OXFORD MATHEMATICAL MONOGRAPHS

OXFORD MATHEMATICAL MONOGRAPHS

For a full list of titles please visit
http://ukcatalogue.oup.com/category/academic/series/science/maths/omm.do

Woodhouse: *Geometric Quantization, Second Edition,* paperback
Hirschfeld: *Projective Geometries over Finite Fields, Second Edition*
Evans and Kawahigashi: *Quantum Symmetries of Operator Algebras*
Klingen: *Arithmetical Similarities: Prime Decomposition and Finite Group Theory*
Matsuzaki and Taniguchi: *Hyperbolic Manifolds and Kleinian Groups*
Macdonald: *Symmetric Functions and Hall Polynomials, Second Edition,* paperback
Catto, Le Bris, and Lions: *Mathematical Theory of Thermodynamic Limits: Thomas-Fermi Type Models*
Mcduff And Salamon: *Introduction to Symplectic Topology,* paperback
Holschneider: Wavelets: *An Analysis Tool,* paperback
Goldman: *Complex Hyperbolic Geometry*
Colbourn and Rosa: *Triple Systems*
Kozlov, Maz'ya and Movchan: *Asymptotic Analysis of Fields in Multi-Structures*
Maugin: *Nonlinear Waves in Elastic Crystals*
Dassios and Kleinman: *Low Frequency Scattering*

Ambrosio, Fusco and Pallara: *Functions of Bounded Variation and Free Discontinuity Problems*
Slavyanov and Lay: *Special Functions: A Unified Theory Based on Singularities*
Joyce: *Compact Manifolds with Special Holonomy*
Carbone and Semmes: *A Graphic Apology for Symmetry and Implicitness*
Boos: *Classical and Modern Methods in Summability*
Higson and Roe: *Analytic K-Homology*
Semmes: *Some Novel Types of Fractal Geometry*
Iwaniec and Martin: *Geometric Function Theory and Nonlinear Analysis*
Johnson and Lapidus: *The Feynman Integral and Feynman's Operational Calculus,* paperback
Lyons and Qian: *System Control and Rough Paths*
Ranicki: *Algebraic and Geometric Surgery*
Ehrenpreis: *The Radon Transform*
Lennox and Robinson: *The Theory of Infinite Soluble Groups*
Ivanov: *The Fourth Janko Group*
Huybrechts: *Fourier-Mukai Transforms in Algebraic Geometry*
Hida: *Hilbert Modular Forms and Iwasawa Theory*
Boffi and Buchsbaum: *Threading Homology through Algebra*
Vazquez: *The Porous Medium Equation*
Benzoni-Gavage and Serre: *Multi-Dimensional Hyperbolic Partial Differential Equations*
Calegari: *Foliations and the Geometry of 3-Manifolds*
Boyer and Galicki: *Sasakian Geometry*
Choquet-Bruhat: *General Relativity and the Einstein Equations*
Ignaczak and Ostoja-Starzewski: *Thermoelasticity with Finite Wave Speeds*
Scott: *Traces and Determinants of Pseudodifferential Operators*
Franchi and Le Jan: *Hyperbolic Dynamics and Brownian Motion: An Introduction*
Jain, Srivastava, and Tuganbaev: *Cyclic Modules and the Structure of Rings*
Ringström: *On the Topology and Future Stability of the Universe*
Johnson, Lapidus and Nielsen: *Feynman's Operational Calculus and Beyond: Noncommutativity and Time-Ordering*
Edmunds and Evans: *Spectral Theory and Differential Operators, Second Edition*
Van Mieghem: *The Mittag-Leffler and Gamma Function*

The Mittag-Leffler and Gamma Function

PIET VAN MIEGHEM

OXFORD
UNIVERSITY PRESS

OXFORD
UNIVERSITY PRESS

Great Clarendon Street, Oxford, OX2 6DP,
United Kingdom

Oxford University Press is a department of the University of Oxford.
It furthers the University's objective of excellence in research, scholarship,
and education by publishing worldwide. Oxford is a registered trade mark of
Oxford University Press in the UK and in certain other countries.

Published in the United States of America by Oxford University Press

198 Madison Avenue, New York, NY 10016, United States of America

British Library Cataloguing in Publication Data

Data available

Library of Congress Control Number: 2026933347

ISBN 9780198993377

DOI: 10.1093/oso/9780198993377.001.0001

Printed and bound by
CPI Group (UK) Ltd., Croydon, CR0 4YY

The manufacturer's authorized representative in the EU for product safety is
Oxford University Press España S.A. of Parque Empresarial San Fernando de Henares,
Avenida de Castilla, 2 – 28830 Madrid (www.oup.es/en or product.safety@oup.com).
OUP España S.A. also acts as importer into Spain of products made by the manufacturer.

MIX
Paper | Supporting responsible forestry
FSC® C013604
FSC
www.fsc.org

Ars Mathematicae

to the loving memory
of Saskia and Nathan

to Huijuan, Vincent and Laurens

to honor Gösta Mittag-Leffler

Contents

Preface

The Mittag-Leffler function does not appear in the classic *Handbook of Mathematical Functions* by Abramowitz and Stegun (1968) and is only briefly mentioned in its successor by Olver *et al.* (2010, 10.46.3). For those who first encounter the Mittag-Leffler function, this de facto omission is surprising, even from a pure mathematical point of view. My belief that the Mittag-Leffler function, although exotic at first glance, deserves a proper place in the theory of functions has motivated me to write this book.

The study of properties of entire functions, which are generalization of polynomials to infinite degree in the complex plane, has led Mittag-Leffler, around 1900, to the study of his function. The Mittag-Leffler function $E_a(z)$ is a primary example of an entire function, tunable in its order (Section 1.4) via the parameter a, and is represented by the simple eyeing Taylor series

$$E_a(z) = \sum_{k=0}^{\infty} \frac{z^k}{\Gamma(ak+1)}$$

in which another basic complex function, the Gamma function $\Gamma(z)$, appears. If $a = 1$ and $\Gamma(ak+1) = k!$, then the Taylor series of the Mittag-Leffler function $E_a(z)$ reduces to the most famous entire function, the exponential function: $E_1(z) = \sum_{k=0}^{\infty} \frac{z^k}{k!} = e^z$. If $a = 0$, then the Mittag-Leffler function represents the geometric series. About 10 years earlier than Mittag-Leffler, Hadamard (Section 6.3) focused on a related entire function $\sum_{k=0}^{\infty} \frac{z^k}{(k!)^a}$, which he compared with the Mittag-Leffler function $E_a(z)$ and which avoids the prominent position of the Gamma function. The reduction of the Gamma function to the factorial in Hadamard's entire function may suggest an easier treatment, however, the opposite seems to hold.

This book mainly targets the mathematical properties of the Mittag-Leffler function, not its applications to fractional analysis nor its numerical evaluation. Since the Gamma function plays a crucial role in the properties of the Mittag-Leffler function, a comprehensive treatment of the Mittag-Leffler function requires the knowledge of the Gamma function. Therefore, the current book consists of two parts: the first and main part is on the

Mittag-Leffler function and the second part discusses the Gamma function, together with its allied functions. The second part on the Gamma function was originally placed in an appendix and can be read before the first part, because it does not depend on the Mittag-Leffler function.

I have tried to write in a readable way, where – I hope – enough intermediate steps are given to ease the reader and spare his frustration by a ban on the abundant "it is easy to see" in current literature.

26 January 2026 PIET VAN MIEGHEM

Acknowledgements

I am very grateful to Professor Francesco Mainardi for very stimulating discussions. After the submission of the first version arXiv:2005.13330, he and Professor Alexander Apelblat have contacted me and pointed me to earlier work. At the time of writing, Professor Mainardi continues to stimulate and to express his deep interest in the Mittag-Leffler function. Moreover, Professor Mainardi has continuously sent me articles about the Mittag-Leffler function and its application to fractional calculus, a wealth of citations that have enriched the book. I also thank Professor Nico Temme profoundly for his numerous comments, pointing me to papers and kindly sharing his great knowledge on the theory of functions. Apart from communicating his vision about "fractional time" (Hilfer, 2000, Chapter II) as the most important driver towards the increasing appearance of the Mittag-Leffler function, Professor Rudolf Hilfer has supplied me with many of his articles on the Mittag-Leffler function. Dr. Rui Ferreira has pointed me to the lower bound (6.27), that has led to the inclusion of **art.** 115.

I would like to thank Johan Dubbeldam, Rui Ferreira and Robin Persoons for valuable comments and pointing me to errors in an earlier version.

Part I

The Mittag-Leffler function

1
Introduction

1.1 Definition of the Mittag-Leffler function

The Mittag-Leffler function is defined for complex numbers z, in the complex parameters a and b, by its Taylor series around $z_0 = 0$,

$$E_{a,b}(z) = \sum_{k=0}^{\infty} \frac{z^k}{\Gamma(b + ak)} \tag{1.1}$$

The Mittag-Leffler function, denoted as $E_a(z) = \sum_{k=0}^{\infty} \frac{z^k}{\Gamma(1+ak)} \triangleq E_{a,1}(z)$, was introduced by Gösta Mittag-Leffler (1903a,b) in 1903 with $b = 1$ in (1.1). The letter "E" refers to "entire" and entire functions[1] were at that time new and challenging, after Weierstrass's pioneering work on entire functions (see e.g. Titchmarsh (1964, Chapter VIII), Sansone and Gerretsen (1960, Chapter 6) and Tannery and Molk (1893, Chapitre III)). Entire functions are the generalizations of polynomials to infinite order and their most characteristic property is that entire functions do not have singularities in the finite complex plane.

We show in Section 1.4 that the Taylor series (1.1) around $z_0 = 0$ converges for any complex number z and b, provided that the complex number a possesses a positive real part, i.e. $\operatorname{Re}(a) > 0$. Immediately, the crucial role of the parameter a over the parameter b is highlighted. While z always denotes a complex number, the scope of the parameters a and b, either real or complex, is specified or is clear in the sequel. We consider the definition $E_{a,b}(z)$ in the parameters a and b, because the functional relations for $E_{a,b}(z)$ are closed and expressed in terms of $E_{a,b}(z)$, whereas confinement to $E_a(z) = E_{a,1}(z)$ with fixed $b = 1$ only, deprives the analysis from a complete and more elegant picture (see e.g. the derivative of $E_{a,b}(z)$ in **art.** 15). There exist generalizations[2] of the Mittag-Leffler function $E_{a,b}(z)$ to more parameters, which are beyond the present scope, but discussed by

[1] In French, we speak about "une fonction entière", while in German, an entire function is "ein ganz holomorphe oder gesamte Funktion". Sometimes, entire functions are also called "integral functions", which might be confusing with the integral of a function.

[2] The generalized Mittag-Leffler function is $E_{a,b,c}(z) = \sum_{k=0}^{\infty} \frac{\Gamma(c+k)}{\Gamma(c)} \frac{z^k}{\Gamma(b+ak)}$, where $E_{a,b,0}(z) = E_{a,b}(z)$.

Haubold *et al.* (2011) and also covered in the book by Gorenflo *et al.* (2020) on Mittag-Leffler functions and their applications. The Mittag-Leffler function $E_a(z)$ is treated by Erdélyi *et al.* (1953a, Sec. 18.1 on pp. 206-211) and by Sansone and Gerretsen (1960, Sec. 6.13 on pp. 345-348). Gorenflo *et al.* (2020) treat nearly all known aspects and results of the Mittag-Leffler function and its multiple parameter generalizations in their encyclopedic work, containing a wealth of citations.

1.1.1 Motivation for the study of the Mittag-Leffler function

I believe that there are, at least, three compelling reasons that justify a study of the Mittag-Leffler function $E_{a,b}(z)$.

First, the Mittag-Leffler function $E_{a,b}(z)$ in (1.1) is an entire function in the complex variable z in two complex parameters a and b. Due to its parameters a and b, the Mittag-Leffler function $E_{a,b}(z)$ constitutes a broad class of entire functions, that contain, for example, the exponential function $E_{1,1}(z) = e^z$ and exponential-like functions such as the cosine $E_{2,1}(-z^2) = \cos z$ and $E_{1,2}(z) = \frac{e^z-1}{z}$ and many more (see Section 2.1). Just as in Mittag-Leffler's time, this first motivation is the main focus in this book.

Second, a Mittag-Leffler random variable (Chapter 5) is heavy-tailed of the power-law type and plays a role in so-called stable distributions. Many observed properties in real-world networks are approximately power-law distributed. The discovery of power-law behavior was a major stimulus that gave birth to Network Science, the theory of dynamics processes on graphs (see e.g. Barabási (2016), Newman (2018), Van Mieghem (2023a)). The Mittag-Leffler random variable may model such power-law like properties, although none of its moments exists, which is a rather complicating, but, at the same time, fascinating factor.

Third, the Mittag-Leffler function $E_{a,b}(z)$ naturally arises in fractional calculus. Fractional calculus is attracting increasing interest over the last years and is penetrating into more application domains. The analytic solution (1.10) in Section 1.3 below is undoubtedly the most important driver towards the increasing appearance of the Mittag-Leffler function $E_{a,b}(z)$. The application of the solution (1.10) of a fractional order integral or differential equation is illustrated in Haubold *et al.* (2011), Gorenflo *et al.* (2020), Mainardi (2021) and Sandev and Tomovski (2019); for example, in the fractional generalization of the heat equation, random walks, Lévy flights, superdiffusive transport and viscoelasticity, and fractional Ohm's Law. Abel's integral equation, whose solution involves $E_{a,b}(z)$, is treated in Gorenflo *et al.* (2014, Chapter 7). A "fractional" generalization of the Poisson renewal process, discussed in Gorenflo *et al.* (2014, Sec. 9.4), consists of replacing the exponential interarrival time between events by a Mittag-Leffler distribution $E_{a,1}(-t)$ with real $t \geq 0$ and $0 < a \leq 1$. Undoubtedly, fractional calculus is currently the main driver to study the Mittag-Leffler function. Therefore, we briefly sketch the relation between the Mittag-Leffler function and fractional calculus in Section 1.3.

1.1.2 This book

The aim here is to collect and deduce the most relevant functional properties of the Mittag-Leffler function $E_{a,b}(z)$ defined in (1.1). In the course of writing, it became clear that the Gamma function $\Gamma(z)$ plays a fundamental role in the theory of the Mittag-Leffler function. Therefore, I have decided to include a second part that treats the Gamma function. That second part dives a bit deeper into properties of the Gamma function than most classical books.

The manuscript is written in the form of articles (**art.**) as in my book *Graph Spectra of Complex Networks* (Van Mieghem, 2023a). As explained in the Preface of the first edition of *Graph Spectra of Complex Networks*, articles are independent blocks that contain a message, concept and/or result. The article-style was common in the early stages of mathematics, when derivations were presented in human language, more natural than in the current theorem-proof style, but, I admit, sometimes less precise. To some extent, I find that the state of the art and the knowledge about the Mittag-Leffler function is rather immature and incomplete. Many results still lie hidden as mentioned in the Epilogue in Chapter 8. A polished and near to complete theory as for the elementary functions is lacking. This observation justifies the older presentation style, used before the rise of abstract formalization of mathematics that culminated in the Bourbaki school. Another advantage is the readability for a group of scientists larger than only mathematicians. The risk of using the article-style is a depreciation of the pure formalists. About half of the results in this book have been established before and are reviewed. Many results are new.

As in any book, there are omissions. There are few numerical evaluations and no graphical plots, because Wolfram's Mathematica can amply fulfill this role. Methods to numerically compute the Mittag-Leffler function are found in the last chapter by Garrappa and Popolizio of the third volume of the series by Karniadakis (2019), in Garrappa (2015) and in the works of Seybold and Hilfer (2005, 2008). Zeng and Chen (2015) focus on Padé approximations for the Mittag-Leffler function. I refer to Ferreira (2025) and references therein for the discrete Mittag-Leffler function. For the history of the Mittag-Leffler function and detailed historical citations, I refer to the great book by Gorenflo *et al.* (2020), where the life of Mittag-Leffler is told and his five papers on "his" function are reviewed. Hardy (1928) praises Mittag-Leffler as the pivotal person in mathematics of his time. Mainardi and Consiglio (2020) honor the pioneers of the Mittag-Leffler function, with a special attention to Davis and Gross.

1.2 Outline of Part I on the Mittag-Leffler function

After introducing a definition of the fractional derivative, Section 1.3 explains the main value of the Mittag-Leffler function for fractional analysis. Section 1.4 briefly summarizes the properties of entire functions that are defined for any complex number z by their Taylor series such as (1.1) for the Mittag-Leffler function $E_{a,b}(z)$.

Chapter 2 deduces direct consequences of the Taylor series in (1.1). Section 2.2 derives the unique cyclotomic property and Section 2.3 presents the progression property of the Mittag-Leffler function $E_{a,b}(z)$. The derivative of $E_{a,b}(z)$, which satisfies an important recursion, is studied in Section 2.4. The Mittag-Leffler function is defined by a Taylor series (1.1) around $z_0 = 0$, but an entire function possesses a Taylor series with infinite radius of convergence around each finite point z_0 in the complex plane, which is derived in Section 2.5. Intimately related to the Taylor series around $z_0 \neq 0$ and to the m-th derivative of the Mittag-Leffler function $E_{a,b}(z)$ is the polynomial $q_j(a, b, m)$ in the parameters a and b of degree $m - j$, novel in the theory of the Mittag-Leffler function and studied in Section 2.6. After searching for a possible differential equation in Section 2.8, a separate Section 2.9 is created, that only focuses on the logarithm of Mittag-Leffler function $E_{a,b}(z)$.

Chapter 3 explores integrals that contain the Mittag-Leffler function $E_{a,b}(z)$ as integrand and that can be evaluated analytically. One of the most important integrals is the Laplace transform (3.13) of the Mittag-Leffler function $E_{a,b}(z)$. Section 3.6 applies the Euler-Maclaurin summation formula (3.40) to the Mittag-Leffler function $E_{a,b}(z)$. The Euler-Maclaurin summation formula is an important formula to evaluate a summation by its associated integral and correction terms. A variant (3.41) of Euler-Maclaurin summation formula already appeared in the asymptotic study of the Gamma function by Stirling and de Moivre (**art**. 195).

We continue in Chapter 4 with complex integrals for the Mittag-Leffler function $E_{a,b}(z)$: some are deduced from the inverse Laplace transform of (3.13) and others are complex representations of the Taylor series (1.1). We treat two different approaches, each containing a different contour integral for $E_{a,b}(z)$: Sections 4.1-4.3 concentrate on Mittag-Leffler's contour integral, while Section 4.4 investigates the contour integral deduced from a general summation method due to Plana.

Chapter 5 is devoted to the Mittag-Leffler function $E_{a,b}(z)$ in probability theory. The complete monotonicity of $E_{a,b}(z)$ for $0 < a < 1$ and for some values of b, apart from $b = 1$, has been and is still studied in depth. **Art**. 89 presents a different proof of the monotonicity of the Mittag-Leffler function $E_{a,b}(z)$, while Section 5.2 and 5.3 focus on the Mittag-Leffler random variable.

Estimates of Hadamard and bounds are found in Chapter 6, while Chapter 7 covers various, unrelated topics such as expansions of the Mittag-

Leffler function $E_{a,b}(z)$ in powers of a in **art.** 116, the deduction of the integral (4.50) from the Taylor series (1.1) and the Taylor series of $\frac{1}{\Gamma(z)}$ in **art.** 117, Mobius inversion in **art.** 118, Apelblat series in **art.** 120 and the Mittag-Leffler function $E_{a,b}(z)$ as a limit in **art.** 122-126. Section 7.5 is new and explores the related integral

$$I_{a,b}(z) = \int_0^\infty \frac{z^u}{\Gamma(b+au)}\,du \tag{1.2}$$

that naturally appears in the Euler-Maclaurin summation for $E_{a,b}(z)$ in **art.** 53. Chapter 8 concludes this work with some open problems.

For self-consistency, we have included a second part on the Gamma function $\Gamma(z)$, whose properties are essential for the Mittag-Leffler function $E_{a,b}(z)$. Part II on the Gamma function, summarized in Section 9.1, can be read independently of part I on the Mittag-Leffler function.

Appendix A briefly introduces the concept of a characteristic coefficient of a complex function. Appendix B investigates the inverse Laplace transform, separated in real and imaginary part.

1.3 Mittag-Leffler function and fractional calculus

In contrast to the theory of the Mittag-Leffler function, there exist many books on fractional calculus (see e.g. Gorenflo *et al.* (2020), Podlubny (1999), Hilfer (2000)).

Following and summarizing the work[3] of Riemann (1876), Hadamard (1893, art. 31-32, p. 54, Troisième partie) introduces the "general" derivative $D_x^\alpha f(x)$, defined, for *negative* real α, as

$$D_x^\alpha f(x) = \frac{1}{\Gamma(-\alpha)} \int_p^x (x-z)^{-\alpha-1} f(z)\,dz \tag{1.3}$$

where p is a number at which $f(z)$ is regular between p and x. If the function is regular at the origin, p is chosen as 0. The case $\alpha = 0$ is defined as $D_x^0 f(x) = f(x)$. If the function is presented by a power series, $f(z) = \sum_{k=0}^\infty f_k z^k$, then Riemann computes

$$x^\alpha D_x^\alpha f(x) = \sum_{k=0}^\infty f_k \frac{\Gamma(k+1)}{\Gamma(k+1-\alpha)} z^k$$

because $D_x^\alpha x^k = \frac{1}{\Gamma(-\alpha)} \int_0^x (x-z)^{-\alpha-1} z^k dz = x^{k-\alpha} \frac{\Gamma(k+1)}{\Gamma(k+1-\alpha)}$ by (3.27). Hadamard transforms Riemann's definition (1.3) with $p = 0$ by the substitution $z = tx$,

$$x^\alpha D_x^\alpha f(x) = \frac{1}{\Gamma(-\alpha)} \int_0^1 (1-t)^{-\alpha-1} f(tx)\,dt \tag{1.4}$$

[3] Weber and Dedekind mention that Riemann's manuscript contains the date 14 January 1847 and originates from Riemann's student time. It was not published by Riemann, but incorporated in his "Nachlass" by Weber and Dedekind as a characteristic work illustrating the depth and development of Riemann's genius.

In the subsequent articles (from 33 up to 38), Hadamard (1893, pp. 57-65) derives properties of the "general" derivative $D_x^{\alpha} f(x)$ and considers the more general form $\varphi(x) = \int_0^1 V(t) f(tx)\,dt$ and the specific coefficient $\mathcal{C}_m = \int_0^1 V(t)\, t^m dt$ for the corresponding "generalized" power series, whose properties and convergence he studies.

The k-th order derivative of a complex function can be deduced from Cauchy's integral as

$$\left.\frac{d^k f(z)}{dz^k}\right|_{z=z_0} = \frac{\Gamma(1+k)}{2\pi i} \int_{C(z_0)} \frac{f(\omega)\, d\omega}{(\omega - z_0)^{k+1}} \tag{1.5}$$

where $C(z_0)$ is a contour around the point z_0 in a region of the complex plane where the function $f(z)$ is analytic. The integer number k at the right-hand side in (1.5) can be *formally* extended (Van Mieghem, 2022) to a complex number α, which then defines the left-hand side as a *complex fractional derivative*,

$$D^{\alpha} f(z_0) \equiv \left.\frac{d^{\alpha} f(z)}{dz^{\alpha}}\right|_{z=z_0} = \frac{\Gamma(1+\alpha)}{2\pi i} \int_{C(z_0)} \frac{f(\omega)\, d\omega}{(\omega - z_0)^{\alpha+1}} \tag{1.6}$$

However, the extension of the integer k towards a complex number α transforms the analytic function ω^k in the integrand to a considerably more complicated function $\omega^a = e^{a \log \omega}$, that is multi-valued and that contains a branch-cut in the complex plane at negative real ω. Consequently, the contour $C(z_0)$ in (1.6) cannot be closed as in (1.5), which makes (1.6) a mathematically ill-posed formula, indicating an inherent challenge to define a "fractional derivative". As a matter of facts, there are many possible, but different definitions as witnessed e.g. by the appearance of p in (1.3) and (1.7) below! By choosing or deforming the contour $C(z_0)$ in an appropriate manner (which is one of the core ingenuities in complex integration and amply illustrated in this book), we can transform (1.6) with the integer m to

$$D_m^{\alpha} f(t) = \left.\frac{d^{\alpha} f(z)}{dz^{\alpha}}\right|_{z=t} = \frac{1}{\Gamma(m-\alpha)} \int_p^t \frac{f^{(m)}(x)}{(t-x)^{\alpha+1-m}} dx \quad \text{for } \operatorname{Re}(\alpha) < m \tag{1.7}$$

which is known as the Caputo fractional derivative and reduces for $m = 0$ to Riemann's definition (1.3). Caputo and Mainardi (1971b, Appendix) derive (1.7) for $m = 0$ and $p = 0$ from the m-fold integral (3.29). As explained in Van Mieghem (2022), the integral in (1.7) is a convolution of the function $g(x) = f^{(m)}(x)$, the m-th derivative of $f(x)$, and the function $x^{-\alpha-1+m}$, which is a "power law", while Hadamard studies a more general convolution $\varphi(x) = \int_0^1 V(t) f(tx)\,dt$. We refer to Gorenflo *et al.* (2020) for the theory of the fractional derivative.

After the extension of the *classical* derivative (1.5) to a *fractional* derivative (1.7), we replace or extend the derivative in the linear matrix differential equation

$$\frac{ds(t)}{dt} = Qs(t) \tag{1.8}$$

where Q is an $N \times N$ matrix and $s(t)$ is an $N \times 1$ vector, to a Caputo fractional derivative (1.7)

$$D_m^{\alpha} s_{\alpha}(t) = Q s_{\alpha}(t) \tag{1.9}$$

The classical solution of (1.8) for *any* $N \times N$ matrix Q is

$$s(t) = e^{Qt} s(0)$$

while, for $0 < \alpha \leq 1$ and[4] $m = 0$, the "fractional α process" is described by

$$s_{\alpha}(t) = E_{\alpha}(Qt^{\alpha})\, s_{\alpha}(0) \tag{1.10}$$

where the Mittag-Leffler matrix $E_{\alpha}(Qt^{\alpha}) = \sum_{k=0}^{\infty} \frac{Q^k t^{ak}}{\Gamma(1+ak)}$ is also defined for any matrix Q.

If the matrix Q is diagonalizable, then the eigenvalue decomposition (Van Mieghem, 2023a) is

$$Q = \sum_{k=1}^{N} \mu_k x_k y_k^T$$

where μ_k is the eigenvalue belonging to the right-eigenvector x_k and the left-eigenvector y_k of the matrix Q. We assume the ordering $|\mu_1| \geq |\mu_2| \geq \cdots \geq \mu_N = 0$, but $\text{Re}(\mu_k) \leq 0$ for reasons of stability and existence, while we suppose eigenvalues with multiplicity 1 for simplicity. Then, after spectral decomposition, the matrix (1.10) transforms (Van Mieghem, 2022) to

$$s_{\alpha}(t) = \sum_{j=1}^{N} \left(y_j^T s_{\alpha}(0) \right) E_{\alpha}(\mu_j t^{\alpha})\, x_j \tag{1.11}$$

If the matrix Q is diagonalizable, then the solution vector $s_{\alpha}(t)$ in (1.11) of the fractional matrix differential equation (1.9) for $m = 0$ and $0 < \alpha \leq 1$ is expressed as a linear combination of eigenvectors $x_1, x_2, \ldots x_N$, where each eigenvector x_j is weighted by the Mittag-Leffler function $E_{\alpha}(\mu_j t^{\alpha})$ and the initial condition scalar product $y_j^T s_{\alpha}(0)$. For $\alpha = 1$ with $E_1(z) = e^z$, the solution (1.11) reduces to the classical vector solution of (1.8)

$$s(t) = \sum_{j=1}^{N} \left(y_j^T s(0) \right) e^{\mu_j t} x_j$$

with exponential, decaying (and possibly oscillating if $\mu_j \in \mathbb{C}$) functions in time t with rates equal to the eigenvalues μ_m of the matrix Q.

[4] The more general form for $m > 0$ is also analytically known (Van Mieghem, 2022, Appendix C).

The major importance of (1.10) is that fractional linear differential equations have an exact mathematical solution. This exact analytic result (1.10) generalizes a large number of physical processes that are described by a linear differential equation of order N. In case the matrix Q is diagonalizable, its solution (1.11) shows that the eigenfunctions are Mittag-Leffler functions, that reduce for $\alpha = 1$ to the classical set of exponential eigenfunctions. Often in physics, a good intuition and understanding of the natural process exist, but a precise mathematical description together with an elegant solution, that allows for predictions and deeper insight, is lacking. The curious fact in fractional analysis is a reverse picture: the mathematical solution (1.11) is known, but the precise physical meaning is still challenging. For example, the extension of Markov theory to a "fractional setting", where the Chapman-Kolmogorov equation (1.8) is replaced by (1.9) and where the matrix Q is the infinitesimal generator of the Markov chain, introduces memory and destroys the typical Markovian property that the present state of the process only depends upon the previous state. The current state of such fractional Markov process depends in a complicated way upon the previous states, which is not yet fully understood as illustrated in Van Mieghem (2022) and D'Alessandro and Van Mieghem (2025).

1.4 Complex function theory: entire functions

1.4.1 Taylor series of a function

The Taylor or power series expansion of a complex function $f(z)$ around a point z_0 is defined as

$$f(z) = \sum_{k=0}^{\infty} f_k(z_0)\,(z - z_0)^k \tag{1.12}$$

with the k-th Taylor coefficient

$$f_k(z_0) = \frac{1}{k!}\left.\frac{d^k f(z)}{dz^k}\right|_{z=z_0} \tag{1.13}$$

$$= \frac{1}{2\pi i}\int_{C(z_0)} \frac{f(\omega)\, d\omega}{(\omega - z_0)^{k+1}} \tag{1.14}$$

where the contour $C(z_0)$ encloses z_0 and $f(z)$ is analytic on and inside the contour $C(z_0)$. The theory of complex functions (Titchmarsh, 1964; Whittaker and Watson, 1996; Sansone and Gerretsen, 1960; Markushevich, 1985) heavily relies on the Taylor series, because the necessary and sufficient condition that a function $f(z)$ should be expansible in a Taylor series is that $f(z)$ is analytic in a region of the complex plane.

The geometric series

$$S_n(z) = \sum_{k=0}^{n} z^k = 1 + z + z^2 + \cdots + z^n \tag{1.15}$$

holds for any complex number z and any integer n, because $S_{-n}(z) = \sum_{k=0}^{n} z^{-k} = S_n\left(\frac{1}{z}\right)$, and satisfies $zS_n(z) - S_n(z) = z^{n+1} - 1$, that establishes the functional equation for any complex z and any integer n,

$$S_n(z) = \frac{z^{n+1} - 1}{z - 1} \tag{1.16}$$

while $S_n(1) = n$. The functional equation (1.16) shows that $\lim_{n\to\infty} S_n(z)$ exists, only if $|z| < 1$. The geometric series converges to $\lim_{n\to\infty} S_n(z) = \frac{1}{1-z}$, provided that $|z| < 1$. Cauchy's integral (1.13) for the k-th derivative emphasizes the intimate relation of a Taylor series and the geometric series, because

$$\begin{aligned}\sum_{k=0}^{n} f_k(z_0)(z - z_0)^k &= \frac{1}{2\pi i}\int_{C(z_0)} \frac{f(\omega)}{(\omega - z_0)} \sum_{k=0}^{n} \left(\frac{z - z_0}{\omega - z_0}\right)^k d\omega \\ &= \frac{1}{2\pi i}\int_{C(z_0)} \frac{f(\omega)\left(1 - \left(\frac{z-z_0}{\omega-z_0}\right)^{n+1}\right)}{\omega - z} d\omega\end{aligned}$$

If $|z - z_0| < |\omega - z_0|$, i.e. z_0 is sufficiently close to z, then the limit $n \to \infty$ returns Cauchy's integral, which is (1.14) with $k = 0$, with z_0 replaced by z, and the left-hand side becomes $f(z)$.

1.4.2 Radius of convergence

Since the Mittag-Leffler function $E_{a,b}(z)$ is defined in (1.1) as a power series in z around $z_0 = 0$, a first concern is its validity range in the complex number z. The radius R of convergence of the Taylor series (1.12) around $z_0 = 0$ with Taylor coefficients denoted by $f_k(z_0) = f_k$ for brevity, i.e. $f(z) = \sum_{k=0}^{\infty} f_k z^k$, satisfies[5] $\frac{1}{R} = \limsup_{k\to\infty} |f_k|^{1/k}$ or $\frac{1}{R} = \lim_{k\to\infty} \left|\frac{f_{k+1}}{f_k}\right|$ when the latter exists (Titchmarsh, 1964).

Using $\frac{\Gamma(z+a)}{\Gamma(z+b)} \sim z^{a-b}$ in **art**. 198 when $z \to \infty$, the radius of convergence of the power series in (1.1) satisfies

$$\frac{1}{R} = \lim_{k\to\infty} \left|\frac{\Gamma(b + ak)}{\Gamma(b + a + ak)}\right| = \lim_{k\to\infty} |ak|^{-a} = 0$$

for $\mathrm{Re}(a) > 0$ and all complex b. Hence, the Taylor series (1.1) of the Mittag-Leffler function $E_{a,b}(z)$ converges for all complex z, provided that $\mathrm{Re}(a) > 0$. For $\mathrm{Re}(a) < 0$, the radius of convergence is $R = 0$ and the Taylor series (1.1) does not converge for any b and any complex number $z \neq 0$. If $\mathrm{Re}(a) = 0$ and $a = i\alpha$ is purely imaginary with $\alpha \in \mathbb{R}$, then the

---

[5] Indeed, writing $f(z) = \sum_{k=0}^{\infty} f_k z^k = \sum_{k=0}^{\infty} \left((f_k)^{\frac{1}{k}} z\right)^k$ leads, in association with the geometric series $\sum_{k=0}^{\infty} w^k = \frac{1}{1-w}$, that converges for $|w| < 1$, to the convergence condition $\left|(f_k)^{\frac{1}{k}} z\right| < 1$ or, equivalently, to $\frac{1}{|z|} > |f_k|^{\frac{1}{k}}$ for all "sufficiently" large k.

Taylor series (1.1) is bounded as

$$|E_{i\alpha,b}(z)| \leq \sum_{k=0}^{\infty} \frac{|z|^k}{|\Gamma(b+i\alpha k)|} = \sum_{k=0}^{K-1} \frac{|z|^k}{|\Gamma(b+i\alpha k)|} + \sum_{k=K}^{\infty} \frac{|z|^k}{|\Gamma(b+i\alpha k)|}$$

The first, finite series $\sum_{k=0}^{K-1} \frac{|z|^k}{|\Gamma(b+i\alpha k)|}$ is finite. The second series behaves, for large K and with the asymptotic formula (14.8) of the Gamma function, as

$$\sum_{k=K}^{\infty} \frac{|z|^k}{|\Gamma(b+i\alpha k)|} = \sum_{k=K}^{\infty} \frac{\left(|z|\, e^{\frac{\pi}{2}|\alpha|}\right)^k}{\sqrt{2\pi}\,(\alpha k)^{b-\frac{1}{2}}} \left(1 + O\left(\frac{1}{k}\right)\right)$$

which only converges, for any complex b, provided that $|z|\, e^{\frac{\pi}{2}|\alpha|} < 1$, implying that the radius of convergence[6] equals $R = e^{-\frac{\pi}{2}|\alpha|} \leq 1$. If $|z| = e^{-\frac{\pi}{2}|\alpha|}$, the series converges for $\mathrm{Re}\,(b) > \frac{3}{2}$, which follows from the Dirichlet series of the Riemann Zeta-function (**art.** 138). If $a = 0$, then the Mittag-Leffler function $E_{0,b}(z)$ reduces to the geometric series and $E_{0,b}(z) = \frac{1}{\Gamma(b)} \frac{1}{1-z}$ for $|z| < 1$. Hence, only if $\mathrm{Re}\,(a) = 0$, then the radius of convergence of the Taylor series (1.1) of the Mittag-Leffler function $E_{a,b}(z)$ is positive and finite. Consequently, $E_{i\alpha,b}(z)$ for real α is not an entire function in z.

Many analytic functions can be expressed in terms of the hypergeometric function, defined by Gauss's series (Abramowitz and Stegun, 1968, 15.1.1),

$$F(a, b; c; z) = \frac{\Gamma(c)}{\Gamma(a)\Gamma(b)} \sum_{k=0}^{\infty} \frac{\Gamma(a+k)\Gamma(b+k)}{\Gamma(c+k)k!} z^k \qquad \text{convergent for } |z| < 1 \tag{1.17}$$

where the argument of the Gamma functions is of the form $\alpha k + \beta$ with $\alpha = 1$, in contrast to the Mittag-Leffler function in (1.1) where $\alpha = a$ is real positive. Just the fact that a is real and not an integer colors the theory of the Mittag-Leffler function $E_{a,b}(z)$ and causes its main challenges.

1.4.3 The order of an entire function

An entire function has a power series with infinite radius of convergence and is, thus, analytic in the entire complex plane with an essential singularity at infinity. Associated to entire complex functions is the concept[7] of the order ρ, which is defined (Titchmarsh, 1964, p. 248) for any $\varepsilon > 0$ as $f(z) = O\left(e^{|z|^{\rho+\varepsilon}}\right)$ when $|z| \to \infty$. A necessary and sufficient condition

---

[6] Gorenflo *et al.* (2020, p. 20) mention, without derivation, that $R = e^{\frac{\pi}{2}|\mathrm{Im}\, a|}$, whereas it should be $R = e^{-\frac{\pi}{2}|\mathrm{Im}\, a|}$.

[7] The concept of the order of an entire function is related to the *extreme value index* of a probability distribution $F_X(x) = \Pr[X \leq x]$ of a random variable, where the limit process may apply to both large and small values of x. For example, the *extreme value index* $\alpha = \lim_{x \downarrow 0} \frac{\log F_X(x)}{\log x}$ of the probability distribution of a non-negative random variable X for small values, characterizes the behavior of the random variable X for small values (see e.g. further Van Mieghem (2014, p. 398) and its applications to the shortest path problem).

(Titchmarsh, 1964, p. 253), that the function $f(z) = \sum_{k=0}^{\infty} f_k z^k$ is an entire function of order ρ, is that

$$\frac{1}{\rho} = \lim_{k\to\infty} \frac{-\log|f_k|}{k\log k}$$

Applied to the power series in (1.1) of the Mittag-Leffler function $E_{a,b}(z)$, yields, after invoking for large k Stirling's asymptotic formula $\Gamma(ak+b) \sim \sqrt{2\pi} e^{-ak} (ak)^{ak+b-\frac{1}{2}}$ in Chapter 14,

$$\frac{1}{\rho} = \lim_{k\to\infty} \frac{\log|\Gamma(b+ak)|}{k\log k} = \lim_{k\to\infty} \frac{\log\sqrt{2\pi} - ak + \left(ak+b-\frac{1}{2}\right)\log(ak)}{k\log k} = a$$

Hence, we find the important property that the Mittag-Leffler function $E_{a,b}(z)$ is an entire complex function in z of order $\rho = \frac{1}{a}$ for $\mathrm{Re}(a) > 0$ and any b.

1.4.4 Zeros of an entire function

The number $n(r)$ of zeros $z_1, z_2, \ldots$ of an entire function $f(z)$ of order ρ, for which $|z_n| \leq r$ is non-decreasing in r, grows as $n(r) = O(r^{\rho+\varepsilon})$ with real, positive r. Roughly stated (Titchmarsh, 1964, p. 249), the higher the order ρ of an entire function, the more zeros it may have in a given region of the complex plane. Moreover, if the modulus of a zero z_n is $r_n = |z_n|$, then

$$\sum_{n=1}^{\infty} \frac{1}{(r_n)^{\alpha}} \text{ converges if } \alpha > \rho$$

and the lower bound of α is the exponent ρ_1 of convergence; thus $\rho_1 \leq \rho$. If $\sum_{n=1}^{\infty} \left(\frac{r}{r_n}\right)^{p+1}$ converges for an integer p, then $p+1 > \rho_1$ and the smallest integer p is called the genus of $f(z)$. In any case, $p \leq \rho_1 \leq \rho$. We may have $\rho_1 < \rho$, for example for $f(z) = e^z$, whose order is $\rho = 1$, but the exponent of convergence $\rho_1 = 0$, because e^z does not have zeros.

Hadamard's factorization theorem (Titchmarsh, 1964, p. 250) states that an entire function $f(z)$ of order ρ with zeros $z_1, z_2, \ldots$, different from zero, can be written for any complex z as

$$f(z) = e^{p_{\lfloor\rho\rfloor}(z)} \prod_{n=1}^{\infty} \left(1 - \frac{z}{z_n}\right) e^{\sum_{j=1}^{\lfloor\rho\rfloor} \frac{1}{j}\left(\frac{z}{z_n}\right)^j} \tag{1.18}$$

where $p_n(z)$ is a polynomial of degree n and $\lfloor\rho\rfloor$ is the largest integer not exceeding ρ, i.e. $\lfloor\rho\rfloor \leq \rho$. In particular, if the order $\rho < 1$, then Hadamard's formula (1.18) simplifies to

$$f(z) = f(0) \prod_{n=1}^{\infty} \left(1 - \frac{z}{z_n}\right) \tag{1.19}$$

Applied to the Mittag-Leffler function $E_{a,b}(z)$ of order $\rho = \frac{1}{a}$, the theory indicates that more zeros are expected for small a than for large a, which seems contradictory to the monotonicity of $E_{a,b}(z)$ for $0 < a < 1$ on the negative real axis in **art**. 89. The determination of the zeros of $E_{a,b}(z)$ is generally difficult (see e.g. Wiman (1905b), Hilfer and Seybold (2006), Gorenflo *et al.* (2014, Sec. 4.6), Popov and Sedletskii (2013)) and omitted here, although **art**. 69 briefly sketches the method of Wiman (1905b).

2

Deductions from the Taylor series

An entire function $f(z)$ is defined for all complex numbers z by its Taylor series. This Chapter 2 studies the Taylor series $E_{a,b}(z) = \sum_{k=0}^{\infty} \frac{z^k}{\Gamma(b+ak)}$ in (1.1) of the entire Mittag-Leffler function around $z_0 = 0$. After algebraic manipulations in Section 2.1, we highlight the cyclotomic property in Section 2.2 and the progression property in Section 2.3, that are inverses of each other (**art.** 13). The cyclotomic property enables the derivation of closed forms of $E_{a,b}(z)$ for integer values of $a = m > 0$ and $b = n$. The combination of the progression and cyclotomic property leads to a closed-form expression of $E_{\frac{m}{n},b}(z)$ for rational values of $a = \frac{m}{n}$ in **art.** 14.

Section 2.4 explores the derivative of the Mittag-Leffler function $E_{a,b}(z)$ with respect to z, resulting in the important recursive formula (2.40), the Taylor series (2.47) of $E_{a,b}(z)$ around an arbitrary point z_0 in Section 2.5 and the discovery of the polynomial $q_j(a,b,m)$ in both a and b in Section 2.6. The polynomial $q_j(a,b,m)$ appears in any Taylor series (2.53) around an arbitrary point $z_0 \neq 0$ of the Mittag-Leffler function $E_{a,b}(z)$, but also of its related integral $I_{a,b}(z)$, as shown in **art.** 135, and of the Wright function in (3.8) in **art.** 41. The complicated form of the Taylor series (2.53) of the Mittag-Leffler function $E_{a,b}(z)$ around an arbitrary point $z_0 \neq 0$ contrasts the simplicity of Taylor series $E_{a,b}(z) = \sum_{k=0}^{\infty} \frac{z^k}{\Gamma(b+ak)}$ around $z_0 = 0$. Integration of $E_{a,b}(z)$ with respect to z is shown in Section 2.7 to be awkward.

Section 2.8 deduces that there is no linear differential equation for $E_{a,b}(z)$ with irrational parameter a. The last Section 2.9 in Chapter 2 explores the logarithm of the Mittag-Leffler function via series expansions, based on characteristic coefficients (Appendix A).

2.1 Direct consequences of the Taylor series

1. *On the real axis.* For real $a > 0$ and real b, the Mittag-Leffler function $E_{a,b}(z)$ is real on the real z-axis. Indeed, with the polar form of the complex number $z = re^{i\theta}$, where the angle $0 \leq \theta \leq 2\pi$ and $r = |z|$, the Taylor series

(1.1) is written, for real parameters $a > 0$ and b, as

$$E_{a,b}\left(re^{i\theta}\right) = \sum_{k=0}^{\infty} \frac{r^k \cos k\theta}{\Gamma(b+ak)} + i \sum_{k=0}^{\infty} \frac{r^k \sin k\theta}{\Gamma(b+ak)} \tag{2.1}$$

The positive real z-axis is reached for the angle $\theta = 0$, whereas the negative real z-axis corresponds to $\theta = \pi$. For both $\theta = 0$ and $\theta = \pi$, the function $\sin k\theta = 0$ for all integers $k \geq 0$ and the imaginary part in (2.1) vanishes. On the other hand, the function $\cos k\theta = 0$ for $k\theta = \left(\frac{1}{2} + m\right)\pi$, where $m \in \mathbb{Z}$, indicates that there does not exist an angle $0 \leq \theta \leq 2\pi$ for which the Mittag-Leffler function $E_{a,b}\left(re^{i\theta}\right)$ is purely imaginary for all r.

Moreover, for real positive a and b, the Mittag-Leffler function $E_{a,b}(z)$ attains its maximum on the real positive axis, because $\left|E_{a,b}\left(re^{i\theta}\right)\right| \leq \sum_{k=0}^{\infty} \frac{r^k\left|e^{ik\theta}\right|}{|\Gamma(b+ak)|} = \sum_{k=0}^{\infty} \frac{r^k}{\Gamma(b+ak)} = E_{a,b}(r)$.

2. *Special values of the parameters a and b.* If $b = 0$, then, since $\lim_{z\to 0} \frac{1}{\Gamma(z)} = 0$, the Taylor series (1.1) reduces to

$$E_{a,0}(z) = \sum_{k=1}^{\infty} \frac{z^k}{\Gamma(ak)} = \sum_{k=0}^{\infty} \frac{z^{k+1}}{\Gamma(ak+a)}$$

and, hence,

$$E_{a,0}(z) = zE_{a,a}(z) \tag{2.2}$$

If $a = 0$, then

$$E_{0,b}(z) = \frac{1}{\Gamma(b)} \frac{1}{1-z} \qquad \text{for } |z| < 1$$

From $E_{a,b}(z) = \frac{1}{\Gamma(b)} + \sum_{k=1}^{\infty} \frac{z^k}{\Gamma(b+ak)}$, we observe that $\lim_{a\to\infty} E_{a,b}(z) = \frac{1}{\Gamma(b)} = E_{a,b}(0)$.

3. *The Mittag-Leffler function with $b = \beta + am$ where $m \in \mathbb{Z}$.* Rewriting the definition (1.1) of $E_{a,b}(z)$,

$$\begin{aligned} E_{a,b}(z) &= \sum_{k=0}^{\infty} \frac{z^k}{\Gamma(b+ak)} = \sum_{k=1}^{\infty} \frac{z^{k-1}}{\Gamma(b-a+ak)} \\ &= \frac{1}{z}\left(\sum_{k=0}^{\infty} \frac{z^k}{\Gamma(b-a+ak)} - \frac{1}{\Gamma(b-a)}\right) \end{aligned}$$

leads to "the shift down of b by a" formula

$$E_{a,b}(z) = \frac{1}{z}\left(E_{a,b-a}(z) - \frac{1}{\Gamma(b-a)}\right) \tag{2.3}$$

or, similarly after replacing b by $a+b$,

$$E_{a,b}(z) = \frac{1}{\Gamma(b)} + zE_{a,b+a}(z)$$

from which, for $b = 0$, we find again (2.2). If $b = \beta + ma$ in (2.3), then we obtain a recursion in m

$$E_{a,\beta+ma}(z) = \frac{1}{z}\left(E_{a,\beta+(m-1)a}(z) - \frac{1}{\Gamma(\beta+(m-1)a)}\right) \tag{2.4}$$

After iteration of (2.4), we find

$$z^m E_{a,\beta+ma}(z) = E_{a,\beta}(z) - \sum_{l=0}^{m-1} \frac{z^l}{\Gamma(\beta+la)} \tag{2.5}$$

The case for $m = 1$ in (2.5) reduces to (2.3).

Relation (2.5) is directly retrieved from the definition (1.1) of $E_{a,b}(z)$,

$$\begin{aligned} E_{a,\beta+ma}(z) &= \sum_{k=0}^{\infty} \frac{z^k}{\Gamma(\beta+a(m+k))} = \sum_{k=m}^{\infty} \frac{z^{k-m}}{\Gamma(\beta+ak)} \\ &= z^{-m}\left(\sum_{k=0}^{\infty} \frac{z^k}{\Gamma(\beta+ak)} - \sum_{k=0}^{m-1} \frac{z^k}{\Gamma(\beta+ak)}\right) \end{aligned}$$

Similarly,

$$\begin{aligned} E_{a,\beta-na}(z) &= \sum_{k=0}^{\infty} \frac{z^k}{\Gamma(\beta+a(k-n))} = \sum_{k=-n}^{\infty} \frac{z^{k+n}}{\Gamma(\beta+ak)} \\ &= z^n\left(\sum_{k=0}^{\infty} \frac{z^k}{\Gamma(\beta+ak)} + \sum_{k=1}^{n} \frac{z^{-k}}{\Gamma(\beta-ak)}\right) \end{aligned}$$

Hence, for a non-negative integer n, it holds that

$$z^{-n} E_{a,\beta-na}(z) = E_{a,\beta}(z) + \sum_{k=1}^{n} \frac{1}{\Gamma(\beta-ak)} \frac{1}{z^k} \tag{2.6}$$

Example If $a = 1$ and $\beta = 1$ in (2.6), then $\frac{1}{\Gamma(\beta-ak)} = \frac{1}{\Gamma(1-k)} = 0$ for $k \geq 1$ and, with $E_{1,1}(z) = e^z$, we find (also from (2.12)), for integers $m \geq 0$, that

$$E_{1,1-m}(z) = z^m e^z \tag{2.7}$$

4. After splitting odd and even indices in the k-sum of (1.1), we obtain

$$E_{a,b}(-z) = \sum_{k=0}^{\infty} \frac{(-1)^k z^k}{\Gamma(b+ak)} = \sum_{k=0}^{\infty} \frac{z^{2k}}{\Gamma(b+2ak)} - \sum_{k=0}^{\infty} \frac{z^{2k+1}}{\Gamma(b+a+2ak)}$$

and

$$E_{a,b}(-z) = E_{2a,b}(z^2) - zE_{2a,b+a}(z^2) \tag{2.8}$$

Property (2.8) cannot be expressed for $E_a(-z) = E_{a,1}(-z)$ in terms of itself and motivates our viewpoint that the complex function theory of the Mittag-Leffler function should focus on the two-parameter function $E_{a,b}(z)$, rather than on the one-parameter function $E_a(z)$. The differentiation rule in **art.** 16 below is the more fundamental motivation.

Adding $E_{a,b}(z) = E_{2a,b}(z^2) + zE_{2a,b+a}(z^2)$ to (2.8) leads to

$$E_{2a,b}(z^2) = \frac{E_{a,b}(z) + E_{a,b}(-z)}{2} \tag{2.9}$$

and, similarly,

$$E_{2a,b+a}(z^2) = \frac{E_{a,b}(z) - E_{a,b}(-z)}{2z} \tag{2.10}$$

Examples From $E_{1,1}(z) = E_1(z) = e^z$, the relation (2.9) indicates that $E_2(z) = \cosh(\sqrt{z})$ and next

$$E_4(z) = \frac{1}{2}\left\{\cosh\left(z^{\frac{1}{4}}\right) + \cos\left(z^{\frac{1}{4}}\right)\right\}$$

The odd variant (2.10) gives $E_{2,2}(z^2) = \frac{\sinh z}{z}$. More generally, (2.7) in (2.9) and (2.10) indicate, for integers $m \geq 0$, that $E_{2,1-2m}(z^2) = z^{2m}\cosh z$ and $E_{2,-2m}(z^2) = z^{2m+1}\sinh z$.

5. *Mittag-Leffler functions of the type* $E_{1,b}(z)$. If the parameter $a = 1$, then the second series (16.18) of the incomplete Gamma function (**art.** 219) indicates that

$$E_{1,b}(z) = \sum_{k=0}^{\infty} \frac{z^k}{\Gamma(b+k)} = e^z \gamma^*(b-1, z) \tag{2.11}$$

With the Taylor series (16.7) of the incomplete Gamma function and the integral representation (16.4), (2.11) is specified as

$$\begin{aligned} E_{1,b}(z) &= \frac{e^z}{\Gamma(b-1)} \sum_{k=0}^{\infty} \frac{(-z)^k}{k!\,(b-1+k)} \\ &= z^{1-b} e^z \left(1 - \frac{1}{\Gamma(b-1)} \int_z^{\infty} e^{-t} t^{b-2} dt\right) \end{aligned} \tag{2.12}$$

If $\mathrm{Re}(b) > 1$, then (2.12) also equals

$$E_{1,b}(z) = \frac{z^{1-b} e^z}{\Gamma(b-1)} \int_0^z e^{-t} t^{b-2} dt \tag{2.13}$$

Moreover, $E_{1,b}(z)$ can also be written in terms of Kummer's confluent hypergeometric function (Abramowitz and Stegun, 1968, 13.1.2)

$$M(a, b; z) = \frac{\Gamma(b)}{\Gamma(a)} \sum_{k=0}^{\infty} \frac{\Gamma(a+k)}{\Gamma(b+k)} \frac{z^k}{k!} \tag{2.14}$$

as

$$E_{1,b}(z) = \sum_{k=0}^{\infty} \frac{z^k}{\Gamma(b+k)} = \frac{M(1, b, z)}{\Gamma(b)}$$

2.2 The cyclotomic property

The cyclotomic property and the progression property in Section 2.3 are remarkable, general features of the Mittag-Leffler function $E_{a,b}(z)$, that unravel its basic functional form, as demonstrated in **art**. 14.

6. *Cyclotomic property.* We apply an instance of the geometric sum (1.15) and (1.16),

$$\sum_{r=0}^{m-1} e^{i\frac{2\pi kr}{m}} = \frac{1-e^{2\pi ki}}{1-e^{i\frac{2\pi k}{m}}} = m1_{m|k}$$

where $m|k$ means that the integer m divides the integer k and where the indicator function 1_x equals 1 if the condition x is true, else $1_x = 0$, to the Taylor series (1.1),

$$\sum_{r=0}^{m-1} E_{a,b}\left(e^{i\frac{2\pi r}{m}}z\right) = \sum_{k=0}^{\infty} \frac{\sum_{r=0}^{m-1} e^{i\frac{2\pi kr}{m}} z^k}{\Gamma(b+ak)} = m\sum_{k=0}^{\infty} \frac{1_{m|k}z^k}{\Gamma(b+ak)}$$
$$= m\sum_{l=0}^{\infty} \frac{z^{ml}}{\Gamma(b+aml)}$$

Hence, we obtain the cyclotomic property of the Mittag-Leffler function

$$E_{am,b}(z^m) = \frac{1}{m}\sum_{r=0}^{m-1} E_{a,b}\left(ze^{i\frac{2\pi r}{m}}\right) \tag{2.15}$$

The right-hand side of (2.15) for $z = re^{i\theta}$ is the average over m Mittag-Leffler function evaluations on a circle in the complex plane, centered at the origin with radius $r = |z|$, starting at angle $\theta, \theta+\frac{2\pi}{m}, \theta+\frac{4\pi}{m}, \ldots, \theta+2\pi-\frac{2\pi}{m}$. The m points $\left\{e^{i\frac{2\pi r}{m}}\right\}_{0\leq r\leq m-1}$ on the unit circle are the roots of $z^m - 1$.

For $m = 2$, the cyclotomic formula (2.15) reduces to

$$2E_{2a,b}\left(z^2\right) = E_{a,b}(z) + E_{a,b}(-z)$$

and we retrieve (2.9), because (2.15) essentially follows by multi-sectioning of a power series (Riordan, 1968, Section 4.3) of which splitting in odd and even terms is the obvious case in $m = 2$ sections.

7. *Representations of the cyclotomic property.* There are several representations of the cyclotomic property. Letting r run in opposite sense over the circle, yields $E_{am,b}(z^m) = \frac{1}{m}\sum_{r=0}^{m-1} E_{a,b}\left(ze^{-i\frac{2\pi r}{m}}\right)$ and another form of the cyclotomic formula (2.15) is

$$E_{am,b}(z^m) = \frac{1}{2m}E_{a,b}(z) + \frac{1}{2m}\sum_{r=-(m-1)}^{m-1} E_{a,b}\left(ze^{i\frac{2\pi r}{m}}\right) \tag{2.16}$$

If $m = 2n+1$ is odd, then (2.15) is split into two sums

$$E_{(2n+1)a,b}\left(z^{2n+1}\right) = \frac{1}{2n+1}\left(\sum_{r=0}^{n} E_{a,b}\left(ze^{i\frac{2\pi r}{2n+1}}\right) + \sum_{r=n+1}^{2n} E_{a,b}\left(ze^{i\frac{2\pi r}{2n+1}}\right)\right)$$

Let $l = 2n+1-r$ in the last sum, then

$$E_{(2n+1)a,b}\left(z^{2n+1}\right) = \frac{1}{2n+1}\left(\sum_{r=0}^{n} E_{a,b}\left(ze^{i\frac{2\pi r}{2n+1}}\right) + \sum_{r=1}^{n} E_{a,b}\left(ze^{-i\frac{2\pi r}{2n+1}}\right)\right)$$

Thus, if $m = 2n+1$ is odd, then

$$E_{(2n+1)a,b}\left(z^{2n+1}\right) = \frac{1}{2n+1}\sum_{r=-n}^{n} E_{a,b}\left(ze^{i\frac{2\pi r}{2n+1}}\right) \tag{2.17}$$

Similarly, if $m = 2n$ is even, then a slight asymmetry appears

$$E_{2na,b}\left(z^{2n}\right) = \frac{1}{2n}\sum_{r=-n}^{n-1} E_{a,b}\left(ze^{i\frac{2\pi r}{2n+1}}\right) = \frac{1}{2n}\sum_{r=-(n-1)}^{n} E_{a,b}\left(ze^{i\frac{2\pi r}{2n+1}}\right) \tag{2.18}$$

Since $E_{a,b}(z)$ is real on the real z-axis for real $a > 0$ and b (**art.** 1), the reflection principle (Titchmarsh, 1964, p. 155) tells that $(E_{a,b}(z))^* = E_{a,b}(z^*)$, where $z^* = x - iy$ is the complex conjugate of $z = x + iy$. Hence, for real $z = x$ and odd $m = 2n+1$, (2.17) reduces, for real $a > 0$ and b, to

$$E_{(2n+1)a,b}\left(x^{2n+1}\right) = \frac{1}{2n+1}\left(E_{a,b}(x) + 2\sum_{r=1}^{n} \operatorname{Re} E_{a,b}\left(xe^{i\frac{2\pi r}{2n+1}}\right)\right)$$

8. *Application of the cyclotomic property.* The case $a = b = 1$ in (2.15)

$$E_{m,1}(z) = E_m(z) = \frac{1}{m}\sum_{r=0}^{m-1} e^{z^{\frac{1}{m}} e^{i\frac{2\pi r}{m}}} \tag{2.19}$$

can be extended to integer values of b. Indeed, using

$$E_{1,n}(z) = \sum_{k=0}^{\infty} \frac{z^k}{(k+n-1)!} = \sum_{k=n-1}^{\infty} \frac{z^{k-(n-1)}}{k!} = z^{-(n-1)}\left(e^z - \sum_{j=0}^{n-2}\frac{z^j}{j!}\right)$$

in $E_{m,n}(z) = \frac{1}{m}\sum_{r=0}^{m-1} E_{1,n}\left(z^{\frac{1}{m}} e^{i\frac{2\pi r}{m}}\right)$ in (2.15) yields

$$E_{m,n}(z) = \frac{z^{-\frac{n-1}{m}}}{m}\sum_{r=0}^{m-1} e^{-i\frac{2\pi r}{m}(n-1)}\left(e^{z^{\frac{1}{m}} e^{i\frac{2\pi r}{m}}} - \sum_{j=0}^{n-2}\frac{z^{\frac{j}{m}} e^{i\frac{2\pi r}{m}j}}{j!}\right)$$

The last double sum

$$\begin{aligned}\sum_{r=0}^{m-1}\sum_{j=0}^{n-2} e^{-i\frac{2\pi r}{m}(n-1-j)}\frac{z^{\frac{j}{m}}}{j!} &= \sum_{q=1}^{n-1}\frac{z^{\frac{n-1-q}{m}}}{(n-1-q)!}\sum_{r=0}^{m-1} e^{-i\frac{2\pi qr}{m}} \\ &= \sum_{q=1}^{n-1}\frac{z^{\frac{n-1-q}{m}} m 1_{m|q}}{(n-1-q)!}\end{aligned}$$

vanishes for all integers $n \leq m$. Thus, for integer $0 \leq n \leq m$, we arrive at

$$E_{m,n}(z) = \frac{z^{-\frac{n-1}{m}}}{m}\sum_{r=0}^{m-1} e^{-i\frac{2\pi r}{m}(n-1)} e^{z^{\frac{1}{m}} e^{i\frac{2\pi r}{m}}}$$

Formula (2.5) with $a = m$ and $\beta = n$ shows, for integers $p \geq 0$, that

$$E_{m,n+pm}(z) = z^{-p}\frac{z^{\frac{1-n}{m}}}{m}\sum_{r=0}^{m-1} e^{-i\frac{2\pi r}{m}(n-1)} e^{z^{\frac{1}{m}} e^{i\frac{2\pi r}{m}}} - z^{-p}\sum_{l=0}^{p-1}\frac{z^l}{\Gamma(n+lm)} \tag{2.20}$$

which indicates that $E_{m,b}(z)$, for integer values of $b = n + pm$ larger than m, additional terms in z need to complement the pure cyclotomic form. The closed form (2.20) for positive integer $a = m$ and any integer values for b can also be derived from the basic complex integral (4.4) as in Wong and Zhao (2002, Theorem 2.1), after a considerably longer computation.

For the case, where $a = 1$ and $b = -n$ is a negative integer and $n \geq 0$, the cyclotomic formula (2.15) reduces, with $E_{1,-n}(z) = z^{n+1}e^z$ in (2.7), to

$$E_{m,-n}(z^m) = \frac{z^{n+1}}{m}\sum_{r=0}^{m-1} e^{i\frac{2\pi r}{m}(n+1)} e^{ze^{i\frac{2\pi r}{m}}}$$

In summary, we find, for positive integers $m > 0$ and all integers $n \in \mathbb{Z}$ constrained by $n \leq m$, that

$$E_{m,n}(z) = \frac{z^{\frac{1-n}{m}}}{m}\sum_{r=0}^{m-1} e^{i\frac{2\pi r}{m}(1-n)} e^{z^{\frac{1}{m}} e^{i\frac{2\pi r}{m}}} \tag{2.21}$$

9. *Iterations of the cyclotomic property.* After replacing am by a and z^m by z, (2.15) becomes

$$E_{a,b}(z) = \frac{1}{m}\sum_{r=0}^{m-1} E_{\frac{a}{m},b}\left(z^{\frac{1}{m}} e^{i\frac{2\pi r}{m}}\right) \tag{2.22}$$

We start a recursion and write the right-hand side r-term by using (2.22),

$$E_{\frac{a}{m},b}\left(z^{\frac{1}{m}} e^{i\frac{2\pi r}{m}}\right) = \frac{1}{m}\sum_{k=0}^{m-1} E_{\frac{a}{m^2},b}\left(z^{\frac{1}{m^2}} e^{i\left(\frac{2\pi r}{m^2}+\frac{2\pi k}{m}\right)}\right)$$

Substituted into (2.22) yields the first iteration

$$E_{a,b}(z) = \frac{1}{m^2}\sum_{k_1=0}^{m-1}\sum_{k_2=0}^{m-1} E_{\frac{a}{m^2},b}\left(z^{\frac{1}{m^2}} e^{i\left(\frac{2\pi k_2}{m^2}+\frac{2\pi k_1}{m}\right)}\right) \tag{2.23}$$

We repeat the recipe, by using (2.22),

$$E_{\frac{a}{m^2},b}\left(z^{\frac{1}{m^2}} e^{i\left(\frac{2\pi k_2}{m^2}+\frac{2\pi k_1}{m}\right)}\right) = \frac{1}{m}\sum_{r=0}^{m-1} E_{\frac{a}{m^3},b}\left(z^{\frac{1}{m^3}} e^{2\pi i\left(\frac{k_2}{m^3}+\frac{k_1}{m^2}+\frac{r}{m}\right)}\right)$$

Substituted again into (2.23) and relabeling the indices yields

$$E_{a,b}(z) = \frac{1}{m^3}\sum_{k_1=0}^{m-1}\sum_{k_2=0}^{m-1}\sum_{k_3=0}^{m-1} E_{\frac{a}{m^3},b}\left(z^{\frac{1}{m^3}} e^{2\pi i\left(\frac{k_3}{m^3}+\frac{k_2}{m^2}+\frac{k_1}{m}\right)}\right)$$

from which we deduce the p-th iteration of the cyclotomic formula (2.22) as

$$E_{a,b}(z) = \frac{1}{m^p} \sum_{k_1=0}^{m-1} \sum_{k_2=0}^{m-1} \cdots \sum_{k_p=0}^{m-1} E_{\frac{a}{m^p},b}\left(z^{\frac{1}{m^p}} e^{2\pi i \sum_{j=1}^{p} \frac{k_j}{m^j}}\right) \tag{2.24}$$

Let $\frac{a}{m^p} = 1$ and $b = 1$, then (2.24) simplifies to

$$E_{m^p}(z) = \frac{1}{m^p} \sum_{k_1=0}^{m-1} \sum_{k_2=0}^{m-1} \cdots \sum_{k_p=0}^{m-1} \exp\left(z^{\frac{1}{m^p}} e^{\frac{2\pi i}{m^p} \sum_{l=0}^{p-1} k_{p-l} m^l}\right)$$

which should be compared with (2.19)

$$E_{m^p}(z) = \frac{1}{m^p} \sum_{r=0}^{m^p-1} \exp\left(z^{\frac{1}{m^p}} e^{i\frac{2\pi r}{m^p}}\right)$$

In particular for $m = 2$, the multiple sum becomes

$$E_{2^p}(z) = \frac{1}{2^p} \sum_{k_1=0}^{1} \sum_{k_2=0}^{1} \cdots \sum_{k_p=0}^{1} \exp\left(z^{\frac{1}{2^p}} e^{\frac{2\pi i}{2^p} \sum_{l=0}^{p-1} k_{p-l} 2^l}\right)$$

where $r = \sum_{l=0}^{p-1} k_{p-l} 2^l$ is the binary representation of the integer r in p digits.

10. *Generalization of the cyclotomic property.* We anticipate the concept of multisectioning of a Taylor series $f(z) = \sum_{k=0}^{\infty} f_k z^k$ in **art.** 12, that leads, for $0 \leq j < m$, to

$$\sum_{r=0}^{\infty} f_{rm+j} z^{rm+j} = \frac{1}{m} \sum_{k=0}^{m-1} e^{\frac{2\pi i k(m-j)}{m}} f\left(e^{\frac{2\pi i k}{m}} z\right) \tag{2.25}$$

If $f_k = \frac{1}{\Gamma(ak+b)}$, then $E_{a,b}(z) = \sum_{k=0}^{\infty} \frac{z^k}{\Gamma(ak+b)}$ and (2.25) shows for $0 \leq j < m$, with $z^j E_{am,aj+b}(z^m) = \sum_{r=0}^{\infty} \frac{z^{mr+j}}{\Gamma(a(mr+j)+b)}$, that

$$z^j E_{am,b+aj}(z^m) = \frac{1}{m} \sum_{k=0}^{m-1} e^{\frac{2\pi i k(m-j)}{m}} E_{a,b}\left(e^{\frac{2\pi i k}{m}} z\right) \tag{2.26}$$

which simplifies for $j = 0$ to (2.15). The generalization (2.26) of the cyclotomic property (2.15) was derived by Srivastava (1994, eq. (26)) and also appeared in the book by Podlubny (1999, eq. (1.98)).

11. *The limit $m \to \infty$ in the cyclotomic property (2.26).* Applying the Riemann sum (12.47) in **art.** 168 to the generalized cyclotomic property (2.26) indicates that

$$\begin{aligned}\int_0^1 e^{-2\pi i j x} E_{a,b}\left(z e^{2\pi i x}\right) dx &= \lim_{m\to\infty} z^j E_{am,b+aj}(z^m) \\ &= \lim_{m\to\infty} z^j \sum_{k=0}^{\infty} \frac{z^{km}}{\Gamma(amk+b+aj)}\end{aligned}$$

Only the $k = 0$ term in the summation is different from zero for $\text{Re}(a) > 0$ in the limit and we obtain, for integer $j \geq 0$, the integral

$$\int_0^1 e^{-2\pi ijx} E_{a,b}\left(e^{2\pi ix} z\right) dx = \frac{z^j}{\Gamma(aj+b)} \tag{2.27}$$

Formula (2.27) is a manifestation of the derivative (1.13) and (1.14). The right-hand side of (2.27) equals $\frac{z^j}{\Gamma(aj+b)} = \frac{1}{j!} \frac{d^j}{dy^j} E_{a,b}(zy)\Big|_{y=0}$ in (1.13) and the integral form (1.14) indicates that

$$\frac{z^j}{\Gamma(aj+b)} = \frac{1}{2\pi i} \int_{C(0)} \frac{E_{a,b}(z\omega)\, d\omega}{\omega^{j+1}}$$

Evaluating the contour around the origin $\omega = 0$ by a circle with unit radius $\omega = e^{2\pi ix}$ leads to (2.27).

2.3 The progression property

12. *Progression property and multisectioning.* A progression is a sequence of the form $\alpha, \alpha+\beta, \alpha+2\beta, \alpha+3\beta, \ldots$, where α and β are numbers. The set of integers $k = \{0, 1, 2, \ldots\}$ can be generated by a progression $k = rm + j$ with $0 \leq j < m$ and $r \geq 0$, so that for any converging Taylor series holds that

$$f(z) = \sum_{k=0}^{\infty} f_k z^k = \sum_{r=0}^{\infty} \sum_{j=0}^{m-1} f_{rm+j} z^{rm+j} = \sum_{j=0}^{m-1} F_j(z; m) \tag{2.28}$$

where the series

$$F_j(z; m) = \sum_{r=0}^{\infty} f_{rm+j} z^{rm+j} \qquad \text{with } 0 \leq j < m \tag{2.29}$$

is called (Riordan, 1968, Section 4.3) the j-th m-section of the series $f(z)$. The j-th m-section $F_j(z; m)$ of the series $f(z)$ also equals

$$F_j(z; m) = \frac{1}{m} \sum_{k=1}^{m} e^{-\frac{2\pi ikj}{m}} f\left(e^{\frac{2\pi ik}{m}} z\right) \tag{2.30}$$

Indeed, introducing the Taylor series $f(z) = \sum_{n=0}^{\infty} f_n z^n$ into (2.30)

$$F_j(z; m) = \frac{1}{m} \sum_{k=1}^{m} e^{-\frac{2\pi ikj}{m}} f\left(e^{\frac{2\pi ik}{m}} z\right) = \frac{1}{m} \sum_{n=0}^{\infty} f_n z^n \left(\sum_{k=1}^{m} e^{\frac{2\pi ik(n-j)}{m}} \right)$$

and using $\sum_{k=1}^{m} e^{\frac{2\pi ik(n-j)}{m}} = m\delta_{m|(n-j)}$, where $m|(n-j)$ implies that $n = rm + j$ for integers $r \geq 0$, leads to (2.29). Chapter 12 on the digamma function presents more examples of the progression property.

13. *Progression property of the Mittag-Leffler function.* The sequence of the integers k can be written as the progression $k = rm + j$ with integer period m, where the integer $0 \leq j < m$ and $0 \leq r$. The Taylor series (1.1) then becomes

$$E_{a,b}(z) = \sum_{k=0}^{\infty} \frac{z^k}{\Gamma(b+ak)} = \sum_{r=0}^{\infty} \sum_{j=0}^{m-1} \frac{z^{rm+j}}{\Gamma(b+arm+aj)}$$

Reversing the summation $E_{a,b}(z) = \sum_{j=0}^{m-1} z^j \sum_{r=0}^{\infty} \frac{(z^m)^r}{\Gamma(b+aj+amr)}$ and invoking the Taylor series (1.1) leads, for any integer $m > 0$, to the progression formula

$$E_{a,b}(z) = \sum_{j=0}^{m-1} z^j E_{am,b+aj}(z^m) \tag{2.31}$$

Hence, (2.31) illustrates the interesting fact that the j-th m-section (2.29) of the Mittag-Leffler function is again a Mittag-Leffler function, but with different parameters a and b. For example, the case $n = 2$ in (2.31) is

$$E_{a,b}(z) = E_{2a,b}(z^2) + zE_{2a,b+a}(z^2)$$

which is (2.8) with z replaced by $-z$. For $n = 3$, (2.31) reduces to

$$E_{a,b}(z) = E_{3a,b}(z^3) + zE_{3a,b+a}(z^3) + z^2 E_{3a,b+2a}(z^3)$$

If $b = a$, then (2.31) simplifies to $E_{a,a}(z) = \frac{1}{z} \sum_{j=1}^{m} z^j E_{am,aj}(z^m)$.

The progression formula (2.31) expresses $E_{a,b}(z)$ as a finite sum of Mittag-Leffler functions with larger a as well as larger b parameter for $j > 0$. In that sense, the progression formula (2.31) can be regarded as kind of a reverse of the cyclotomic formula (2.15). Introducing the cyclotomic property (2.15) in (2.31) yields

$$E_{a,b}(z) = \frac{1}{n} \sum_{j=0}^{n-1} \sum_{r=0}^{n-1} z^j E_{a,b+aj}\left(ze^{i\frac{2\pi r}{n}}\right) \tag{2.32}$$

As a verification, by introducing the Taylor series (1.1), we show that the relation (2.32) is an identity. Indeed,

$$\begin{aligned} E_{a,b}(z) &= \frac{1}{n} \sum_{j=0}^{n-1} \sum_{r=0}^{n-1} z^j \sum_{k=0}^{\infty} \frac{\left(ze^{i\frac{2\pi r}{n}}\right)^k}{\Gamma(ak+b+aj)} \\ &= \frac{1}{n} \sum_{k=0}^{\infty} \left(\sum_{r=0}^{n-1} e^{i\frac{2\pi k}{n} r} \right) \sum_{j=0}^{n-1} \frac{z^{j+k}}{\Gamma(ak+b+aj)} \end{aligned}$$

Formula $\sum_{r=0}^{n-1} e^{i\frac{2\pi r}{n} k} = n 1_{n|k}$ requests that $k = pn$, i.e. k must be a multiplier of n, where p runs over all integers,

$$E_{a,b}(z) = \sum_{p=0}^{\infty} \sum_{j=0}^{n-1} \frac{z^{j+pn}}{\Gamma(a(np+j)+b)} = \sum_{l=0}^{\infty} \frac{z^l}{\Gamma(al+b)}$$

The last step again stems from summing over all integers $l = np + j$ in progressions of n.

Substitution of the generalized cyclotomic property (2.26) into the progression formula (2.31) leads, with $\sum_{j=0}^{n-1} e^{i\frac{2\pi rj}{n}} = n1_{n|r}$, to an identity. Hence, just as (2.28) and (2.30), the generalized cyclotomic property (2.26) and the progression formula (2.31) are inverse formulae, for integers $0 \leq j < m$ and $m > 0$:

$$E_{a,b}(z) = \sum_{j=0}^{m-1} z^j E_{am,b+aj}(z^m)$$
$$\iff \quad z^j E_{am,b+aj}(z^m) = \frac{1}{m}\sum_{r=0}^{m-1} e^{-2\pi i\frac{rj}{m}} E_{a,b}\left(e^{\frac{2\pi ir}{m}} z\right)$$

14. *Rational order* $a = \frac{m}{n} \in \mathbb{Q}$. The progression formula (2.31) for $a = \frac{m}{n}$, where m and n are integers, becomes[1]

$$E_{\frac{m}{n},b}(z) = \sum_{j=0}^{n-1} z^j E_{m,b+\frac{m}{n}j}(z^n) \tag{2.33}$$

Introducing the cyclotomic property (2.15) with $a = 1$ in (2.33) leads to

$$E_{\frac{m}{n},b}(z) = \frac{1}{m}\sum_{j=0}^{n-1} z^j \sum_{r=0}^{m-1} E_{1,b+\frac{m}{n}j}\left(z^{\frac{n}{m}} e^{i\frac{2\pi r}{m}}\right) \tag{2.34}$$

Art. 5 shows that $E_{1,b}(z)$ can be written in terms of the incomplete Gamma function as $E_{1,b}(z) = e^z \gamma^*(b-1, z)$ in (2.11) and other known functions. Hence, formula (2.34) underlines the intimate connection of the Mittag-Leffler function at rational parameter $a = \frac{m}{n}$ and any complex b with the incomplete Gamma function. Since an irrational number a can be approximated arbitrarily close (Hardy and Wright, 2008, Chapter IX) by a fraction $\frac{m}{n}$, (2.34) approximates the Mittag-Leffler function $E_{a,b}(z)$ for any real $a > 0$ as a finite double sum in terms of known analytic functions.

For example, using (2.13) in (2.34) gives, for $\text{Re}(b) > 1$,

$$E_{\frac{m}{n},b}(z) = \frac{z^{\frac{n}{m}(1-b)}}{m}\sum_{r=0}^{m-1} e^{i\frac{2\pi r}{m}(1-b)} e^{z^{\frac{n}{m}} e^{i\frac{2\pi r}{m}}}$$
$$\times \int_0^{z^{\frac{n}{m}} e^{i\frac{2\pi r}{m}}} u^{b-2} e^{-u} \sum_{j=0}^{n-1} \frac{\left(u^{\frac{m}{n}} e^{-i\frac{2\pi r}{n}}\right)^j}{\Gamma\left(\frac{m}{n}j + b - 1\right)} du \tag{2.35}$$

The finite sum formula (2.35), that can approximate any irrational number a arbitrarily close by a rational number $\frac{m}{n}$, contains the functional essence of a Mittag-Leffler function $E_{a,b}(z)$ with real positive a and any complex z and complex b with $\text{Re}(b) > 1$. In Section 4.2 on the basic integral for the Mittag-Leffler function, we will again encounter the characteristic functional forms of the type $\frac{z^{\frac{1-b}{a}}}{a} e^{z^{\frac{1}{a}}}$.

[1] If $m = n$ and $b = 1$ in (2.33), then we find, after substituting (2.21) and invoking $\sum_{j=0}^{n-1} e^{-i\frac{2\pi r}{n}j} = n1_{r=0}$, an identity.

If $m = 1$, then (2.35) reduces, for $\mathrm{Re}(b) > 1$, to

$$E_{\frac{1}{n},b}(z) = z^{n(1-b)} e^{z^n} \int_0^{z^n} e^{-u} \sum_{j=0}^{n-1} \frac{u^{b-2+\frac{j}{n}}}{\Gamma\left(\frac{j}{n}+b-1\right)} du$$

which is the generalization (2.44) for any b of Wiman's formula (2.45) for $b = 1$, differently derived below in **art.** 17. If $n = 1$, then (2.35) reduces, for $\mathrm{Re}(b) > 1$, to

$$E_{m,b}(z) = \frac{z^{\frac{1-b}{m}}}{m} \sum_{r=0}^{m-1} e^{i\frac{2\pi r}{m}(1-b)} e^{z^{\frac{1}{m}} e^{i\frac{2\pi r}{m}}} \frac{\int_0^{z^{\frac{1}{m}} e^{i\frac{2\pi r}{m}}} u^{b-2} e^{-u} du}{\Gamma(b-1)}$$

which is the generalization of (2.21) for non-integer b with $\mathrm{Re}(b) > 1$.

2.4 Derivatives of the Mittag-Leffler function

2.4.1 Direct differentiation of $E_{a,b}(z)$

15. *Differentiation with respect to* z. The derivative of the definition (1.1) with respect to z is

$$\begin{aligned}\frac{d}{dz} E_{a,b}(z) &= \sum_{k=1}^{\infty} \frac{k z^{k-1}}{\Gamma(b+ak)} = \sum_{k=0}^{\infty} \frac{(k+1) z^k}{\Gamma(b+a+ak)} \\ &= \frac{1}{a} \sum_{k=0}^{\infty} \frac{(ak+b+a-1) z^k}{\Gamma(b+a+ak)} - \frac{b-1}{a} \sum_{k=0}^{\infty} \frac{z^k}{\Gamma(b+a+ak)} \\ &= \frac{1}{a} \sum_{k=0}^{\infty} \frac{z^k}{\Gamma(b+a-1+ak)} - \frac{b-1}{a} \sum_{k=0}^{\infty} \frac{z^k}{\Gamma(b+a+ak)}\end{aligned}$$

and, with the definition (1.1),

$$\frac{d}{dz} E_{a,b}(z) = \frac{1}{a} E_{a,b+a-1}(z) - \frac{b-1}{a} E_{a,b+a}(z) \tag{2.36}$$

Using $E_{a,b+a}(z) = \frac{1}{z}\left(E_{a,b}(z) - \frac{1}{\Gamma(b)}\right)$ in (2.3) simplifies (2.36) to

$$az \frac{d}{dz} E_{a,b}(z) = E_{a,b-1}(z) - (b-1) E_{a,b}(z) \tag{2.37}$$

The m-th derivative

$$a \frac{d^m}{dz^m} E_{a,b}(z) = \frac{d^{m-1}}{dz^{m-1}} E_{a,b+a-1}(z) - (b-1) \frac{d^{m-1}}{dz^{m-1}} E_{a,b+a}(z)$$

has a recursive structure, when denoting $h_m(b) = \frac{d^m}{dz^m} E_{a,b}(z)$ and $h_0(b) = E_{a,b}(z)$,

$$a h_m(b) = h_{m-1}(b+a-1) - (b-1) h_{m-1}(b+a)$$

which can be iterated resulting in

$$a^m \frac{d^m}{dz^m} E_{a,b}(z) = \sum_{j=0}^{m} q_j(a,b,m) E_{a,b+ma-j}(z)$$

For example,

$$a^2 \frac{d^2 E_{a,b}(z)}{dz^2} = E_{a,b+2a-2}(z) - (2b+a-3) E_{a,b+2a-1}(z) + (b-1)(b+a-1) E_{a,b+2a}(z)$$

$$a^3 \frac{d^3 E_{a,b}(z)}{dz^3} = E_{a,b+3a-3}(z) - (3b+3a-6) E_{a,b+3a-2}(z) + \{(2b+a-3)(b+2a-2) + (b-1)(b+a-1)\} E_{a,b+3a-1}(z) - (b-1)(b-1+a)(b-1+2a) E_{a,b+3a}(z)$$

With (2.5), we find that

$$(az)^m \frac{d^m}{dz^m} E_{a,b}(z) = \sum_{j=0}^{m} q_j(a,b,m) \left(E_{a,b-j}(z) - \sum_{l=0}^{m-1} \frac{z^l}{\Gamma(b-j+la)} \right) \tag{2.38}$$

with coefficients

$$q_m(a,b,m) = 1$$

$$q_{m-1}(a,b,m) = -\left(mb + \frac{m(m-1)}{2} a - \frac{m(m+1)}{2} \right)$$

$$q_0(a,b,m) = (-1)^m \prod_{k=0}^{m-1} (b-1+ka)$$

In general, $q_j(a,b,m)$ are polynomials in a and b of order $m-j$. Closed forms of the polynomial $q_j(a,b,m)$ will be deduced in Section 2.6, after we have derived the Taylor expansion of $E_{a,b}(z)$ around an arbitrary point z_0 in art. 18.

2.4.2 Recursive formulas for derivatives

16. *Differential recursion.* Using the functional equation of the Gamma function, $\Gamma(z+1) = z\Gamma(z)$, we write

$$E_{a,b}(xz) = \sum_{k=0}^{\infty} \frac{x^k z^k}{\Gamma(b+ak)} = \sum_{k=0}^{\infty} \frac{x^k z^k}{(b-1+ak)\Gamma(b-1+ak)}$$

so that

$$z^{b-1} E_{a,b}(xz^a) = \sum_{k=0}^{\infty} \frac{x^k z^{b-1+ak}}{(b-1+ak)\Gamma(b-1+ak)}$$

Differentiation with respect to z gives

$$\frac{d}{dz}\left\{ z^{b-1} E_{a,b}(xz^a) \right\} = \sum_{k=0}^{\infty} \frac{x^k z^{b-2+ak}}{\Gamma(b-1+ak)} = z^{b-2} \sum_{k=0}^{\infty} \frac{(xz^a)^k}{\Gamma(b-1+ak)}$$

With the definition (1.1), we arrive for any x at the differential recursion in b,

$$\frac{d}{dz}\left\{z^{b-1}E_{a,b}\left(xz^{a}\right)\right\} = z^{b-2}E_{a,b-1}\left(xz^{a}\right) \tag{2.39}$$

Differentiating (2.39) again m-times and using the recursion (2.39) yields

$$\frac{d^m}{dz^m}\left\{z^{b-1}E_{a,b}\left(xz^{a}\right)\right\} = z^{b-1-m}E_{a,b-m}\left(xz^{a}\right) \tag{2.40}$$

The differential recursion (2.39) in b holds for a broader class of functions. Indeed, the function $f_{a,b}(z)$ with Taylor series $f_{a,b}(z) = \sum_{k=0}^{\infty}\frac{f_k z^k}{\Gamma(b+ak)}$ as well as its associated integral $\mathcal{F}_{a,b}(z) = \int_0^{\infty}\frac{f(u)z^u}{\Gamma(b+au)}du$ obey the same differential recursion in b,

$$\frac{d}{dz}\left\{z^{b-1}f_{a,b}\left(xz^{a}\right)\right\} = z^{b-2}f_{a,b-1}\left(xz^{a}\right) \tag{2.41}$$

17. *Fractional values of a.* Let $a = \frac{m}{n}$ where $m \leq n$ are integers, then the differentiation formula (2.40) becomes, for $x = 1$,

$$\begin{aligned}
\frac{d^m}{dz^m}\left\{z^{b-1}E_{\frac{m}{n},b}\left(z^{\frac{m}{n}}\right)\right\} &= z^{b-1}\sum_{k=0}^{\infty}\frac{z^{\frac{m}{n}k-m}}{\Gamma\left(b-m+\frac{m}{n}k\right)} \\
&= z^{b-1}\sum_{k=0}^{\infty}\frac{z^{\frac{m}{n}(k-n)}}{\Gamma\left(b+\frac{m}{n}(k-n)\right)} \\
&= z^{b-1}\sum_{k=-n}^{\infty}\frac{z^{\frac{m}{n}k}}{\Gamma\left(b+\frac{m}{n}k\right)} \\
&= z^{b-1}\sum_{k=1}^{n}\frac{z^{-\frac{m}{n}k}}{\Gamma\left(b-\frac{m}{n}k\right)} + z^{b-1}\sum_{k=0}^{\infty}\frac{z^{\frac{m}{n}k}}{\Gamma\left(b+\frac{m}{n}k\right)}
\end{aligned}$$

and we find the differential equation for $z^{b-1}E_{\frac{m}{n},b}\left(z^{\frac{m}{n}}\right)$,

$$\frac{d^m}{dz^m}\left\{z^{b-1}E_{\frac{m}{n},b}\left(z^{\frac{m}{n}}\right)\right\} - z^{b-1}E_{\frac{m}{n},b}\left(z^{\frac{m}{n}}\right) = z^{b-1}\sum_{k=1}^{n}\frac{z^{-\frac{m}{n}k}}{\Gamma\left(b-\frac{m}{n}k\right)} \tag{2.42}$$

For $n = 1$, (2.42) simplifies to

$$\frac{d^m}{dz^m}\left\{z^{b-1}E_{m,b}\left(z^{m}\right)\right\} = \frac{z^{b-1-m}}{\Gamma(b-m)} + z^{b-1}E_{m,b}\left(z^{m}\right)$$

In particular, when $b = 1$, then it holds for $m \geq 1$ that $\frac{d^m}{dz^m}E_m\left(z^m\right) = E_m\left(z^m\right)$, which illustrates that $y = E_m\left(z^m\right)$, explicitly given in (2.19), is a solution of the differential equation $\frac{d^m y}{dz^m} = y$.

For $m = 1$, (2.42) reduces to

$$\frac{d}{dz}\left\{z^{b-1}E_{\frac{1}{n},b}\left(z^{\frac{1}{n}}\right)\right\} = z^{b-1}E_{\frac{1}{n},b}\left(z^{\frac{1}{n}}\right) + \sum_{k=1}^{n}\frac{z^{b-1-\frac{1}{n}k}}{\Gamma\left(b-\frac{1}{n}k\right)}$$

from which

$$\frac{d}{dz}\left\{e^{-z}z^{b-1}E_{\frac{1}{n},b}\left(z^{\frac{1}{n}}\right)\right\} = e^{-z}\sum_{k=1}^{n}\frac{z^{b-1-\frac{1}{n}k}}{\Gamma\left(b-\frac{1}{n}k\right)}$$

Integrating both sides from z_0 to z yields, for all complex z, z_0 and b,

$$e^{-z}z^{b-1}E_{\frac{1}{n},b}\left(z^{\frac{1}{n}}\right) = e^{-z_0}z_0^{b-1}E_{\frac{1}{n},b}\left(z_0^{\frac{1}{n}}\right) + \sum_{k=1}^{n}\frac{\int_{z_0}^{z} t^{b-1-\frac{1}{n}k}e^{-t}dt}{\Gamma\left(b-\frac{1}{n}k\right)}$$

If we choose $z_0 = 0$, then we must require[2] that $\operatorname{Re} b \geq 1$ and obtain

$$E_{\frac{1}{n},b}\left(z^{\frac{1}{n}}\right) - z^{1-b}e^{z}\left\{\frac{1_{\{b=1\}}}{\Gamma(b)} + \sum_{k=1}^{n}\frac{1}{\Gamma\left(b-\frac{1}{n}k\right)}\int_0^z t^{b-1-\frac{1}{n}k}e^{-t}dt\right\} \tag{2.43}$$

where the indicator function 1_x is defined in **art.** 6. After replacing z by $z^{\frac{1}{n}}$ in (2.43) and letting $j = n-k$ in the summation, we obtain

$$E_{\frac{1}{n},b}(z) = z^{(1-b)n}e^{z^n}\left\{\frac{1_{\{b=1\}}}{\Gamma(b)} + \sum_{j=0}^{n-1}\frac{1}{\Gamma\left(b-1+\frac{j}{n}\right)}\int_0^{z^n} t^{\left(b-1+\frac{j}{n}\right)-1}e^{-t}dt\right\} \tag{2.44}$$

which is presented without indicator $1_{\{b=1\}}$ in (4.1). The progression formula (2.31) for $a = \frac{1}{n}$ with the integral $E_{1,b}(z) = \frac{z^{1-b}e^z}{\Gamma(b-1)}\int_0^z e^{-t}t^{b-2}dt$ in (2.13), valid for $\operatorname{Re}(b) > 1$, again reproduces (2.44). Formula (2.44) reduces, for $n = 1$, to (2.12) and for $b = 1$, to Wiman's form in Wiman (1905a)

$$E_{\frac{1}{n}}(z) = e^{z^n}\left\{1 + \int_0^{z^n} e^{-t}\sum_{j=1}^{n-1}\frac{t^{\frac{j}{n}-1}}{\Gamma\left(\frac{j}{n}\right)}dt\right\} \tag{2.45}$$

For positive real $x = \operatorname{Re}(z)$, it holds that

$$\int_0^{x^n} t^{\left(b-1+\frac{j}{n}\right)-1}e^{-t}dt < \int_0^{\infty} t^{\left(b-1+\frac{j}{n}\right)-1}e^{-t}dt = \Gamma\left(b-1+\frac{j}{n}\right)$$

[2] Less confining for b is the choice $z_0 = |z_0|\,e^{i\arg z_0} \to \infty$ at an angle $\arg z_0$ for which $E_{\frac{1}{n},b}\left(z_0^{\frac{1}{n}}\right) \to 0$, which is studied in Chapter 4. Then, $\int_z^{z_0} t^{b-1-\frac{1}{n}k}e^{-t}dt = \Gamma\left(b-\frac{1}{n}k, z\right)$ is the incomplete Gamma function, which is defined in **art.** 220 for all complex b and z, except for real negative z. Thus, for $|\arg z| > \frac{\pi}{n}$ (**art.** 69), we find

$$E_{\frac{1}{n},b}\left(z^{\frac{1}{n}}\right) = -z^{1-b}e^{z}\sum_{k=1}^{n}\frac{\Gamma\left(b-\frac{1}{n}k, z\right)}{\Gamma\left(b-\frac{1}{n}k\right)}$$

Introducing the recursion (16.11)

$$E_{\frac{1}{n},b}\left(z^{\frac{1}{n}}\right) = -\sum_{k=1}^{n}\sum_{l=0}^{m-1}\frac{1}{\Gamma\left(b-\frac{1}{n}k-l\right)}\frac{1}{z^{l+\frac{k}{n}}} - z^{1-b}e^{z}\sum_{k=1}^{n}\frac{\Gamma\left(b-\frac{1}{n}k-m, z\right)}{\Gamma\left(b-\frac{1}{n}k-m\right)}$$

indicates, for large $|z|$ and $|\arg z| > \frac{\pi}{n}$, that $E_{\frac{1}{n},b}\left(z^{\frac{1}{n}}\right) = -\frac{z^{-\frac{1}{n}}}{\Gamma\left(b-\frac{1}{n}\right)} + O\left(z^{-\frac{2}{n}}\right)$ agreeing with (4.16).

and (2.44) shows that $E_{\frac{1}{n},b}(x) < x^{(1-b)n} e^{x^n} \left\{1_{\{b=1\}} + n - 1_{\{b=1\}}\right\}$. Thus,

$$E_{\frac{1}{n},b}(x) < n x^{(1-b)n} e^{x^n}$$

illustrates for $a = \frac{1}{n}$ that the entire function $E_{\frac{1}{n},b}(z)$ has order $\rho = \frac{1}{a} = n$. This bound also reappears in **art.** 75. Moreover, it is interesting to compare (2.21) formally, written for an integer $b < m$,

$$E_{m,b}(z) = \frac{z^{\frac{1-b}{m}}}{m} \left\{ e^{z^{\frac{1}{m}}} + \sum_{r=1}^{m-1} e^{-i\frac{2\pi r}{m}(b-1)} e^{z^{\frac{1}{m}} e^{i\frac{2\pi r}{m}}} \right\}$$

with Bieberbach's integral (7.41) for $E_{a,b}(z)$ and with the bound, where $\frac{1}{n}$ is replaced by m,

$$E_{m,b}(z) < \frac{z^{\frac{1-b}{m}}}{m} e^{z^{\frac{1}{m}}}$$

We return to the relation between $E_{a,b}(z)$ and $E_{\frac{1}{a},b}(z)$ later in **art.** 75.

Example If $n = 1$ in (2.43), then we retrieve (2.12). If $n = 2$ in (2.43), then we find for $b = 1$

$$E_{\frac{1}{2}}\left(z^{\frac{1}{2}}\right) = e^z \left\{ 1 + \frac{1}{\Gamma\left(\frac{1}{2}\right)} \int_0^z t^{-\frac{1}{2}} e^{-t} dt \right\} = e^z \left\{ 1 + \frac{2}{\sqrt{\pi}} \int_0^{\sqrt{z}} e^{-u^2} du \right\}$$

With the definition (Abramowitz and Stegun, 1968, 7.1.1) of the error function $\operatorname{erf}(x) = \frac{2}{\sqrt{\pi}} \int_0^z e^{-u^2} du$, we arrive at

$$E_{\frac{1}{2}}(z) = e^{z^2} \left\{ 1 + \frac{2}{\sqrt{\pi}} \int_0^z e^{-u^2} du \right\} = e^{z^2} \left\{ 1 + \operatorname{erf}(z) \right\} \tag{2.46}$$

2.5 Taylor series of $E_{a,b}(z)$ around an arbitrary point z_0

18. *Taylor series around* z_0. Since the Mittag-Leffler function $E_{a,b}(z)$ is an entire function, any Taylor series around an arbitrary, finite point z_0 has infinite radius of convergence,

$$E_{a,b}(z) = \sum_{m=0}^{\infty} \frac{1}{m!} \left. \frac{d^m E_{a,b}(z)}{dx^m} \right|_{x=z_0} (z - z_0)^m$$

With the m-th derivative (2.38), we find

$$E_{a,b}(z) = \sum_{m=0}^{\infty} \left\{ \sum_{j=0}^{m} q_j(a,b,m) \left(E_{a,b-j}(z_0) - \sum_{l=0}^{m-1} \frac{z_0^l}{\Gamma(b-j+la)} \right) \right\} \frac{\left(\frac{z-z_0}{az_0}\right)^m}{m!}$$

but, unfortunately (see **art.** 15), the coefficients $q_j(a,b,m)$ are not available in closed form yet. However, the closed form (2.40) of the m-th derivative, which is a fundamental property of $E_{a,b}(z)$, opens a new avenue. The

Taylor series around z_0 of

$$z^{b-1}E_{a,b}\left(z^a\right)=\sum_{m=0}^{\infty}\frac{1}{m!}\frac{d^m}{dx^m}\left\{x^{b-1}E_{a,b}\left(x^a\right)\right\}\Big|_{x=z_0}\left(z-z_0\right)^m$$

becomes with the differential recursion (2.40)

$$z^{b-1}E_{a,b}\left(z^a\right)=z_0^{b-1}\sum_{m=0}^{\infty}\frac{E_{a,b-m}\left(z_0^a\right)}{m!}z_0^{-m}\left(z-z_0\right)^m$$

The function $z^{b-1}E_{a,b}\left(z^a\right)$ has a branch cut at the negative real axis for real a and b, implying that the radius R of convergence equals $|z_0|$.

We proceed by removing real powers of $z^\beta=e^{\beta\log z}$ that destroy analyticity in the complex plane and introduce formally the Taylor series $\left(\frac{z}{z_0}\right)^{1-b}=\left(\frac{z}{z_0}-1+1\right)^{1-b}=\sum_{k=0}^{\infty}\binom{1-b}{k}\left(\frac{z}{z_0}-1\right)^k$, valid for any b and $|z|<|z_0|$, in

$$E_{a,b}\left(z^a\right)=\left(\frac{z}{z_0}\right)^{1-b}\sum_{m=0}^{\infty}\frac{E_{a,b-m}\left(z_0^a\right)}{m!}\left(\frac{z}{z_0}-1\right)^m$$

yielding, after executing the Cauchy product,

$$E_{a,b}\left(z^a\right)=\sum_{m=0}^{\infty}\left\{\sum_{k=0}^{m}\binom{1-b}{m-k}\frac{E_{a,b-k}\left(z_0^a\right)}{k!}\right\}\left(\frac{z}{z_0}-1\right)^m\qquad\text{for } |z|<|z_0|$$

Next, letting $y=z^a$ and $y_0=z_0^a$, we first expand $\left(\left(\frac{y}{y_0}\right)^{\frac{1}{a}}-1\right)^m$ in a Taylor series around y_0

$$\begin{aligned}\left(\left(\frac{y}{y_0}\right)^{\frac{1}{a}}-1\right)^m&=\sum_{q=0}^{m}\binom{m}{q}\left(\frac{y}{y_0}\right)^{\frac{q}{a}}(-1)^{m-q}\\&=\sum_{q=0}^{m}\binom{m}{q}\left(\frac{y-y_0}{y_0}+1\right)^{\frac{q}{a}}(-1)^{m-q}\\&=\sum_{j=0}^{\infty}\left\{\sum_{q=0}^{m}\binom{m}{q}\binom{\frac{q}{a}}{j}(-1)^{m-q}\right\}\left(\frac{y-y_0}{y_0}\right)^j\end{aligned}$$

Substitution and reversing the m- and k-sum yields, with $Y=\frac{y-y_0}{y_0}$,

$$\begin{aligned}E_{a,b}(y)&=\sum_{j=0}^{\infty}\left[\sum_{m=0}^{\infty}\sum_{k=0}^{m}\binom{1-b}{m-k}\frac{E_{a,b-k}(y_0)}{k!}\left\{\sum_{q=0}^{m}\binom{m}{q}\binom{\frac{q}{a}}{j}(-1)^{m-q}\right\}\right]Y^j\\&=\sum_{j=0}^{\infty}\left[\sum_{k=0}^{\infty}\frac{E_{a,b-k}(y_0)}{k!}\sum_{m=k}^{\infty}\binom{1-b}{m-k}\sum_{q=0}^{m}\binom{m}{q}\binom{\frac{q}{a}}{j}(-1)^{m-q}\right]Y^j\end{aligned}$$

We observe from (A.25) in Appendix A that the q-sum is the characteristic coefficient of $(1+z)^{\frac{1}{a}}$,

$$E_{a,b}(y) = \sum_{j=0}^{\infty} \left[\sum_{k=0}^{\infty} \frac{E_{a,b-k}(y_0)}{k!} \sum_{m=k}^{\infty} \binom{1-b}{m-k} s[m,j]|_{(1+z)^{\frac{1}{a}}} \right] \left(\frac{y-y_0}{y_0} \right)^j$$

Since a characteristic coefficient $s[k,m]_{f(z)}(z_0) = 0$ for $k > m$, as follows from its definition (A.2), we obtain

$$E_{a,b}(y) = \sum_{j=0}^{\infty} \left[\sum_{k=0}^{j} \frac{E_{a,b-k}(y_0)}{k!} \sum_{m=k}^{j} \binom{1-b}{m-k} s[m,j]|_{(1+z)^{\frac{1}{a}}} \right] \left(\frac{y-y_0}{y_0} \right)^j$$

Finally, after replacing $y_0 = w$, we arrive, with (A.25) and (1.12), at the Taylor series of $E_{a,b}(y)$ around w,

$$E_{a,b}(y) = \sum_{j=0}^{\infty} \frac{1}{j!} \left. \frac{d^j E_{a,b}(z)}{dz^j} \right|_{z=w} (y-w)^j \tag{2.47}$$

with the closed form of the j-th derivative of the Mittag-Leffler function

$$\left. \frac{d^j E_{a,b}(z)}{dz^j} \right|_{z=w} = w^{-j} \sum_{k=0}^{j} E_{a,b-k}(w) \sum_{m=k}^{j} \binom{m}{k} \frac{\Gamma(2-b) \sum_{q=m}^{j} S_j^{(q)} \mathcal{S}_q^{(m)} \frac{1}{a^q}}{\Gamma(2-b-m+k)} \tag{2.48}$$

Since the Mittag-Leffler function $E_{a,b}(z)$ is an entire function, the j-th Taylor coefficient around w decreases faster than any power of $(z-w)^j$ for large j and any z, we deduce that $\left. \frac{d^j E_{a,b}(z)}{dz^j} \right|_{z=w} = o\left(j! w^j\right)$. Thus, the j-th Taylor coefficient $\frac{1}{j!} \left. \frac{d^j E_{a,b}(z)}{dz^j} \right|_{z=w}$ increases at most as a polynomial in w of order j. In contrast to the relatively simple Taylor series of $E_{a,b}(z)$ in (1.1) around the origin, the general form (2.48) emphasizes the complicated nature of the Mittag-Leffler function $E_{a,b}(z)$ elsewhere in the complex plane.

19. *Verification of (2.48) for* $a = b = 1$. The first orthogonality condition of the Stirling numbers – the second is (2.63) below –

$$\sum_{k=n}^{m} S_k^{(n)} \mathcal{S}_m^{(k)} = \delta_{nm} \tag{2.49}$$

shows that the Taylor series (2.47) simplifies to

$$E_{a,b}(y) = \sum_{j=0}^{\infty} \left[\sum_{k=0}^{j} E_{1,1-k}(w) \sum_{m=k}^{j} \binom{m}{k} \frac{\Gamma(2-b)}{\Gamma(2-b-m+k)} \delta_{jm} \right] \frac{\left(\frac{y-w}{w}\right)^j}{j!}$$

With $E_{1,1-k}(z) = z^k e^z$ in (2.7) and $b = 1$, we establish correctness of

$$E_{1,1}(y) = e^w \sum_{j=0}^{\infty} \left[\sum_{k=0}^{j} \frac{\binom{j}{k} w^{k-j}}{\Gamma(1-j+k)} \right] \frac{(y-w)^j}{j!} = e^w \sum_{j=0}^{\infty} \frac{(y-w)^j}{j!} = e^y$$

2.6 The polynomial $q_j(a,b,m)$

20. *Polynomial $q_j(a,b,m)$ defined in (2.38).* We recall the polynomial $q_j(a,b,m)$ defined in the m-th derivative (2.38),

$$(az)^m \frac{d^m}{dz^m} E_{a,b}(z) = \sum_{j=0}^{m} q_j(a,b,m) \left(E_{a,b-j}(z) - \sum_{l=0}^{m-1} \frac{z^l}{\Gamma(b-j+la)} \right)$$

The same derivative follows from (2.48) as

$$(az)^m \frac{d^m E_{a,b}(z)}{dz^m} = \sum_{j=0}^{m} E_{a,b-j}(z) \sum_{n=j}^{m} \binom{n}{j} \frac{\Gamma(2-b) \sum_{q=n}^{m} S_m^{(q)} \mathcal{S}_q^{(n)} a^{m-q}}{\Gamma(2-b-n+j)}$$

Equating corresponding terms in $E_{a,b-j}(z)$ indicates that

$$q_j(a,b,m) = \sum_{n=j}^{m} \binom{n}{j} \frac{\Gamma(2-b)}{\Gamma(2-b-n+j)} \sum_{l=n}^{m} S_m^{(l)} \mathcal{S}_l^{(n)} a^{m-l} \tag{2.50}$$

and that

$$\sum_{j=0}^{m} q_j(a,b,m) \sum_{l=0}^{m-1} \frac{z^l}{\Gamma(b-j+la)} = 0 \tag{2.51}$$

Rewriting the latter as $\sum_{l=0}^{m-1} \left(\sum_{j=0}^{m} \frac{q_j(a,b,m)}{\Gamma(b-j+la)} \right) z^l = 0$ indicates, after equating corresponding powers of z, that, for $0 \le l \le m-1$,

$$\sum_{j=0}^{m} \frac{q_j(a,b,m)}{\Gamma(b-j+la)} = 0$$

If $l = m$, then we obtain from (2.38), for $z = 0$, with $\left.\frac{d^m}{dz^m} E_{a,b}(z)\right|_{z=0} = \frac{m!}{\Gamma(b+am)}$,

$$\sum_{j=0}^{m} \frac{q_j(a,b,m)}{\Gamma(b-j+ma)} = \frac{m! a^m}{\Gamma(b+am)}$$

which bears resemblance to $\frac{x^m}{\Gamma(x+1)} = \sum_{j=0}^{m} \frac{\mathcal{S}_m^{(j)}}{\Gamma(x+1-j)}$, the generating function (A.6) of the Stirling numbers $\mathcal{S}_m^{(j)}$ of the second kind.

As a consequence of (2.51), (2.38) simplifies to

$$(az)^m \frac{d^m}{dz^m} E_{a,b}(z) = \sum_{j=0}^{m} q_j(a,b,m) E_{a,b-j}(z) \tag{2.52}$$

With (2.52), the Taylor series (2.47) around z_0 becomes

$$E_{a,b}(z) = \sum_{m=0}^{\infty} \left\{ \sum_{j=0}^{m} q_j(a,b,m) E_{a,b-j}(z_0) \right\} \frac{\left(\frac{z-z_0}{az_0}\right)^m}{m!} \tag{2.53}$$

Art. 16 and the differential recursion (2.41) show that a general class

$f_{a,b}(z)$ of functions possesses the same Taylor series (2.53) with formally $E_{a,b}$ replaced by $f_{a,b}$.

The polynomial $q_j(a,b,m)$ in (2.50) equals, for a few values of j,

$$q_m(a,b,m) = 1$$

$$q_{m-1}(a,b,m) = \binom{m}{2}(1-a) + m(1-b)$$

$$q_{m-2}(a,b,m) = \binom{m}{2}\left(\frac{(m-2)(3m-1)}{12}a^2 - \binom{m-1}{2}a - (1-b)b\right) + (m-1)(1-b)\binom{m}{2}(1-a) + \binom{m}{3}\frac{3m-5}{4}$$

illustrating the rather irregular structure of $q_j(a,b,m)$ in a,b and m.

Another form of the polynomial $q_j(a,b,m)$ arises after reversion of the summations in (2.50) as

$$q_j(a,b,m) = \sum_{l=j}^{m} S_m^{(l)} a^{m-l} \sum_{n=j}^{l} \binom{n}{j} \frac{\Gamma(2-b)}{\Gamma(2-b-n+j)} \mathcal{S}_l^{(n)} \tag{2.54}$$

Since $\frac{\Gamma(2-b)}{\Gamma(2-b-n+j)} = \prod_{k=0}^{n-j-1}(1-b-k)$, we observe that $q_j(a,b,m)$ is a polynomial in a and b of degree $m-j$ with integer coefficients.

The last sum in (2.54) equals $\Gamma(2-b)\sum_{n=0}^{l-j}\binom{n+j}{j}\frac{\mathcal{S}_l^{(n+j)}}{\Gamma(2-b-n)}$, which only for $j=0$ reduces to $(1-b)^l$ by the above generating function of the Stirling numbers $\mathcal{S}_m^{(j)}$. Thus, we verify

$$q_0(a,b,m) = (-1)^m \prod_{k=0}^{m-1}(b-1+ka) \tag{2.55}$$

in **art.** 15 as

$$q_0(a,b,m) = \sum_{l=0}^{m} S_m^{(l)} a^{m-l} \sum_{n=0}^{l} \frac{\Gamma(2-b)}{\Gamma(2-b-n)} \mathcal{S}_l^{(n)} = a^m \sum_{l=0}^{m} S_m^{(l)} \left(\frac{1-b}{a}\right)^l$$

$$= a^m \prod_{k=0}^{m-1}\left(\frac{1-b}{a} - k\right) = (-1)^m \prod_{k=0}^{m-1}(b-1+ka)$$

where the generating function $\prod_{k=0}^{m-1}(x-k) = \sum_{k=0}^{m} S_m^{(k)} x^k$ of the Stirling numbers $S_m^{(k)}$ of the first kind in (A.5) has been used.

21. *A differential form for the polynomial* $q_j(a,b,m)$. We introduce the characteristic coefficient (A.25) of the function $(1+z)^{\frac{1}{a}}$ into the polynomial (2.50)

$$q_j(a,b,m) = a^m \sum_{n=j}^{m} \binom{n}{j} \frac{\Gamma(2-b)}{\Gamma(2-b-n+j)} \frac{m!}{n!} s[n,m]|_{(1+z)^{\frac{1}{a}}}$$

$$= \frac{m! a^m}{j!} \sum_{n=j}^{m} \frac{\Gamma(2-b)}{\Gamma(2-b-n+j)(n-j)!} s[n,m]|_{(1+z)^{\frac{1}{a}}}$$

With the definition (A.2) of the characteristic coefficient

$$s[n,m]|_{(1+z)^{\frac{1}{a}}} = \frac{1}{m!}\frac{d^m}{dz^m}\left(\left((1+z)^{\frac{1}{a}}-1\right)^n\right)\Big|_{z=0} \tag{2.56}$$

we have

$$\begin{aligned} q_j(a,b,m) &= \frac{a^m}{j!}\frac{d^m}{dz^m}\left(\sum_{n=j}^{m}\frac{\Gamma(2-b)\left((1+z)^{\frac{1}{a}}-1\right)^n}{\Gamma(2-b-n+j)(n-j)!}\right)\Bigg|_{z=0} \\ &= \frac{a^m}{j!}\frac{d^m}{dz^m}\left(\sum_{n=0}^{m-j}\frac{\Gamma(2-b)}{\Gamma(2-b-n)\,n!}\left((1+z)^{\frac{1}{a}}-1\right)^{n+j}\right)\Bigg|_{z=0} \end{aligned}$$

The finite sum

$$\begin{aligned} S &= \sum_{n=0}^{m-j}\frac{\Gamma(2-b)}{\Gamma(2-b-n)\,n!}\left((1+z)^{\frac{1}{a}}-1\right)^{n+j} \\ &= \left((1+z)^{\frac{1}{a}}-1\right)^j\sum_{n=0}^{m-j}\binom{1-b}{n}\left((1+z)^{\frac{1}{a}}-1\right)^n \end{aligned}$$

can be rewritten as a difference of infinite sums. Indeed,

$$\begin{aligned} \widetilde{S} &= \sum_{n=0}^{m-j}\binom{1-b}{n}\left((1+z)^{\frac{1}{a}}-1\right)^n \\ &= \sum_{n=0}^{\infty}\binom{1-b}{n}\left((1+z)^{\frac{1}{a}}-1\right)^n - \sum_{n=m-j+1}^{\infty}\binom{1-b}{n}\left((1+z)^{\frac{1}{a}}-1\right)^n \\ &= (1+z)^{\frac{1-b}{a}} - \sum_{n=m-j+1}^{\infty}\binom{1-b}{n}\left((1+z)^{\frac{1}{a}}-1\right)^n \end{aligned}$$

is valid, because $z \to 0$ implies that the convergence requirement $|w| < 1$ for the binomial series $(1+w)^\alpha = \sum_{k=0}^{\infty}\binom{\alpha}{k}w^k$, i.e. $\left|(1+z)^{\frac{1}{a}}-1\right| < 1$, can be satisfied. The result is

$$\begin{aligned} q_j(a,b,m) = \frac{a^m}{j!}\Bigg(&\frac{d^m}{dz^m}\left((1+z)^{\frac{1-b}{a}}\left((1+z)^{\frac{1}{a}}-1\right)^j\right)\Bigg|_{z=0} \\ &- \sum_{n=m+1}^{\infty}\binom{1-b}{n-j}\frac{d^m}{dz^m}\left(\left((1+z)^{\frac{1}{a}}-1\right)^n\right)\Bigg|_{z=0}\Bigg) \end{aligned}$$

The contour integral (1.14) representation of the derivative illustrates that

$$\begin{aligned} \frac{d^m}{dz^m}\left(\left((1+z)^{\frac{1}{a}}-1\right)^n\right)\Bigg|_{z=0} &= \frac{m!}{2\pi i}\int_{C(0)}\frac{\left((1+z)^{\frac{1}{a}}-1\right)^n}{z^{m+1}}dz \\ &= \frac{m!}{2\pi i}\int_{C(0)}\left(\frac{(1+z)^{\frac{1}{a}}-1}{z}\right)^n z^{n-m-1}dz \end{aligned}$$

where $\frac{(1+z)^{\frac{1}{a}}-1}{z}$ is analytic in a region sufficiently close around $z = 0$.

Hence, if $n > m$, then the derivative is zero. Thus, we find the differential form of the polynomial,

$$q_j(a,b,m) = \frac{a^m}{j!} \frac{d^m}{dz^m} \left((1+z)^{\frac{1-b}{a}} \left((1+z)^{\frac{1}{a}} - 1 \right)^j \right) \Big|_{z=0} \tag{2.57}$$

which is not a characteristic coefficient due to the appearance of the factor $(1+z)^{\frac{1-b}{a}}$. The polynomial $q_j(a,b,m)$ in (2.57) equals the m-th Taylor coefficient (1.13) of the function $\frac{m!a^m}{j!}(1+z)^{\frac{1-b}{a}} \left((1+z)^{\frac{1}{a}} - 1 \right)^j$ around $z_0 = 0$.

22. *Deductions from the differential form (2.57).* We expand the binomial in the differential form (2.57)

$$q_j(a,b,m) = \frac{a^m}{j!} \sum_{k=0}^{j} \binom{j}{k} (-1)^{j-k} \frac{d^m}{dz^m} \left((1+z)^{\frac{1-b+k}{a}} \right) \Big|_{z=0}$$

and obtain

$$q_j(a,b,m) = \frac{m!a^m}{j!} \sum_{k=0}^{j} (-1)^{j-k} \binom{j}{k} \binom{\frac{1-b+k}{a}}{m} \tag{2.58}$$

Example For $j=1$ and $j=2$, the binomial form (2.58) becomes

$$q_1(a,b,m) = m!a^m \left(\binom{\frac{2-b}{a}}{m} - \binom{\frac{1-b}{a}}{m} \right)$$

$$q_2(a,b,m) = \frac{m!a^m}{2} \left(\binom{\frac{3-b}{a}}{m} - 2\binom{\frac{2-b}{a}}{m} + \binom{\frac{1-b}{a}}{m} \right)$$

Evaluating the binomial coefficient in (2.58)

$$q_j(a,b,m) = \frac{a^m}{j!} \sum_{k=0}^{j} (-1)^{j-k} \binom{j}{k} \frac{\Gamma\left(\frac{1-b+k}{a}+1\right)}{\Gamma\left(\frac{1-b+k}{a}+1-m\right)}$$

becomes, with $\frac{\Gamma(x+1)}{\Gamma(x+1-m)} = \prod_{k=0}^{m-1}(x-k)$ in (A.5),

$$q_j(a,b,m) = \sum_{k=0}^{j} \frac{(-1)^{j-k}}{k!\,(j-k)!} \prod_{r=0}^{m-1} (1-b+k-ra) \tag{2.59}$$

Leibniz' rule tells us that the differential form (2.57) can be written as

$$q_j(a,b,m) = \frac{a^m}{j!} \sum_{k=0}^{m} \binom{m}{k} \frac{d^{m-k}}{dz^{m-k}} (1+z)^{\frac{1-b}{a}} \frac{d^k}{dz^k} \left((1+z)^{\frac{1}{a}} - 1 \right)^j \Big|_{z=0}$$

The Taylor series $(1+z)^{\frac{1-b}{a}} = \sum_{n=0}^{\infty} \binom{\frac{1-b}{a}}{n} z^n$ shows that

$$\frac{d^{m-k}}{dz^{m-k}} \left((1+z)^{\frac{1-b}{a}} \right) \Big|_{z=0} = (m-k)! \binom{\frac{1-b}{a}}{m-k}$$

With the definition of the characteristic coefficient (2.56), we find

$$q_j(a,b,m) = \frac{a^m m!}{j!} \sum_{k=0}^{m} \binom{\frac{1-b}{a}}{m-k} s[j,k]|_{(1+z)^{\frac{1}{a}}}$$

Finally, with $s[j,k]|_{(1+z)^{\frac{1}{a}}} = \frac{j!}{k!} \sum_{l=j}^{k} S_k^{(l)} \mathcal{S}_l^{(j)} \alpha^{-l}$ in (A.25), we arrive at the alternative form, which looks similar to (2.50),

$$q_j(a,b,m) = \sum_{n=j}^{m} \binom{m}{n} \frac{\Gamma\left(\frac{1-b}{a}+1\right)}{\Gamma\left(\frac{1-b}{a}-m+n+1\right)} \sum_{l=j}^{n} S_n^{(l)} \mathcal{S}_l^{(j)} a^{m-l} \tag{2.60}$$

With $\frac{\Gamma(x+1)}{\Gamma(x+1-m)} = \prod_{k=0}^{m-1}(x-k)$, (2.60) becomes

$$\begin{aligned} q_j(a,b,m) &= \sum_{n=j}^{m} \binom{m}{n} \prod_{k=0}^{m-n-1} \left(\frac{1-b}{a}-k\right) \sum_{l=j}^{n} S_n^{(l)} \mathcal{S}_l^{(j)} a^{m-l} \\ &= \sum_{n=j}^{m} \binom{m}{n} \prod_{k=0}^{m-n-1} (1-b-ka) \sum_{l=j}^{n} S_n^{(l)} \mathcal{S}_l^{(j)} a^{n-l} \end{aligned} \tag{2.61}$$

which, again, indicates that the polynomial $q_j(a,b,m)$ in a and b has integer coefficients.

23. *Three-term recursion for the polynomial* $q_j(a,b,m)$. The binomial form (2.58) directly leads to the three-term recursion for all $j \geq 0$,

$$(j+1)\, q_{j+1}(a,b,m) + q_j(a,b,m) - q_j(a,b-1,m) = 0 \tag{2.62}$$

Indeed, adding $(j+1)!q_{j+1}(a,b,m)$ and $j!q_j(a,b,m)$ in (2.58)

$$\begin{aligned} R &= \frac{(j+1)!q_{j+1}(a,b,m) + j!q_j(a,b,m)}{m!a^m} \\ &= \sum_{k=0}^{j} (-1)^{j-k} \binom{j}{k} \binom{\frac{1-b+k}{a}}{m} - \sum_{k=0}^{j+1} (-1)^{j-k} \binom{j+1}{k} \binom{\frac{1-b+k}{a}}{m} \\ &= \binom{\frac{2-b+j}{a}}{m} - \sum_{k=1}^{j} (-1)^{j-k} \left\{ \binom{j+1}{k} - \binom{j}{k} \right\} \binom{\frac{1-b+k}{a}}{m} \end{aligned}$$

Since $\binom{j+1}{k} - \binom{j}{k} = \binom{j}{k-1}$, we have

$$\begin{aligned} R &= \binom{\frac{2-b+j}{a}}{m} - \sum_{k=1}^{j} (-1)^{j-k} \binom{j}{k-1} \binom{\frac{1-b+k}{a}}{m} \\ &= \binom{\frac{2-b+j}{a}}{m} + \sum_{k=0}^{j-1} (-1)^{j-k} \binom{j}{k} \binom{\frac{2-b+k}{a}}{m} \\ &= \sum_{k=0}^{j} (-1)^{j-k} \binom{j}{k} \binom{\frac{1-(b-1)+k}{a}}{m} = \frac{j!}{m!a^m} q_j(a,b-1,m) \end{aligned}$$

which proves (2.62).

24. *Special cases of the polynomial $q_j(a,b,m)$.* For $a=1$, the second orthogonality formula of the Stirling numbers (Olver *et al.*, 2010, Sec. 26.8.39)

$$\sum_{l=n}^{m} S_m^{(l)} \mathcal{S}_l^{(n)} = \delta_{nm} \tag{2.63}$$

indicates that

$$q_j(1,b,m) = \sum_{n=j}^{m} \binom{n}{j} \frac{\Gamma(2-b)}{\Gamma(2-b-n+j)} \delta_{nm} = \binom{m}{j} \frac{\Gamma(2-b)}{\Gamma(2-b-m+j)} \tag{2.64}$$

For $a=0$, (2.59) and (2.54) simplify to

$$q_j(0,b,m) = \sum_{k=0}^{j} \frac{(-1)^{j-k}}{k!\,(j-k)!} (1-b+k)^m = \sum_{n=j}^{m} \binom{n}{j} \frac{\Gamma(2-b)}{\Gamma(2-b-n+j)} \mathcal{S}_m^{(n)}$$

For $b=1$, the factor $\frac{\Gamma(2-b)}{\Gamma(2-b-n+j)} = \frac{1}{(j-n)!}$ is only non-zero for $n=j$ and the polynomial $q_j(a,b,m)$ in (2.50) simplifies to

$$q_j(a,1,m) = \sum_{l=j} S_m^{(l)} \mathcal{S}_l^{(j)} \, a^{m-l} \tag{2.65}$$

for which $q_0(a,1,m) = 0$ for $m>0$, due to $\mathcal{S}_l^{(0)} = S_l^{(0)} = \delta_{0,l}$, while $q_m(a,1,m) = 1$ for $m \geq 0$. Moreover, the differential form (2.57), (2.56) as well as (A.25) show that

$$q_j(a,1,m) = \frac{a^m}{j!} \frac{d^m}{dz^m} \left(\left((1+z)^{\frac{1}{a}} - 1 \right)^j \right) \Big|_{z=0} = a^m \frac{m!}{j!}\, s[j,m] \big|_{(1+z)^{\frac{1}{a}}} \tag{2.66}$$

With (2.65), the polynomial $q_j(a,b,m)$ in (2.61) is recursively written as

$$q_j(a,b,m) = \sum_{n=j}^{m} \binom{m}{n} \prod_{k=0}^{m-n-1} (1-b-ka)\, q_j(a,1,n)$$

Similarly, (2.50) results, with (2.65), in an "a from b separated" recursion

$$q_j(a,b,m) = \sum_{n=j}^{m} \binom{n}{j} \frac{\Gamma(2-b)}{\Gamma(2-b-n+j)} q_n(a,1,m)$$

from which we find with $\binom{-z}{j} = (-1)^j \binom{z-1+j}{j}$, valid for any complex z and positive integer j,

$$j!\,q_j(a,b,m) = \sum_{n=j}^{m} \binom{1-b}{n-j} n!\,q_n(a,1,m)$$

If $a=b=1$, then (2.65) and the orthogonality condition (2.63) of the Stirling numbers shows that

$$q_j(1,1,m) = \delta_{jm} \tag{2.67}$$

which directly follows from (2.66) for $a = 1$.

If $a = b$, then (2.61) becomes

$$q_j(a,a,m) = \sum_{n=j}^{m} \binom{m}{n} \prod_{k=1}^{m-n} (1-ka) \sum_{l=j}^{n} S_n^{(l)} \mathcal{S}_l^{(j)} a^{n-l}$$

which illustrates that $q_j(a,a,m)$ has a zero at $a = 1$ for $j < m$. The polynomial $q_j(a,a,m)$ in a will reappear in **art.** 95.

2.7 Integral of the Mittag-Leffler function

25. *Integration of the Taylor series around* $z_0 = 0$. Integration of the Taylor series (1.1) leads to

$$\int_0^z E_{a,b}(u)\,du = a \sum_{k=0}^{\infty} \frac{z^{k+1}}{(ak+a)\,\Gamma(b+ak)} \tag{2.68}$$

Clearly, if $b = a$, then (2.68) simplifies to

$$\int_0^z E_{a,a}(u)\,du = a \sum_{k=0}^{\infty} \frac{z^{k+1}}{\Gamma(a+1+ak)} = azE_{a,a+1}(z) \tag{2.69}$$

If b is different from a, the evaluation of the integral is considerably more complicated.

The identity, valid for any number c,

$$\frac{1}{ak+a} = \frac{1}{ak+b+c} + \frac{c-(a-b)}{(ak+b+c)}\frac{1}{(ak+a)}$$

introduced into the integrated Taylor series (2.68) leads, for $c = 0$, to

$$\begin{aligned}\int_0^z E_{a,b}(u)\,du &= a \sum_{k=0}^{\infty} \left(1 + \frac{-(a-b)}{ak+a}\right) \frac{z^{k+1}}{\Gamma(ak+b+1)} \\ &= azE_{a,b+1}(z) + a(b-a) \sum_{k=0}^{\infty} \frac{z^{k+1}}{(ak+a)\,\Gamma(ak+b+1)}\end{aligned}$$

Let

$$R(b) = \frac{1}{a}\int_0^z E_{a,b}(u)\,du = \sum_{k=0}^{\infty} \frac{z^{k+1}}{(ak+a)\,\Gamma(b+ak)} \tag{2.70}$$

and the above becomes a recursion

$$R(b) = zE_{a,b+1}(z) + (b-a)\,R(b+1) \tag{2.71}$$

After p iterations of the recursion (2.71), we find that

$$R(b) = z\sum_{j=1}^{p} E_{a,b+j}(z)\prod_{l=0}^{j-2}(l+b-a) + R(b+p)\prod_{l=0}^{p-1}(l+b-a)$$
$$= z\sum_{j=1}^{p} E_{a,b+j}(z)\frac{\Gamma(b-a+j-1)}{\Gamma(b-a)} + R(b+p)\frac{\Gamma(b-a+p)}{\Gamma(b-a)}$$

For any positive integer p, we arrive with the definition (2.70) at

$$\int_0^z E_{a,b}(u)\,du = az\sum_{j=1}^{p} E_{a,b+j}(z)\frac{\Gamma(b-a+j-1)}{\Gamma(b-a)} + \frac{\sum_{k=0}^{\infty}\frac{\Gamma(p+b-a)}{\Gamma(p+b+ak)}\frac{z^{k+1}}{(k+1)}}{\Gamma(b-a)} \tag{2.72}$$

If p tends to infinity, then the asymptotic formula (14.12) indicates that $\frac{\Gamma(p+b-a)}{\Gamma(p+b+ak)} = p^{-a(k+1)}\left(1+O(\frac{1}{p})\right)$. Thus, for $\operatorname{Re}(a) > 0$ and any finite z, it holds that

$$\int_0^z E_{a,b}(u)\,du = az\sum_{j=0}^{\infty} E_{a,b+j+1}(z)\frac{\Gamma(b-a+j)}{\Gamma(b-a)} \tag{2.73}$$

Unfortunately, the infinite series in (2.73) converges rather slowly. If $b = a$, then (2.73) reduces to (2.69). For $a = 0$, (2.72) simplifies, for any p while restricting $|z| < 1$, to

$$\int_0^z E_{0,b}(u)\,du = \frac{-\log(1-z)}{\Gamma(b)}$$

26. *The Taylor series (2.53) around z_0 revisited.* The differential form (2.57) of the polynomial $q_j(a,b,m)$ also opens interesting perspectives towards the Taylor series (2.53) around $z_0 \neq 0$.

We reverse the summations in the Taylor series (2.53) around $z_0 \neq 0$,

$$E_{a,b}(z) = \sum_{m=0}^{\infty}\sum_{j=0}^{m} q_j(a,b,m)\,E_{a,b-j}(z_0)\frac{1}{m!}\left(\frac{z-z_0}{az_0}\right)^m$$
$$= \sum_{j=0}^{\infty} E_{a,b-j}(z_0)\sum_{m=j}^{\infty}\frac{q_j(a,b,m)}{m!}\left(\frac{z-z_0}{az_0}\right)^m$$

Since $q_j(a,b,m) = 0$ if $m < j$, the summation index can be lowered from $m = j$ to $m = 0$ and the last m-sum becomes, with (2.57) and with $Z = \frac{z-z_0}{az_0}$,

$$\sum_{m=j}^{\infty}\frac{q_j(a,b,m)}{m!}Z^m = \frac{1}{j!}\sum_{m=0}^{\infty}\frac{d^m}{dz^m}\left((1+z)^{\frac{1-b}{a}}\left((1+z)^{\frac{1}{a}}-1\right)^j\right)\Bigg|_{z=0}\frac{Z^m}{m!}$$

The infinite sum is a Taylor series (1.12) equal to $\left(\frac{z}{z_0}\right)^{\frac{1-b}{a}}\left(\left(\frac{z}{z_0}\right)^{\frac{1}{a}}-1\right)^j$.

Thus, we retrieve

$$E_{a,b}(z) = \left(\frac{z}{z_0}\right)^{\frac{1-b}{a}} \sum_{j=0}^{\infty} \frac{E_{a,b-j}(z_0)}{j!} \left(\left(\frac{z}{z_0}\right)^{\frac{1}{a}} - 1\right)^j \tag{2.74}$$

which was a starting point in **art.** 18.

27. *Companion of (2.73) for* $\int_0^z E_{a,b}(u)\,du$. We integrate the series (2.74),

$$\int_0^z E_{a,b}(u)\,du = \sum_{j=0}^{\infty} \frac{E_{a,b-j}(z_0)}{j!} \int_0^z \left(\frac{u}{z_0}\right)^{\frac{1-b}{a}} \left(\left(\frac{u}{z_0}\right)^{\frac{1}{a}} - 1\right)^j du$$

Substituting of the integration variable $w = \left(\frac{u}{z_0}\right)^{\frac{1}{a}}$,

$$\int_0^z \left(\frac{u}{z_0}\right)^{\frac{1-b}{a}} \left(\left(\frac{u}{z_0}\right)^{\frac{1}{a}} - 1\right)^j du = az_0 \int_0^{\left(\frac{z}{z_0}\right)^{\frac{1}{a}}} w^{(a-b+1)-1} (w-1)^j\,dw$$

suggests us to choose the arbitrary expansion point $z_0 = z \neq 0$, in which case the integral reduces to the Beta-integral (10.11). Thus, with $z_0 = z$ and provided that $\mathrm{Re}(a-b) > -1$, we find that

$$\int_0^z \left(\frac{u}{z}\right)^{\frac{1-b}{a}} \left(\left(\frac{u}{z}\right)^{\frac{1}{a}} - 1\right)^j du = (-1)^j\, az \frac{\Gamma(a-b+1)\, j!}{\Gamma(a-b+j+2)}$$

Thus, we arrive at the companion of (2.73)

$$\int_0^z E_{a,b}(u)\,du = az \sum_{j=0}^{\infty} E_{a,b-j}(z)\,(-1)^j \,\frac{\Gamma(a-b+1)}{\Gamma(a-b+j+2)} \tag{2.75}$$

28. *A remarkable identity.* Both series (2.73) and (2.75) of $\int_0^z E_{a,b}(u)\,du$ produce a remarkable identity for Mittag-Leffler functions.

The reflection formula (11.4) shows that $\frac{\Gamma(b-a+j)}{\Gamma(b-a)} = \frac{(-1)^j \Gamma(a-b+1)}{\Gamma(a-b+1-j)}$ and (2.73) can be rewritten as

$$\begin{aligned}
\int_0^z E_{a,b}(u)\,du &= az \sum_{j=0}^{\infty} E_{a,b+j+1}(z) \frac{(-1)^j \Gamma(a-b+1)}{\Gamma(a-b+1-j)} \\
&= -az \sum_{j=1}^{\infty} E_{a,b+j}(z) \frac{(-1)^j \Gamma(a-b+1)}{\Gamma(a-b+2-j)} \\
&= -az \sum_{j=-\infty}^{-1} E_{a,b-j}(z) \frac{(-1)^j \Gamma(a-b+1)}{\Gamma(a-b+2+j)}
\end{aligned}$$

Equated to (2.75) results in a remarkable null-formula

$$\sum_{j=-\infty}^{\infty} \frac{(-1)^j E_{a,b-j}(z)}{\Gamma(a-b+2+j)} = \sum_{j=-\infty}^{\infty} E_{a,b+j}(z)\,\Gamma(b-a+j-1) = 0$$

where the last equation follows similarly from (2.73). Thus, the Mittag-Leffler function

$$E_{a,b}(z) = -\sum_{j=-\infty;j\neq 0}^{\infty} \frac{\Gamma(b-a+j-1)}{\Gamma(b-a-1)} E_{a,b+j}(z)$$

can be written as a linear combination of the set $\{E_{a,b+j}(z)\}_{-\infty\leq j\neq 0\leq\infty}$.

29. *Integration of the differential recursion.* Integration of the differential recursion (2.39) yields

$$\begin{aligned} z^b E_{a,b+1}(xz^a) - z_0^b E_{a,b+1}(xz_0^a) &= \int_{z_0}^{z} u^{b-1} E_{a,b}(xu^a)\, du \\ &= \frac{1}{a} x^{-\frac{b}{a}} \int_{xz_0^a}^{xz^a} t^{\frac{b}{a}-1} E_{a,b}(t)\, dt \end{aligned}$$

If we replace xz^a by w, then we obtain

$$\int_{w_0}^{w} t^{\frac{b}{a}-1} E_{a,b}(t)\, dt = a\left(w^{\frac{b}{a}} E_{a,b+1}(w) - (w_0)^{\frac{b}{a}} E_{a,b+1}(w_0)\right) \tag{2.76}$$

which is a special case of the more general integral (3.28). For $b = a$ in (2.76), we find

$$\int_{w_0}^{w} E_{a,a}(t)\, dt = a\left(wE_{a,a+1}(w) - w_0 E_{a,a+1}(w_0)\right)$$

from which (2.69) follows. The factor $t^{\frac{b}{a}-1}$ in the integral of the closed formula (2.76) explains why “pure” integration of the Mittag-Leffler function as in (2.73) and (2.75) is so complicated.

2.8 Differential equations

30. *Differential equation for the Mittag-Leffler function $E_{a,b}(z)$.* We rewrite the differentiation formula (2.52)

$$(az)^m \frac{d^m}{dz^m} E_{a,b}(z) - q_0(a,b,m)\, E_{a,b}(z) = \sum_{j=1}^{m} q_j(a,b,m)\, E_{a,b-j}(z)$$

for m consecutive integers as a linear set,

$$\begin{bmatrix} 1 & 0 & \cdots & 0 \\ q_1(a,b,2) & 1 & \cdots & 0 \\ \vdots & \vdots & \ddots & \vdots \\ q_1(a,b,m) & q_2(a,b,m) & \cdots & 1 \end{bmatrix} \begin{bmatrix} E_{a,b-1}(z) \\ E_{a,b-2}(z) \\ \vdots \\ E_{a,b-m}(z) \end{bmatrix} = \begin{bmatrix} \tau(z;1) \\ \tau(z;2) \\ \vdots \\ \tau(z;m) \end{bmatrix}$$

where

$$\tau(z;m) = (az)^m \frac{d^m}{dz^m} E_{a,b}(z) - q_0(a,b,m)\, E_{a,b}(z)$$

In matrix form, the linear set is written as

$$Q_m(a,b)\, v_m(z) = w_m(z) \tag{2.77}$$

where

$$Q_m(a,b) = \begin{bmatrix} 1 & 0 & \cdots & 0 \\ q_1(a,b,2) & 1 & \cdots & 0 \\ \vdots & \vdots & \ddots & \vdots \\ q_1(a,b,m) & q_2(a,b,m) & \cdots & 1 \end{bmatrix}$$

is an $m \times m$ lower triangular matrix with determinant $\det Q_m(a,b) = 1$ and where the $m \times 1$ vectors are

$$v_m(z) = \begin{bmatrix} E_{a,b-1}(z) \\ E_{a,b-2}(z) \\ \vdots \\ E_{a,b-m}(z) \end{bmatrix} \text{ and } w_m(z) = \begin{bmatrix} \tau(z;1) \\ \tau(z;2) \\ \vdots \\ \tau(z;m) \end{bmatrix}$$

Since $\det Q_m(a,b) = 1$, the matrix $Q_m(a,b)$ is invertible and (2.77) becomes

$$v_m(z) = Q_m^{-1}(a,b)\, w_m(z) \tag{2.78}$$

where the $m \times m$ matrix $Q_m^{-1}(a,b)$ is again a lower triangular matrix, that is exactly solvable (Van Mieghem, 2023b) in terms of the polynomials $q_j(a,b,m)$. The vector $v_m(z)$, with j-th vector component $(v_m(z))_j = E_{a,b-j}(z)$, is expressed in (2.78) in terms of the vector $w_m(z)$, whose components only consist of the function $E_{a,b}(z)$ and its derivatives of $E_{a,b}(z)$.

If there exist an $m \times 1$ non-zero vector $u_m(z)$ that is orthogonal to the vector $v_m(z)$, satisfying $u_m^T(z)\, v_m(z) = 0$, then combining (2.78) and $u_m^T(z)\, v_m(z) = 0$ leads to the quadratic form

$$0 = u_m^T(z)\, Q_m^{-1}(a,b)\, w_m(z) = \sum_{j=1}^{m} \sum_{l=1}^{m} \left(Q_m^{-1}(a,b)\right)_{jl} (u_m(z))_j\, (w_m(z))_l \tag{2.79}$$

which formally represents a set of m linear differential equations for $E_{a,b}(z)$ of maximum order m with non-constant coefficients.

We proceed by finding an $m \times 1$ vector $u_m(z)$ that is orthogonal to the vector $v_m(z)$. We investigate the orthogonality condition $u_m^T(z)\, v_m(z) = 0$. We assume that each component of the vector $u_m(z)$ is analytic in z around the origin and, thus, that there exist a converging Taylor series $(u_m(z))_j = \sum_{k=0}^{\infty} v_k(j;m)\, z^k$ around $z_0 = 0$. Then, the orthogonality condition $u_m^T(z)\, v_m(z) = 0$ becomes, with the Taylor series (1.1) of $E_{a,b}(z)$,

$$0 = \sum_{j=1}^{m} (u_m(z))_j\, E_{a,b-j}(z) = \sum_{j=1}^{m} \sum_{k=0}^{\infty} v_k(j;m)\, z^k \sum_{k=0}^{\infty} \frac{z^k}{\Gamma(ak+b-j)}$$

After executing the product of Taylor series,

$$0 = \sum_{k=0}^{\infty} \left(\sum_{j=1}^{m} \sum_{l=0}^{k} \frac{v_{k-l}(j;m)}{\Gamma(al+b-j)} \right) z^k$$

and equating corresponding powers in z, we find, for any integer $k \geq 0$, that

$$0 = \sum_{l=0}^{k} \sum_{j=1}^{m} \frac{v_{k-l}(j;m)}{\Gamma(al+b-j)} \tag{2.80}$$

The orthogonality condition (2.80) for the Taylor coefficients $v_{k-l}(j,m)$ of the vector components $(u_m(z))_j$ indicates that, if $al+b-j = n \in \mathbb{Z}^-$ for which $\frac{1}{\Gamma(n)} = 0$, then the corresponding Taylor coefficient $v_{k-l}(j,m)$ can be chosen at will. In other words, if $a \in \mathbb{Q}$ is a rational number and b is an integer, then a linear differential equation for $E_{a,b}(z)$ exists. Examples are $\frac{d^m}{dz^m} E_m(z^m) = E_m(z^m)$ and, more general, the differential equation (2.42) in **art.** 17 (see also Podlubny (1999, p. 23)). On the other hand, if $al+b \notin \mathbb{Z}$ for any integer l, implying that a is non-rational, but a real number and/or b is not an integer, then we will show below that there does not exist a linear differential equation for any complex z. If the complex number z is confined, for example to real numbers, then linear differential equations exist, e.g. (5.3) in **art.** 17.

31. *No linear differential equation for $E_{a,b}(z)$ with constant u_m vector in (2.79).* The orthogonality condition (2.80) in **art.** 30 simplifies most, if the $m \times 1$ non-zero vector $u_m(z) = u_m$ is independent of z, to $u_m^T v_m(z) = 0$ or

$$0 = \sum_{j=1}^{m} (u_m)_j E_{a,b-j}(z) = \sum_{k=0}^{\infty} \left(\sum_{j=1}^{m} \frac{(u_m)_j}{\Gamma(ak+b-j)} \right) z^k$$

which must hold for any number z, leading to the case $l = k$ in (2.80) where $v_0(j;m) = (u_m(0))_j = (u_m)_j$,

$$0 = \sum_{j=1}^{m} \frac{v_0(j;m)}{\Gamma(ak+b-j)} \tag{2.81}$$

Substituting $\Gamma(ak+b-j) = \Gamma(ak+b) \left(\prod_{n=1}^{j} (ak+b-n) \right)^{-1}$, obtained after iterating the functional equation $\Gamma(z+1) = z\Gamma(z)$ of the Gamma function in (10.1), yields for $ak+b \notin \mathbb{Z}$ for any integer k, to

$$0 = \sum_{j=1}^{m} v_0(j;m) \prod_{n=1}^{j} (ak+b-n)$$

Using the generating function $\prod_{n=0}^{m-1}(x-n) = \sum_{k=0}^{m} S_m^{(k)} x^k$ in (A.5) of the Stirling numbers $S_m^{(k)}$ of the first kind

$$\prod_{n=1}^{j}(ak+b-n)=\prod_{n=0}^{j-1}(ak+b-1-n)=\sum_{n=0}^{j}S_j^{(n)}\,(ak+b-1)^n$$

shows, with $S_j^{(0)}=\delta_{0j}$ equal to zero because $j\geq 1$, that

$$0=\sum_{j=1}^{m}v_0(j;m)\sum_{n=1}^{j}S_j^{(n)}(ak+b-1)^n=\sum_{n=1}^{m}\left(\sum_{j=n}^{m}v_0(j;m)S_j^{(n)}\right)(ak+b-1)^n$$

Since this set of equations in m unknowns $v_0(1;m), v_0(2;m), \ldots, v_0(m;m)$ must hold for any integer $k\geq 0$, for which $ak+b-1\neq 0$, it is only satisfied if

$$\sum_{j=n}^{m}v_0(j;m)\,S_j^{(n)}=0$$

After multiplying both sides by $\mathcal{S}_n^{(r)}$, summing over n from $0<r$ to m,

$$\sum_{n=r}^{m}\sum_{j=n}^{m}v_0(j;m)\,\mathcal{S}_n^{(r)}S_j^{(n)}=0$$

reversing the sums

$$\sum_{j=r}^{m}v_0(j;m)\sum_{n=r}^{j}\mathcal{S}_n^{(r)}S_j^{(n)}=0$$

and invoking the second orthogonality formula $\sum_{n=r}^{j}\mathcal{S}_n^{(r)}S_j^{(n)}=\delta_{jr}$ in (2.63) shows that $v_0(r;m)=0$ for any $1\leq r\leq m$. In conclusion, for non-rational a and non-integer b, there does not exist a non-zero constant vector u_m that produces a zero quadratic form (2.79).

As a corollary, for non-rational $a>0$ and non-integer b, the set of m function $E_{a,b-1}(z)$, $E_{a,b-2}(z)$, ..., $E_{a,b-m}(z)$ is linearly independent (Courant and Hilbert, 1953, Chapter II), i.e. for all z, the homogeneous linear relation

$$0=\sum_{j=1}^{m}c_jE_{a,b-j}(z)$$

only holds for a set of m constant coefficients $c_1,c_2,\ldots,c_m$ if each coefficient $c_j=0$ for $1\leq j\leq m$. Moreover, from any given system of infinitely linearly independent functions $v_1,v_2,\ldots$, an orthonormal system $\varphi_1,\varphi_2,\ldots$ can be constructed (see e.g. Van Mieghem (2023a, art. 352) for orthogonal polynomials). The series (2.73) and (2.75) exemplify linear expansions of the function $\int_0^x E_{a,b}(t)\,dt$ into a infinite set of linearly independent functions $\{E_{a,b-j}(z)\}_{j\geq 0}$ and $\{E_{a,b+j}(z)\}_{j\geq 1}$.

32. *No linear differential equation for $E_{a,b}(z)$ with non-rational a.* Incorporating that $(u_m)_j=v_0(j,m)=0$ for all $1\leq j\leq m$, the orthogonality condition (2.80) becomes, for all $k \geq 1$,

$$0 = \sum_{l=0}^{k-1} \sum_{j=1}^{m} \frac{v_{k-l}(j;m)}{\Gamma(al+b-j)} = \sum_{l=1}^{k} \sum_{j=1}^{m} \frac{v_{k+1-l}(j;m)}{\Gamma(al+b-j)}$$

The first-order Taylor coefficients $v_1(j;m)$ for $1 \leq j \leq m$ of the vector $u_m(z)$ must satisfy, taken $l = k$,

$$0 = \sum_{j=1}^{m} \frac{v_1(j;m)}{\Gamma(ak+b-j)}$$

However, this equation is algebraically the same as (2.81), where now the unknowns $v_0(j;m)$ are replaced by $v_1(j;m)$. **Art.** 31 demonstrates that only the zero solution for $v_1(j;m)$ exist. Continuing the recipe for the second-order $v_2(j;m)$ and higher-order Taylor coefficients $v_l(j;m)$ with $l > 2$, iteratively leads to the zero solution.

In summary, there does not exist a linear differential equation in the derivatives $\frac{d^l E_{a,b}(z)}{dz^l}$, where $1 \leq l$, for Mittag-Leffler functions $E_{a,b}(z)$ with non-rational parameter a and non-integer parameter b, valid for all complex z. The non-existence of a linear differential equation in the "classical" derivatives $\frac{d^l E_{a,b}(z)}{dz^l}$ for all complex z suggests that an extension to "fractional" derivatives as in (1.7) might be more successful.

33. *Derivation of $E_{a,b}(z)$ with respect to the parameters a and b.* From the definition (1.1), partial differentiating yields

$$\begin{aligned}\frac{\partial}{\partial a} E_{a,b}(z) &= \frac{\partial}{\partial a}\left(\frac{1}{\Gamma(b)} + \sum_{k=1}^{\infty} \frac{z^k}{\Gamma(b+ak)}\right) = \sum_{k=1}^{\infty} \frac{d}{dy} \frac{1}{\Gamma(y)}\bigg|_{y=b+ak} \frac{dy}{da} z^k \\ &= -\sum_{k=1}^{\infty} \frac{\psi(y)}{\Gamma(y)}\bigg|_{y=b+ak} k z^k\end{aligned}$$

where $\psi(y)$ is the digamma function (Chapter 12). Thus,

$$\frac{\partial}{\partial a} E_{a,b}(z) = -\sum_{k=1}^{\infty} \frac{\psi(b+ak)}{\Gamma(b+ak)} k z^k$$

Similarly, we find

$$\frac{\partial}{\partial b} E_{a,b}(z) = -\sum_{k=0}^{\infty} \frac{\psi(b+ak)}{\Gamma(b+ak)} z^k$$

Since the digamma function $\psi(y) > 0$ for real $y > 1.46163$, the Mittag-Leffler function $E_{a,b}(x)$ is strictly decreasing in a and b for positive real x and a, provided that $b > 1.46163$. Furthermore, invoking the functional equations (10.1) and (12.1) of the Gamma and digamma functions, respectively,

$$\frac{\partial}{\partial a}E_{a,b}(z) = -\frac{1}{a}\sum_{k=0}^{\infty}\frac{\psi(b+ak)}{\Gamma(b+ak)}(ak+b-1-(b-1))\,z^k$$
$$= -\frac{1}{a}\sum_{k=0}^{\infty}\frac{\psi(b-1+ak)}{\Gamma(b-1+ak)}z^k - \frac{1}{a}\sum_{k=0}^{\infty}\frac{z^k}{\Gamma(b+ak)}$$
$$+\frac{b-1}{a}\sum_{k=0}^{\infty}\frac{\psi(b+ak)}{\Gamma(b+ak)}z^k$$

shows that

$$a\frac{\partial}{\partial a}E_{a,b}(z) = \frac{\partial}{\partial b}E_{a,b-1}(z) - (b-1)\frac{\partial}{\partial b}E_{a,b}(z) - E_{a,b}(z) \tag{2.82}$$

We rewrite (2.82) as

$$a\frac{\partial}{\partial a}E_{a,b}(z) = \frac{\partial}{\partial b}\left(E_{a,b-1}(z) - (b-1)\,E_{a,b}(z)\right)$$

use (2.37) and find the second-order partial differential equation

$$\frac{\partial}{\partial a}E_{a,b}(z) = z\frac{\partial^2}{\partial z\partial b}E_{a,b}(z)$$

The latter is readily generalized. Partial differentiating m times gives

$$\frac{\partial^m}{\partial a^m}E_{a,b}(z) = \sum_{k=0}^{\infty}\left.\frac{d^m}{dy^m}\frac{1}{\Gamma(y)}\right|_{y=b+ak} k^m z^k$$

which suggests us to let $z = ye^w$ so that

$$\frac{\partial^m}{\partial a^m}E_{a,b}(ye^w) = \frac{\partial^m}{\partial w^m}\sum_{k=0}^{\infty}\left.\frac{d^m}{dy^m}\frac{1}{\Gamma(y)}\right|_{y=b+ak} y^k e^{kw}$$

leading to the partial differentiation equation for any integer $m \geq 0$ and any y (independent of a, b and w),

$$\frac{\partial^m}{\partial a^m}E_{a,b}(ye^w) = \frac{\partial^{2m}}{\partial w^m\partial b^m}E_{a,b}(ye^w) \tag{2.83}$$

2.9 The logarithm of the Mittag-Leffler function

34. *Logarithmic derivative.* The logarithmic derivative $\frac{d}{dz}\log E_{a,b}(z) = \frac{\frac{d}{dz}E_{a,b}(z)}{E_{a,b}(z)}$ follows directly from (2.37) as

$$\frac{d\log E_{a,b}(z)}{dz} = \frac{1}{az}\left(\frac{E_{a,b-1}(z)}{E_{a,b}(z)} - (b-1)\right) \tag{2.84}$$

Similarly, invoking (2.39), we find the companion logarithmic derivative

$$\frac{d}{dz}\log\left(z^{b-1}E_{a,b}(xz^a)\right) = \frac{\frac{d}{dz}\left\{z^{b-1}E_{a,b}(xz^a)\right\}}{z^{b-1}E_{a,b}(xz^a)} = \frac{1}{z}\frac{E_{a,b-1}(xz^a)}{E_{a,b}(xz^a)} \tag{2.85}$$

Since $b-1+ak > b-1+a$ for $k \geq 1$ and real $a > 0$, we have, for positive real x and for $b > 1-a$, because then it holds that $\Gamma(b-1+ak) > 0$ for $k \geq 1$,

$$E_{a,b}(x) = \frac{1}{\Gamma(b)} + \sum_{k=1}^{\infty} \frac{x^k}{(b-1+ak)\Gamma(b-1+ak)}$$
$$< \frac{1}{\Gamma(b)} + \frac{1}{b-1+a}\left(\sum_{k=0}^{\infty} \frac{x^k}{\Gamma(b-1+ak)} - \frac{1}{\Gamma(b-1)}\right)$$

and

$$E_{a,b}(x) < \frac{1}{b-1+a}\left(\frac{a}{\Gamma(b)} + E_{a,b-1}(x)\right)$$

Thus, for real $x > 0$ and for $b > 1-a$, the inequality is equivalent to

$$a\left(1 - \frac{1}{\Gamma(b) E_{a,b}(x)}\right) < \frac{E_{a,b-1}(x)}{E_{a,b}(x)} - (b-1)$$

so that the logarithmic derivative (2.84) is lower bounded by

$$\frac{d \log E_{a,b}(x)}{dx} > \frac{1}{x}\left(1 - \frac{1}{\Gamma(b) E_{a,b}(x)}\right) \tag{2.86}$$

Example If $b = 1$ and $a = 1$, then $E_{1,1}(z) = e^z$ and the above inequality for positive real x results in the well-known bound (see e.g. Van Mieghem (2014, p. 103)) that $e^{-x} > 1 - x$, which holds for all real x.

35. *Second-order logarithmic derivative.* Similarly as in **art.** 34, we differentiate (2.84) again,

$$\frac{d^2 \log E_{a,b}(z)}{dz^2} = -\frac{1}{az^2}\left(\frac{E_{a,b-1}(z)}{E_{a,b}(z)} - (b-1)\right) + \frac{1}{az}\frac{d}{dz}\left(\frac{E_{a,b-1}(z)}{E_{a,b}(z)}\right)$$

The derivative at the right-hand side is computed by using (2.37),

$$\frac{d}{dz}\left(\frac{E_{a,b-1}(z)}{E_{a,b}(z)}\right) = \frac{\frac{E_{a,b-2}(z)}{E_{a,b}(z)} + \frac{E_{a,b-1}(z)}{E_{a,b}(z)} - \left(\frac{E_{a,b-1}(z)}{E_{a,b}(z)}\right)^2}{az} \tag{2.87}$$

After substitution of the latter into the former, we arrive at the second-order logarithmic derivative

$$\frac{d^2 \log E_{a,b}(z)}{dz^2} = \frac{a(b-1) + \frac{E_{a,b-2}(z)}{E_{a,b}(z)} - (a-1)\frac{E_{a,b-1}(z)}{E_{a,b}(z)} - \left(\frac{E_{a,b-1}(z)}{E_{a,b}(z)}\right)^2}{(az)^2} \tag{2.88}$$

As already observed in **art.** 15, higher-order derivatives will become involved, which suggests us to consider the companion differential rule (2.39) in **art.** 38.

36. *The Taylor series of* $\log E_{a,b}(z)$ *around* $z_0 = 0$. Substituting the power series definition (1.1) of $E_{a,b}(z)$ into (2.84), indicates that

$$\lim_{z\to 0}\frac{d\log E_{a,b}(z)}{dz} = \lim_{z\to 0}\frac{1}{\sum_{k=0}^{\infty}\frac{z^k}{\Gamma(b+ak)}}\lim_{z\to 0}\frac{\sum_{k=0}^{\infty}\frac{z^k}{\Gamma(b-1+ak)}-\sum_{k=0}^{\infty}\frac{(b-1)z^k}{\Gamma(b+ak)}}{az}$$
$$= \Gamma(b)\lim_{z\to 0}\sum_{k=1}^{\infty}\frac{kz^{k-1}}{\Gamma(b+ak)}$$

We find that

$$\lim_{z\to 0}\frac{d\log E_{a,b}(z)}{dz} = \frac{\Gamma(b)}{\Gamma(b+a)}$$

Proceeding in this way to determine higher-order derivatives becomes cumbersome. The Taylor series of $\log E_{a,b}(z)$ follows from the general Taylor series (A.26) in Appendix A, based upon characteristic coefficients. Confining to the expansion point $z_0 = 0$, the characteristic coefficients (A.3) of the Mittag-Leffler function $E_{a,b}(z)$ are

$$s[k,m]|_{E_{a,b}(z)} = \sum_{\sum_{i=1}^{k} j_i = m; j_i > 0}\prod_{i=1}^{k}\frac{1}{\Gamma(b+j_i a)}$$

The Taylor series of $\log E_{a,b}(z)$ around $z_0 = 0$ follows from (A.26) as

$$\log E_{a,b}(z) = \sum_{m=0}^{\infty} c_m\, z^m$$

where we define $c_0 = \log E_{a,b}(0) = -\log(\Gamma(b))$ and the coefficients c_m for $m > 0$,

$$c_m = -\sum_{k=1}^{m}\frac{(-\Gamma(b))^k}{k}\, s[k,m]|_{E_{a,b}(z)}(0)$$

Explicitly, the coefficient c_m is

$$c_m = \frac{\Gamma(b)}{\Gamma(b+ma)} + \frac{(-1)^{m-1}}{m}\left(\frac{\Gamma(b)}{\Gamma(b+a)}\right)^m$$
$$+\sum_{k=2}^{m-1}\frac{(-1)^{k-1}}{k}\sum_{\sum_{i=1}^{k} j_i = m; j_i > 0}\prod_{i=1}^{k}\frac{\Gamma(b)}{\Gamma(b+j_i a)}$$

The list of the coefficients c_m for $m = 1$ up to $m = 5$ is

$$c_1 = \frac{\Gamma(b)}{\Gamma(b+a)}$$
$$c_2 = \frac{\Gamma(b)}{\Gamma(b+2a)} - \frac{1}{2}\left(\frac{\Gamma(b)}{\Gamma(b+a)}\right)^2$$
$$c_3 = \frac{\Gamma(b)}{\Gamma(b+3a)} + \frac{\Gamma^2(b)}{\Gamma(b+2a)\Gamma(b+a)} + \frac{1}{3}\left(\frac{\Gamma(b)}{\Gamma(b+a)}\right)^3$$

$$c_4 = \frac{\Gamma(b)}{\Gamma(b+4a)} - \frac{\Gamma^2(b)}{\Gamma(b+3a)\Gamma(b+a)} - \frac{1}{2}\left(\frac{\Gamma(b)}{\Gamma(b+2a)}\right)^2$$
$$+ \frac{\Gamma^3(b)}{\Gamma(b+2a)\Gamma^2(b+a)} - \frac{1}{4}\left(\frac{\Gamma(b)}{\Gamma(b+a)}\right)^4$$
$$c_5 = \frac{\Gamma(b)}{\Gamma(b+5a)} - \frac{\Gamma^2(b)}{\Gamma(b+4a)\Gamma(b+a)} - \frac{\Gamma^2(b)}{\Gamma(b+3a)\Gamma(b+2a)}$$
$$+ \frac{\Gamma^3(b)}{\Gamma^2(b+2a)}\left(\frac{1}{\Gamma(b+a)} + \frac{1}{\Gamma(b+3a)}\right) - \frac{\Gamma^4(b)}{\Gamma^3(b+a)\Gamma(b+2a)}$$
$$+ \frac{1}{5}\left(\frac{\Gamma(b)}{\Gamma(b+a)}\right)^5$$

Unfortunately, it seems difficult to further sum the terms in the Taylor coefficient c_m.

37. *An expansion of* $\log E_{a,b}(z)$ *in powers of* $\log z$. Integration of (2.84) leads to

$$\log \frac{E_{a,b}(z)}{E_{a,b}(z_0)} = \frac{(1-b)}{a} \log \frac{z}{z_0} + \frac{1}{a} \int_{z_0}^{z} \frac{E_{a,b-1}(w)}{E_{a,b}(w)} \frac{dw}{w}$$

Using (2.87), partial integration shows that

$$\int_{z_0}^{z} \frac{E_{a,b-1}(w)}{E_{a,b}(w)} \frac{dw}{w} = \frac{E_{a,b-1}(w)}{E_{a,b}(w)} \log w \Big|_{z_0}^{z} - \frac{1}{a} \int_{z_0}^{z} \frac{\log w}{w}$$
$$\times \left\{ \frac{E_{a,b-2}(w)}{E_{a,b}(w)} + \frac{E_{a,b-1}(w)}{E_{a,b}(w)} - \left(\frac{E_{a,b-1}(w)}{E_{a,b}(w)}\right)^2 \right\} dw$$

leading to

$$\log \frac{E_{a,b}(z)}{E_{a,b}(z_0)} = \frac{(1-b)}{a} \log \frac{z}{z_0} + \frac{1}{a} \left(\frac{E_{a,b-1}(z)}{E_{a,b}(z)} \log z - \frac{E_{a,b-1}(z_0)}{E_{a,b}(z_0)} \log z_0 \right)$$
$$- \int_{z_0}^{z} \frac{\log w}{a^2 w} \left\{ \frac{E_{a,b-2}(w)}{E_{a,b}(w)} + \frac{E_{a,b-1}(w)}{E_{a,b}(w)} - \left(\frac{E_{a,b-1}(w)}{E_{a,b}(w)}\right)^2 \right\} dw$$

Since $\frac{\log w}{w} dw = \frac{1}{2} d\left(\log^2 w\right)$, again partial integration of the right-hand side integral R leads to

$$R = \int_{z_0}^{z} \frac{\log w}{w} \left\{ \frac{E_{a,b-2}(w)}{E_{a,b}(w)} + \frac{E_{a,b-1}(w)}{E_{a,b}(w)} - \left(\frac{E_{a,b-1}(w)}{E_{a,b}(w)}\right)^2 \right\} dw$$
$$= \left\{ \frac{E_{a,b-2}(w)}{E_{a,b}(w)} + \frac{E_{a,b-1}(w)}{E_{a,b}(w)} - \left(\frac{E_{a,b-1}(w)}{E_{a,b}(w)}\right)^2 \right\} \frac{\log^2 w}{2} \Bigg|_{z_0}^{z}$$
$$- \int_{z_0}^{z} \frac{\log^2 w}{2} \frac{d}{dw} \left(\left\{ \frac{E_{a,b-2}(w)}{E_{a,b}(w)} + \frac{E_{a,b-1}(w)}{E_{a,b}(w)} - \left(\frac{E_{a,b-1}(w)}{E_{a,b}(w)}\right)^2 \right\} \right) dw$$

Each derivative of a fraction of Mittag-Leffler functions can be computed with (2.87), which indicates that again a factor $\frac{1}{aw}$ will appear so that $\frac{\log^2 w}{2w} dw = \frac{1}{3!} d\left(\log^3 w\right)$ which enables a next integration.

Continuing partial integration will lead to an expansion of the form

$$\log \frac{E_{a,b}(z)}{E_{a,b}(z_0)} = \frac{1-b}{a} \log\left(\frac{z}{z_0}\right) - \sum_{k=1}^{K} \frac{(-1)^k \left(\mathcal{F}_k(z) \log^k z - \mathcal{F}_k(z_0) \log^k z_0\right)}{k!} \tag{2.89}$$

$$+ \frac{(-1)^K}{K!} \int_{z_0}^{z} \frac{\log^K w}{w} \mathcal{F}_{K+1}(w)\, dw$$

where $K \geq 1$ is an integer and $\mathcal{F}_k(z)$ consists of a sum of fractions of Mittag-Leffler functions, that satisfy the recursion, for integer $k > 1$,

$$\mathcal{F}_k(w) = w \frac{d}{dw} \mathcal{F}_{k-1}(w)$$

and $\mathcal{F}_1(z) = \frac{1}{a} \frac{E_{a,b-1}(z)}{E_{a,b}(z)}$. We have computed above or in (2.87),

$$\mathcal{F}_2(z) = \frac{1}{a^2} \left\{ \frac{E_{a,b-2}(z)}{E_{a,b}(z)} + \frac{E_{a,b-1}(z)}{E_{a,b}(z)} - \left(\frac{E_{a,b-1}(z)}{E_{a,b}(z)}\right)^2 \right\}$$

but we omit computations for $k > 2$. The expansion (2.89) is valid for all complex z and z_0, except for negative real numbers (due to the branch cut in $\log z$). Clearly, the choice $z_0 = 1$ simplifies (2.89).

The companion of (2.89), expressed as an expansion in $\log z$ follows from the general expansion in (A.30) as

$$E_{a,b}(z) = E_{a,b}\left(\frac{1}{\xi}\right) + \sum_{m=1}^{\infty} \left(\sum_{k=1}^{m} \left.\frac{d^k E_{a,b}(p)}{dp^k}\right|_{p=\frac{1}{\xi}} \frac{\mathcal{S}_m^{(k)}}{\xi^k} \right) \frac{\log^m(\xi z)}{m!}$$

where (2.52) leads to

$$E_{a,b}(z) = E_{a,b}\left(\frac{1}{\xi}\right) + \sum_{m=1}^{\infty} \left(\sum_{k=1}^{m} \frac{\mathcal{S}_m^{(k)}}{a^k} \sum_{j=0}^{k} q_j(a,b,k)\, E_{a,b-j}\left(\frac{1}{\xi}\right) \right) \frac{\log^m(\xi z)}{m!}$$

38. *Higher-order logarithmic derivatives.* The second-order logarithmic derivative of the companion in (2.85) is

$$\frac{d^2}{dz^2} \log\left(z^{b-1} E_{a,b}(xz^a)\right) = \frac{d}{dz}\left(\frac{1}{z} \frac{E_{a,b-1}(xz^a)}{E_{a,b}(xz^a)}\right) = \frac{d}{dz}\left(\frac{z^{b-2} E_{a,b-1}(xz^a)}{z^{b-1} E_{a,b}(xz^a)}\right)$$

Invoking (2.39) yields

$$\frac{d}{dz}\left(\frac{z^{b-2} E_{a,b-1}(xz^a)}{z^{b-1} E_{a,b}(xz^a)}\right) = \frac{z^{b-3} E_{a,b-2}(xz^a)}{z^{b-1} E_{a,b}(xz^a)} - \left(\frac{z^{b-2} E_{a,b-1}(xz^a)}{z^{b-1} E_{a,b}(xz^a)}\right)^2$$

and

$$\frac{d^2}{dz^2} \log\left(z^{b-1} E_{a,b}(xz^a)\right) = \frac{1}{z^2}\left(\frac{E_{a,b-2}(xz^a)}{E_{a,b}(xz^a)} - \left(\frac{E_{a,b-1}(xz^a)}{E_{a,b}(xz^a)}\right)^2\right) \tag{2.90}$$

whose structure is more pleasing than that of (2.88).

A next differentiation yields

$$\frac{d^3}{dz^3}\log\left(z^{b-1}E_{a,b}\left(xz^a\right)\right)=\frac{d}{dz}\left(\frac{z^{b-3}E_{a,b-2}\left(xz^a\right)}{z^{b-1}E_{a,b}\left(xz^a\right)}-\left(\frac{z^{b-2}E_{a,b-1}\left(xz^a\right)}{z^{b-1}E_{a,b}\left(xz^a\right)}\right)^2\right)$$

The first derivative equals

$$\frac{d}{dz}\left(\frac{z^{b-3}E_{a,b-2}\left(xz^a\right)}{z^{b-1}E_{a,b}\left(xz^a\right)}\right)=\frac{1}{z^3}\left(\frac{E_{a,b-3}\left(xz^a\right)}{E_{a,b}\left(xz^a\right)}-\frac{E_{a,b-2}\left(xz^a\right)E_{a,b-1}\left(xz^a\right)}{\left(E_{a,b}\left(xz^a\right)\right)^2}\right)$$

The second is

$$\frac{d}{dz}\left(\frac{z^{b-2}E_{a,b-1}\left(xz^a\right)}{z^{b-1}E_{a,b}\left(xz^a\right)}\right)^2=\frac{2}{z^3}\left(\frac{E_{a,b-1}\left(xz^a\right)E_{a,b-2}\left(xz^a\right)}{\left(E_{a,b}\left(xz^a\right)\right)^2}-\frac{E^3_{a,b-1}\left(xz^a\right)}{\left(E_{a,b}\left(xz^a\right)\right)^3}\right)$$

Combining results in

$$\frac{d^3}{dz^3}\log\left(z^{b-1}E_{a,b}\left(xz^a\right)\right)=\frac{1}{z^3}\left(\begin{array}{c}\frac{E_{a,b-3}(xz^a)}{E_{a,b}(xz^a)}-\frac{3E_{a,b-1}(xz^a)E_{a,b-2}(xz^a)}{(E_{a,b}(xz^a))^2}\\+2\left(\frac{E_{a,b-1}(xz^a)}{E_{a,b}(xz^a)}\right)^3\end{array}\right) \tag{2.91}$$

Rather than continuing step-wise differentiation, we consider

$$\frac{d^{m+1}}{dz^{m+1}}\log\left(z^{b-1}E_{a,b}\left(xz^a\right)\right)=\frac{d^m}{dz^m}\left(\frac{z^{b-2}E_{a,b-1}\left(xz^a\right)}{z^{b-1}E_{a,b}\left(xz^a\right)}\right)$$

and invoke Leibniz' rule

$$\begin{aligned}L&=\frac{d^m}{dz^m}\left(\frac{z^{b-2}E_{a,b-1}\left(xz^a\right)}{z^{b-1}E_{a,b}\left(xz^a\right)}\right)\\&=\sum_{k=0}^{m}\binom{m}{k}\frac{d^k}{dz^k}\left(\frac{1}{z^{b-1}E_{a,b}\left(xz^a\right)}\right)\frac{d^{m-k}}{dz^{m-k}}\left(z^{b-2}E_{a,b-1}\left(xz^a\right)\right)\end{aligned}$$

With (2.40), we obtain

$$\frac{d^{m-k}}{dz^{m-k}}\left\{z^{b-2}E_{a,b-1}\left(xz^a\right)\right\}=z^{b-2-m+k}E_{a,b-1-m+k}\left(xz^a\right)$$

Applying (A.27) in Appendix A and replacing z_0 by z specifies, for $k\geq 1$,

$$\frac{1}{k!}\frac{d^k}{dz^k}\left(\frac{1}{z^{b-1}E_{a,b}\left(xz^a\right)}\right)=\sum_{n=1}^{k}\frac{(-1)^n}{\left(z^{b-1}E_{a,b}\left(xz^a\right)\right)^{n+1}}\,s[n,k]|_{z^{b-1}E_{a,b}(xz^a)}(z)$$

where the explicit form (A.3) of the characteristic coefficient indicates that

$$\begin{aligned}s[n,k]|_{\frac{E_{a,b}(xz^a)}{z^{1-b}}}(z)&=\sum_{\sum_{i=1}^{n}j_i=k;j_i>0}\prod_{i=1}^{n}\frac{\frac{d^{j_i}}{dz^{j_i}}\left\{z^{b-1}E_{a,b}\left(xz^a\right)\right\}}{j_i!}\\&=\sum_{\sum_{i=1}^{n}j_i=k;j_i>0}\prod_{i=1}^{n}\frac{z^{b-1-j_i}E_{a,b-j_i}\left(xz^a\right)}{j_i!}\end{aligned}$$

Since $\prod_{i=1}^{n} \frac{z^{b-1-j_i} E_{a,b-j_i}(xz^a)}{j_i!} = z^{n(b-1)-\sum_{i=1}^{n} j_i} \prod_{i=1}^{n} \frac{E_{a,b-j_i}(xz^a)}{j_i!}$, we find the characteristic coefficient

$$s[n,k]|_{z^{b-1}E_{a,b}(xz^a)}(z) = z^{n(b-1)-k} \sum_{\sum_{i=1}^{n} j_i = k; j_i > 0} \prod_{i=1}^{n} \frac{E_{a,b-j_i}(xz^a)}{j_i!}$$

so that

$$\frac{1}{k!}\frac{d^k}{dz^k}\left(\frac{1}{z^{b-1}E_{a,b}(xz^a)}\right) = \sum_{n=1}^{k} \frac{(-1)^n \sum_{\sum_{i=1}^{n} j_i = k; j_i > 0} \prod_{i=1}^{n} \frac{E_{a,b-j_i}(xz^a)}{j_i!}}{z^{(b-1)+k}\left(E_{a,b}(xz^a)\right)^{n+1}}$$

for $k \geq 1$. Substituting these results into Leibniz's formula yields the logarithmic derivatives

$$l_m = \frac{d^m}{dz^m} \log\left(z^{b-1}E_{a,b}(xz^a)\right)$$

explicitly, for any integer $m \geq 0$, as

$$l_{m+1} = \frac{E_{a,b-1-m}(xz^a)}{z^{m+1}E_{a,b}(xz^a)} + \frac{m!\sum_{k=1}^{m}\left\{\sum_{n=1}^{k} \frac{\sum_{\sum_{i=1}^{n} j_i = k; j_i > 0} \prod_{i=1}^{n} \frac{E_{a,b-j_i}(xz^a)}{j_i!}}{(-1)^n (E_{a,b}(xz^a))^{n+1}}\right\} \frac{E_{a,b-1-m+k}(xz^a)}{(m-k)!}}{z^{m+1}} \tag{2.92}$$

Reversing the k- and n sum in (2.92) shows that the *scaled* logarithmic derivatives $l'_m = z^m l_m$,

$$l'_{m+1} = \frac{E_{a,b-1-m}(xz^a)}{E_{a,b}(xz^a)} + \sum_{n=1}^{m} \frac{(-1)^n m!}{\left(E_{a,b}(xz^a)\right)^{n+1}} \sum_{k=n}^{m} \frac{E_{a,b-1-m+k}(xz^a)}{(m-k)!} \sum_{\sum_{i=1}^{n} j_i = k; j_i > 0} \prod_{i=1}^{n} \frac{E_{a,b-j_i}(xz^a)}{j_i!}$$

is a sum of $m+1$ fractions of Mittag-Leffler functions, alternating in sign.

In addition to (2.85), (2.90) and (2.91), we list the scaled logarithmic derivatives $l'_m = z^m l_m$ for $m = 4$ and 5,

$$l'_4 = \frac{E_{a,b-4}(xz^a)}{E_{a,b}(xz^a)} - \frac{4E_{a,b-3}(xz^a)E_{a,b-1}(xz^a) + 3\left(E_{a,b-2}(xz^a)\right)^2}{\left(E_{a,b}(xz^a)\right)^2} + \frac{12E_{a,b-2}(xz^a)\left(E_{a,b-1}(xz^a)\right)^2}{\left(E_{a,b}(xz^a)\right)^3} - 6\left(\frac{E_{a,b-1}(xz^a)}{E_{a,b}(xz^a)}\right)^4$$

$$l'_5 = \frac{E_{a,b-5}(xz^a)}{E_{a,b}(xz^a)} - \frac{10E_{a,b-3}(xz^a)E_{a,b-2}(xz^a) + 5E_{a,b-4}(xz^a)E_{a,b-1}(xz^a)}{\left(E_{a,b}(xz^a)\right)^2} + \frac{20E_{a,b-3}(xz^a)\left(E_{a,b-1}(xz^a)\right)^2 + 30E_{a,b-1}(xz^a)\left(E_{a,b-2}(xz^a)\right)^2}{\left(E_{a,b}(xz^a)\right)^3} - \frac{60E_{a,b-2}(xz^a)\left(E_{a,b-1}(xz^a)\right)^3}{\left(E_{a,b}(xz^a)\right)^4} + 24\left(\frac{E_{a,b-1}(xz^a)}{E_{a,b}(xz^a)}\right)^5$$

The recursion of the characteristic coefficients (A.4) enables exact computation to any desired, finite m by symbolic software.

In summary, the derivatives $l_m = \frac{d^m}{dz^m} \log \left(z^{b-1} E_{a,b}\left(xz^a\right)\right)$ in (2.92), for any z, a and b, give the Taylor series of $L_{a,b}\left(xz^a\right) \overset{\text{def}}{=} \log\left(\frac{E_{a,b}(xz^a)}{E_{a,b}(xz_0^a)}\right)$,

$$L_{a,b}\left(xz^a\right) = \log\left(\frac{z}{z_0}\right)^{(1-b)} + \sum_{m=1}^{\infty} \left.\frac{d^m \log\left(z^{b-1}E_{a,b}\left(xz^a\right)\right)}{dz^m}\right|_{z=z_0} \frac{\left(z-z_0\right)^m}{m!}$$

around any complex z_0 with $\arg z_0 \neq \pi$, explicitly as,

$$\begin{aligned} L_{a,b}\left(xz^a\right) &= (1-b)\log\left(\frac{z}{z_0}\right) + \sum_{m=0}^{\infty} \frac{E_{a,b-1-m}\left(xz_0^a\right)}{E_{a,b}\left(xz_0^a\right)} \frac{\left(\frac{z-z_0}{z_0}\right)^{m+1}}{(m+1)!} \\ &+ \sum_{m=1}^{\infty} \left\{ \sum_{k=1}^{m} \frac{E_{a,b-1-m+k}\left(xz_0^a\right)}{(m-k)!} \sum_{n=1}^{k} \frac{\sum_{\sum_{i=1}^{n} j_i = k;\ j_i>0} \prod_{i=1}^{n} \frac{E_{a,b-j_i}\left(xz_0^a\right)}{j_i!}}{(-1)^n \left(E_{a,b}\left(xz_0^a\right)\right)^{n+1}} \right\} \\ &\times \frac{\left(\frac{z-z_0}{z_0}\right)^{m+1}}{m+1} \end{aligned} \tag{2.93}$$

The Taylor series around $z_0 = 0$ in **art.** 36 is limited to a region around the origin in the complex plane. Taylor series (2.93) possesses a radius $R\left(z_0\right)$ of convergence around z_0 that equals the distance between z_0 and the nearest zero of $E_{a,b}\left(xz^a\right)$ to z_0.

3

Integrals containing $E_{a,b}(z)$

Chapter 3 focuses on integrals, whose integrand includes the Mittag-Leffler function and that can be evaluated exactly. Section 3.1 outlines classes of such computable integrals, while Section 3.2 confines to a special class that produces Laplace transforms of the Mittag-Leffler function. Section 3.3 extends the Laplace transforms in Section 3.2 by incorporating Gauss's multiplication formula (11.8) of the Gamma function. By relying on an extension (3.27) of the Beta- integral, Section 3.4 considers another instance of the general class, that relates to the fractional derivative of the Mittag-Leffler function. Section 3.5 is devoted to Berberan-Santos' integral (3.36) for the double argument in the parameter a and presents a generalization (3.37). The final Section 3.6 explores applications of the famous Euler-Maclaurin summation formula (3.40) to the Mittag-Leffler function: **Art**. 57 illustrates how the *basic complex integral* (4.4) can be deduced from the Euler-Maclaurin summation formula and **art**. 57 derives a representation (3.50) of the Mittag-Leffler function $E_{a,b}(z)$, that exhibits the dominant asymptotic behavior $\frac{z^{\frac{1-b}{a}}}{a}e^{z^{\frac{1}{a}}}$ for large $|z|$.

3.1 Classes of integrals

39. *A class of Mittag-Leffler integrals.* We will investigate here integrals, whose integrand contains the Mittag-Leffler function and another arbitrary function $g(u)$ defined on the interval $u \in [\alpha, \beta]$,

$$\int_{\alpha}^{\beta} E_{a,b}(\lambda u^c)\, g(u)\, du = \int_{\alpha}^{\beta} g(u) \sum_{k=0}^{\infty} \frac{(\lambda u^c)^k}{\Gamma(ak+b)} du$$

Since a Taylor series can be integrated term-wise within its domain of convergence, provided the resulting integral exist and also lies with the domain of convergence, we can interchange the sum and integral,

$$\int_{\alpha}^{\beta} E_{a,b}(\lambda u^c)\, g(u)\, du = \sum_{k=0}^{\infty} \frac{\lambda^k}{\Gamma(ak+b)} \int_{\alpha}^{\beta} g(u)\, u^{ck} du \qquad (3.1)$$

The Mittag-Leffler function is an entire function, whose radius of convergence is infinite (Section 1.4), which allows us to extend $\beta \to \infty$ and $\alpha \to -\infty$. However, this extension to an infinitely large interval needs a little more attention as shown by Titchmarsh (1964, Chapter 1). Most integrals in this Chapter 3 are of the class

$$\int_0^\infty E_{a,b}(\lambda u^c)\, g(u)\, du = \sum_{k=0}^\infty \frac{\lambda^k}{\Gamma(ak+b)} \int_0^\infty g(u)\, u^{ck} du \tag{3.2}$$

where $\mathrm{Re}(c) > 0$ and $\int_0^\infty g(u)\, u^{s-1} du$ is the Mellin transform (**art.** 245) of the function $g(u)$.

40. *Borel summation.* A specific example of the class in (3.2) with $g(u) = e^{-su}$, further evaluated in **art.** 42, involves the integral

$$\int_0^\infty g(u)\, u^{ck} du = \int_0^\infty e^{-su} u^{ck} du = \frac{\Gamma(ck+1)}{s^{ck+1}}$$

where the last line follows from Euler's integral (10.2) for the Gamma function for $\mathrm{Re}(c) > -1$. Another class of integrals than considered in **art.** 39 concerns Borel's class (Hardy, 1948, Section 8.5). If the entire function $f(z) = \sum_{k=0}^\infty f_k z^k$ is integrated and reversal of integration and summation is allowed, then

$$\int_0^\infty e^{-su} f(u)\, du = \sum_{k=0}^\infty f_k \int_0^\infty e^{-su} z^k du$$

Euler's integral (10.2) for $\mathrm{Re}(s) > 0$ then leads to

$$\int_0^\infty e^{-su} f(u)\, du = \frac{1}{s} \sum_{k=0}^\infty k! f_k \left(\frac{1}{s}\right)^k \tag{3.3}$$

which is, formally, again a Taylor series in $\frac{1}{s}$ with Taylor coefficient $k! f_k$. The precise summability conditions, under which the formal reversal is allowed, is treated by Hardy (1948, Section 8.5). Two examples are here immediate: (a) $f_k = \frac{1}{\Gamma(ak+b)}$, investigated in **art.** 42, and (b) $f_k = \frac{1}{k!}\frac{1}{\Gamma(ak+b)}$.

41. *The Wright function $W_{a,b}(z)$.* In case (b) in **art.** 40, the entire function

$$W_{a,b}(z) = \sum_{k=0}^\infty \frac{1}{\Gamma(ak+b)} \frac{z^k}{k!} \tag{3.4}$$

is called the Wright function and is studied in Gorenflo *et al.* (2020). The Taylor series (3.4) converges for all complex z and b, provided that $\mathrm{Re}(a) > -1$ (computed as in Section 1.4.1).

The derivative $\frac{dW_{a,b}(z)}{dz} = \sum_{k=0}^\infty \frac{k}{\Gamma(ak+b)} \frac{z^{k-1}}{k!}$ can be expressed in two different ways. Firstly and similar to (2.37) in **art.** 15, it holds that

$$az\frac{dW_{a,b}(z)}{dz} = W_{a,b-1}(z) - (b-1)\, W_{a,b}(z) \tag{3.5}$$

Secondly and unlike the Mittag-Leffler function $E_{a,b}(z)$, we find

$$\frac{dW_{a,b}(z)}{dz} = \sum_{k=1}^{\infty} \frac{1}{\Gamma(ak+b)} \frac{z^{k-1}}{(k-1)!} = \sum_{k=0}^{\infty} \frac{1}{\Gamma(ak+a+b)} \frac{z^k}{k!} = W_{a,a+b}(z) \tag{3.6}$$

Combining both (3.5) and (3.6) results in a recursion in the parameter b

$$azW_{a,a+b}(z) = W_{a,b-1}(z) - (b-1)W_{a,b}(z) \tag{3.7}$$

The Wright function also obeys the same differential recursion (2.39) as the Mittag-Leffler function,

$$\frac{d}{dz}\left\{z^{b-1}W_{a,b}(xz^a)\right\} = z^{b-2}W_{a,b-1}(xz^a) \tag{3.8}$$

The Wright function possesses by (3.3), as Laplace transform for $\mathrm{Re}(s) > 0$, the Mittag-Leffler function,

$$\int_0^{\infty} e^{-su} W_{a,b}(u)\, du = \frac{1}{s} E_{a,b}\left(\frac{1}{s}\right) \tag{3.9}$$

The inverse Laplace transform (B.2)

$$W_{a,b}(u) = \frac{1}{2\pi i} \int_{c-i\infty}^{c+i\infty} E_{a,b}\left(\frac{1}{s}\right) \frac{e^{su}}{s} ds \qquad \text{with } c > -1$$

is less manageable than the complex integral (4.9) deduced from Hankel's integral (15.1) of the Gamma function in **art.** 64.

Except for the deduction of a complex integral for $W_{a,b}(z)$ in **art.** 64 and its appearance in the distribution of the stable random variable R_a in **art.** 96, we do not treat the Wright function $W_{a,b}(z)$ and refer to the books by Gorenflo *et al.* (2020) and Mainardi (2021, Appendix F).

3.2 Laplace transforms

42. *Laplace transform of $z^{B-1}E_{a,b}(xz^A)$.* The Laplace transform of $t^{B-1}E_{a,b}(xt^A)$ is an instance of the general class (3.2) of Mittag-Leffler integrals in **art.** 39,

$$\int_0^{\infty} e^{-st} t^{B-1} E_{a,b}(xt^A)\, dt = \int_0^{\infty} e^{-st} \sum_{k=0}^{\infty} \frac{x^k t^{B-1+Ak}}{\Gamma(b+ak)} dt$$

Fubini's theorem[1] states that the summation and integration can be reversed

$$\int_0^{\infty} e^{-st} t^{B-1} E_{a,b}(xt^A)\, dt = \sum_{k=0}^{\infty} \frac{x^k}{\Gamma(b+ak)} \int_0^{\infty} e^{-st} t^{B+Ak-1} dt$$

provided the integrals $\int_0^{\infty} e^{-st} t^{B-1} E_{a,b}(xt^A)\, dt$ and $\int_0^{\infty} e^{-st} t^{B+Ak-1} dt =$

[1] Another argument is that a Taylor series can be integrated within its range of convergence.

$\frac{\Gamma(B+Ak)}{s^{B+kA}}$ exist and the resulting series at the right-hand side of (3.10) converges. The Euler integral (10.2) requires that $\text{Re}(B+kA) > 0$ for all integers $k \geq 0$, which is satisfied if $\text{Re}(B) > 0$ and $\text{Re}(A) > 0$, leading to

$$\int_0^\infty e^{-st} t^{B-1} E_{a,b}\left(xt^A\right) dt = \sum_{k=0}^{\infty} \frac{\Gamma(B+Ak)}{\Gamma(b+ak)} \frac{x^k}{s^{B+kA}} \tag{3.10}$$

Stirling's formula $\Gamma(z) = \sqrt{\frac{2\pi}{z}} z^z e^{-z} \left(1 + O\left(\frac{1}{z}\right)\right)$ in (14.17) indicates for large k that, to first order,

$$\frac{\Gamma(B+Ak)}{\Gamma(b+ak)} \sim k^{B-b} e^{k((a-A)+A\log A - a\log a)} k^{(A-a)k}$$

Hence, the series in (3.10) diverges if $A > a$ and converges for all $\frac{x}{s^A}$ if $A < a$. If $A = a$, then (14.9) shows, for any complex number $z = re^{i\theta}$ with large modulus $r = |z|$, that $\frac{|\Gamma(a+z)|}{|\Gamma(b+z)|} = r^{a-b}\left(1 + O\left(\frac{1}{r}\right)\right)$ and thus that $\frac{\Gamma(B+ak)}{\Gamma(b+ak)} \sim (ak)^{B-b}$ for large k. For a fixed, large integer k_0, the convergence of the series $\frac{1}{s^B}\sum_{k=k_0}^{\infty} \frac{\Gamma(B+ak)}{\Gamma(b+ak)} \left(\frac{x}{s^a}\right)^k \sim \frac{a^{B-b}}{s^B} \sum_{k=k_0}^{\infty} k^{B-b} \left(\frac{x}{s^a}\right)^k$ in (3.10) requires that $\left|\frac{x}{s^a}\right| < 1$, for any finite B and b.

43. *Particular A and B values in the Laplace transform of $z^{B-1}E_{a,b}\left(xz^A\right)$.* After introducing the duplication formula $\Gamma(2z) = \frac{2^{2z-1}}{\sqrt{\pi}} \Gamma(z)\Gamma\left(z + \frac{1}{2}\right)$ in (11.9) of the Gamma function

$$\frac{\Gamma(B+Ak)}{\Gamma(b+ak)} = \frac{\sqrt{\pi}}{2^{b-1}} \frac{\Gamma(B+Ak)}{2^{ak}\Gamma\left(\frac{b}{2} + \frac{a}{2}k\right)\Gamma\left(\frac{b+1}{2} + \frac{a}{2}k\right)}$$

two choices in (3.10) follow as $B = \frac{b}{2}$ and $A = \frac{a}{2}$ and $B = \frac{b+1}{2}$ and $A = \frac{a}{2}$, respectively. If $B = \frac{b}{2}$ and $A = \frac{a}{2}$, then we find

$$\int_0^\infty e^{-st} t^{\frac{b}{2}-1} E_{a,b}\left(xt^{\frac{a}{2}}\right) dt = \frac{\sqrt{\pi}}{2^{b-1}s^{\frac{b}{2}}} \sum_{k=0}^{\infty} \frac{1}{\Gamma\left(\frac{b+1}{2} + \frac{a}{2}k\right)} \left(\frac{x}{2^a s^{\frac{a}{2}}}\right)^k$$

resulting, with the definition (1.1), into

$$\int_0^\infty e^{-st} t^{\frac{b}{2}-1} E_{a,b}\left(xt^{\frac{a}{2}}\right) dt = \frac{\sqrt{\pi}}{2^{b-1}s^{\frac{b}{2}}} E_{\frac{a}{2},\frac{b+1}{2}}\left(\frac{x}{(4s)^{\frac{a}{2}}}\right) \tag{3.11}$$

while the choice $B = \frac{b+1}{2}$ and $A = \frac{a}{2}$ leads to

$$\int_0^\infty e^{-st} t^{\frac{b+1}{2}-1} E_{a,b}\left(xt^{\frac{a}{2}}\right) dt = \frac{\sqrt{\pi}}{2^{b-1}s^{\frac{b+1}{2}}} E_{\frac{a}{2},\frac{b}{2}}\left(\frac{x}{(4s)^{\frac{a}{2}}}\right) \tag{3.12}$$

The more important choice of the parameters is $A = a$ and $B = b$ in (3.10), which restricts s to $|s^a| > |x|$, leads, for $\text{Re}(b) > 0$ and $\text{Re}(a) > 0$, to

$$\int_0^\infty e^{-st} t^{b-1} E_{a,b}\left(xt^a\right) dt = \frac{1}{s^b} \sum_{k=0}^{\infty} \left(\frac{x}{s^a}\right)^k = \frac{s^{a-b}}{s^a - x} \tag{3.13}$$

The Laplace transform (3.13) plays a key role in the theory of the Mittag-Leffler function.

Example After substituting $t = u^2$ in the integral in (3.11), then replacing $s = \frac{1}{4t}$, we arrive at an integral that plays a role in diffusion theory (Podlubny, 1999, eq. (1.109)),

$$\int_0^\infty u^{b-1} e^{-\frac{u^2}{4t}} E_{a,b}\left(xu^a\right) du = \sqrt{\pi} t^{\frac{b}{2}} E_{\frac{a}{2},\frac{b+1}{2}}\left(xt^{\frac{a}{2}}\right) \tag{3.14}$$

Example After choosing $b = 1$, (3.11) also applies to $E_a(z)$,

$$\int_0^\infty e^{-st} t^{-\frac{1}{2}} E_a\left(xt^{\frac{a}{2}}\right) dt = \frac{\sqrt{\pi}}{\sqrt{s}} E_{\frac{a}{2}}\left(\frac{x}{(4s)^{\frac{a}{2}}}\right)$$

which further reduces for $a = 1$ to

$$\frac{\sqrt{\pi}}{\sqrt{s}} E_{\frac{1}{2}}\left(\frac{x}{2\sqrt{s}}\right) = \int_0^\infty e^{-st+x\sqrt{t}} t^{-\frac{1}{2}} dt = 2\int_0^\infty e^{-su^2+xu} du$$

With $su^2 - xu = s\left(u^2 - \frac{x}{s}u\right) = s\left(u - \frac{x}{2s}\right)^2 - \frac{x^2}{4s}$, we have

$$\int_0^\infty e^{-su^2+xu} du = \int_0^\infty e^{-s\left(u-\frac{x}{2s}\right)^2 + \frac{x^2}{4s}} du = \frac{e^{\frac{x^2}{4s}}}{\sqrt{s}} \int_{-\frac{x}{2\sqrt{s}}}^\infty e^{-y^2} dy$$

Replacing $z = \frac{x}{2\sqrt{s}}$ leads to

$$E_{\frac{1}{2}}(z) = e^{z^2} \frac{2}{\sqrt{\pi}} \int_{-z}^\infty e^{-y^2} dy$$

With the error function $\text{erfc}(z) = \frac{2}{\sqrt{\pi}} \int_z^\infty e^{-y^2} dy = 1 - \text{erf}(z)$ and $\text{erf}(-z) = \frac{2}{\sqrt{\pi}} \int_0^{-z} e^{-y^2} dy = -\text{erf}(z)$, we find again (2.46).

44. *Alternative and extended derivation of the Laplace transform (3.13).* We denote

$$L_{a,b}(s;x) = \int_0^\infty e^{-st} t^{b-1} E_{a,b}\left(xt^a\right) dt = \lim_{T\to\infty} L_{a,b}(s,T;x)$$

with

$$L_{a,b}(s,T;x) = \int_0^T e^{-st} t^{b-1} E_{a,b}\left(xt^a\right) dt = \int_0^T e^{-st} \frac{d}{dt}\left\{t^b E_{a,b+1}\left(xt^a\right)\right\} dt$$

where the derivative (2.39) is introdued. Partial integration yields, for $\text{Re}(b) \geq 0$ and real $T > 0$,

$$L_{a,b}(s,T;x) = e^{-sT} T^b E_{a,b+1}\left(xT^a\right) - 1_{\{b=0\}} + s\int_0^T e^{-st} t^b E_{a,b+1}\left(xt^a\right) dt \tag{3.15}$$

because the lower limit $\lim_{t\to 0} e^{-st} t^b E_{a,b+1}\left(xt^a\right) = \frac{1}{\Gamma(b+1)} \lim_{t\to 0} t^b = 0$ only if $\text{Re}(b) > 0$, but $\lim_{t\to 0} e^{-st} t^b E_{a,b+1}\left(xt^a\right) = 1$ for $b = 0$, provided

$\mathrm{Re}(a) > 0$ (see (2.2)), else divergence occurs. For real $T > 0$, (3.15) represents a recursion in $\Lambda[b] = L_{a,b}(s, T; x)$,

$$\Lambda[b] = e^{-sT} T^b E_{a,b+1}(xT^a) - 1_{\{b=0\}} + s\Lambda[b+1]$$

Iterating n times yields

$$\Lambda[b] = e^{-sT} T^b \sum_{j=0}^{n-1} (sT)^j E_{a,b+j}(xT^a) - \sum_{j=0}^{n-1} s^j 1_{\{b=-j\}} + s^n \Lambda[b+n] \quad (3.16)$$

We will show in the remainder that $\lim_{n\to\infty} s^n \Lambda[b+n] = 0$, so that (3.16) produces, for real $T > 0$, $\mathrm{Re}(a) > 0$ and $\mathrm{Re}(b) \geq 0$,

$$\int_0^T e^{-st} t^{b-1} E_{a,b}(xt^a)\, dt = e^{-sT} T^b \sum_{j=0}^{\infty} (sT)^j E_{a,b+j}(xT^a) - \sum_{j=0}^{\infty} s^j 1_{\{b=-j\}} \quad (3.17)$$

The "finite T" Laplace transform (3.17) is again an example of an expansion (**art.** 31) into an infinite set of linearly independent functions $\{E_{a,b+j}(z)\}_{j\geq 1}$.

The Taylor series (1.1) of the Mittag-Leffler function shows that

$$\int_0^T e^{-st} t^{b-1} E_{a,b}(xt^a)\, dt = \sum_{k=0}^{\infty} \frac{\int_0^T e^{-st} t^{b+ak-1} dt}{\Gamma(ak+b)} x^k$$

After writing the coefficient of x^k in terms of the incomplete Gamma function $\Gamma(a, z)$ in (16.1) and entire incomplete Gamma function $\gamma^*(a, z)$ in (16.3)

$$\begin{aligned}\frac{\int_0^T e^{-st} t^{b+ak-1} dt}{\Gamma(ak+b)} &= \frac{\int_0^\infty e^{-st} t^{b+ak-1} dt - \int_T^\infty e^{-st} t^{b+ak-1} dt}{\Gamma(ak+b)} \\ &= \frac{1}{s^{ak+b}} \left(1 - \frac{\int_{sT}^\infty e^{-u} u^{b+ak-1} du}{\Gamma(ak+b)}\right) \\ &= \frac{1}{s^{ak+b}} \left(1 - \frac{\Gamma(ak+b, sT)}{\Gamma(ak+b)}\right) = T^{ak+b} \gamma^*(ak+b, sT)\end{aligned}$$

we obtain

$$\Lambda[b] = \int_0^T e^{-st} t^{b-1} E_{a,b}(xt^a)\, dt = T^b \sum_{k=0}^{\infty} \gamma^*(ak+b, sT)(xT^a)^k \quad (3.18)$$

Thus, (3.18) gives $s^n \Lambda[b+n] = (sT)^n \sum_{k=0}^{\infty} T^{ak+b} \gamma^*(ak+b+n, sT) x^k$. The series $\gamma^*(a, z) = \frac{1}{\Gamma(a)} \sum_{k=0}^{\infty} \frac{(-1)^k z^k}{k!(a+k)}$ in (16.7) demonstrates, for finite T, that $\lim_{n\to\infty} s^n \Lambda[b+n] = 0$, which established (3.17).

On the other hand, for fixed $ak+b$ and large T, the asymptotic formula $\gamma^*(ak+b, sT) = (sT)^{-ak+b} - \frac{e^{-sT}}{sT\Gamma(ak+b)}\left(1 + O\left(\frac{1}{T}\right)\right)$ in (16.23) shows in (3.18) that

$$\int_0^T e^{-st} t^{b-1} E_{a,b}(xt^a)\, dt = \sum_{k=0}^{\infty} \frac{x^k}{s^{ak-b}} - \frac{e^{-sT}}{s} T^{b-1} E_{a,b}(xT^a)\left(1 + O\left(\frac{1}{T}\right)\right)$$

which for $T \to \infty$ again leads to (3.13), provided the last term vanishes. The asymptotic expansion (4.19) illustrates that the fastest possible growth in z of $E_{a,b}(z) = \frac{C}{a}|z|^{\frac{1-b}{a}} e^{|z|^{\frac{1}{a}}}$, where C is the number terms in the k-sum in (4.19). Thus, $\frac{e^{-sT}}{s}T^{b-1}E_{a,b}(xT^a) \sim \frac{C}{as}|x|^{\frac{1-b}{a}} e^{\left(|x|^{\frac{1}{a}} - \mathrm{Re}(s)\right)T}$ vanishes if $\mathrm{Re}(s) > |x|^{\frac{1}{a}}$, which amounts to the same validity conditions as in (3.13).

3.3 Integral multiplication formulae

45. *Integral multiplication formula for $E_{a,b}(z)$.* We generalize the "duplication" integrals (3.11) and (3.12). Invoking Gauss's multiplication formula (11.8) into the Taylor series (1.1) yields

$$E_{a,b}(z) = (2\pi)^{\frac{1}{2}(1-n)} n^{nb-\frac{1}{2}} \sum_{k=0}^{\infty} \frac{\prod_{j=1}^{n-1}\Gamma\left(b + ak + \frac{j}{n}\right)}{\Gamma(nb + nak)} (n^{na}z)^k$$

We introduce the Mellin transform (16.36) of a product of Gamma functions for $\mathrm{Re}(s) > 0$,

$$\prod_{j=1}^{n-1}\Gamma\left(s + \frac{j}{n}\right) = \int_0^\infty u^{s-1} h_n(u)\, du$$

where inverse Mellin transformation specifies the function $h_n(u)$ in (16.39) in **art.** 232 as a Taylor series in u, and we obtain

$$E_{a,b}(z) = (2\pi)^{\frac{1}{2}(1-n)} n^{nb-\frac{1}{2}} \int_0^\infty h_n(u)\, u^{b-1} \sum_{k=0}^{\infty} \frac{(u^a n^{na} z)^k}{\Gamma(nb + nak)} du$$

Thus, for an integer $n \geq 2$, we arrive at an integral multiplication formula for the Mittag-Leffler functions,

$$E_{a,b}(z) = (2\pi)^{\frac{1}{2}(1-n)} n^{nb-\frac{1}{2}} \int_0^\infty h_n(u)\, u^{b-1} E_{na,nb}(u^a n^{na} z)\, du \tag{3.19}$$

From $E_{a,b}(z) = \sum_{k=0}^{\infty} \frac{z^k}{(b-1+ak)\Gamma(b-1+ak)}$, the companion of (3.19) follows similarly as

$$E_{a,b}(z) = (2\pi)^{\frac{1}{2}(1-n)} n^{n(b-1)+\frac{1}{2}} \int_0^\infty h_n(u)\, u^{b-2} E_{na,n(b-1)+1}(u^a n^{na} z)\, du \tag{3.20}$$

which directly reduces for $b = 1$ to the Mittag-Leffler function

$$E_a(z) = (2\pi)^{\frac{1}{2}(1-n)} n^{\frac{1}{2}} \int_0^\infty h_n(u)\, u^{-1} E_{na}(u^a n^{na} z)\, du$$

46. *Special cases of the integral multiplication formula for $E_{a,b}(z)$.* The case $n = 3$ in (3.19) with $h_3(u)$ in (16.43) expressed in terms of the modified Bessel function $K_\nu(z)$, defined in (16.41), becomes

$$E_{a,b}(z) = \frac{3^{3b-\frac{1}{2}}}{\pi} \int_0^\infty K_{\frac{1}{3}}\left(2\sqrt{u}\right) u^{b-\frac{1}{2}} E_{3a,3b}\left(3^{3a} u^a z\right) du \tag{3.21}$$

The companion of (3.21) follows from (3.20) as

$$E_{a,b}(z) = \frac{3^{3(b-1)+\frac{1}{2}}}{\pi} \int_0^\infty K_{\frac{1}{3}}\left(2\sqrt{u}\right) u^{b-\frac{3}{2}} E_{3a,3b-2}\left(3^{3a} u^a z\right) du \tag{3.22}$$

Apelblat (2020, eq. (25), (28) and (31)) has presented three remarkable integral functional relations,

$$E_a\left(t^a\right) = \frac{1}{\sqrt{\pi t}} \int_0^\infty e^{-\frac{u^2}{4t}} E_{2a}\left(u^{2a}\right) du \tag{3.23}$$

$$E_a\left(t^a\right) = \frac{1}{\pi} \int_0^\infty \sqrt{\frac{u}{t}} K_{\frac{1}{3}}\left(\frac{2u^{\frac{3}{2}}}{\sqrt{27t}}\right) E_{3a}\left(u^{3a}\right) du \tag{3.24}$$

$$\frac{E_a\left(t^a\right) - 1}{\sqrt{t}} = \int_0^\infty \frac{J_1\left(2\sqrt{tu}\right)}{\sqrt{u}} E_a\left(u^a\right) du \tag{3.25}$$

which he has skillfully derived by manipulations of Laplace transforms. Apart from the last relation (3.25), where $J_p(z)$ is the Bessel function of order p, the first two are special cases of the multiplication formula's for $E_{a,b}(z)$ in **art.** 45. Indeed, the first integral (3.23) is a special case of the fractional diffusion integral (3.14) for $b = 1$ and a replaced by $2a$. With $z = t^a$ and after substituting $x = \frac{u^3}{27t}$ in (3.22), we obtain

$$E_{a,b}\left(t^a\right) = \frac{t^{\frac{1}{2}-b}}{\pi} \int_0^\infty K_{\frac{1}{3}}\left(2\sqrt{\frac{u^3}{27t}}\right) u^{3b-\frac{5}{2}} E_{3a,3b-2}\left(u^{3a}\right) dx$$

which leads to (3.24) after choosing $b = 1$.

3.4 Convolution integrals

This Section 3.4 considers convolution integrals, defined by

$$f(x) = \int_0^x h(x-u)\, g(u)\, du \tag{3.26}$$

47. *Generalized integration.* By using a variation of the integral of the Beta-function in (10.11),

$$\int_0^x u^{z-1}(x-u)^{w-1} du = x^{z+w-1} \frac{\Gamma(z)\Gamma(w)}{\Gamma(z+w)} \tag{3.27}$$

for real $x > 0$, $\mathrm{Re}(z) > 0$ and $\mathrm{Re}(w) > 0$, we obtain another instance of the general class (3.1) of Mittag-Leffler integrals in **art.** 39,

$$\int_0^x \frac{(x-u)^{w-1}}{\Gamma(w)} u^{\gamma-1} E_{a,b}\left(\lambda u^\beta\right) du = \sum_{k=0}^\infty \lambda^k \frac{\int_0^x (x-u)^{w-1} u^{\beta k+\gamma-1}}{\Gamma(b+ak)\Gamma(w)} du$$
$$= x^{\gamma+w-1} \sum_{k=0}^\infty \frac{\lambda^k x^{\beta k}}{\Gamma(b+ak)} \frac{\Gamma(\gamma+\beta k)}{\Gamma(\gamma+w+\beta k)}$$

which reduces for $\beta = a$, $\gamma = b$ to

$$\frac{1}{\Gamma(w)} \int_0^x (x-u)^{w-1} u^{b-1} E_{a,b}(\lambda u^a)\, du = x^{b-1+w} E_{a,b+w}(\lambda x^a) \qquad (3.28)$$

valid for $\mathrm{Re}(a) > 0$, $\mathrm{Re}(b) > 0$, $\mathrm{Re}(w) > 0$, real positive x and any complex λ.

The m-fold integral[2] (3.28) for $w = m$ possesses the same form as the m-fold differentiation in (2.40)

$$\frac{d^m}{dx^m}\left\{x^{b-1} E_{a,b}(\lambda x^a)\right\} = x^{b-1-m} E_{a,b-m}(\lambda x^a)$$

that is better recognized with Cauchy's integral (1.5) for the m-th derivative of an analytic function, $\left.\frac{d^m f(z)}{dz^m}\right|_{z=z_0} = \frac{m!}{2\pi i} \int_{C(z_0)} \frac{f(\omega)\, d\omega}{(\omega - z_0)^{m+1}}$, where $C(z_0)$ is a closed contour around z_0,

$$\frac{(-1)^{m+1} m!}{2\pi i} \int_{C(x)} (x-\omega)^{-m-1} \omega^{b-1} E_{a,b}(\lambda \omega^a)\, d\omega = x^{b-1-m} E_{a,b-m}(\lambda x^a)$$

Hence, (3.28) written with the reflection formula (11.4), suggests that

$$\frac{\frac{\sin \pi w}{\pi} \Gamma(1-w)}{2\pi i} \int_{C(x)} (x-\omega)^{w-1} \omega^{b-1} E_{a,b}(\lambda \omega^a)\, d\omega = x^{b-1+w} E_{a,b+w}(\lambda x^a)$$

holds for any negative real w, leading to a fractional derivative (1.6) after modifying the contour, due to the branch in z^w for $w \notin \mathbb{Z}$. The Laplace transform (3.13) easier connects to fractional derivatives, avoiding the contour integral. Applications of the Mittag-Leffler function to fractional calculus are amply illustrated in Gorenflo *et al.* (2014).

Example For $a = b = 1$ in (3.28) and, next, $x = 1$, $\lambda = z$ and $w+1 \to b$, we obtain, for $\mathrm{Re}(b) > 1$,

$$E_{1,b}(z) = \frac{1}{\Gamma(b-1)} \int_0^1 (1-u)^{b-2} e^{zu} du$$

Let $t = 1 - u$, then the incomplete Gamma function (16.5) appears

$$E_{1,b}(z) = \frac{e^z \int_0^1 t^{b-2} e^{-zt} dt}{\Gamma(b-1)} = \frac{z^{1-b} e^z}{\Gamma(b-1)} \int_0^z u^{b-2} e^{-u} du = e^z \gamma^*(b-1, z)$$

which is again equal to (2.13), while $E_{1,b}(z) = e^z \gamma^*(b-1, z)$ in (2.11) holds for any b.

---

[2] The m-fold integral is

$$F_m(x,a) = \int_a^x du_1 \int_a^{u_1} du_2 \ldots \int_a^{u_{m-1}} du_m f(u_m) = \int_a^x \frac{(x-u)^{m-1}}{(m-1)!} f(u)\, du \qquad (3.29)$$

The generalization towards fractional calculus, where the integer m is extended to a real number, is treated in Gorenflo and Mainardi (1997).

Example For $b=1$, $w=a$ and any λ, (3.28) simplifies to

$$\frac{1}{\Gamma(a)}\int_0^x (x-u)^{a-1} E_a(\lambda u^a)\,du = x^a E_{a,1+a}(\lambda x^a)$$

Formula (2.5) reduces for $\beta=m=1$ to $zE_{a,1+a}(z)=E_a(z)-1$ and leads to a functional equation for the Mittag-Leffler function $E_a(z)$,

$$E_a(\lambda x^a) = 1+\lambda\int_0^x \frac{(x-u)^{a-1}}{\Gamma(a)} E_a(\lambda u^a)\,du \tag{3.30}$$

Iterating the functional equation (3.30) again produces the defining Taylor series (1.1) around $z_0=0$. Taking the Laplace transform of both sides of (3.30) and invoking (3.13) results in an identity.

48. *Integration of a product of Mittag-Leffler functions.* We extend the idea in **art.** 47 and consider, for $x\geq 0$,

$$\begin{aligned} L &= \frac{1}{\Gamma(w)}\int_0^x (x-u)^{w-1} u^{\beta-1} E_{a,b}(\lambda u^\alpha) E_{c,d}(\mu(x-u)^\gamma)\,du \\ &= \frac{1}{\Gamma(w)}\sum_{k=0}^\infty\sum_{m=0}^\infty \frac{\lambda^k}{\Gamma(b+ak)}\frac{\mu^m}{\Gamma(d+cm)}\int_0^x (x-u)^{w+m\gamma-1} u^{\beta+k\alpha-1}du \end{aligned}$$

Introducing the integral (3.27) of the Beta-function results in

$$L = \frac{x^{\beta+w-1}}{\Gamma(w)}\sum_{k=0}^\infty\sum_{m=0}^\infty \frac{(\lambda x^\alpha)^k}{\Gamma(b+ak)}\frac{(\mu x^\gamma)^m}{\Gamma(d+cm)}\frac{\Gamma(\beta+k\alpha)\Gamma(w+m\gamma)}{\Gamma(\beta+k\alpha+w+m\gamma)}$$

After the choice $\alpha=a$, $\beta=b$, $\gamma=c$ and $d=w$, the double sum simplifies to

$$L = \frac{x^{b+w-1}}{\Gamma(w)}\sum_{k=0}^\infty\sum_{m=0}^\infty \frac{(\lambda x^a)^k(\mu x^c)^m}{\Gamma(b+w+ka+mc)}$$

which equals

$$\begin{aligned} L &= \frac{x^{b+w-1}}{\Gamma(w)}\sum_{k=0}^\infty (\lambda x^a)^k E_{c,b+w+ka}(\mu x^c) \\ &= \frac{x^{b+w-1}}{\Gamma(w)}\sum_{k=0}^\infty (\mu x^c)^k E_{a,b+w+kc}(\lambda x^a) \end{aligned}$$

Further computations require the choice $c=a$,

$$L = \frac{x^{b+w-1}}{\Gamma(w)}\sum_{k=0}^\infty\sum_{m=0}^\infty \frac{(\lambda x^a)^k(\mu x^a)^m}{\Gamma(b+w+a(k+m))}$$

Let $q=k+m$, then $0\leq q$ and $m=q-k\geq 0$, while $k\geq 0$, thus

$$L = \frac{x^{b+w-1}}{\Gamma(w)}\sum_{q=0}^\infty\sum_{k=0}^q \frac{\lambda^k\mu^{q-k}x^{aq}}{\Gamma(b+w+aq)} = \frac{x^{b+w-1}}{\Gamma(w)}\sum_{q=0}^\infty \frac{\mu^q x^{aq}}{\Gamma(b+w+aq)}\sum_{k=0}^q\left(\frac{\lambda}{\mu}\right)^k$$

Executing the finite geometric series $\sum_{k=0}^{q}\left(\frac{\lambda}{\mu}\right)^k = \frac{\left(\frac{\lambda}{\mu}\right)^{q+1}-1}{\frac{\lambda}{\mu}-1}$ in (1.16) leads to

$$L = \frac{x^{b+w-1}}{\Gamma(w)(\lambda-\mu)}\left\{\lambda\sum_{q=0}^{\infty}\frac{(\lambda x^a)^q}{\Gamma(b+w+aq)} - \mu\sum_{q=0}^{\infty}\frac{(\mu x^a)^q}{\Gamma(b+w+aq)}\right\}$$

Finally, we arrive for $\mathrm{Re}(b) > 0$ and $\mathrm{Re}(w) > 0$ at

$$P = \int_0^x (x-u)^{w-1} u^{b-1} E_{a,b}(\lambda u^a) E_{a,w}(\mu(x-u)^a)\, du = \frac{\lambda E_{a,b+w}(\lambda x^a) - \mu E_{a,b+w}(\mu x^a)}{x^{1-b-w}(\lambda-\mu)} \tag{3.31}$$

Clearly[3], since $\lim_{\mu\to 0} E_{a,w}(\mu(x-u)^a) = \frac{1}{\Gamma(w)}$, then (3.31) reduces to (3.28).

49. *A convolution integral containing the Mittag-Leffler function.* Let $h(x) = x^{b-1}E_{a,b}(\lambda x^a)$ in the definition (3.26) of a convolution integral, then a convolution transform, for $\mathrm{Re}(b) > 0$, appears

$$f(x) = \int_0^x (x-u)^{b-1} E_{a,b}(\lambda(x-u)^a)\, g(u)\, du \tag{3.32}$$

which was studied by Bharatiya (1967) for $b = 1$. It follows from (3.32) that $f(0) = 0$. We show that the convolution transform (3.32) can be inverted or solved for $g(x)$.

Taking the Laplace transform $F(s) = \mathcal{L}_0[f(x)] = \int_0^\infty e^{-sx} f(x)dx$ of the convolution (3.32) and using (3.13) for $\mathrm{Re}(b) > 0$, $\mathrm{Re}(a) > 0$ and $|s^a| > |\lambda|$ yields $F(s) = \frac{s^{a-b}}{s^a-\lambda}G(s)$, from which we find

$$G(s) = s^b F(s) - \lambda s^{b-a} F(s) \tag{3.33}$$

The Laplace transform of the generalized fractional derivative $D_m^\alpha f(t)$ in (1.7) with $p = 0$ and integer $m \geq 0$ is (Van Mieghem, 2022, Sec. II.G)

$$\mathcal{L}_s\left[\left.\frac{d^\alpha f(z)}{dz^\alpha}\right|_{z=x}\right] = s^\alpha\left\{\mathcal{L}_s[f(t)] - \sum_{l=0}^{m-1} s^{-l-1} f^{(l)}(0)\right\} \quad \text{for } \mathrm{Re}(\alpha) \leq m \tag{3.34}$$

Euler's integral (10.2) shows, for $\mathrm{Re}(z) > 0$, that $s^{-z} = \int_0^\infty e^{-sx}\frac{x^{z-1}}{\Gamma(z)}dx$. After taking the inverse Laplace transform (see Appendix B) of (3.33)

[3] In case $\mu \to \lambda$, then after using de l'Hospital's rule,

$$\lim_{\mu\to\lambda}\frac{\lambda E_{a,b+w}(\lambda x^a) - \mu E_{a,b+w}(\mu x^a)}{\lambda-\mu} = E_{a,b+w}(\lambda x^a) + \lambda x^a \left.\frac{dE_{a,b+w}(z)}{dz}\right|_{z=\lambda x^a}$$

and the differential rule (2.37), formula (3.31) becomes

$$P = \int_0^x (x-u)^{w-1} u^{b-1} E_{a,b}(\lambda u^\alpha) E_{a,w}(\lambda(x-u)^a)\, du = \frac{(a+1-(b+w))E_{a,b+w}(\lambda x^a) + E_{a,b+w-1}(\lambda x^a)}{a x^{1-b-w}}$$

and recalling that $f(0) = 0$, the inverse of the transformation (3.32), for $\mathrm{Re}(a-b) > 0$ and $0 < \mathrm{Re}(b) \leq m$, is formally

$$g(x) = \left.\frac{d^b f(z)}{dz^b}\right|_{z=x} + \mathcal{L}_s^{-1}\left[\sum_{l=1}^{m-1} s^{b-l-1} f^{(l)}(0)\right] - \lambda \int_0^x \frac{(x-u)^{a-b-1}}{\Gamma(a-b)} f(u)\, du$$

which simplifies for $b = 1$ with $m = 1$ to Bharatiya (1967, eq. (3.2)) in the usual derivative

$$g(x) = \left.\frac{df(z)}{dz}\right|_{z=x} - \frac{\lambda}{\Gamma(a-1)} \int_0^x (x-u)^{a-2} f(u)\, du \tag{3.35}$$

3.5 The Berberan-Santos integral

50. *Berberan-Santos' integral for the double argument.* Berberan-Santos (2005b) applied a simplified form (B.6) of the inverse Laplace contour integral (B.2). First, let the Laplace transform of a real function $g(u)$ be equal to $E_{a,b}(-s)$, hence, $\int_0^\infty e^{-su} g(u)\, du = E_{a,b}(-s)$. From the Laplace transform with $s = \sigma + iT$

$$|E_{a,b}(-s)| = \left|\int_0^\infty e^{-\sigma u} e^{-iTu} g(u)\, du\right| \leq \int_0^\infty e^{-\sigma t} g(t)\, dt$$

it follows that $\lim_{s\to\infty} \int_0^\infty e^{-st} g(t)\, dt = 0$ for any direction in which s with $\mathrm{Re}(s) > 0$ tends to infinity. Since $E_{a,b}(z)$ is entire function of order $\frac{1}{a}$, implying (Section 1.4.3) that $E_{a,b}(z) = O\left(e^{z^{\frac{1}{a}}}\right)$ for any complex b, it holds that $E_{a,b}(-z) = O\left(e^{r^{\frac{1}{a}} e^{i\frac{\theta+\pi}{a}}}\right) = O\left(e^{r^{\frac{1}{a}} \cos\frac{\theta+\pi}{a}}\right)$ and $|E_{a,b}(-s)| \to 0$ provided that $\cos\frac{\theta+\pi}{a} < 0$ or $\frac{\pi a}{2} - \pi < \theta < \frac{3\pi a}{2} - \pi$. With $s = re^{i\theta}$ and $-\frac{\pi}{2} \leq \theta \leq \frac{\pi}{2}$ because $\mathrm{Re}(s) > 0$, the limit $\lim_{s\to\infty} |E_{a,b}(-s)| = 0$ requires that $0 \leq a \leq 1$. Incidentally, the definition of c in Appendix B indicates that $c = 0$ and Berberan-Santos' inverse Laplace transform (B.6) then yields

$$g(t) = \frac{2}{\pi} \int_0^\infty \mathrm{Re}\,(E_{a,b}(-iw)) \cos tw dw \qquad \text{for } t > 0$$

Next, Berberan-Santos observed from (2.8) that

$$E_{a,b}(-iw) = E_{2a,b}\left(-w^2\right) - iwE_{2a,b+2a}\left(-w^2\right)$$

so that, for real w,

$$\mathrm{Re}\,(E_{a,b}(-iw)) = E_{2a,b}\left(-w^2\right)$$

Hence, the inverse Laplace transform becomes

$$g(t) = \frac{2}{\pi} \int_0^\infty E_{2a,b}\left(-w^2\right) \cos tw dw \qquad \text{for } t > 0$$

Finally, taking the Laplace transform of both sides and reversing the integrals

$$E_{a,b}\left(-s\right) = \frac{2}{\pi}\int_0^\infty E_{2a,b}\left(-w^2\right)\left(\int_0^\infty e^{-st}\cos twdt\right)dw$$

results in Berberan-Santos' remarkable integral for the double argument in a

$$E_{a,b}\left(-s\right) = \frac{2s}{\pi}\int_0^\infty \frac{E_{2a,b}\left(-w^2\right)}{s^2+w^2}dw \qquad \text{for } 0 \leq a \leq 1,\ \operatorname{Re}\left(s\right) > 0 \tag{3.36}$$

Example For $a = \frac{1}{2}$, $b = 1$ and with $E_1(z) = e^z$, Berberan-Santos' integral (3.36) yields again (2.46),

$$E_{\frac{1}{2}}\left(-x\right) = \frac{2x}{\pi}\int_0^\infty \frac{e^{-t^2}}{t^2+x^2}dt = e^{x^2}\operatorname{erfc}\left(x\right)$$

51. *A generalization of Berberan-Santos' integral (3.36).* The integral $\int_C \frac{E^p_{a,b}(-iw)}{w^2+s^2}dw$ for real $p > 0$ and $\operatorname{Re}(s) > 0$, where the contour C is a semi-circle in the positive $\operatorname{Im} w$-plane, centered at the origin $w = 0$ with radius $R > |s|$, equals, by Cauchy's theorem,

$$\int_C \frac{E^p_{a,b}\left(iw\right)}{w^2+s^2}dw = \pi\frac{E^p_{a,b}\left(-s\right)}{s}$$

because the contour C only encloses the simple pole at $w = is$. Evaluation of the integral along the contour C results in

$$\int_C \frac{E^p_{a,b}\left(iw\right)}{w^2+s^2}dw = \int_{-R}^R \frac{E^p_{a,b}\left(ix\right)}{x^2+s^2}dx + i\int_0^\pi \frac{E^p_{a,b}\left(Re^{i\left(\theta+\frac{\pi}{2}\right)}\right)}{R^2e^{2i\theta}+s^2}Re^{i\theta}d\theta$$

For $0 \leq a \leq 1$,

$$\left|\int_0^\pi \frac{E^p_{a,b}\left(Re^{i\left(\theta+\frac{\pi}{2}\right)}\right)}{R^2e^{2i\theta}+s^2}Re^{i\theta}d\theta\right| \leq \frac{1}{R}\int_0^\pi \frac{\left|E_{a,b}\left(Re^{i\left(\theta+\frac{\pi}{2}\right)}\right)\right|^p}{\left|e^{2i\theta}+\frac{s^2}{R^2}\right|}d\theta$$

vanishes for $R \to \infty$, as shown in **art.** 50. Hence, for any positive real p, for $0 \leq a \leq 1$ and for $\operatorname{Re}(s) > 0$, it holds that

$$E^p_{a,b}\left(-s\right) = \frac{s}{\pi}\int_{-\infty}^\infty \frac{E^p_{a,b}\left(ix\right)}{x^2+s^2}dx \tag{3.37}$$

After invoking $E_{a,b}(ix) = E_{2a,b}\left(-x^2\right) + ixE_{2a,b+a}\left(-x^2\right)$ in (2.8), the fact that $xE_{2a,b+2a}\left(-x^2\right)$ is odd and $E_{2a,b}\left(-x^2\right)$ is even and Newton's binomium, we find from (3.37) for integer values of $p \geq 1$, the generalization of the Berberan-Santos integral in the double a argument,

$$E^p_{a,b}\left(-s\right) = \frac{2s}{\pi}\sum_{k=0}^{\left[\frac{p}{2}\right]}(-1)^k\binom{p}{2k}\int_0^\infty \frac{E^{p-2k}_{2a,b}\left(-x^2\right)x^{2k}E^{2k}_{2a,b+a}\left(-x^2\right)}{x^2+s^2}dx \tag{3.38}$$

If $p = 1$ in (3.38), then we find Berberan-Santos's integral (3.36). If $p = 2$, then (3.38) becomes

$$E_{a,b}^2(-s) = \frac{2s}{\pi}\int_0^\infty \frac{E_{2a,b}^2(-x^2) - x^2 E_{2a,b+a}^2(-x^2)}{x^2+s^2}dx$$

52. *Berberan-Santos' integral and Laplace transforms.* Using $\frac{1}{s^2+w^2} = \int_0^\infty e^{-t(s^2+w^2)}dt$ in (3.36) and reversing the integrals, justified by absolute convergence, gives

$$\begin{aligned} E_{a,b}(-s) &= \frac{2s}{\pi}\int_0^\infty e^{-ts^2}\left\{\int_0^\infty e^{-tw^2}E_{2a,b}(-w^2)\,dw\right\}dt \\ &= \frac{s}{\pi}\int_0^\infty e^{-ts^2}\left\{\int_0^\infty e^{-tx}x^{-\frac{1}{2}}E_{2a,b}(-x)\,dx\right\}dt \end{aligned}$$

Thus, Berberan-Santos' integral (3.36) is equivalent to the statement that the Laplace transform of the Laplace transform of $x^{-\frac{1}{2}}E_{2a,b}(-x)$ equals $\frac{\pi E_{a,b}(-\sqrt{s})}{\sqrt{s}}$. The latter property follows from the Stieltjes transform (B.7), which is an iteration of the Laplace transform and which is treated in the book by Widder (1946, Chapter VIII). Appendix B discusses the Laplace transform pair of Gross (1950), in which the inverse Laplace transform (B.2) is of the same form as the Laplace transform (B.1) itself.

3.6 Euler-Maclaurin summation applied to $E_{a,b}(z)$

53. *Euler-Maclaurin summation.* Titchmarsh and Heath-Brown (1986, p. 13), and also Rainville (1971, pp. 27-28) and Hardy (1948, pp. 318-319), prove the basic summation formula due to Euler,

$$\begin{aligned} \sum_{\alpha<k\leq\beta} f(k) &= \int_\alpha^\beta f(x)\,dx + \int_\alpha^\beta \left(x - \lfloor x\rfloor - \frac{1}{2}\right)\frac{df(x)}{dx}dx \\ &\quad + \left(\alpha - \lfloor\alpha\rfloor - \frac{1}{2}\right)f(\alpha) - \left(\beta - \lfloor\beta\rfloor - \frac{1}{2}\right)f(\beta) \end{aligned} \tag{3.39}$$

valid for any function $f(x)$ with continuous derivative in the interval $[\alpha, \beta]$ and $\lfloor x\rfloor$ denotes the greatest integer not exceeding x. Titchmarsh and Heath-Brown (1986, p. 13) observe that (3.39) is additive with respect to the interval $[\alpha, \beta]$ so that it suffices to suppose that $m \leq \alpha < \beta \leq m+1$, where $m \in \mathbb{Z}$, in which case (3.39) is readily verified to be correct.

The basic summation formula (3.39) lies at the basis of the Euler-Maclaurin summation formula (3.40). Indeed, by additivity, we can limit the interval to $[0, 1]$ and further assume that α and β are integers. We recognize (Section A.2) that $B_1(x) = x - \frac{1}{2}$ is a Bernoulli polynomial. Continued partial integration of the second integral in (3.39), invoking $\frac{dB_n(x)}{dx} = nB_{n-1}(x)$ in (A.16) and $B_n(1) = B_n(0) = B_n$, for integer $n \geq 2$

that follows from the reflection formula (A.17) of the Bernoulli polynomials, leads to the Euler-Maclaurin summation formula, for integer α and β,

$$\sum_{k=\alpha+1}^{\beta} f(k) = \int_{\alpha}^{\beta} f(t)\,dt + \sum_{n=1}^{N} (-1)^n \frac{B_n}{n!} \left[f^{(n-1)}(\beta) - f^{(n-1)}(\alpha) \right] + R_N \tag{3.40}$$

with remainder term

$$R_N = \frac{(-1)^{N-1}}{N!} \int_{\alpha}^{\beta} B_N(u - \lfloor u \rfloor)\, f^{(N)}(u)\,du$$

Rademacher (1973, p.14) presents another derivation of the Euler-Maclaurin summation formula (3.40). A detailed study of the Euler-Maclaurin formula (3.40), with a convergence analysis of the remainder R_N and with alternative derivations e.g. due to Poisson and Fourier, is given by Hardy (1948, Chapter XIII). Euler (1755, Pars posterior, Chapt. 6) deduces and gives many examples of the Euler-Maclaurin summation formula (3.40) for the (integer) variable x in the form (in our notation)

$$\sum_{k=1}^{x} f(k) = \int_{1}^{x} f(t)\,dt + C + \sum_{n=1}^{N} (-1)^n \frac{B_n}{n!} f^{(n-1)}(x) \tag{3.41}$$

where the constant $C = -f(1) - \sum_{n=1}^{N} (-1)^n \frac{B_n}{n!} f^{(n-1)}(1)$ is determined from (3.41) for an appropriate value of x and N, although Euler realized that for some functions f the series $\sum_{n=1}^{N} (-1)^n \frac{B_n}{n!} f^{(n-1)}(x)$ diverges for $N \to \infty$.

54. *Application of the basic summation formula (3.39) to the Mittag-Leffler function.* Substitution of the function $f(w) = \frac{z^w}{\Gamma(aw+b)}$ in the basic summation formula (3.39) yields

$$\sum_{k=1}^{\infty} \frac{z^k}{\Gamma(ak+b)} = \int_0^{\infty} \frac{z^x}{\Gamma(ax+b)} dx - \frac{1}{2} \frac{1}{\Gamma(b)} + \int_0^{\infty} \left(x - \lfloor x \rfloor - \frac{1}{2} \right) \frac{z^x (\log z - a\, \psi(ax+b))}{\Gamma(ax+b)} dx$$

where $\psi(z)$ is the digamma function in **art.** 154. Hence, for non-negative real z, we obtain an integral for the Mittag-Leffler function

$$E_{a,b}(z) = \int_0^{\infty} \frac{\left(1 + \left(x - \lfloor x \rfloor - \frac{1}{2}\right) (\log z - a\, \psi(ax+b))\right) z^x}{\Gamma(ax+b)} dx + \frac{1}{2} \frac{1}{\Gamma(b)} \tag{3.42}$$

The integral (3.42) simplifies for $z = 1$ to

$$E_{a,b}(1) = \int_0^{\infty} \frac{\left(1 - a\left(x - \lfloor x \rfloor - \frac{1}{2}\right)\, \psi(ax+b)\right)}{\Gamma(ax+b)} dx + \frac{1}{2} \frac{1}{\Gamma(b)}$$

55. *Application of Euler-Maclaurin summation to the Mittag-Leffler function.* The right-hand side summation in the Euler-Maclaurin summation formula (3.40) requires higher-order derivatives for the function $f(w) = \frac{z^w}{\Gamma(b+aw)}$. We invoke Leibniz' rule

$$f^{(k)}(x) = \frac{d^k}{dw^k} \frac{e^{w \log z}}{\Gamma(b+aw)}\bigg|_{w=x} = \sum_{j=0}^{k} \binom{k}{j} (\log z)^{k-j} z^x \frac{d^j}{dw^j} \frac{1}{\Gamma(b+aw)}\bigg|_{w=x}$$

which simplifies, with $\frac{d^j}{dw^j}\frac{1}{\Gamma(b+aw)} = a^j \frac{d^j}{dy^j}\frac{1}{\Gamma(y)}\Big|_{y=b+ax}$, to

$$\frac{f^{(k)}(x)}{k!} = z^x \sum_{j=0}^{k} \frac{a^j}{j!} \frac{d^j}{dy^j} \frac{1}{\Gamma(y)}\bigg|_{y=b+ax} \frac{(\log z)^{k-j}}{(k-j)!}$$

Since $\frac{1}{\Gamma(z)}$ is an entire function, the j-sum converges when $k \to \infty$. Application of the Euler-Maclaurin summation formula (3.40) to the definition of the Mittag-Leffler function $E_{a,b}(z) = \sum_{k=0}^{\infty} \frac{z^k}{\Gamma(b+ak)}$ in (1.1) shows that $\beta \to \infty$, but that we can choose the integer $\alpha = m$. The asymptotic form (14.9) of the Gamma function indicates for all j that $\lim_{\beta\to\infty} \frac{d^j}{dw^j}\frac{1}{\Gamma(b+aw)}\Big|_{w=\beta} = 0$ and the Euler-Maclaurin summation formula (3.40), for non-negative real z, becomes

$$\sum_{k=m+1}^{\infty} \frac{z^k}{\Gamma(b+ak)} = \int_m^{\infty} \frac{z^t}{\Gamma(b+at)} dt + z^m S_N(z) + R_N \tag{3.43}$$

where the Euler-Maclaurin sum is

$$S_N(z) = \sum_{n=0}^{N-1} (-1)^n \frac{B_{n+1}}{n+1} \sum_{j=0}^{n} \frac{a^j}{j!} \frac{d^j}{dy^j} \frac{1}{\Gamma(y)}\bigg|_{y=b+ma} \frac{(\log z)^{n-j}}{(n-j)!} \tag{3.44}$$

The integral in (3.43) is written in terms of the integral $I_{a,b}(z)$, defined in (1.2) and studied in Section 7.5, as

$$\int_m^{\infty} \frac{z^t}{\Gamma(b+at)} dt = z^m \int_0^{\infty} \frac{z^u}{\Gamma(b+am+au)} du = z^m I_{a,b+ma}(z)$$

In summary, with $\sum_{k=m+1}^{\infty} \frac{z^k}{\Gamma(b+ak)} = \sum_{k=0}^{\infty} \frac{z^{k+m+1}}{\Gamma(b+a(m+1)+ak)}$, the Euler-Maclaurin expansion (3.43) becomes

$$z^{m+1} E_{a,b+(m+1)a}(z) = z^m I_{a,b+ma}(z) + z^m S_N(z) + R_N \tag{3.45}$$

56. *The Euler-Maclaurin sum $S_N(z)$.* We concentrate on the sum $S_N(z)$ in (3.44) and reverse the summations,

$$S_N(z) = \sum_{j=0}^{N-1} \frac{a^j}{j!} \frac{d^j}{dy^j} \frac{1}{\Gamma(y)}\bigg|_{y=b+ma} \sum_{n=j}^{N-1} (-1)^n \frac{B_{n+1}}{n+1} \frac{(\log z)^{n-j}}{(n-j)!}$$

Let $x = \log z$ and use $\frac{d^j}{dx^j}x^n = \frac{n!}{(n-j)!}x^{n-j}$, then

$$\begin{aligned}\sum_{n=j}^{N-1}(-1)^n \frac{B_{n+1}}{n+1}\frac{x^{n-j}}{(n-j)!} &= \sum_{n=j}^{N-1}(-1)^n \frac{B_{n+1}}{n+1}\frac{1}{n!}\frac{d^j}{dx^j}x^n \\ &= \frac{d^j}{dx^j}\sum_{n=j}^{N-1}(-1)^n \frac{B_{n+1}}{(n+1)!}x^n \\ &= \frac{d^j}{dx^j}\sum_{n=0}^{N-1}(-1)^n \frac{B_{n+1}}{(n+1)!}x^n\end{aligned}$$

because $\sum_{n=0}^{j-1}(-1)^n \frac{B_{n+1}}{(n+1)!}x^n$ is a polynomial of order $j-1$ in x, which vanishes after j-times differentiation. Hence, we obtain

$$\sum_{n=j}^{N-1}(-1)^n \frac{B_{n+1}}{n+1}\frac{x^{n-j}}{(n-j)!} = -\frac{d^j}{dx^j}\left(\frac{1}{x}\sum_{n=1}^{N}\frac{B_n}{n!}(-x)^n\right)$$

and an alternative form of the Euler-Maclaurin sum (3.44)

$$S_N(z) = \sum_{j=0}^{N-1}\frac{a^j}{j!}\left.\frac{d^j}{dy^j}\frac{1}{\Gamma(y)}\right|_{y=b+ma}\frac{d^j}{dx^j}\left.\left(\sum_{n=1}^{N}\frac{B_n}{n!}(-x)^{n-1}\right)\right|_{x=\log z} \tag{3.46}$$

The generating function $\frac{x}{e^x-1} = 1+\sum_{n=1}^{\infty}B_n\frac{x^n}{n!}$ in (A.15) of the Bernoulli numbers B_n indicates that, only for $|x| < 2\pi$, we can proceed to the limit $N \to \infty$,

$$\sum_{n=j}^{\infty}(-1)^n \frac{B_{n+1}}{n+1}\frac{x^{n-j}}{(n-j)!} = -\frac{d^j}{dx^j}\left(\frac{1}{x}\left(\frac{-x}{e^{-x}-1}-1\right)\right) = \frac{d^j}{dx^j}\left(\frac{1}{x}-\frac{1}{1-e^{-x}}\right)$$

and convergence implies that the remainder $R_\infty = 0$. Thus, for $|\log z| < 2\pi$ or equivalently $|z| \le e^{\sqrt{(2\pi)^2-(\arg z)^2}}$, we find that the Euler-Maclaurin sum (3.46) converges, with $R_\infty = 0$, to

$$S_\infty(z) = \sum_{j=0}^{\infty}\frac{a^j}{j!}\left.\frac{d^j}{dy^j}\frac{1}{\Gamma(y)}\right|_{y=b+ma}\frac{d^j}{dx^j}\left.\left(x^{-1}-\frac{1}{1-e^{-x}}\right)\right|_{x=\log z}$$

Invoking the Fermi-Dirac integrals $F_{-j-1}(y) = \frac{d^j}{dy^j}\left(\frac{1}{1+e^{-y}}\right)$ in **art.** 116 and $\frac{d^j}{dx^j}x^{-1} = (-1)^j j! x^{-1-j}$ gives

$$\begin{aligned}S_\infty(z) = &\frac{1}{\log z}\sum_{j=0}^{\infty}\left.\frac{d^j}{dy^j}\frac{1}{\Gamma(y)}\right|_{y=b+ma}\left(-\frac{a}{\log z}\right)^j \\ &-\sum_{j=0}^{\infty}\frac{a^j}{j!}\left.\frac{d^j}{dy^j}\frac{1}{\Gamma(y)}\right|_{y=b+ma}F_{-j-1}(\log z + i\pi)\end{aligned}$$

We recognize from (7.5) that, for $|z| < 1$,

$$E_{a,b}\left(-ze^{i\pi}\right) = E_{a,b}(z) = \frac{1}{\Gamma(b)} - \sum_{j=0}^{\infty} \frac{a^j}{j!} \frac{d^j}{dy^j} \frac{1}{\Gamma(y)}\bigg|_{y=b} F_{-j-1}(\log z + i\pi)$$

and from the series (7.42) for $I_{a,b}(z)$, we obtain

$$S_\infty(z) = E_{a,b+ma}(z) - \frac{1}{\Gamma(b+ma)} - I_{a,b+ma}(z) \qquad (3.47)$$

which is valid for sufficiently small $|z|$. The Euler-Maclaurin expansion (3.45) indeed shows, for $N \to \infty$, with (3.47) and (2.5), that the remainder $R_\infty = 0$.

In particular, for $z = 1$, the Euler-Maclaurin expansion (3.43) simplifies to

$$\sum_{k=m+1}^{\infty} \frac{1}{\Gamma(b+ak)} = \int_m^\infty \frac{dt}{\Gamma(b+at)} + \sum_{n=0}^{\infty} \frac{(-1)^n B_{n+1}}{(n+1)!} \frac{d^n}{dy^n} \frac{1}{\Gamma(y)}\bigg|_{y=b+am} a^n$$

where the latter sum, studied in **art.** 116, converges for $|a| < 1$.

57. *An integral for the Euler-Maclaurin sum $S_N(z)$.* Introducing the Hankel integral (15.1) for $\frac{1}{\Gamma(z)}$ into the Euler-Maclaurin sum (3.44) yields

$$S_N(z) = \sum_{n=0}^{N-1} \frac{(-1)^n B_{n+1}}{n+1} \sum_{j=0}^{n} \frac{a^j}{j!} \left(\frac{(-1)^j}{2\pi i} \int_C \frac{e^w \log^j w}{w^{b+ma}} dw \right) \frac{(\log z)^{n-j}}{(n-j)!}$$

$$= \sum_{n=0}^{N-1} \frac{(-1)^n B_{n+1}}{(n+1)!} \frac{1}{2\pi i} \int_C \sum_{j=0}^{n} \binom{n}{j} (-a\log w)^j \; (\log z)^{n-j} \frac{e^w dw}{w^{b+ma}}$$

The binomium $\sum_{j=0}^n \binom{n}{j} (-a\log w)^j \; (\log z)^{n-j} = (-a\log w + \log z)^n$ simplifies to

$$S_N(z) = \frac{1}{2\pi i} \int_C \left(\sum_{n=0}^{N-1} (-1)^n \frac{B_{n+1}}{(n+1)!} (-a\log w + \log z)^n \right) w^{-b-ma} e^w dw$$

and we obtain an integral representation for the Euler-Maclaurin sum

$$S_N(z) = \frac{1}{2\pi i} \int_C \left(\sum_{n=1}^{N} \frac{B_n}{n!} (a\log w - \log z)^n \right) \frac{w^{-b-ma} e^w}{a\log w - \log z} dw$$

Only if $\left|\log\left(\frac{w^a}{z}\right)\right| < 2\pi$, the generating function (A.15) of the Bernoulli numbers tells us that the inner series converges for $N \to \infty$ to $\frac{a\log w - \log z}{e^{a\log w - \log z} - 1} - 1 = z\frac{a\log w - \log z}{w^a - z} - 1$, so that

$$S_\infty(z) = \frac{1}{2\pi i} \int_C \left(z\frac{a\log w - \log z}{w^a - z} - 1 \right) \frac{w^{-b-ma} e^w}{a\log w - \log z} dw$$

$$= \frac{z}{2\pi i} \int_C \frac{w^{-b-ma} e^w}{w^a - z} dw - \frac{1}{2\pi i} \int_C \frac{w^{-b-ma} e^w}{a\log w - \log z} dw$$

Art. 61 and 75 explain that we can deform the contour C to always obey the convergence constraint $\left|\log\left(\frac{w^a}{z}\right)\right| < 2\pi$. After proper modification of the contour C into C_ϕ in the second integral, then $I_{a,b}(z) = \frac{1}{2\pi i}\int_{C_\phi} \frac{w^{-b}e^w}{a\log w - \log z}dw$ in (7.39) indicates that the second integral equals $I_{a,b+ma}(z)$. The Euler-Maclaurin expansion (3.45) with remainder $R_\infty = 0$ reduces to

$$E_{a,b+(m+1)a}(z) = \frac{1}{2\pi i}\int_{C_\phi} \frac{w^{-b-ma}e^w}{w^a - z}dw = \frac{1}{2\pi i}\int_{C_\phi} \frac{w^{a-(b+(m+1)a)}e^w}{w^a - z}dw$$

which is, after the same contour modification from C into C_ϕ, precisely the *basic complex integral* (4.4). Hence, we have shown that the basic complex integral (4.4) can also be deduced from the Euler-Maclaurin summation formula, but the classical demonstration in **art.** 61 is shorter.

58. *Transforming the Taylor series of the Mittag-Leffler function.* A fundamental property in the theory of integration is the principle of substitution of the integration variable. In summations as opposed to integrals, a substitution of the summation variable is, in general, less flexible or powerful. Here, we will essentially transform the summation variable in the Taylor series (1.1) of the Mittag-Leffler function by making a detour to integrals via the basic summation formula (3.39).

We start from (3.39) for non-negative real z, written as

$$\sum_{k=0}^{\infty} \frac{z^k}{\Gamma(ak+b)} = \int_0^\infty \frac{z^x}{\Gamma(ax+b)}dx + R_1 \tag{3.48}$$

where the remainder is

$$R_1 = \int_0^\infty \left(x - \lfloor x \rfloor - \frac{1}{2}\right) \frac{z^x\left(\log z - a\,\psi(ax+b)\right)}{\Gamma(ax+b)}dx + \frac{1}{2}\frac{1}{\Gamma(b)}$$

Confining to real a and b, we make the substitution $u = ax + b - 1$ in the integrals,

$$\int_0^\infty \frac{z^x}{\Gamma(ax+b)}dx = \frac{z^{\frac{1-b}{a}}}{a}\int_{b-1}^\infty \frac{z^{\frac{u}{a}}}{\Gamma(u+1)}du \tag{3.49}$$

and

$$R_1 = \frac{z^{\frac{1-b}{a}}}{a}\int_{b-1}^\infty \frac{\left((u+1-b) - a\left\lfloor \frac{1}{a}(u+1-b)\right\rfloor - \frac{a}{2}\right)\left(\frac{\log z}{a} - \psi(u+1)\right)}{z^{-\frac{u}{a}}\Gamma(u+1)}du$$
$$+ \frac{1}{2}\frac{1}{\Gamma(b)}$$

Again using the basic summation formula (3.39) with $f(x) = \frac{z^{\frac{x}{a}}}{\Gamma(x+1)}$, we return from the integral $\int_{b-1}^\infty \frac{z^{\frac{u}{a}}}{\Gamma(u+1)}du$ in (3.49) to its corresponding summation, that can be evaluated exactly,

$$\int_{b-1}^{\infty}\frac{z^{\frac{u}{a}}}{\Gamma(u+1)}du=\sum_{k=\lfloor b-1\rfloor+1}^{\infty}\frac{z^{\frac{k}{a}}}{k!}-\left(b-\lfloor b\rfloor-\frac{1}{2}\right)\frac{z^{\frac{b-1}{a}}}{\Gamma(b)}$$
$$-\int_{b-1}^{\infty}\left(x-\lfloor x\rfloor-\frac{1}{2}\right)\frac{z^{\frac{x}{a}}\left(\frac{\log z}{a}-\psi(x+1)\right)}{\Gamma(x+1)}dx$$

With $\sum_{k=\lfloor b-1\rfloor+1}^{\infty}\frac{z^{\frac{k}{a}}}{k!}=e^{z^{\frac{1}{a}}}-\sum_{k=0}^{\lfloor b-1\rfloor}\frac{z^{\frac{k}{a}}}{k!}$, substitution of (3.49) into (3.48) yields

$$\sum_{k=0}^{\infty}\frac{z^k}{\Gamma(ak+b)}=\frac{z^{\frac{1-b}{a}}}{a}e^{z^{\frac{1}{a}}}-\frac{z^{\frac{1-b}{a}}}{a}\sum_{k=0}^{\lfloor b-1\rfloor}\frac{z^{\frac{k}{a}}}{k!}+\frac{\left(\frac{1}{2}+\frac{1}{2a}-\frac{b-1-\lfloor b-1\rfloor}{a}\right)}{\Gamma(b)}$$
$$+\frac{z^{\frac{1-b}{a}}}{a}\int_{b-1}^{\infty}\left(\frac{3}{2}-b-\frac{a}{2}\right)\frac{z^{\frac{x}{a}}\left(\frac{\log z}{a}-\psi(x+1)\right)}{\Gamma(x+1)}dx$$
$$+\frac{z^{\frac{1-b}{a}}}{a}\int_{b-1}^{\infty}\frac{\left(\lfloor x\rfloor-a\left\lfloor\frac{(x+1-b)}{a}\right\rfloor\right)\left(\frac{\log z}{a}-\psi(x+1)\right)}{z^{-\frac{x}{a}}\Gamma(x+1)}dx$$

We evaluate the first integral with $\int_{b-1}^{\infty}\frac{d}{dx}\left(\frac{z^{\frac{x}{a}}}{\Gamma(x+1)}\right)dx=-\frac{z^{\frac{b-1}{a}}}{\Gamma(b)}$ and arrive, for real a and b and any complex z with $|\arg z|<\pi$ (i.e. avoiding real negative z), at

$$E_{a,b}(z)=\frac{z^{\frac{1-b}{a}}}{a}e^{z^{\frac{1}{a}}}-\frac{z^{\frac{1-b}{a}}}{a}\sum_{k=0}^{\lfloor b-1\rfloor}\frac{z^{\frac{k}{a}}}{k!}+\left(1+\frac{\lfloor b-1\rfloor}{a}\right)\frac{1}{\Gamma(b)}+R_{a,b}(z) \tag{3.50}$$

where the remainder is

$$R_{a,b}(z)=z^{\frac{1-b}{a}}\int_{b-1}^{\infty}\left(\frac{\lfloor x\rfloor}{a}-\left\lfloor\frac{x+1-b}{a}\right\rfloor\right)\frac{z^{\frac{x}{a}}\left(\frac{\log z}{a}-\psi(x+1)\right)}{\Gamma(x+1)}dx \tag{3.51}$$

For $b=1$, (3.50) simplifies to the elegant relation

$$E_a(z)=\frac{e^{z^{\frac{1}{a}}}}{a}+1-\frac{1}{a}+\int_0^{\infty}\left(\frac{\lfloor x\rfloor}{a}-\left\lfloor\frac{x}{a}\right\rfloor\right)\frac{d}{dx}\left(\frac{z^{\frac{x}{a}}}{\Gamma(x+1)}\right)dx \tag{3.52}$$

whose remainder $R_a(x)$ is evaluated for real $x>0$ in **art.** 115. If $b<1$, then the k-sum in (3.50) disappears.

After splitting the integral in (3.51) into two terms, subdividing the integration interval according to the integral part $\lfloor x\rfloor$, which results into integrals of the type $\int_k^{k+1}\frac{z^{\frac{x}{a}}\left(\frac{1}{a}\log z-\psi(x+1)\right)}{\Gamma(x+1)}dx=\frac{z^{\frac{k+1}{a}}}{\Gamma(k+2)}-\frac{z^{\frac{k}{a}}}{\Gamma(k+1)}$, the remainder $R_{a,b}(z)$ can be evaluated exactly again in the appearing terms on the right-hand side, illustrating that (3.50) is an identity. Although $R_{a,b}(z)$ can be evaluated exactly, standard numerical integration packages (such as Mathematica) perform rather poorly in computing the integral in $R_{a,b}(z)$ without proper fine-tuning. Relation (3.50) exhibits the same dominant term $\frac{z^{\frac{1-b}{a}}}{a}e^{z^{\frac{1}{a}}}$ for large $|z|$ as in Mittag-Leffler's integral in (4.28)

$$E_{a,b}(z) = \frac{z^{\frac{1-b}{a}}}{a}e^{z^{\frac{1}{a}}} + \frac{1}{2\pi i a}\int_{C'_{a\phi}} \frac{t^{\frac{1-b}{a}}e^{t^{\frac{1}{a}}}}{t-z}dt$$

However, for moderate and small z, none of the terms in (3.50) dominates.

4

Integral representations for $E_{a,b}(z)$

We present several integral analyses, due to Mittag-Leffler, Wiman, Barnes, Paris and Kaminski, that aim to unravel the behavior of the Mittag-Leffler function $E_{a,b}(z)$ in the complex z-plane.

Two different types of complex integrals for the Mittag-Leffler function $E_{a,b}(z)$ are discussed. The basic complex integral (4.7) in Section 4.2 follows from Hankel's contour integral for $\frac{1}{\Gamma(z)}$ in Section 15.1 and also from the inverse Laplace transform of (3.13) and the Euler-Maclaurin summation formula (**art.** 57). The second complex integral (4.33) in Section 4.4 is an instance of a Plana-like summation formula. Section 4.4.4 derives further forms of the integral (4.55) for $E_a(-z)$, which is the special for $b = 1$ and $0 < a < 1$ of the integral (4.52).

We believe that relation (4.13) is the key result, derived from the generalization (4.12) of the basic integral (4.4) of the Mittag-Leffler function $E_{a,b}(z)$, from which the asymptotic behavior for different a-regimes and different sectors in the complex z-plane is deduced.

4.1 Explorations of Wiman

We start with the analysis of Anders Wiman (1905a), but generalize his computations of $E_a(z)$ to $E_{a,b}(z)$ with arbitrary b.

59. *Asymptotic formula for rational* $a = \frac{1}{n}$. We write (2.44), for $\text{Re}(b) \geq 1$ and for complex z with $\text{Re}(z^n) > 0$, but without indicator $1_{\{b=1\}}$,

$$E_{\frac{1}{n},b}(z) = z^{(1-b)n} e^{z^n} \left\{ n - \sum_{j=0}^{n-1} \frac{1}{\Gamma\left(b-1+\frac{j}{n}\right)} \int_{z^n}^{\infty} t^{\left(b-1+\frac{j}{n}\right)-1} e^{-t} dt \right\} \tag{4.1}$$

in terms of the incomplete Gamma function $\Gamma(a,z) = \int_z^\infty e^{-x} x^{a-1} dx$ in (16.1) as

$$E_{\frac{1}{n},b}(z) = z^{(1-b)n} e^{z^n} \left\{ n - \sum_{j=0}^{n-1} \frac{\Gamma\left(b-1+\frac{j}{n}, z^n\right)}{\Gamma\left(b-1+\frac{j}{n}\right)} \right\}$$

The recursion $\Gamma(a,z) = \frac{z^{a-1}}{e^z}\sum_{k=0}^{m-1}\frac{\prod_{j=1}^{k}(a-j)}{z^k} + \Gamma(a-m,z)\prod_{j=1}^{m}(a-j)$ in (16.10) for a fixed integer m and for a complex z with $|\arg z| < \frac{\pi}{2n}$ leads to

$$E_{\frac{1}{n},b}(z) = \frac{e^{z^n}\left\{n - \sum_{j=0}^{n-1}\frac{\Gamma\left(b-1+\frac{j}{n}-m,z^n\right)}{\Gamma\left(b-1+\frac{j}{n}-m\right)}\right\}}{z^{(b-1)n}} - \sum_{j=0}^{n-1}\sum_{k=1}^{m}\frac{z^{j-nk}}{\Gamma\left(b+\frac{j}{n}-k\right)}$$

Its integral form

$$E_{\frac{1}{n},b}(z) = \frac{ne^{z^n}}{z^{(b-1)n}} - \sum_{j=0}^{n-1}\left\{\sum_{k=1}^{m}\frac{z^{j-nk}}{\Gamma\left(b+\frac{j}{n}-k\right)} + \frac{e^{z^n}\int_{z^n}^{\infty}e^{-x}x^{\left(b-1+\frac{j}{n}-m\right)-1}dx}{z^{(b-1)n}\Gamma\left(b-1+\frac{j}{n}-m\right)}\right\} \tag{4.2}$$

reduces for $b = 1$ to Wiman's expression (Wiman, 1905a, p. 194),

$$E_{\frac{1}{n}}(z) = ne^{z^n} - \sum_{j=1}^{n-1}\left\{\sum_{k=1}^{m}\frac{z^{j-nk}}{\Gamma\left(\frac{j}{n}-k+1\right)} + \frac{ne^{z^n}\int_z^{\infty}e^{-u^n}u^{-nm+j-1}du}{\Gamma\left(\frac{j}{n}-m\right)}\right\}$$

Expansion (4.2) is a key relation, from which interesting insights will be deduced.

60. *Deductions from formula (4.2).* First, for large z and complex a, the asymptotic formula for $\Gamma(a,z) = \int_z^{\infty}e^{-x}x^{a-1}dx \sim e^{-z}z^{a-1}\left(1+O\left(\frac{1}{z}\right)\right)$ in (16.21) indicates that the integral in (4.2)

$$\frac{z^{(1-b)n}e^{z^n}}{\Gamma\left(b-1+\frac{j}{n}-m\right)}\int_{z^n}^{\infty}e^{-x}x^{\left(b-1+\frac{j}{n}-m\right)-1}dx = \frac{z^{j-n(m+1)}\left(1+O\left(\frac{1}{z^n}\right)\right)}{\Gamma\left(b+\frac{j}{n}-(m+1)\right)}$$

equals approximately the next term $k = m+1$ in the sum $\sum_{k=1}^{m}\frac{z^{j-nk}}{\Gamma\left(\frac{j}{n}-k+1\right)}$. Second, the reflection formula (11.4) of the Gamma function shows that

$$\sum_{k=1}^{m}\frac{z^{j-nk}}{\Gamma\left(\frac{j}{n}-k+1\right)} = \frac{\sin\pi\left(b-1+\frac{j}{n}\right)}{\pi}z^j\sum_{k=1}^{m}(-1)^k\frac{\Gamma\left(2-b-\frac{j}{n}+k\right)}{(z^n)^k}$$

If $m \to \infty$, then the series does not converge. For sufficiently large z, the initial terms in the series decrease in absolute value and $\sum_{k=1}^{m}\frac{z^{j-nk}}{\Gamma\left(\frac{j}{n}-k+1\right)}$ is thus an asymptotic series, just as Stirling's asymptotic series in **art.** 196. The best choice for m is the minimizer in k of $\frac{z^{j-nk}}{\Gamma\left(\frac{j}{n}-k+1\right)}$. Wiman (1905a, p. 194) argues (without demonstration) that, if we choose m so that $\frac{z^{j-nm}}{\Gamma\left(b+\frac{j}{n}-m\right)}$ is minimal, then $\frac{ne^{z^n}}{\Gamma\left(\frac{j}{n}-m\right)}\int_z^{\infty}e^{-u^n}u^{-nm+j-1}du = O\left(z^{\frac{n}{2}}e^{-z^n}\right)$ for sufficiently large z. Indeed, the ratio $r_k = \frac{u_{k+1}}{u_k}$ of two consecutive terms, with $u_k = \frac{z^{j-nk}}{\Gamma\left(\frac{j}{n}-k+1\right)}$, in the series $\sum_{k=1}^{m}\frac{z^{j-nk}}{\Gamma\left(\frac{j}{n}-k+1\right)}$ is $r_k = \frac{\frac{j}{n}-k}{z^n}$, from which $|r_k| < \frac{|1-k|}{|z|^n}$, because $0 \le \frac{j}{n} \le 1$. Moreover, for $k > 1$ and real, positive z, the ratio r_k is negative, indicating that the terms u_k are alternating in k (i.e. have different sign). Further, $|r_k|$ is linearly increasing k, in agreement with the divergent nature of $\sum_{k=1}^{\infty}\frac{z^{j-nk}}{\Gamma\left(\frac{j}{n}-k+1\right)}$. As long as

$|r_k| < 1$, the series converges[1]; if $|r_k| > 1$, the terms $|u_k|$ start increasing and divergence sets in. Now, $|r_k| = 1$ occurs if $|1-k| < k < |z|^n$ and $|u_{[z^n]}|$ is about the smallest (in absolute value) term, equal to

$$\left|u_{[z^n]}\right| = \left|\frac{z^{j-nz^n}}{\Gamma\left(\frac{j}{n} - z^n + 1\right)}\right| = \left|z^j\right| \left|\frac{\Gamma\left(2 - b - \frac{j}{n} + z^n\right)}{\pi\left(z^n\right)^{z^n}}\right|$$

For $2-b-\frac{j}{n}=1$ (to simplify the analysis, see also **art.** 69), the asymptotic series (14.17), $\Gamma(z+1) \simeq \sqrt{2\pi} z^{z+\frac{1}{2}} e^{-z}$, indicates that

$$\left|u_{[z^n]}\right| \approx \left|z^j\right| \left|\frac{\Gamma\left(1+z^n\right)}{\pi\left(z^n\right)^{z^n}}\right| \approx \sqrt{\frac{2}{\pi}} \left|z^j\right| \left|z^{\frac{n}{2}} e^{-z^n}\right|$$

In summary, the main result in Wiman (1905a) is the accurate asymptotic formula, for large z with $|\arg z| < \frac{\pi}{2n}$,

$$E_{\frac{1}{n},b}(z) = nz^{(1-b)n} e^{z^n} - \sum_{j=0}^{n-1} \sum_{k=1}^{\lfloor z^n \rfloor} \frac{z^{j-nk}}{\Gamma\left(b + \frac{j}{n} - k\right)} + O\left(\left|z^{\frac{3n}{2}-1} e^{-z^n}\right|\right) \quad (4.3)$$

If only K terms, instead of the huge number $\lfloor z^n \rfloor$ of terms, are evaluated in the k-sum, the remaining error is of the order $O\left(z^{-nK}\right)$, roughly equal to first omitted term.

Although the case $a = \frac{1}{n}$ in (4.3) is a rather special case, **art.** 14 demonstrates that its analysis can be generalized to any rational number $a = \frac{m}{n} \in \mathbb{Q}$, and as such, the properties of $E_{a,b}(z)$ for any positive real a can be arbitrarily closely approximated. Wiman (1905a) further shows for $E_{\frac{1}{n},1}(z)$ with $b=1$ that (4.3) holds for $-\frac{\pi}{n} < \arg z < \frac{\pi}{n}$ and that, for complex z with $\frac{\pi}{n} \le \arg z \le 2\pi - \frac{\pi}{n}$, the result equals (4.3) without the first term $nz^{(1-b)n}e^{z^n}$. Since Wiman's derivation is specific for $a = \frac{1}{n}$, where the involved integral $\int_L e^{-z} z^\nu dz$ with integer ν encloses an analytic region, void from the branch cut at real negative z for $\nu \notin \mathbb{Z}$, we omit Wiman's further analysis and treat the general case for real order a in **art.** 66.

4.2 Basic complex integral for $E_{a,b}(z)$

Besides the Taylor series (1.1), the basic complex integral (4.4) is the most prominent representation of the Mittag-Leffler function $E_{a,b}(z)$ from which its complex behavior is deduced.

---

[1] The alternating series $S_A = \sum_{k=0}^{\infty} (-1)^k c_k = \sum_{k=0}^{m} (-1)^k c_k + r_m$ is convergent if all numbers $c_0, c_1, \ldots$ are positive and decrease monotonously to zero. The remainder is $(-1)^{m+1} r_m = \sum_{k=m+1}^{m} (-1)^{m+1+k} c_k = c_{m+1} - (c_{m+2} - c_{m+3}) - \cdots < c_{m+1}$, which shows that S_A is less than the first term omitted. That property also holds for alternatingly divergent series, that initially converge. Such series are called semi-convergent (Sansone and Gerretsen, 1960, p. 446).

4.2.1 Constructions of the basic complex integral

Three different derivations of the basic complex integral are given: the first, based on Euler-Maclaurin summation, already appeared in **art.** 57, the second in **art.** 61 relies on Hankel's integral for $\frac{1}{\Gamma(z)}$ and the third in **art.** 62 follows from the Laplace transform (3.13) of the Mittag-Leffler function.

61. *Basic complex integral for $E_{a,b}(z)$ via Hankel's integral.* We substitute Hankel's deformed integral (15.3) in the truncated power series of the Mittag-Leffler function (1.1)

$$\sum_{k=0}^{m-1} \frac{z^k}{\Gamma(b+ak)} = \frac{1}{2\pi i} \sum_{k=0}^{m-1} \int_{C_\phi} \left(zw^{-a}\right)^k w^{-b} e^w dw$$
$$= \frac{1}{2\pi i} \int_{C_\phi} \sum_{k=0}^{m-1} \left(zw^{-a}\right)^k w^{-b} e^w dw$$

Since the sum is finite, reversal of sum and integral is allowed. The geometric series (1.15) can be summed by (1.16) and we arrive, for any finite integer m and any complex a and b, to

$$\sum_{k=0}^{m-1} \frac{z^k}{\Gamma(b+ak)} = \frac{1}{2\pi i} \int_{C_\phi} \frac{\left(zw^{-a}\right)^m - 1}{zw^{-a} - 1} w^{-b} e^w dw$$

The limit process of $m \to \infty$ requires, at the left-hand side, that $\mathrm{Re}(a) > 0$ in order that the Taylor series (1.1) converges. At the right-hand side, $\left(zw^{-a}\right)^m$ only tends to zero if $m \to \infty$, provided that $|zw^{-a}| < 1$. The requirement $|zw^{-a}| < 1$ means that $|z| < |w^a|$ for any point w on the contour C_ϕ, which is described in **art.** 204. In other words, the contour C_ϕ with $\frac{\pi}{2} < \phi < \pi$, in particular the radius ρ of the circle at $w = 0$, must be appropriately chosen to satisfy $|z| < \rho^{\mathrm{Re}\, a}$ or $|z|^{\frac{1}{\mathrm{Re}\, a}} < \rho$ for any z. Then, we obtain, for all complex b and $\mathrm{Re}(a) > 0$, the *basic complex integral*

$$E_{a,b}(z) = \frac{1}{2\pi i} \int_{C_\phi} \frac{w^{a-b} e^w}{w^a - z} dw \tag{4.4}$$

If we choose the angle $\phi = \frac{\pi}{2}$ and $c = \rho > |z|^{\frac{1}{\mathrm{Re}\, a}}$ in (4.4), then we must restrict $\mathrm{Re}(b-a) > 0$ by (15.8), because e^w does not tend to zero along the imaginary w-axis. The contour $C_{\frac{\pi}{2}}$ is depicted in Caputo and Mainardi (1971a, Fig. 19).

62. *Basic complex integral for $E_{a,b}(z)$ via inverse Laplace transform.* Inverse Laplace transformation (B.2) of (3.13) yields, with $c' > |x|^{\frac{1}{a}}$ for $\mathrm{Re}(a) > 0$ and $\mathrm{Re}(b) > 0$,

$$t^{b-1} E_{a,b}(xt^a) = \frac{1}{2\pi i} \int_{c'-i\infty}^{c'+i\infty} \frac{s^{a-b} e^{st}}{s^a - x} ds \tag{4.5}$$

where the integrand is analytic for $\mathrm{Re}(s) > c'$. Evaluating the integral along the line $s = c' + iu$ yields

$$t^{b-1} E_{a,b}(xt^a) = \frac{1}{2\pi} \int_{-\infty}^{\infty} \frac{(c'+iu)^{a-b} e^{tc'} e^{itu}}{(c'+iu)^a - x} du$$

After substituting the real variable $v = tu$ for real $t > 0$ and simplifying the relation,

$$E_{a,b}(xt^a) = \frac{1}{2\pi} \int_{-\infty}^{\infty} \frac{(c't+iv)^{a-b} e^{(c't+iv)}}{(c't+iv)^a - xt^a} dv$$

we replace $z = xt^u$, denote $c = c't$, which must obey $c > t|x|^{\frac{1}{a}} = |z|^{\frac{1}{a}}$, and find

$$E_{a,b}(z) = \frac{1}{2\pi} \int_{-\infty}^{\infty} \frac{(c+iv)^{a-b} e^{(c+iv)}}{(c+iv)^a - z} dv \qquad c > |z|^{\frac{1}{a}} \tag{4.6}$$

Returning to the complex $w = c+it$, we arrive, for $\mathrm{Re}(b) > 0$ and $\mathrm{Re}(a) > 0$, at the *basic complex integral*

$$E_{a,b}(z) = \frac{1}{2\pi i} \int_{c-i\infty}^{c+i\infty} \frac{w^{a-b} e^w}{w^a - z} dw \qquad c > |z|^{\frac{1}{a}} \tag{4.7}$$

Although the path of integration in (4.7) is easier, the contour C_ϕ with $\frac{\pi}{2} < \phi < \pi$ and $c = \rho > |z|^{\frac{1}{\mathrm{Re}\, a}}$ in (4.4) holds more generally for all b and $\mathrm{Re}(a) > 0$.

63. *Evaluation of the basic complex integral (4.6) along the line $w = c+it$.* We evaluate the integral in (4.6), valid for $\mathrm{Re}(a) > 0$ and $\mathrm{Re}(b) > 0$. Using $c + it = \sqrt{c^2+t^2} e^{i \arctan \frac{t}{c}} = c\sqrt{1+\left(\frac{t}{c}\right)^2} e^{i \arctan \frac{t}{c}}$ in (4.6), we find, after substitution $u = \frac{t}{c}$,

$$E_{a,b}(z) = \frac{e^c c^{1-b}}{2\pi} \int_{-\infty}^{\infty} \frac{(1+u^2)^{\frac{a-b}{2}} e^{i(cu+(a-b)\arctan u)}}{(1+u^2)^{\frac{a}{2}} e^{ia \arctan u} - \frac{z}{c^a}} du$$

We split the integration interval into two parts and, after tedious manipulations, we arrive, for $\mathrm{Re}(b) > 0$, $\mathrm{Re}(a) > 0$ and $c > |z|^{\frac{1}{a}}$, at

$$\begin{aligned} E_{a,b}(z) = & \frac{c^{1-b}}{\pi e^{-c}} \int_0^{\infty} \frac{(1+u^2)^{a-\frac{b}{2}} \cos(cu - b \arctan u)}{(1+u^2)^a - \frac{2z}{c^a}(1+u^2)^{\frac{a}{2}} \cos(a \arctan u) + \left(\frac{z}{c^a}\right)^2} du \\ & - \frac{zc^{1-b-a}}{\pi e^{-c}} \int_0^{\infty} \frac{(1+u^2)^{\frac{a-b}{2}} \cos(cu + (a-b)\arctan u)}{(1+u^2)^a - \frac{2z}{c^a}(1+u^2)^{\frac{a}{2}} \cos(a \arctan u) + \left(\frac{z}{c^a}\right)^2} du \end{aligned} \tag{4.8}$$

For $z = 0$ and $c > 0$, (4.8) simplifies to (15.11) in **art.** 207. The positive real number $c > |z|^{\frac{1}{a}}$ can be regarded as a tunable parameter.

64. *A complex integral for the Wright function $W_{a,b}(z)$.* Similarly as in **art.** 61, substitution of Hankel's deformed integral (15.3) in the Taylor series $W_{a,b}(z) = \sum_{k=0}^{\infty} \frac{1}{\Gamma(ak+b)} \frac{z^k}{k!}$ of the Wright function (3.4) yields, after reversal of sum and integral,

$$W_{a,b}(z) = \frac{1}{2\pi i} \int_{C_\phi} w^{-b} e^w dw \sum_{k=0}^{\infty} \frac{\left(w^{-a} z\right)^k}{k!}$$

The function $\sum_{k=0}^{\infty} \frac{\left(w^{-a} z\right)^k}{k!} = e^{w^{-a} z}$ is analytic in the complex w-plane, except for the negative real w-axis, which is a branch cut for $w^a = e^{a \log w}$ for non-integer values of a. Since the Hankel contour C_ϕ already avoids the branch cut, there is no need to alter the contour C_ϕ as in **art.** 61 and we obtain the contour integral representation of the Wright function

$$W_{a,b}(z) = \frac{1}{2\pi i} \int_{C_\phi} w^{-b} e^{w + w^{-a} z} dw \tag{4.9}$$

valid for all $\text{Re}(a) > -1$ and all complex b and z.

If we choose $\phi = \pi$, then $C_\phi = C$ is the usual Hankel contour (**art.** 204). Let ε be the radius of a circle at the origin, then evaluation of the integral $\mathbb{I} = \int_C w^{-b} e^{w + w^{-a} z} dw$ along the Hankel contour C yields

$$\begin{aligned}
\mathbb{I} = &-\int_{\infty}^{\varepsilon} \left(xe^{-i\pi}\right)^{-b} e^{-x + z\left(xe^{-i\pi}\right)^{-a}} dx - \int_{\varepsilon}^{\infty} \left(xe^{i\pi}\right)^{-b} e^{-x + z\left(xe^{i\pi}\right)^{-a}} dx \\
&+ i \int_{-\pi}^{\pi} \left(\varepsilon e^{i\theta}\right)^{1-b} e^{\varepsilon e^{i\theta} + z\left(\varepsilon e^{i\theta}\right)^{-a}} d\theta \\
= &\int_{\varepsilon}^{\infty} \frac{e^{bi\pi + ze^{ai\pi} x^{-a}} - e^{-bi\pi + ze^{-ai\pi} x^{-a}}}{x^b e^x} dx + i\varepsilon^{1-b} \int_{-\pi}^{\pi} \frac{e^{\varepsilon e^{i\theta} + z\varepsilon^{-a} e^{-ai\theta}}}{e^{i\theta(b-1)}} d\theta
\end{aligned}$$

The limit $\varepsilon \to 0$ exists, provided $\text{Re}(b) \leq 1$ and $-1 < \text{Re}(a) \leq 0$, so that

$$W_{a,b}(z) = \frac{1}{\pi} \int_0^{\infty} x^{-b} e^{-x} \,\text{Im}\left(e^{bi\pi + ze^{ai\pi} x^{-a}}\right) dx + 1_{\{b=1\}}$$

For real a and b,

$$\text{Im}\left(e^{bi\pi + ze^{ai\pi} x^{-a}}\right) = e^{|z| \cos(\arg z + a\pi) x^{-a}} \sin\left(b\pi + |z| \sin(\arg z + a\pi) x^{-a}\right)$$

we finally obtain, for real $b \leq 1$ and real $-1 < a \leq 0$, the integral

$$W_{a,b}(z) = \frac{1}{\pi} \int_0^{\infty} e^{-x + \frac{|z| \cos(\arg z + a\pi)}{x^a}} \frac{\sin\left(b\pi + \frac{|z| \sin(\arg z + a\pi)}{x^a}\right)}{x^b} dx + 1_{\{b=1\}} \tag{4.10}$$

Convergence of (4.10) for $|z| \to \infty$ requires that $\cos(\arg z + a\pi) < 0$ or $\pi > \arg z + a\pi > \frac{\pi}{2}$, implying that $1 - \frac{\arg z}{\pi} > a > \frac{1}{2} - \frac{\arg z}{\pi}$. If $\arg z = 0$, then the integral (4.10), valid for $-1 < a \leq 0$, diverges for $|z| \to \infty$, while if $\arg z = \pi$, then (4.10) converges for $-\frac{1}{2} < a < 0$.

65. *Mittag-Leffler function $E_{a,b}(z)$ for negative real a.* Although $E_{a,b}(z)$, defined by the power series (1.1), is not valid for negative $\text{Re}(a)$ values, the

Taylor series (7.5) of $E_{a,b}(z)$ in a around $a_0 = 0$, for $|z| < 1$,

$$E_{a,b}(-z) = \frac{1}{\Gamma(b)} - \sum_{m=0}^{\infty} \frac{1}{m!} \frac{d^m}{du^m} \frac{1}{\Gamma(u)}\bigg|_{u=b} F_{-m-1}(\log z) a^m$$

and its companion (7.7)

$$E_{a,b}(-z) = \sum_{m=0}^{\infty} \frac{1}{m!} \frac{d^m}{du^m} \frac{1}{\Gamma(u)}\bigg|_{u=b} F_{-m-1}(-\log z)(-a)^m$$

indicate, for $|z| < 1$, that

$$E_{a,b}\left(\frac{1}{z}\right) = \frac{1}{\Gamma(b)} - E_{-a,b}(-z)$$

which may be used, by analytic continuation, as a definition of $E_{a,b}(z)$ for $\mathrm{Re}(a) < 0$. Thus, for $\mathrm{Re}(a) < 0$, the Mittag-Leffler function $E_{a,b}(z)$ possesses an essential singularity at $z = 0$ and is not an entire function.

The same result is also formally deduced in Gorenflo *et al.* (2014, Sec. 4.8, pp. 80-82) from the basic complex integral (4.4) in **art.** 61

$$E_{a,b}(z) = \frac{1}{2\pi i} \int_{C_\phi} \frac{w^{a-b} e^w}{w^a - z} dw$$

which is valid for $\mathrm{Re}(a) > 0$ and any complex b and z. Introducing the identity

$$\frac{w^{a-b}}{w^a - z} = \frac{1}{w^b - z w^{b-a}} = \frac{1}{w^b} - \frac{1}{w^b - \frac{1}{z} w^{a+b}}$$

formally leads to

$$E_{a,b}(z) = \frac{1}{2\pi i} \int_{C_\phi} \frac{e^w}{w^b} dw - \frac{1}{2\pi i} \int_{C_\phi} \frac{w^{-a-b} e^w}{w^{-a} - \frac{1}{z}} dw$$

Invoking Hankel's integral (15.4) to the first integral and the basic complex integral (4.4) to the last integral formally – thus ignoring the $\mathrm{Re}(a) > 0$ restriction – leads again to

$$E_{a,b}(z) = \frac{1}{\Gamma(b)} - E_{-a,b}\left(\frac{1}{z}\right) \tag{4.11}$$

The Taylor series (1.1) of $E_{a,b}(z)$ then shows that

$$E_{-a,b}(z) = \frac{1}{\Gamma(b)} - E_{a,b}\left(\frac{1}{z}\right) = -\sum_{k=1}^{\infty} \frac{z^{-k}}{\Gamma(b + ak)}$$

Based upon (4.11), Hanneken *et al.* (2009) further study $E_{-a,b}(z)$. For example, after combining (4.11) and $E_{a,b+a}(z) = \frac{1}{z}\left(E_{a,b}(z) - \frac{1}{\Gamma(b)}\right)$ in (2.4) for $m = 1$, they list

$$E_{-a,b}(z) = -\frac{1}{z} E_{a,a+b}\left(\frac{1}{z}\right)$$

Besides plots of $E_{-a,b}(z)$ versus z for several values of a and b, Hanneken *et al.* (2009) also present tables with examples of explicit evaluations of both $E_{a,b}(z)$ and $E_{-a,b}(z)$ for integer values of a.

66. *General series for the Mittag-Leffler function.* Wiman (1905b) derives the general asymptotic expansion from Mittag-Leffler's contour integral (4.26) by substituting the finite geometric series (1.16) for $\omega \neq z$ and $\omega \neq z_0$,

$$\frac{1}{\omega - z} = \frac{1}{\omega - z}\left(\frac{\omega - z_0}{z - z_0}\right)^m - \frac{1}{z - z_0}\sum_{k=0}^{m-1}\left(\frac{\omega - z_0}{z - z_0}\right)^k$$

We follow Wiman's idea, but we substitute the finite geometric series, with $\omega = w^a$ and $z_0 = 0$, in the general integral (4.4), for all b and $\text{Re}(a) > 0$, with contour $\frac{\pi}{2} < \phi < \pi$ and $c > |z|^{\frac{1}{a}}$, to obtain,

$$E_{a,b}(z) = -\sum_{k=0}^{m-1}\frac{z^{-k-1}}{2\pi i}\int_{C_\phi} w^{(k+1)a-b}e^w dw + \frac{z^{-m}}{2\pi i}\int_{C_\phi}\frac{w^{a(m+1)-b}e^w}{w^a - z}dw$$

After using Hankel's integral (15.3), we arrive, for all b and $\text{Re}(a) > 0$, $\frac{\pi}{2} < \phi < \pi$ and $c > |z|^{\frac{1}{a}}$ and any finite integer $m \geq 0$, at

$$E_{a,b}(z) = -\sum_{k=1}^{m}\frac{1}{z^k}\frac{1}{\Gamma(b - ka)} + \frac{1}{z^m}\frac{1}{2\pi i}\int_{C_\phi}\frac{w^{a(m+1)-b}e^w}{w^a - z}dw \tag{4.12}$$

Relation (4.12) thus generalizes the Taylor series (1.1) and also includes negative powers of z,

$$\sum_{k=-m}^{\infty}\frac{z^k}{\Gamma(b + ka)} = \frac{1}{z^m}\frac{1}{2\pi i}\int_{C_\phi}\frac{w^{a(m+1)-b}e^w}{w^a - z}dw$$

4.2.2 Residue evaluation of the basic integral

67. *Evaluation of Wiman's general integral in (4.12).* Our main purpose is the evaluation of the complex integral in (4.12) by Cauchy's powerful residue theorem, which requires that the path C_ϕ of integration in (4.12) is deformed into a contour or closed loop. As demonstrated in **art.** 61 and **art.** 206, the path C_ϕ in the complex w-plane consists of an arc of a circle at the origin with radius $\rho > |z|^{\frac{1}{a}}$, ranging from an angle 0 to an angle $\frac{\pi}{2} < \phi < \pi$ and the straight line $re^{i\phi}$ with $r \geq \rho$ and its reflection around the real w-axis, but the path C_ϕ is oriented from $\lim_{r\to\infty} re^{-i\phi}$ along the arc towards $\lim_{r\to\infty} re^{i\phi}$. Due to the branch cut w^a for non-integer a, the path cannot be closed over an arc of a circle at the origin with infinite radius, ranging from angle ϕ to angle $-\phi$. Instead, the closure must avoid crossing the branch cut at the negative real w-axis.

Hence, we consider the closed loop-integral

$$\mathbb{L} = \int_L \frac{w^{a(m+1)-b}e^w}{w^a - z} dw$$

where the contour L consists of the straight line at $w = \rho e^{i\phi}t$ with $\frac{\pi}{2} < \phi < \pi$ and starting at $t = 1$ towards infinity, the circular arc at infinity turning from ϕ towards $\pi - \varepsilon$, the line above the negative real axis, the small circle with arbitrary small radius $\varepsilon > 0$ turning around the origin from $\pi - \varepsilon$ to $-\pi + \varepsilon$, the line from the origin just below the negative real axis, over the circular arc with infinite radius from $-\pi + \varepsilon$ to $-\phi$ and following at the negative side of the straight line at $w = \rho e^{-i\phi}t$ down to the intersection with circle at $w = \rho e^{-i\phi}$ and ending along that circle, centered at the origin with radius $\rho > |z|^{\frac{1}{a}}$, at $w = \rho e^{i\phi}$. The contour L thus encloses the singularities of $(w^a - z)^{-1}$, which are simple poles. Indeed, the zeros of $w^a - z = 0$ are all simple and lie at $w_k = |z|^{\frac{1}{a}} e^{i\frac{\arg z}{a}} e^{i\frac{2\pi k}{a}}$ for each $k \in \mathbb{Z}$ and $-\pi < \arg z \leq \pi$. If a is irrational, there are infinitely many zeros in separate Riemann sheets. The contour L only encloses the poles $w_k = |w_k| e^{i\theta_k}$ with argument $\theta_k = \frac{\arg z}{a} + \frac{2\pi k}{a}$ between $-\pi < \theta_k \leq \pi$. Thus, the integer k obeys $-\pi a < \arg z + 2\pi k \leq \pi a$ or ranges between $-\left(\frac{a}{2} + \frac{\arg z}{2\pi}\right) < k \leq \frac{a}{2} - \frac{\arg z}{2\pi}$. There does not seem to exist a simple closed formula for the number $K(a, \arg z)$ of singularities enclosed as a function of a and $\arg z$. The number $K(a, \arg z)$ is zero if $0 < a < 1$ or if $\arg z > \pi a$. Hence, a must be at least equal to 1, else no singularities are enclosed.

Cauchy's residue theorem shows, with $w_k - |z|^{\frac{1}{a}} e^{i\frac{\arg z}{a}} e^{i\frac{2\pi k}{a}}$, that

$$\begin{aligned}
\mathbb{L} &= 2\pi i \sum_{-\left(\frac{a}{2}+\frac{\arg z}{2\pi}\right)<k\leq\frac{a}{2}-\frac{\arg z}{2\pi}} \lim_{w\to w_k} \frac{w - w_k}{w^a - z} w^{(m+1)a-b} e^w \\
&= \frac{2\pi i}{a} \sum_{-\left(\frac{a}{2}+\frac{\arg z}{2\pi}\right)<k\leq\frac{a}{2}-\frac{\arg z}{2\pi}} w_k^{1+ma-b} e^{w_k}
\end{aligned}$$

Since $w_k^a = z$, we obtain

$$\mathbb{L} = 2\pi i \frac{z^m}{a} \sum_{-\left(\frac{a}{2}+\frac{\arg z}{2\pi}\right)<k\leq\frac{a}{2}-\frac{\arg z}{2\pi}} w_k^{1-b} e^{w_k}$$

We can always choose the radius $0 < \varepsilon < r$ small enough, provided that $\text{Re}(a(m+1) - b) > -1$ as shown in **art.** 204. Evaluation of the contour L in the integral $\mathbb{L}$ yields

$$\begin{aligned}
\mathbb{L} &= \int_{C_\phi} \frac{w^{a(m+1)-b}e^w}{w^a - z} dw + \int_\infty^0 \frac{\left(ye^{i\pi}\right)^{a(m+1)-b} e^{-y}}{\left(ye^{i\pi}\right)^a - z} d\left(ye^{i\pi}\right) \\
&\quad + \int_0^\infty \frac{\left(ye^{-i\pi}\right)^{a(m+1)-b} e^{-y}}{\left(ye^{-i\pi}\right)^a - z} d\left(ye^{-i\pi}\right) \\
&= \int_{C_\phi} \frac{w^{a(m+1)-b}e^w}{w^a - z} dw + \int_0^\infty \left(\frac{e^{i\pi(a(m+1)-b)}}{y^a e^{i\pi a} - z} - \frac{e^{-i\pi(a(m+1)-b)}}{y^a e^{-i\pi a} - z} \right) \frac{y^{a(m+1)-b}}{e^y} dy
\end{aligned}$$

Equating, for $\operatorname{Re}(a(m+1)-b) > -1$, to the two evaluations of the integral $\mathbb{L}$ leads to

$$\int_{C_\phi} \frac{w^{a(m+1)-b}e^w}{w^a - z}dw = \int_0^\infty \left(\frac{e^{-(a(m+1)-b)i\pi}}{e^{-ai\pi}y^a - z} - \frac{e^{(a(m+1)-b)i\pi}}{e^{ai\pi}y^a - z} \right) \frac{y^{a(m+1)-b}}{e^y}dy + 2\pi i \frac{z^m}{a} \sum_{-\left(\frac{a}{2}+\frac{\arg z}{2\pi}\right)<k\leq\frac{a}{2}-\frac{\arg z}{2\pi}} w_k^{1-b}e^{w_k}$$

In summary, from (4.12) and with

$$\frac{e^{-(a(m+1)-b)i\pi}}{e^{-ai\pi}y^a - z} - \frac{e^{(a(m+1)-b)i\pi}}{e^{ai\pi}y^a - z} = \frac{z\sin\pi(a(m+1)-b) - \sin\pi(am-b)y^a}{\frac{1}{2i}\left(y^{2a} - 2zy^a\cos(a\pi) + z^2\right)}$$

we find for any finite integer $m \geq 0$, for complex z with $-\pi < \arg z \leq \pi$, for $\operatorname{Re}(a) > 0$ and $\operatorname{Re}(a(m+1)) > \operatorname{Re}(b) - 1$, the *general form* for the Mittag-Leffler function,

$$E_{a,b}(z) = \frac{|z|^{\frac{1-b}{a}}}{a} \sum_{-\left(\frac{a}{2}+\frac{\arg z}{2\pi}\right)<k\leq\frac{a}{2}-\frac{\arg z}{2\pi}} e^{i\frac{\arg z+2\pi k}{a}(1-b)}e^{|z|^{\frac{1}{a}}e^{i\frac{\arg z+2\pi k}{a}}} - \sum_{k=1}^{m} \frac{1}{z^k}\frac{1}{\Gamma(b-ka)}$$
$$+ \int_0^\infty \left(\frac{\sin\pi(b-am)y^a + z\sin\pi(a(m+1)-b)}{y^{2a} - 2zy^a\cos(a\pi) + z^2} \right) \frac{y^{a(m+1)-b}}{\pi z^m e^y}dy \tag{4.13}$$

The general form (4.13) is undoubtedly one of the most important and most general representations of the Mittag-Leffler function $E_{a,b}(z)$.

If $a = l$ and $b = n$ are integers, then $\sin(a-b)\pi = \sin\pi b = 0$, the integral in (4.13) disappears[2] and (4.13) simplifies to

$$E_{l,n}(z) = \frac{z^{\frac{1-n}{l}}}{l} \sum_{-\left(\frac{l}{2}+\frac{\arg z}{2\pi}\right)<k\leq\frac{l}{2}-\frac{\arg z}{2\pi}} e^{i\frac{2\pi k}{l}(1-n)}e^{z^{\frac{1}{l}}e^{i\frac{2\pi k}{l}}}$$

which equals (2.21). The integral in (4.13), studied further in **art.** 82-84, can be regarded for a real pair (a, b) as an interpolator between integer pairs (l, n).

68. *Discussion of the general form (4.13) for* $m = 0$. We distinguish between two cases: real negative and real positive values of z.

a. For real $z = -x^a = x^a e^{i\pi}$, $m = 0$ and $w_k = xe^{i\frac{(2k+1)\pi}{a}}$, the general

[2] Indeed, the branch cut of w^a disappears and the contour in the basic integral (4.7) can be closed over the negative w-axis.

form (4.13) becomes

$$E_{a,b}(-x^a) = -x^a \frac{\sin(a-b)\pi}{\pi} \int_0^\infty \frac{y^{a-b} e^{-y} dy}{y^{2a} + 2x^a y^a \cos(a\pi) + x^{2a}} + \frac{\sin \pi b}{\pi} \int_0^\infty \frac{y^{2a-b} e^{-y}}{y^{2a} + 2x^a y^a \cos(a\pi) + x^{2a}} dy + \frac{x^{1-b}}{a} e^{i\left\{\frac{\pi(1-b)}{a}\right\}} \sum_{k=0}^{\lfloor a \rfloor} e^{x \cos \frac{(2k-1)\pi}{a}} e^{i\left\{-x \sin \frac{(2k-1)\pi}{a} - \frac{2k\pi(1-b)}{a}\right\}} \tag{4.14}$$

For negative real x, $b = 1$ and $0 < a < 1$, the condition to enclose a pole is $-a\pi < \arg z + 2\pi k \leq a\pi$. Since $\arg z = \pi \notin (-a\pi, a\pi]$ for $0 < a < 1$, the residue sum disappears and (4.14) simplifies to

$$E_a(-x^a) = x^a \frac{\sin a\pi}{\pi} \int_0^\infty \frac{y^{a-1} e^{-y} dy}{y^{2a} + 2x^a y^a \cos(a\pi) + x^{2a}} \qquad \text{for } 0 < a < 1 \tag{4.15}$$

from which the asymptotic behavior for large x, with $y^{2a} + 2x^a y^a \cos a\pi + x^{2a} \sim x^{2a}$, is

$$E_a(-x^a) \sim \frac{\sin a\pi}{\pi} \frac{\Gamma(a)}{x^a} \tag{4.16}$$

Art. 82 presents another derivation in (4.55) that is equivalent to (4.15). **Art.** 117 verifies the integral in (4.13), only for $0 < a < 1$, via a series approach. For small, real x, the Taylor series (1.1) gives

$$\begin{aligned} E_a(-x^a) &= \sum_{k=0}^\infty \frac{(-1)^k x^{ak}}{\Gamma(1+ak)} - 1 - \frac{x^a}{\Gamma(1+a)} + O\left(x^{2a}\right) \\ &= \exp\left(-\frac{x^a}{\Gamma(1+a)}\right) + O\left(x^{2a}\right) \end{aligned}$$

Hence, $E_a(-x^a)$ for $0 < a < 1$ is said to "interpolate" between an exponential (for small x) and a power law (for large x). These two regimes have been studied by Mainardi (2014), who illustrated their accuracy with several plots for $a = 0.25, 0.5, 0.75, 0.9$ and 0.99.

b. Positive real $z = x$ have $\arg z = 0$. The requirement $-\pi a < \arg z + 2\pi k \leq \pi a$ for the enclosed poles boils down to $k = 0$ for $a \in (0, 2)$ and (4.13) becomes, for $m = 0$,

$$E_{a,b}(x) = \frac{x^{\frac{1-b}{a}}}{a} e^{x^{\frac{1}{a}}} + \frac{1}{\pi} \int_0^\infty \left(\frac{x \sin\pi(a-b) + y^a \sin \pi b}{y^{2a} - 2xy^a \cos(a\pi) + x^2} \right) y^{a-b} e^{-y} dy \tag{4.17}$$

For $b = 1$, $a \in (0, 2)$ and real $x > 0$, (4.17) simplifies to

$$E_a(x) = \frac{e^{x^{\frac{1}{a}}}}{a} - \frac{\sin \pi a}{\pi} \int_0^\infty \left(\frac{x}{y^{2a} - 2\cos(a\pi) x y^a + x^2} \right) y^{a-1} e^{-y} dy$$

After replacing x by x^a and $t = \frac{y}{x}$, we obtain, for real $x > 0$,

$$E_a(x^a) = \frac{e^x}{a} - \frac{\sin \pi a}{\pi} \int_0^\infty \frac{t^{a-1} e^{-xt}}{t^{2a} - 2t^a \cos(a\pi) + 1} dt \qquad \text{for } 0 < a < 2 \tag{4.18}$$

which is the companion of (4.15) with a larger validity range in a.

4.2.3 Asymptotics

In the residue sum of (4.13), the sign of $w_k = |z|^{\frac{1}{a}} e^{i\frac{\arg z}{a}} e^{i\frac{2\pi k}{a}}$ is determined by $\cos\left(\frac{\arg z + 2\pi k}{a}\right)$. Consequently for large $|z|$, in the sectors $-\frac{\pi a}{2} - 2\pi k < \arg z < \frac{\pi a}{2} - 2\pi k$, the Mittag-Leffler function $E_{a,b}(z)$ tends to infinity, while in the complementary sectors, $\frac{\pi a}{2} - 2\pi k < \arg z < \frac{3\pi a}{2} - 2\pi k$, the residue sum tends to zero.

69. *Asymptotic behavior for large z.* The general form (4.13) generalizes Wiman's formula (4.2) for $a = \frac{1}{n}$ to all complex a with $\mathrm{Re}(a) > 0$ and allows us to unravel the behavior of the Mittag-Leffler function $E_{a,b}(z)$ for large z.

Analogously to Wiman's analysis in **art.** 60, the sum $\sum_{k=1}^{m} \frac{1}{z^k} \frac{1}{\Gamma(b-ka)}$ starts to diverge for $m > m_0$ and the value m_0 is close to the minimizer of $|u_k| = \left|\frac{z^{-k}}{\Gamma(b-ka)}\right|$ in k, obeying $\frac{d|u_k|}{dk} = \frac{d\log|u_k|}{dk} = 0$. The reflection formula (11.4) of the Gamma function indicates that $u_k = \frac{\sin \pi(b-ak)}{\pi z^k} \Gamma(1 - b + ka)$. Hence, $|u_k| \le \frac{1}{\pi} |z|^{-k} \Gamma(1 - b + ka)$ and $\log|u_k| \simeq -\log \pi - k \log|z| + \log \Gamma(1 - b + ka)$. The derivative

$$\frac{d\log|u_k|}{dk} \simeq -\log|z| + a\psi(1 - b + ka) \simeq -\log|z| + a\log(ka - b) + O\left(\frac{1}{k}\right)$$

where the last line stems from **art.** 166 and Section 13.5, showing that $\psi(1 + z) = \log z + \frac{1}{2z} + O\left(\frac{1}{z^2}\right)$ for sufficiently large z. Thus, the minimizer of $|u_k|$ obeying $\frac{d\log|u_k|}{dk} = 0$ is about $m_0 \simeq \frac{1}{a}\left(|z|^{\frac{1}{a}} + b\right)$ and $|u_{m_0}| \simeq \frac{\Gamma\left(1+|z|^{\frac{1}{a}}\right)}{z^{\frac{1}{a}\left(|z|^{\frac{1}{a}}+b\right)}}$. The asymptotic series (14.17), $\Gamma(z + 1) \simeq \sqrt{2\pi} z^{z+\frac{1}{2}} e^{-z}$, finally indicates that $|u_{m_0}| = O\left(|z|^{\frac{1-2b}{2a}} e^{-|z|^{\frac{1}{a}}}\right)$.

In summary, the optimal choice of $m_0 \simeq \frac{1}{a}\left(|z|^{\frac{1}{a}} + b\right)$ in (4.13) leads to the asymptotic expansion for large z with smallest error term,

$$E_{a,b}(z) = \frac{|z|^{\frac{1-b}{a}}}{a} \sum_{-\left(\frac{a}{2} + \frac{\arg z}{2\pi}\right) < k \le \frac{a}{2} - \frac{\arg z}{2\pi}} e^{i\frac{\arg z + 2\pi k}{a}(1-b)} e^{|z|^{\frac{1}{a}} e^{i\frac{\arg z + 2\pi k}{a}}}$$
$$- \sum_{k=1}^{m_0} \frac{1}{z^k} \frac{1}{\Gamma(b - ka)} + O\left(|z|^{\frac{1-2b}{2a}} e^{-|z|^{\frac{1}{a}}}\right) \tag{4.19}$$

70. *Asymptotic regimes.* Recall from **art**. 67 that the index k of each enclosed pole w_k must satisfy $-\pi a < \arg z + 2\pi k \le \pi a$. We explore here a few regimes in the complex number z.

a. If $-\pi a < \arg z \le \pi a$, then only $k = 0$ is possible in the residue sum and (4.19) reduces, for $|\arg z| < \pi a$, to

$$E_{a,b}(z) = \frac{z^{\frac{1-b}{a}}}{a} e^{z^{\frac{1}{a}}} - \sum_{k=1}^{\left\lfloor \frac{1}{a}\left(|z|^{\frac{1}{a}}+b\right)\right\rfloor} \frac{1}{z^k} \frac{1}{\Gamma(b-ka)} + O\left(|z|^{\frac{1-2b}{2a}} e^{-|z|^{\frac{1}{a}}}\right) \quad (4.20)$$

which generalizes (4.3) to real a.

b. For $\arg z = \pm\frac{\pi}{2}a$ and $0 < a < 2$, it holds that $|\arg z| < \pi$, $z^{\frac{1}{a}} = |z|^{\frac{1}{a}} e^{i\frac{\arg z}{a}} = |z|^{\frac{1}{a}} e^{\pm i\frac{\pi}{2}}$ and $e^{z^{\frac{1}{a}}} = e^{\pm i|z|^{\frac{1}{a}}}$. Hence, for large $|z|$, (4.20) indicates that

$$\left|E_{a,b}\left(|z| e^{\pm i\frac{\pi}{2}a}\right)\right| = \left|\frac{z^{\frac{1-b}{a}}}{a} e^{z^{\frac{1}{a}}}\right| + O\left(\frac{1}{z}\right) = \frac{e^{\frac{\pi}{2}\operatorname{Im}(b)}}{a} |z|^{\frac{1-\operatorname{Re}(b)}{a}} + O\left(\frac{1}{z}\right)$$

For $b = 1$ and $0 < a < 2$, we observe that $\left|E_a\left(|z| e^{\pm i\frac{\pi}{2}a}\right)\right| = \frac{1}{a} + O\left(\frac{1}{z}\right)$, a result already found by Mittag-Leffler as mentioned in Barnes (1906, p. 285).

c. If $|\arg z| > \pi a$, then no poles are enclosed and the residue sum in (4.19) vanishes, resulting, for $|\arg z| > \pi a$, in

$$E_{a,b}(z) = -\sum_{k=1}^{\left\lfloor \frac{1}{a}\left(|z|^{\frac{1}{a}}+b\right)\right\rfloor} \frac{1}{z^k} \frac{1}{\Gamma(b-ka)} + O\left(|z|^{\frac{1-2b}{2a}} e^{-|z|^{\frac{1}{a}}}\right) \quad (4.21)$$

illustrating large different behaviors of $E_{a,b}(z)$ in the complex plane z. The region $\arg z = \pi a \pm \delta$ for $0 < \delta < \frac{\pi}{2}a$ marks a transition region; the lines $\arg z = \pi a \pm \frac{\pi a}{2}$ are known as Stokes lines, explained in **art**. 71.

d. If $\arg z = \pi$, then $-\frac{a+1}{2} < k \le \frac{a-1}{2}$. Hence, if $2 \le a < 3$ and b is real, then k runs through $0, -1$ and the main contribution $\mathcal{M}$ in (4.19) for sufficiently large $|z|$ consists of the first term

$$\begin{aligned}
\mathcal{M} &= \frac{|z|^{\frac{1-b}{a}}}{a}\left(e^{i\frac{\pi}{a}(1-b)} e^{|z|^{\frac{1}{a}} e^{i\frac{\pi}{a}}} + e^{-i\frac{\pi}{a}(1-b)} e^{|z|^{\frac{1}{a}} e^{-i\frac{\pi}{a}}}\right) \\
&= \frac{2|z|^{\frac{1-b}{a}} e^{|z|^{\frac{1}{a}}\cos\frac{\pi}{a}}}{a} \cos\left(|z|^{\frac{1}{a}} \sin\frac{\pi}{a} + \frac{\pi}{a}(1-b)\right)
\end{aligned}$$

that possesses zeros at $|z|^{\frac{1}{a}} \sin\frac{\pi}{a} + \frac{\pi}{a}(1-b) = \pi\left(l + \frac{1}{2}\right)$ for integers $l \in \mathbb{Z}$. Hence, for sufficiently large $|z|$ and $2 \le a < 3$, (4.19) indicates that $E_{a,b}(-|z|)$ oscillates and the zeros of $E_{a,b}(-|z|)$ have approximately the modulus $|z|_l = \left(\frac{\pi\left(l+\frac{1}{2}+\frac{b-1}{a}\right)}{\sin\frac{\pi}{a}}\right)^a$, with all integers l such that $\frac{\pi\left(l+\frac{1}{2}+\frac{b-1}{a}\right)}{\sin\frac{\pi}{a}} \ge 0$. Wiman (1905b) shows, for $b = 1$ and $a \ge 2$, that $E_a(z)$ has no complex zeros and that all zeros lie on the negative real z-axis. Wiman (1905b) continues his beautiful, though involved, investigations for $0 < a < 2$ and find, based on the asymptotic expansion (4.19) for $b=1$, similar estimates for the zeros.

71. *Stokes lines.* Stokes (1857) observed that asymptotic expansions of certain complex functions can seemingly discontinuously change at certain $\arg z$ in $z=|z|\,e^{i\arg z}$ for large $|z|$. Such lines of rapid change at constant angle $\arg z$ are called Stokes lines. Stokes exemplified the phenomenon by studying the solution $u(t)$ of the differential equation $\frac{du(t)}{dt}+2tu(t)=2$ with initial condition $u(0)=0$. The solution of the homogeneous differential equation $\frac{du_h(t)}{dt}+2tu_h(t)=0$ follows by integration as $u_h(t)=Ce^{-t^2}$. By the method of the variation of a constant, a proposal $u(t)=C(t)\,e^{-t^2}$ is substituted and the solution to the particular differential equation is, valid for all t,

$$u(t)=2e^{-t^2}\int_0^t e^{x^2}dx=2e^{-t^2}\left(t+\frac{t^3}{3}+\frac{t^5}{2.5}+\sum_{k=3}^{\infty}\frac{t^{2k+1}}{k!\,(2k+1)}\right)$$

However, for large t, the series is slowly converging, which encourages us to find a faster converging series. Stokes (1857) writes

$$u(t)=e^{-t^2}\left(2\int_0^1 e^{x^2}dx+2\int_1^t e^{x^2}dx\right)=e^{-t^2}\left(\tilde{c}+\int_1^{t^2}y^{-\frac{1}{2}}e^y dy\right)$$

Partial integration of $I(\alpha)=\int_1^{t^2}y^{\alpha}e^y dy$ results in the recursion $I(\alpha)=t^{2\alpha}e^{t^2}-e-\alpha I(\alpha-1)$. Iterating K times,

$$I(\alpha)=e^{t^2}t^{2\alpha}\sum_{k=0}^{K-1}\frac{(-1)^k\Gamma(\alpha+1)}{\Gamma(\alpha-k+1)}\frac{1}{t^{2k}}-e\sum_{k=0}^{K-1}\frac{(-1)^k\Gamma(\alpha+1)}{\Gamma(\alpha-k+1)}$$
$$+\frac{(-1)^p\Gamma(\alpha+1)}{\Gamma(\alpha-K+1)}I(\alpha-K)$$

leads, for $\alpha=-\frac{1}{2}$, after omitting the integral $I\left(-\frac{1}{2}-K\right)$ and terms of $O\left(e^{-t^2}\right)$, to an asymptotic solution

$$u(t)=ce^{-t^2}+\sum_{k=1}^{K}\frac{1.3.5.\cdots.(2k-1)}{2^{k-1}}\frac{1}{t^{2k-1}} \tag{4.22}$$

where the constant c is non-zero, as shown by Stokes. Of course, if $K\to\infty$, then the series in (4.22) is diverging, similarly as $m_0\to\infty$ in (4.19). Stokes (1857) considers in great detail the behavior of $u\left(|t|\,e^{i\arg t}\right)$ as a function of $\arg t$ for large $|t|$, where the exponential e^{-t^2} in (4.22) can be either large or very small for certain $\arg t$ in a narrow region, while the series in (4.22) is $1+O\left(\frac{1}{t}\right)$.

After reviewing the Stokes phenomenon, Berry (1989) points out that Stokes discontinuity is an artifact of "poor resolution" or series truncation: with the appropriate magnification, the function changes smoothly across the Stokes line. Furthermore, Berry demonstrates that the change across the Stokes line is universal in form and characterized by the error function $\int_{-\infty}^{t} e^{-x^2} dx$.

Paris (2002) starts from (2.6) with $\beta = 1$,

$$E_a(z) = -\sum_{k=1}^{m} \frac{1}{\Gamma(1-ak)} \frac{1}{z^k} + z^{-m} E_{a,1-ma}(z)$$

and truncates the series close at its optimal point m_0 based on the ratio of $k+1$-th over the k-th term (as in **art.** 60) leading to $am_0 = |z|^{\frac{1}{a}} + \nu$, where ν is bounded for large z. Paris then investigates the asymptotics for large z of the remainder $E_{a,1-m_0 a}(z)$, which reduces to studying the integral in (4.17) for $0 < a < 1$ and for large $b = 1 - |z|^{\frac{1}{a}} - \nu$. After invoking the principle of the steepest descent (**art.** 200) and skillful approximations, Paris (2002, eq. (2.10)) finds, for the $\arg z$ in the vicinity of the Stokes lines at $\pm\pi a$, that the leading term for large $|z|$ and $am_0 = |z|^{\frac{1}{a}} + \nu$, is

$$z^{-m_0} E_{a,1-m_0 a}(z) \sim \frac{e^{z^{\frac{1}{a}}}}{2a} \left(1 + \operatorname{erf}\left(\frac{(\pi a \mp \arg z)}{a} \frac{|z|^{\frac{1}{2a}}}{\sqrt{2}}\right)\right)$$

illustrating that the exponential factor $\frac{1}{a} e^{z^{\frac{1}{a}}}$ switches off in the familiar smooth manner as $\arg z$ crosses the Stokes lines at $\pm\pi a$ in the sense of increasing $|\arg z|$. Paris (2002) further studies the case for $a > 1$ as well as the Stokes phenomenon in the parameter a when z (both $|z|$ and $\arg z$) are held fixed with $|z|$.

Wong and Zhao (2002) compute in detail different asymptotic z-regimes of $E_{a,b}(z)$ for (a) $0 < a < 1$, (b) $a \in (2l-1, 2l)$ for a positive integer l, (c) $a \in (2l, 2l+1)$ and (e) $a \in \mathbb{N}_0$, but b is not an integer. Away from the Stokes lines, they derived exponentially improved uniform asymptotic expansions. Near the Stokes lines, Berry-type smooth transitions are obtained. Wong and Zhao (2002) also show that for complex a with $\operatorname{Re}(a) > 0$, the Stokes curves are spirals.

4.2.4 Hankel's contour for the Mittag-Leffler function

Art. 72 first generalizes Hankel's integral for $\frac{1}{\Gamma(z)}$ in Chapter 15, while **art.** 73 then derives a simplification (4.25) that exhibits the characteristic integrand, which already appeared in the key formula (4.13).

72. *Generalization of Hankel's integral for the Gamma function.* We mimic Hankel's method in **art.** 204 and evaluate the integral

$$\mathcal{I} = \int_C E_{a,b}(\lambda w^a)\, w^z e^{sw} dw$$

around the Hankel contour C, where ε is the radius of a circle at the origin,

$$\begin{aligned}\mathcal{I} =& -\int_{\infty}^{\varepsilon}\left(xe^{-i\pi}\right)^{z} E_{a,b}\left(\lambda\left(xe^{-i\pi}\right)^{a}\right)\frac{dx}{e^{sx}} - \int_{\varepsilon}^{\infty}\left(xe^{i\pi}\right)^{z} E_{a,b}\left(\lambda\left(xe^{i\pi}\right)^{a}\right)\frac{dx}{e^{sx}} \\ &+ i\int_{-\pi}^{\pi}\left(\varepsilon e^{i\theta}\right)^{z+1} E_{a,b}\left(\lambda\left(\varepsilon e^{i\theta}\right)^{a}\right) e^{s\varepsilon e^{i\theta}}d\theta \\ =& \int_{\varepsilon}^{\infty}\left(e^{-i\pi z}E_{a,b}\left(\lambda e^{-ai\pi}x^{a}\right) - e^{i\pi z}E_{a,b}\left(\lambda e^{ai\pi}x^{a}\right)\right)x^{z}e^{-sx}dx \\ &+ i\varepsilon^{1+z}\int_{-\pi}^{\pi}e^{i\theta(1+z)}E_{a,b}\left(\lambda\varepsilon^{a}e^{i\theta a}\right)e^{s\varepsilon e^{i\theta}}d\theta\end{aligned}$$

If $\operatorname{Re}(z) > -1$, then we take the limit $\varepsilon \to 0$ and find

$$\mathcal{I} = e^{-i\pi z}\int_{0}^{\infty}E_{a,b}\left(\lambda e^{-ai\pi}x^{a}\right)x^{z}e^{-sx}dx - e^{i\pi z}\int_{0}^{\infty}E_{a,b}\left(\lambda e^{ai\pi}x^{a}\right)x^{z}e^{-sx}dx$$

The Laplace transform $\int_0^\infty e^{-sx}x^{\gamma-1}E_{a,b}\left(\lambda x^{\beta}\right)dx = \sum_{k=0}^{\infty}\frac{\Gamma(\gamma+\beta k)}{\Gamma(b+ak)}\frac{\lambda^k}{s^{\gamma+k\beta}}$ in (3.10) indicates that

$$\begin{aligned}\mathcal{I} &= e^{-i\pi z}\sum_{k=0}^{\infty}\frac{\Gamma((1+z)+ak)}{\Gamma(b+ak)}\frac{\lambda^k e^{-ai\pi k}}{s^{1+z+ka}} - e^{i\pi z}\sum_{k=0}^{\infty}\frac{\Gamma((1+z)+ak)}{\Gamma(b+ak)}\frac{\lambda^k e^{ai\pi k}}{s^{1+z+ka}} \\ &= -\frac{1}{s^{1+z}}\sum_{k=0}^{\infty}\frac{\Gamma((1+z)+ak)}{\Gamma(b+ak)}\left(\frac{\lambda}{s^a}\right)^k\left(e^{i\pi(z+ak)} - e^{-i\pi(z+ak)}\right)\end{aligned}$$

In summary, for $\operatorname{Re}(z) > -1$, $\left|\frac{\lambda}{s^a}\right| < 1$ and any b (see **art.** 42), we arrive at the generalization of Hankel's integral,

$$\int_C E_{a,b}\left(\lambda w^a\right)w^z e^{sw}dw = -\frac{2i}{s^{1+z}}\sum_{k=0}^{\infty}\frac{\Gamma(1+z+ak)\sin\pi(z+ak)}{\Gamma(b+ak)}\left(\frac{\lambda}{s^a}\right)^k \tag{4.23}$$

The generalization (4.23) of Hankel's integral also generalizes the Laplace transforms (3.10)-(3.12) to a complex path. We rewrite (4.23) with the reflection formula (11.4) of the Gamma function as

$$\frac{1}{2i\pi}\int_C E_{a,b}\left(\lambda w^a\right)w^z e^{sw}dw = \frac{1}{s^{1+z}}\sum_{k=0}^{\infty}\frac{1}{\Gamma(b+ak)\Gamma(-z-ak)}\left(\frac{\lambda}{s^a}\right)^k \tag{4.24}$$

Example If $a = b = 1$, then (4.23) reduces to

$$\int_C w^z e^{(s+\lambda)w}dw = -\frac{2iz!}{s^{1+z}}\sum_{k=0}^{\infty}\frac{(k+z)!\sin\pi(z+k)}{k!z!}\left(\frac{\lambda}{s}\right)^k$$

Invoking $\frac{(k+z)!}{k!z!} = \binom{k+z}{k} = (-1)^k\binom{-z-1}{k}$, $\sin\pi(z+k) = (-1)^k\sin\pi(z)$ and the reflection formula (11.4) of the Gamma function leads to

$$\int_C w^z e^{(s+1)w} dw = -\frac{2iz!}{s^{1+z}} \sin \pi z \sum_{k=0}^{\infty} \binom{-1-z}{k} \left(\frac{\lambda}{s}\right)^k$$
$$= -\frac{2i\Gamma(z+1)}{s^{1+z}} \sin \pi z \left(1+\frac{\lambda}{s}\right)^{-1-z} = 2i\pi \frac{(s+\lambda)^{-1-z}}{\Gamma(-z)}$$

which is, indeed, the variant (15.5) of Hankel's contour for the Gamma function.

73. *Special case of the generalization (4.23) of Hankel's integral.* As observed in **art.** 42, the series in (4.23) can be evaluated by choosing $z = b-1$,

$$\int_C E_{a,b}(\lambda w^a) w^{b-1} e^{sw} dw = -\frac{2i}{s^b} \sum_{k=0}^{\infty} \sin \pi (b-1+ak) \left(\frac{\lambda}{s^a}\right)^k$$
$$= \frac{e^{i\pi b} \sum_{k=0}^{\infty} \left(\frac{\lambda e^{i\pi a}}{s^a}\right)^k - e^{-i\pi b} \sum_{k=0}^{\infty} \left(\frac{\lambda e^{-i\pi a}}{s^a}\right)^k}{s^b}$$
$$= \frac{1}{s^b} \left(\frac{e^{i\pi b}}{1-\frac{\lambda e^{i\pi a}}{s^a}} - \frac{e^{-i\pi b}}{1-\frac{\lambda e^{-i\pi a}}{s^a}} \right)$$

Rewriting the right-hand side shows, for $\operatorname{Re}(a) > 0$, all complex b and $\left|\frac{\lambda}{s^a}\right| < 1$, that

$$\frac{1}{2i} \int_C E_{a,b}(\lambda w^a) w^{b-1} e^{sw} dw = \frac{1}{s^b} \frac{\sin \pi b - \frac{\lambda}{s^a} \sin \pi (b-a)}{1 - 2\frac{\lambda}{s^a} \cos \pi a + \left(\frac{\lambda}{s^a}\right)^2} \tag{4.25}$$

If a and b are integers, then the integrand in (4.25) is analytic everywhere; thus, also at the negative w-axis so that the left-hand side integral vanishes by Cauchy's integral theorem.

After deforming the Hankel contour in (4.25) as in **art.** 206, we obtain an inverse Laplace transform with $c > 0$,

$$\frac{1}{2\pi i} \int_{c-i\infty}^{c+i\infty} E_{a,b}(\lambda w^a) w^{b-1} e^{sw} dw = \frac{s^{a-b}}{\pi} \frac{s^a \sin \pi b + \lambda \sin(\pi(a-b))}{s^{2a} - 2\lambda s^a \cos(\pi a) + \lambda^2}$$

provided that the integrals over circular arcs at infinity (as explained in **art.** 206) vanish, which happens if $\lim_{R\to\infty} E_{a,b}\left(R^a e^{ia\phi}\right) R^{b-1} e^{sRe^{i\phi}} = 0$ for all $\pi \leq \phi \leq \frac{\pi}{2}$ and which involves a study of the sectors in terms of a (see Section 4.2.3). It turns out that deforming the contour is only possible for $0 < a < 1$ in which case inverse Laplace transformation leads, for $\operatorname{Re}(a-b) > -1$, to

$$E_{a,b}(\lambda z^a) = \frac{z^{1-b}}{\pi} \int_0^{\infty} \frac{s^a \sin \pi b + \lambda \sin(\pi(a-b))}{s^{2a} - 2\lambda s^a \cos(\pi a) + \lambda^2} s^{a-b} e^{-sz} ds$$

which reduces, for $\lambda = -1$, to (4.52) below.

4.3 Mittag-Leffler's contour integral

74. *Mittag-Leffler's contour integral.* In a rather long article, Mittag-Leffler (1905, at pp. 133-135) proceeds one step further and substitutes $w = t^{\frac{1}{a}}$ in the basic complex integral (4.4). The map $w \to t^{\frac{1}{a}}$ is multi-valued. For $w = |w| e^{i\theta_w}$ with $-\pi \leq \theta_w \leq \pi$, the inverse map $t \to w^a$ shows that $t = |t| e^{i\theta_t} = |w|^a e^{ia\theta_w + 2\pi iak}$, from which the argument $\theta_t = a\theta_w + 2\pi ak$ for any integer k. The contour C_ϕ in the w-plane requires that $\frac{\pi}{2} < |\phi| < \pi$, because $e^w \to 0$ for large w at the contour C_ϕ. Similarly, the transformed contour $C_{a\phi}$ in the t-plane requires that $\frac{a\pi}{2} < |\arg t| < a\pi$ in order that $e^{t^{\frac{1}{a}}} \to 0$ along the straight lines towards infinity. Thus, the map $w \to t^{\frac{1}{a}}$ changes the angles from ϕ to $a\phi$ of the straight lines in the contour (see Fig. 43 in Bieberbach (1945, p. 273) or the figure in Mittag-Leffler (1905, p. 134)). Moreover, we must choose the branch (i.e. the appropriate integer k in $\theta_t = a\theta_w + 2\pi ak$) of the function $t^{\frac{1}{a}}$ that is positive for positive t, i.e. along the positive real t-axis, because the same holds along the positive real w-axis. Performing the substitution $w = t^{\frac{1}{a}}$ in (4.4) leads, for $|z| < |t|$ and $|\arg z| < \frac{a\pi}{2}$, to Mittag-Leffler's contour integral

$$E_{a,b}(z) = \frac{1}{2\pi ia} \int_{C_{a\phi}} \frac{t^{\frac{1-b}{a}} e^{t^{\frac{1}{a}}}}{t-z} dt \qquad \text{with } \frac{\pi}{2} < |\phi| < \pi \tag{4.26}$$

Mittag-Leffler's integral (4.26) for $b = 1$ is actually more elegant than the basic complex integral (4.4) at the expense of a more complicated contour $C_{a\phi}$.

75. *Deductions from Mittag-Leffler's contour integral (4.26) for $E_{a,b}(z)$.* Based on (4.26), we follow Bieberbach (1945, p. 273), who deduces two bounds for $0 < a < 2$ and $b = 1$. First, using $\frac{1}{t-z} = -\frac{1}{z} + \frac{t}{z(t-z)}$ in (4.26) yields

$$E_{a,b}(z) = -\frac{1}{z}\frac{1}{2\pi ia} \int_{C_{a\phi}} t^{\frac{1-b}{a}} e^{t^{\frac{1}{a}}} dt + \frac{1}{2\pi ia} \int_{C_{a\phi}} \frac{t^{\frac{1-b+a}{a}} e^{t^{\frac{1}{a}}}}{z(t-z)} dt$$

Hankel's contour integral (15.3), $\frac{1}{2\pi ia} \int_{C_{a\phi}} t^{\frac{1-b}{a}} e^{t^{\frac{1}{a}}} dt = \frac{1}{2\pi i} \int_{C_\phi} w^{a-b} e^w dw = \frac{1}{\Gamma(b-a)}$, reduces to

$$E_{a,b}(z) = -\frac{1}{z}\frac{1}{\Gamma(b-a)} + \frac{1}{2\pi ia} \int_{C_{a\phi}} \frac{t^{\frac{a-b+1}{a}} e^{t^{\frac{1}{a}}}}{z(t-z)} dt$$

The remaining integral is upper bounded by

$$\left| \frac{1}{2\pi ia} \int_{C_{a\phi}} \frac{t^{\frac{a-b+1}{a}} e^{t^{\frac{1}{a}}}}{z(t-z)} dt \right| < \frac{1}{2\pi a |z|} \int_{C_{a\phi}} \frac{|t|^{\frac{a-b+1}{a}} \left| e^{t^{\frac{1}{a}}} \right|}{|z| \left| 1 - \frac{t}{z} \right|} |dt| = \frac{c}{|z|^2}$$

because the integral $\frac{1}{2\pi a} \int_{C_{a\phi}} \frac{|t|^{\frac{a-b-1}{a}} \left| e^{t^{\frac{1}{a}}} \right|}{\left| 1 - \frac{t}{z} \right|} |dt|$ converges for $t \to \infty$ if $\left| e^{t^{\frac{1}{a}}} \right| =$

$e^{|t|^{\frac{1}{a}} \cos \frac{\arg t}{a}} \to 0$ for large t, which requires that $\frac{\arg t}{a} \geq \frac{\pi}{2}$. In addition, we must prevent that $\frac{t}{z}$ for large z can tend arbitrarily close to 1, which is guaranteed if $|\arg z| \notin \left(\frac{a\pi}{2}, a\pi\right)$, because $|\arg t| \in \left(\frac{a\pi}{2}, a\pi\right)$. Hence, for $|\arg z| \notin \left(\frac{a\pi}{2}, a\pi\right)$, we arrive at

$$\left| E_{a,b}(z) + \frac{1}{z\Gamma(b-a)} \right| < \frac{c}{|z|^2} \tag{4.27}$$

Bieberbach's first bound (4.27) follows from (4.21), if the sum is expanded up to the first term.

For the second bound, Mittag-Leffler (1905, p. 136) cleverly observes that a similar integration path $C'_{a\phi}$ as in Mittag-Leffler's integral (4.26) can be followed with the only difference that the circular part of the path now has a radius smaller than $|z|$. In other words, while $|z| < |t|$ in Mittag-Leffler's integral (4.26), the path $C'_{a\phi}$ now turns over an angle $-a\phi$ to $a\phi$ with the radius smaller than $|z|$. The closed contour (see also Sansone and Gerretsen (1960, p. 346, Fig. 6.13-2)), that first follows the Mittag-Leffler path $C_{a\phi}$ and returns via the path $C'_{a\phi}$, encloses the point $t = z$ as the only singularity, provided $-a\phi < \arg z < a\phi$. Hence, by Cauchy's residue theorem, it holds that

$$\frac{1}{2\pi i a} \int_{C_{a\phi}} \frac{t^{\frac{1-b}{a}} e^{t^{\frac{1}{a}}}}{t-z} dt - \frac{1}{2\pi i a} \int_{C'_{a\phi}} \frac{t^{\frac{1-b}{a}} e^{t^{\frac{1}{a}}}}{t-z} dt = \frac{1}{a} z^{\frac{1-b}{a}} e^{z^{\frac{1}{a}}}$$

and

$$E_{a,b}(z) = \frac{1}{a} z^{\frac{1-b}{a}} e^{z^{\frac{1}{a}}} + \frac{1}{2\pi i a} \int_{C'_{a\phi}} \frac{t^{\frac{1-b}{a}} e^{t^{\frac{1}{a}}}}{t-z} dt \tag{4.28}$$

from which

$$\left| E_{a,b}(z) - \frac{1}{a} z^{\frac{1-b}{a}} e^{z^{\frac{1}{a}}} \right| \leq \frac{1}{2\pi a |z|} \int_{C'_{a\phi}} \frac{\left| t^{\frac{1-b}{a}} \right| \left| e^{t^{\frac{1}{a}}} \right|}{\left| \frac{t}{z} - 1 \right|} dt = \frac{c'}{|z|}$$

by the same convergence argument as above, where now on the circular segment $|t| < |z|$ can be chosen small enough. In summary, the second bound for $|\arg z| \in [0, a\pi)$ is

$$\left| E_{a,b}(z) - \frac{1}{a} z^{\frac{1-b}{a}} e^{z^{\frac{1}{a}}} \right| < \frac{c'}{|z|} \tag{4.29}$$

The derivation illustrates why Mittag-Leffler considers $0 < a < 2$, because $|\arg z| \in (0, 2\pi)$. Bieberbach's second bound (4.29) follows from (4.20), if the sum is omitted.

The second bound (4.29) shows that $aE_{a,b}(z^a) \approx z^{1-b} e^z$ is independent of a so that $aE_{a,b}(z^a) \approx \frac{1}{a} E_{\frac{1}{a},b}\left(z^{\frac{1}{a}}\right)$. Hence, we are led for non-negative real z and $|z| > \varepsilon$ to

$$E_{\frac{1}{a},b}(z) \approx a^2 E_{a,b}\left(z^{a^2}\right)$$

whose exact corresponding relation (7.33) for the associated integral $I_{a,b}(z) = \int_0^\infty \frac{z^u}{\Gamma(b+au)}du$, explored in Section 7.5, is $I_{\frac{1}{a},b}(z) = a^2 I_{a,b}\left(z^{a^2}\right)$.

4.4 Complex integral for $E_{a,b}(z)$ from Cauchy's residue theorem

76. *Deductions from Cauchy's residue theorem.* If $f(z)$ is an entire function and $\lim_{r\to\infty}\left|\frac{f(re^{i\theta})}{\sin(\pi re^{i\theta})}\right| = 0$ in a semicircle with either $-\frac{\pi}{2} < \theta < \frac{\pi}{2}$ or $\frac{\pi}{2} < \theta < \frac{3\pi}{2}$, then it follows directly from Cauchy's residue theorem (Titchmarsh, 1964) for $0 < c < 1$ that

$$\frac{1}{2\pi i}\int_{c-i\infty}^{c+i\infty} \frac{\pi}{\sin \pi w} f(w)\,dw = \begin{cases} -\sum_{k=1}^{\infty}(-1)^k f(k) & \text{if } -\frac{\pi}{2} < \theta < \frac{\pi}{2} \\ \sum_{k=0}^{\infty}(-1)^k f(-k) & \text{if } \frac{\pi}{2} < \theta < \frac{3\pi}{2} \end{cases} \tag{4.30}$$

where, if $-\frac{\pi}{2} < \theta < \frac{\pi}{2}$, then the contour is closed over the positive $\text{Re}(w)$-plane, else over the negative $\text{Re}(w)$-plane. Equation (4.30) is similar to Plana's summation formula (Sansone and Gerretsen, 1960, p. 438),

$$\frac{1}{2\pi i}\int_{c-i\infty}^{c+i\infty} \pi f(w)\cot \pi w\,dw = \begin{cases} -\sum_{k=1}^{\infty} f(k) & \text{if } -\frac{\pi}{2} < \theta < \frac{\pi}{2} \\ \sum_{k=0}^{\infty} f(-k) & \text{if } \frac{\pi}{2} < \theta < \frac{3\pi}{2} \end{cases} \tag{4.31}$$

but the condition $\lim_{r\to\infty}\left|f\left(re^{i\theta}\right)\right| = 0$ on the entire function $f(z)$ is more restrictive.

The Taylor series (1.1) of $E_{a,b}(z) = \sum_{k=0}^{\infty}\frac{z^k}{\Gamma(b+ak)}$ contains $f(w) = \frac{z^w}{\Gamma(b+aw)}$, which is an entire function in w for any complex a and b. The asymptotic behavior (14.9) of

$$\frac{1}{\left|\Gamma\left(b+are^{i\theta}\right)\right|} = \frac{(ar)^{\frac{1}{2}-b}}{\sqrt{2\pi}} e^{-ar(\ln(ar)-1)\cos\theta+\theta ar\sin\theta}\left(1+O\left(\frac{1}{r}\right)\right)$$

in **art.** 197 shows that the above contour can be closed over the positive $\text{Re}(w)$-plane, resulting in[3]

$$E_{a,b}(z) = -\frac{1}{2\pi i}\int_{c+\infty e^{-i\theta}}^{c+\infty e^{i\theta}} \frac{\pi}{\sin \pi w}\frac{(-z)^w}{\Gamma(b+aw)}dw \qquad \text{for } -1 < c < 0 \tag{4.33}$$

where the path above and below the real w-axis follows the lines $c + re^{\pm i\theta}$, where $0 < \theta < \frac{\pi}{2}$ and $r \geq 0$. Thus, the line of integration cannot be parallel with the imaginary axis (i.e. $\theta = \frac{\pi}{2}$), unless $a < 2$. If $a = b = 1$, the reflection formula (11.4) leads to Mellin's integral $\frac{1}{2\pi i}\int_{c+i\infty}^{c+i\infty}\Gamma(w)(-z)^w\,dw =$

[3] The integral (4.33) is rewritten with the reflection formula of the Gamma function as a Barnes-Mellin type integral

$$E_{a,b}(-z) = -\frac{1}{2\pi i}\int_{c+\infty e^{-i\theta}}^{c+\infty e^{i\theta}} \frac{\Gamma(1-w)\Gamma(w)}{\Gamma(b+aw)}z^w dw \qquad \text{for } -1 < c < 0 \text{ and } 0 < \theta < \frac{\pi}{2} \tag{4.32}$$

e^z. Confining to $0 < a < 2$, we rewrite (4.33) with $\widetilde{c} = -c$, as

$$E_{a,b}(-z) = -\frac{1}{2\pi i}\int_{-\widetilde{c}-i\infty}^{-\widetilde{c}+i\infty} \frac{\pi}{\sin \pi w}\frac{z^w}{\Gamma(b+aw)}dw \qquad \text{for } 0 < \widetilde{c} < 1$$

Similar to **art.** 62, evaluation of the contour along the line $w = -\widetilde{c} + it$, followed by substitution $u = -t$ and transforming back as a contour integral along the line $\omega = \widetilde{c} + iu$ results in

$$E_{a,b}(-z) = \frac{1}{2\pi i}\int_{\widetilde{c}-i\infty}^{\widetilde{c}+i\infty} \frac{\pi}{\sin \pi\omega}\frac{z^{-\omega}}{\Gamma(b-a\omega)}d\omega \qquad \text{for } 0 < \widetilde{c} < 1$$

which is recognized as an inverse Mellin transform (B.4). Hence, we find the Mellin transform

$$\int_0^\infty t^{z-1}E_{a,b}(-t)\,dt = \frac{\pi}{\sin \pi z}\frac{1}{\Gamma(b-az)} = \frac{\Gamma(z)\Gamma(1-z)}{\Gamma(b-az)} \tag{4.34}$$

valid for $0 < a < 2$, all b and $0 < \text{Re}(z) < 1$, which could not be found by the Taylor series-based methods of the class (3.2) in Section 3.1.

The companion of (4.33) follows from Plana's summation formula (4.31) as

$$E_{a,b}(z) = -\frac{1}{2\pi i}\int_{c+\infty e^{-i\theta}}^{c+\infty e^{i\theta}} \frac{\pi\cos\pi w}{\sin \pi w}\frac{z^w}{\Gamma(b+aw)}dw \qquad \text{for } -1 < c < 0 \tag{4.35}$$

77. *Shifting the line of integration in (4.33).* Let us consider the contour C, consisting of the line $c + re^{i\theta}$ with $0 \leq r \leq T$, the line parallel to the real axis from $c + Te^{i\theta}$ to the left at $c - m + Te^{i\theta}$, the line back to the real axis at $c - m$ and the reflection of this parallelogram around the real axis. The paths parallel to the real axis

$$\int_{c+Te^{\pm i\theta}}^{c-m+Te^{\pm i\theta}} \frac{\pi}{\sin \pi w}\frac{(-z)^w\,dw}{\Gamma(b+aw)} = \int_c^{c-m}\frac{\pi}{\sin\pi\left(x+Te^{\pm i\theta}\right)}\frac{(-z)^{x+Te^{\pm i\theta}}\,dx}{\Gamma\left(b+ax+aTe^{\pm i\theta}\right)}$$

vanish, for all real $a > 0$ and any integer m, by (14.9) for $T \to \infty$ provided $0 < \theta < \frac{\pi}{2}$. The contour C encloses the poles $\frac{\pi}{\sin \pi w}$ at $w = -k$ from $k = 1$ to m with residue $(-1)^k$. Hence, by shifting the lines $c + re^{\pm i\theta}$ to $c - m + re^{\pm i\theta}$, while maintaining the angle $0 < \theta < \frac{\pi}{2}$, we deform the integral (4.33), for $-(m+1) < c' < -m$, into (left-shifting)

$$E_{a,b}(z) = -\sum_{k=1}^{m}\frac{1}{\Gamma(b-ka)}\frac{1}{z^k} - \frac{1}{2\pi i}\int_{c'+\infty e^{-i\theta}}^{c'+\infty e^{i\theta}}\frac{\pi}{\sin\pi w}\frac{(-z)^w}{\Gamma(b+aw)}dw \tag{4.36}$$

and, for $m - 1 < c'' < m$, into (right-shifting)

$$E_{a,b}(z) = \sum_{k=0}^{m-1}\frac{z^k}{\Gamma(b+ka)} - \frac{1}{2\pi i}\int_{c''+\infty e^{-i\theta}}^{c''+\infty e^{i\theta}}\frac{\pi}{\sin\pi w}\frac{(-z)^w}{\Gamma(b+aw)}dw \tag{4.37}$$

Invoking the reflection formula (11.4) of the Gamma function, the integral (4.36) becomes, with $-(m+1) < c' < -m$ and $0 < \theta < \frac{\pi}{2}$,

$$E_{a,b}(z) = -\sum_{k=1}^{m} \frac{1}{\Gamma(b-ka)} \frac{1}{z^k} \tag{4.38}$$
$$-\frac{1}{2\pi i}\int_{c'+\infty e^{-i\theta}}^{c'+\infty e^{i\theta}} \frac{\sin(\pi(b+aw))}{\sin \pi w}\Gamma(1-b-aw)(-z)^w dw$$

which holds for $\mathrm{Re}(a) > 0$ and all complex b. The contour in (4.38) can be closed over the positive $\mathrm{Re}(w)$-plane and encloses the simples poles of $\frac{1}{\sin \pi w}$ at $w = k$, for integers $k \geq 0$. When $1 - b - aw = -k$ with integer $k \geq 0$, simple poles of $\Gamma(1-b-aw)$ occur at $w_k = \frac{k+1-b}{a}$, but are cancelled by the zeros of $\sin(\pi(b+aw))$ at $w_\nu = \frac{\nu-b}{a}$, where $\nu \in \mathbb{Z}$. Indeed, $\frac{1}{\Gamma(b+aw)}$ in (4.37) is an entire function.

4.4.1 Barnes' study of $E_a(z)$

78. *Barnes' representation of $E_a(z)$.* Barnes (1906, pp. 285-289) presented a different method than Mittag-Leffler (1903a) and aimed to produce the same results as Mittag-Leffler for $E_{a,b}(z)$ with $b = 1$ and real $a > 0$. Barnes (1906, pp. 285-289) slightly modified the integral in (4.38) to

$$I_B = \frac{1}{2\pi i}\int_{C_B} \frac{\sin(\pi(q-a)w)}{\sin \pi w}\Gamma(-aw) z^w dw$$

where the contour C_B embraces the positive half $\mathrm{Re}(s)$-plane and encloses the poles of $\Gamma(-as)$ and of $\sin \pi s$. By introducing the number q, the zeros of $\sin \pi(q-a)s$ at $s_\nu = \frac{\nu}{q-a}$ do not cancel the simple poles of $\Gamma(-as)$ at $s_m = \frac{m}{a}$, provided a does not divide q. Under the non-cancellation restriction for a, Cauchy's integral theorem indicates that

$$I_B = \sum_{m=0}^{\infty} \frac{(-1)^{m-1}}{a\, m!} \frac{\sin \pi \left(q\frac{m}{a} - m\right)}{z^{-\frac{m}{a}} \sin \pi \frac{m}{a}} - \sum_{m=0}^{\infty} (-z)^m \frac{\Gamma(-am)\sin(\pi(q-a)m)}{\pi}$$

The reflection formula (11.4) indicates that

$$\frac{\Gamma(-am)\sin(\pi(q-a)m)}{\pi} = -\frac{\sin \pi(q-a)m}{\sin \pi am \Gamma(1+am)}$$

Barnes now chooses q as an odd integer, i.e. $q = 2p+1$ and $p \in \mathbb{Z}$, such that $\sin \pi(q-a)m = (-1)^m \sin \pi am$ and

$$I_B = \sum_{m=0}^{\infty} \frac{(-1)^{m-1} z^{\frac{m}{a}}}{a\, m!} \frac{\sin \pi \left(q\frac{m}{a} - m\right)}{\sin \pi \frac{m}{a}} + \sum_{m=0}^{\infty} \frac{z^m}{\Gamma(1+am)}$$

By combining the integral expression and the residue expansion, we obtain with the Mittag-Leffler Taylor series (1.1),

$$E_a(z) = \frac{1}{a}\sum_{m=0}^{\infty} \frac{z^{\frac{m}{a}}}{m!} \frac{\sin \pi q\left(\frac{m}{a}\right)}{\sin \pi \frac{m}{a}} + \frac{1}{2\pi i}\int_{C_B} \frac{\sin \pi(q-a)w}{\sin \pi w}\Gamma(-aw) z^w dw$$

Only if $q = 2p+1$, then $\frac{\sin \pi q \frac{m}{a}}{\sin \pi \frac{m}{a}} = \sum_{j=-p}^{p} e^{2\pi i \frac{m}{a} j}$, that follows from $\sin\theta = \frac{e^{i\theta}-e^{-i\theta}}{2i}$ and the geometric series $\sum_{k=0}^{m-1} z^k = \frac{z^m-1}{z-1}$ in (1.16) for any complex z and any positive integer m. Reversing the finite j-sum and the infinite m-sum leads to Barnes' representation of the Mittag-Leffler function for $b=1$, real $a>0$ and integers $p \geq 0$,

$$E_a(z) = \frac{1}{a}\sum_{j=-p}^{p} e^{z^{\frac{1}{a}} e^{2\pi i \frac{j}{a}}} + \frac{1}{2\pi i}\int_{C_B} \Gamma(-aw)\frac{\sin \pi (2p+1-a) w}{\sin \pi w} z^w dw \tag{4.39}$$

which assumes that a does not divide $2p+1$.

If $a = 2p+1$ is an integer, then the integral in (4.39) vanishes and

$$E_{2p+1}(z) = \frac{1}{2p+1}\sum_{j=-p}^{p} e^{z^{\frac{1}{2p+1}} e^{2\pi i \frac{j}{2p+1}}}$$

is an instance of the cyclotomic property (2.17), while if $a = 2p+2$, then $\sin \pi (2p+1-a)s = -\sin \pi s$ and

$$\frac{1}{2\pi i}\int_{C_B} \Gamma(-aw)\frac{\sin \pi (2p+1-a) w}{\sin \pi w} z^w dw = -\frac{1}{2\pi i}\int_{C_B} \Gamma(-as) z^s ds$$
$$= \frac{1}{a}\sum_{m=0}^{\infty} \frac{(-1)^m z^{\frac{m}{a}}}{m!} = \frac{1}{a} e^{-z^{\frac{1}{a}}}$$

so that (4.39) becomes

$$E_{2(p+1)}(z) = \frac{1}{2(p+1)}\sum_{j=-p}^{p+1} e^{z^{\frac{1}{2(p+1)}} e^{2\pi i \frac{j}{2(p+1)}}}$$

which is the companion special case of the cyclotomic property (2.18).

79. *Barnes' contour integral in (4.39).* Barnes chooses the contour C_B as that in (4.38) with $\theta = \frac{\pi}{2}$. Using the asymptotic (14.8) of $\Gamma(b \pm ir)$ and defining $z = |z| e^{i \arg z}$ with $-\pi \leq \arg z \leq \pi$, the integrand $\mathcal{B}$ in (4.39) at a point w along the line $w = c + it$ behaves, for large t, as

$$\begin{aligned}
\mathcal{B} &= \Gamma(-ac-iat)\frac{\sin \pi (2p+1-a)(c+it)}{\sin \pi (c+it)} z^{c+it} \\
&= \sqrt{2\pi}\,(a|t|)^{-ac-\frac{1}{2}-iat} e^{-\frac{\pi}{2}a|t|} e^{\pm i\left(\left(-ac-\frac{1}{2}\right)\frac{\pi}{2}-at\right)}\left(1+O\left(\frac{1}{t}\right)\right) \\
&\quad\times \frac{z^c e^{it(\log|z|+i\arg z)}\left(e^{i\pi(2p+1-a)(c+it)} - e^{-i\pi(2p+1-a)(c+it)}\right)}{e^{i\pi(c+it)} - e^{-i\pi(c+it)}} \\
&= A\,|t|^{-ac} e^{\left(-\frac{\pi}{2}a|t| - t\arg z + \pi|2p+1-a||t| - \pi|t|\right)}\left(1+O\left(\frac{1}{t}\right)\right)
\end{aligned}$$

where A is a constant. Thus, for any finite real c, the integrand

$$|\mathcal{B}| = \left|\Gamma(-aw)\frac{\sin \pi (2p+1-a) w}{\sin \pi w} z^w\right|$$

tends, for $t \to \pm\infty$, to zero provided that

$$-\frac{\pi}{2}a + \pi\,|2p+1-a| + |\arg z| - \pi < 0 \tag{4.40}$$

for real $a > 0$ and integers $p \geq 0$. If $2p+1 > a$, then $|2p+1-a| = 2p+1-a$ and Barnes' condition (4.40) becomes $|\arg z| < \pi\left(\frac{3}{2}a - 2p\right)$ otherwise, if $2p+1 < a$, then $|2p+1-a| = a-2p-1$ and Barnes' condition (4.40) requires that $|\arg z| < \pi\left(2(p+1) - \frac{a}{2}\right)$. Thus, if $p = 0$ and $0 < a < 1$, then the contour integral in (4.39) exists for $|\arg z| < \frac{3\pi}{2}a$. If $p = 0$ and $a > 1$, then the contour integral in (4.39) exists for $|\arg z| < \pi\left(2 - \frac{a}{2}\right)$. These regimes agree with **art.** 70.

For larger integer p, Barnes' formula (4.39) becomes complicated. As shown in (4.13), some terms $e^{z^{\frac{1}{a}}e^{2\pi i\frac{j}{a}}}$ in the first sum are compensated by the contour integral and determining, which $j \in [-p,p]$ cancel, boils down to the same condition as in the first sum of (4.13). In any case, the determination of the asymptotic behavior of $E_a(z)$ in the sector $-\pi < \arg z \leq \pi$ requires a careful study for which we believe that the Wiman and Mittag-Leffler approach in **art.** 70 is more adequate.

4.4.2 The method of Paris and Kaminski for $E_a(z)$

80. Paris and Kaminski (2001, Section 5.1.4) start from Plana's form[4] (4.35) with $-1 < c < 0$ and apply the reflection formula (11.4) of the Gamma function,

$$E_{a,b}(z) = -\frac{1}{2\pi i}\int_{c+\infty e^{-i\theta}}^{c+\infty e^{i\theta}} \Gamma(1-(b+aw))\sin\pi(b+aw)\cot\pi w\; z^w dw \tag{4.41}$$

Paris and Kaminski (2001, Section 5.1.4) mention that their analysis is partly based on Barnes (1906, pp. 285-289), covered in **art.** 78 and **art.** 79. However, instead of skillfully considering a contour integral as Barnes did (**art.** 78), their idea is to use the trigonometric identity for arbitrary ξ and θ,

$$\frac{\sin\xi\theta}{\sin\theta} = 2\sum_{j=0}^{n-1}\cos(\xi-2j-1)\theta + \frac{\sin(\xi-2n)\theta}{\sin\theta} \tag{4.42}$$

Since we extend here the scope of b to all complex numbers, while Barnes (1906) as well as Paris and Kaminski (2001, Section 5.1.4) confined them-

[4] In fact, they alter the contour C_P over the negative $\operatorname{Re}(w)$-plane and limit themselves to $b = 1$,

$$E_a(z) = \frac{1}{2\pi i}\int_{C_P}\frac{z^{-w}}{\Gamma(1-aw)}\pi\cot\pi w\; dw$$

selves to $b = 1$, we also consider the companion

$$\frac{\cos \xi\theta}{\sin \theta} = \frac{\cos (\xi - 2n)\theta}{\sin \theta} - 2\sum_{j=0}^{n-1} \sin (\xi - 2j - 1)\theta \tag{4.43}$$

Formula (4.42), and similarly for (4.43), is deduced from

$$\frac{\sin (\xi + 2)\theta}{\sin \theta} = \frac{\sin \xi\theta \cos 2\theta + \sin 2\theta \cos \xi\theta}{\sin \theta}$$

After using $\cos 2\theta = 1 - 2\sin^2 \theta$ and $\sin 2\theta = 2\sin \theta \cos \theta$, we obtain

$$\frac{\sin (\xi + 2)\theta}{\sin \theta} = 2\cos (1 + \xi)\theta + \frac{\sin \xi\theta}{\sin \theta}$$

which is a recursion[5] in $U_\xi(w) = \frac{\sin \xi\theta}{\sin \theta}$. After n iterations of the recursion $U_\xi(w) = 2\cos (\xi - 1)\theta + U_{\xi-2}(w)$, we arrive at (4.42). Adding (4.42), after multiplication by $\cos \varphi$, and (4.43), after multiplication by $\sin \varphi$, results into the generalization of (4.42),

$$\frac{\sin (\varphi + \xi\theta)}{\sin \theta} = \frac{\sin (\varphi + (\xi - 2n)\theta)}{\sin \theta} + 2\sum_{j=0}^{n-1} \cos (\varphi + (\xi - 2j - 1)\theta) \tag{4.44}$$

while

$$\tfrac{\sin(\xi+1)\theta}{\sin \theta} = \sin \xi\theta \cot \theta + \cos \xi\theta \quad \text{and} \quad \tfrac{\cos(\xi+1)\theta}{\sin \theta} = \cos \xi\theta \cot \theta - \sin \xi\theta$$

similarly combine to

$$\frac{\sin (\varphi + (\xi + 1)\theta)}{\sin \theta} = \sin (\varphi + \xi\theta) \cot \theta + \cos (\varphi + \xi\theta)$$

Our first step is to convert the $\cot \pi w$ into a $\frac{1}{\sin \pi w}$ integrand by using the above identity

$$\sin (\varphi + \xi\theta) \cot \theta = \frac{\sin (\varphi + (\xi + 1)\theta)}{\sin \theta} - \cos (\varphi + \xi\theta)$$

followed by (4.44), after augmenting n and ξ by one,

$$\sin (\varphi + \xi\theta) \cot \theta = \frac{\sin (\varphi + (\xi - 2n - 1)\theta)}{\sin \theta} + 2\sum_{j=1}^{n} \cos (\varphi + (\xi - 2j)\theta) + \cos (\varphi + \xi\theta)$$

Substitution with $\varphi = \pi b$, $\theta = \pi w$ and $\xi = a$ into the contour integral

[5] If $\xi = m$ is a positive integer and $\pi w = \arccos x$, then $U_m(x)$ denotes the Chebyshev orthogonal polynomial of the second kind (Van Mieghem, 2023a, art. 378, 380).

(4.41) yields

$$E_{a,b}(z) = -\frac{1}{2\pi i}\int_{c+\infty e^{-i\theta}}^{c+\infty e^{i\theta}} \Gamma(1-(b+aw))\cos(\pi(b+aw))\ z^w dw$$
$$-\frac{2}{2\pi i}\sum_{j=1}^{n}\int_{c+\infty e^{-i\theta}}^{c+\infty e^{i\theta}} \Gamma(1-(b+aw))\cos(\pi(b+(a-2j)w))\, z^w dw$$
$$-\frac{1}{2\pi i}\int_{c+\infty e^{-i\theta}}^{c+\infty e^{i\theta}} \Gamma(1-(b+aw))\frac{\sin(\pi(b+(a-2n-1)w))}{\sin\pi w} z^w dw$$

The integral

$$J_j = \frac{1}{2\pi i}\int_{c+\infty e^{-i\theta}}^{c+\infty e^{i\theta}} \Gamma(1-b-aw)\cos\pi(b+(a-2j)w)\, z^w dw$$

is rewritten with $\cos\theta = \frac{e^{i\theta}+e^{-i\theta}}{2}$ as

$$J_j = \frac{e^{i\pi b}}{2\pi i}\int_{c+\infty e^{-i\theta}}^{c+\infty e^{i\theta}} \Gamma(1-b-aw)\left(ze^{i\pi(a-2j)}\right)^w dw$$
$$+\frac{e^{-i\pi b}}{2\pi i}\int_{c+\infty e^{-i\theta}}^{c+\infty e^{i\theta}} \Gamma(1-b-aw)\left(ze^{-i\pi(a-2j)}\right)^w dw$$

Each integral can be evaluated by closing the contour over the positive $\mathrm{Re}(w)$-plane, where simple poles of $\Gamma(1-b-aw)$ at $1-b-aw=-k$ with integer $k \geq 0 > c > -1$, i.e. at $\frac{k+1-b}{a} = w_k$ that obey $\mathrm{Re}(w_k) > c$, are enclosed in clockwise direction. Cauchy's residue theorem then states that

$$\frac{1}{2\pi i}\int_{c+\infty e^{-i\theta}}^{c+\infty e^{i\theta}} \Gamma(1-b-aw)\, y^w dw = -\sum_{k>ac+\mathrm{Re}\,b-1}^{\infty} y^{\frac{1-b+k}{a}} r_k$$

where the residue r_k is computed, using the functional equation (10.1) of the Gamma function, as

$$r_k = \lim_{w\to\frac{1-b+k}{a}}\left(w-\frac{1-b+k}{a}\right)\Gamma(1-b-aw)$$
$$= \frac{1}{a}\lim_{w\to\frac{1-b+k}{a}}(aw+b-(1+k))\frac{\Gamma(k+2-(b+aw))}{(-1)^{k+1}\prod_{l=1}^{k+1}((b+aw)-l)} = \frac{1}{a}\frac{(-1)^{k+1}}{k!}$$

and we obtain

$$\frac{1}{2\pi i}\int_{c+\infty e^{-i\theta}}^{c+\infty e^{i\theta}} \Gamma(1-b-aw)\, y^w dw = \frac{y^{\frac{1-b}{a}}}{a}\sum_{k>ac+\mathrm{Re}\,b-1}^{\infty}\frac{(-1)^k y^{\frac{k}{a}}}{k!}$$
$$= \frac{y^{\frac{1-b}{a}}}{a}\left(e^{-y^{\frac{1}{a}}} - \sum_{k=0}^{ac+\mathrm{Re}\,b-2}\frac{(-1)^k y^{\frac{k}{a}}}{k!}\right)$$

At this moment, our derivation deviates from Barnes (1906), who confined himself to $b=1$. In order to avoid the finite k-sum, we consider the more general and flexible integral in (4.36), where the intersection point c' of the path of integration with the $\mathrm{Re}(w)$-axis can be chosen at will. If we choose c' in such a way that all simples poles of $\Gamma(1-b-aw)$ are enclosed, then the above finite k-sum disappears. We enclose all simple poles provided $\mathrm{Re}\, w_0 > c'$, thus, $\frac{1-\mathrm{Re}(b)}{a} > c'$. If $b=1$ as in Barnes, then $c' = c$ and $-1 < c < 0$ and the above finite k-sum vanishes. If $\mathrm{Re}(b) < 0$, then the line at c already encloses all poles. Only if $\mathrm{Re}(b) > 1$, then we shift the line from c towards $-(m+1) < c' < -m$, where $m = \left\lfloor \frac{\mathrm{Re}(b)-1}{a} \right\rfloor$ and $\lfloor x \rfloor$ is the integer part of x. If $\mathrm{Re}(b) > 1$, then we can always choose the line of integration at $-\left\lfloor \frac{\mathrm{Re}(b)-1}{a} \right\rfloor - 1 < c' < -\left\lfloor \frac{\mathrm{Re}(b)-1}{a} \right\rfloor$ to enclose all simple poles of $\Gamma(1-b-aw)$, in which case the integral is

$$J_j = -\frac{z^{\frac{1-b}{a}}}{a}\left(e^{-i\frac{2\pi j(1-b)}{a}} e^{z^{\frac{1}{a}} e^{-i\frac{2\pi j}{a}}} + e^{i\frac{2\pi j(1-b)}{a}} e^{z^{\frac{1}{a}} e^{i\frac{2\pi j}{a}}}\right)$$

In summary, with $-(m+1) < c' < -m$ and $m \geq \left\lfloor \frac{\mathrm{Re}(b)-1}{a} \right\rfloor 1_{\{\mathrm{Re}(b)>1\}}$, where the indicator $1_{\{X\}} = 1$ if the condition or event X is true, else $1_{\{X\}} = 0$, we arrive at the generalization of Paris and Kaminski and of Barnes towards complex b,

$$E_{a,b}(z) = \frac{z^{\frac{1-b}{a}}}{a}\sum_{j=-n}^{n} e^{i\frac{2\pi j(1-b)}{a}} e^{z^{\frac{1}{a}} e^{\frac{2\pi i j}{a}}} - \sum_{k=1}^{m}\frac{1}{\Gamma(b-ka)}\frac{1}{z^k}$$
$$\frac{1}{2\pi i}\int_{c'+\infty e^{-i\theta}}^{c'+\infty e^{i\theta}} \frac{\sin(\pi(b+(a-2n-1)w))}{\sin \pi w}\Gamma(1-b-aw)\, z^w dw \tag{4.45}$$

The integral (4.45) is another representation of Wiman's generalization in (4.13) for any complex b. The integral (4.45) reduces for $b=1$ to (4.39) with $n=p$ and to Paris and Kaminski (2001, (5.1.21),(5.1.22)).

Similarly as Barnes (**art.** 79), Paris and Kaminski (2001, pp. 188-190) proceed to estimate the remaining integral (4.45) and provide a slightly more rigorous analysis than Wiman's in **art.** 60. They also explain the behavior of $E_a(z)$ around the Stokes lines (**art.** 71) at the rays $\arg z = \pm\pi a$.

81. *Verification of the contour integral in (4.45).* The remaining integral in (4.45) encloses the simple poles of $\frac{1}{\sin \pi w}$ at $w = k$, for integers $k \geq 0$ and the simple poles of $\Gamma(1-b-aw)$ at $\frac{k+1-b}{a} = w_k$, provided those poles are not zeros of $\sin(\pi(b+(a-2n)w))$, that appear at $b+(a-2n)w = \nu \in \mathbb{Z}$, i.e. at $w_\nu = \frac{b-\nu}{2n-a}$. Now, $w_k = w_\nu$ if $\frac{k+1-b}{a} = \frac{b-\nu}{2n-a}$ or

$$\frac{2n(b-(k+1))}{a} + (k+1) = \nu$$

which is only possible if $a | (b-(k+1))$. For example, if $b=1$, then $a|k$ implying that a must be an integer. If simple poles at w_k are different from simple zeros at w_ν and recalling that c' is chosen in such a way that all poles of $\Gamma(1-b-aw)$ are enclosed, then by Cauchy's residue theorem,

$$I_0 = -\frac{1}{2\pi i}\int_{c'+\infty e^{-i\theta}}^{c'+\infty e^{i\theta}} \frac{\sin(\pi(b+(a-2n-1)w))}{\sin \pi w}\Gamma(1-b-aw)z^w dw$$

$$= \frac{1}{a}\sum_{k=0}^{\infty} \frac{\sin\left(\pi\left(b+(a-2n-1)\frac{1-b+k}{a}\right)\right)}{\sin \pi\left(\frac{1-b+k}{a}\right)} z^{\frac{1-b+k}{a}} \frac{(-1)^{k+1}}{k!}$$

$$+\frac{1}{\pi}\sum_{j=-m}^{\infty} (-1)^j \sin(\pi(b+(a-2n-1)j))\Gamma(1-b-aj)z^j$$

After simplification and invoking the reflection formula (11.4) of the Gamma function, we find

$$I_0 = -\frac{z^{\frac{1-b}{a}}}{a}\sum_{k=0}^{\infty} \frac{\sin\pi\left((2n+1)\frac{1-b+k}{a}\right)}{\sin\pi\left(\frac{1-b+k}{a}\right)}\frac{z^{\frac{k}{a}}}{k!} + \sum_{j=-m}^{-1}\frac{z^j}{\Gamma(b+aj)} + \sum_{j=0}^{\infty}\frac{z^j}{\Gamma(b+aj)}$$

Substituted into (4.45) leads to

$$E_{a,b}(z) = \frac{z^{\frac{1-b}{a}}}{a}\left(\sum_{j=-n}^{n} e^{i\frac{2\pi j(1-b)}{a}} e^{z^{\frac{1}{a}} e^{\frac{2\pi i j}{a}}} - \sum_{k=0}^{\infty}\frac{\sin\pi\left((2n+1)\frac{1-b+k}{a}\right)}{\sin\pi\left(\frac{1-b+k}{a}\right)}\frac{z^{\frac{k}{a}}}{k!}\right)$$

$$+E_{a,b}(z)$$

implying that

$$\sum_{j=-n}^{n} e^{i\frac{2\pi j(1-b)}{a}} e^{z^{\frac{1}{a}} e^{\frac{2\pi i j}{a}}} = \sum_{k=0}^{\infty}\frac{\sin\pi\left((2n+1)\frac{1-b+k}{a}\right)}{\sin\pi\left(\frac{1-b+k}{a}\right)}\frac{z^{\frac{k}{a}}}{k!}$$

which is with $\frac{\sin\pi(2p+1)\frac{m}{a}}{\sin\pi\frac{m}{a}} = \sum_{j=-p}^{p} e^{2\pi i\frac{m}{a}j}$, indeed, an identity.

4.4.3 Mittag-Leffler function $E_{a,b}(z)$ *and* $0<a<1$

82. *Confinement of (4.33) to* $0<a<1$. We illustrate that the integral (4.33) can be evaluated under certain restriction of the parameter a. We assume here complex z and b, but a is real and positive.

If we choose $\theta = \frac{\pi}{2}$ in (4.33), then we must restrict $0<a<2$ due to (14.8) as explained in **art.** 76. In the case that $0<a<2$, we evaluate the contour in (4.33) along the line $w=c+it$,

$$E_{a,b}(-z) = -\frac{1}{2}\int_{-\infty}^{\infty}\frac{1}{\sin\pi(c+it)}\frac{z^{c+it}}{\Gamma(b+ac+iat)}dt \qquad \text{for } -1<c<0$$

Since $c<0$ and $0<a<2$, the reflection formula (11.4) yields

$$\frac{1}{\Gamma(b-a|c|+ait)} = -\Gamma(1-b+a|c|-ait)\frac{\sin\pi(a|c|-b-ait)}{\pi}$$

and

$$E_{a,b}(-z) = \frac{1}{2\pi}\int_{-\infty}^{\infty}\frac{\sin\pi(a|c|-b-ait)}{\sin\pi(-|c|+it)}\Gamma(a|c|+1-b-ait)z^{-|c|+it}dt$$

We change the sign of c and obtain, for $0 < c < 1$,

$$E_{a,b}(-z) = -\frac{1}{2\pi}\int_{-\infty}^{\infty} \frac{\sin\pi(ac-b-ait)}{\sin\pi(c-it)} \frac{\Gamma(ac+1-b-ait)}{z^{c-it}}dt$$

Similarly as Barnes's condition (4.40), convergence of the integral follows from (14.8) provided that $|\arg z| < \pi\left(1-\frac{a}{2}\right)$.

In order to replace $\Gamma(ac+1-b-ait)$ by Euler's integral (10.2), we must require that $\operatorname{Re}(ac+1-b) > 0$ or $ac+1 > \operatorname{Re}(b)$ for $0 < c < 1$,

$$E_{a,b}(-z) = -\frac{1}{2\pi}\int_{-\infty}^{\infty} dt\, \frac{\sin\pi(a(c-it)-b)}{\sin\pi(c-it)} \frac{1}{z^{c-it}}\int_0^{\infty} u^{a(c-it)-b}e^{-u}du$$

If $|\arg z| < \pi\left(1-\frac{a}{2}\right)$ in addition to $\operatorname{Re}(ac+1-b) > 0$, then both integrals converge absolutely. We can interchange the integrals by absolute convergence, provided that $0 < a < 1$ and that $|\arg z| < \pi\left(1-\frac{a}{2}\right)$,

$$E_{a,b}(-z) = -\int_0^{\infty} u^{-b}e^{-u}du\left(\frac{1}{2\pi}\int_{-\infty}^{\infty} \frac{\sin\pi(a(c-it)-b)}{\sin\pi(c-it)}\left(\frac{u^a}{z}\right)^{c-it}dt\right) \tag{4.46}$$

The integral between brackets in (4.46) can be recasted with $y = \frac{u^a}{z}$ as

$$Q_{a,b}(y) = \frac{1}{2\pi}\int_{-\infty}^{\infty} \frac{\sin\pi(a(c-it)-b)}{\sin\pi(c-it)} y^{c-it}dt \tag{4.47}$$

$$= \frac{1}{2\pi i}\int_{c-i\infty}^{c+i\infty} \frac{\sin\pi(aw-b)}{\sin\pi w} y^{w}dw$$

for $0 < c < 1$ and (4.46) becomes

$$E_{a,b}(-z) = -\int_0^{\infty} u^{-b}e^{-u}Q_{a,b}\left(\frac{u^a}{z}\right)du \tag{4.48}$$

In **art.** 83 below, we prove for complex y with $|\arg y| < \pi(1-a)$, any b and $0 < a < 1$ that

$$Q_{a,b}(y) = \frac{1}{\pi}\left(\frac{y\sin\pi(a-b) - y^2\sin\pi b}{1+2y\cos\pi a + y^2}\right) \tag{4.49}$$

Substitution of (4.49) into (4.46) yields, for $0 < a < 1$, $a+1 > \operatorname{Re}(b)$ and complex z with $|\arg z| < \pi\left(1-\frac{a}{2}\right)$,

$$E_{a,b}(-z) = \frac{1}{\pi}\int_0^{\infty} u^{-b}e^{-u}\left(\frac{-\frac{u^a}{z}\sin\pi(a-b) + \left(\frac{u^a}{z}\right)^2\sin\pi b}{1+2\frac{u^a}{z}\cos\pi a + \left(\frac{u^a}{z}\right)^2}\right)du \tag{4.50}$$

The integral (4.50) is essentially the same as the integral in (4.14) without residue sum.

We present several reformulations of the integral (4.50). A first rewriting of (4.50)

$$E_{a,b}(-z) = \frac{1}{\pi}\int_0^{\infty} u^{a-b}e^{-u}\left(\frac{-z\sin\pi(a-b) + u^a\sin\pi b}{(z\sin\pi a)^2 + (z\cos\pi a + u^a)^2}\right)du \tag{4.51}$$

illustrates that the denominator is always positive for real z. Second, let $u = z^{\frac{1}{a}}t$ in (4.50), then we must additionally require that $\operatorname{Re}\left(z^{\frac{1}{a}}\right) > 0$, thus $\operatorname{Re}\left(r^{\frac{1}{a}}e^{i\frac{\theta}{a}}\right) = r^{\frac{1}{a}}\cos\left(\frac{\theta}{a}\right) > 0$. We obtain, for $\operatorname{Re}\left(z^{\frac{1}{a}}\right) > 0$, $0 < a < 1$ and $a + 1 > \operatorname{Re}(b)$,

$$E_{a,b}(-z) = \frac{z^{\frac{1-b}{a}}}{\pi}\int_0^\infty t^{-b}e^{-z^{\frac{1}{a}}t}\left(\frac{-t^a\sin\pi(a-b) + t^{2a}\sin\pi b}{1 + 2t^a\cos\pi a + t^{2a}}\right)dt \quad (4.52)$$

83. *The integral $Q_{a,b}(y)$.* The contour of the integral $Q_{a,b}(y)$ in (4.47) for complex $y = |y|\,e^{i\arg y}$ can be closed over either half-plane, because the limit

$$\lim_{w\to\infty}\frac{\sin\pi(aw-b)}{\sin\pi w}y^w = \lim_{|w|\to\infty}Ae^{\{\pi(a-1)+|\arg y|\}|w||\sin\varphi|}\,|y|^{|w|\cos\varphi}$$

where $A = e^{i(|w|(\{\arg y - \pi(a-1)\}\cos\varphi + \log|y|\sin\varphi) + \pi b)}$ has modulus a constant $|A|$, vanishes if $|y| < 1$, $\cos\varphi > 0$, $a < 1$ and $|\arg y| < \pi(1-a)$ and for all complex, finite b. If $|y| < 1$, then we close the contour with $0 < c < 1$ in (4.47) over $\operatorname{Re}(w) > 0$-plane and obtain

$$\frac{1}{2\pi i}\int_{c-i\infty}^{c+i\infty}\frac{\sin\pi(aw-b)}{\sin\pi w}y^w dw = -\frac{1}{\pi}\sum_{k=1}^\infty(-1)^k\sin\pi(ak-b)\,y^k$$

Replacing $\sin\pi(ak-b) = \frac{e^{i\pi(ak-b)}}{2i} - \frac{e^{-i\pi(ak-b)}}{2i}$ and executing the geometric sums for $|y| < 1$ results in

$$\begin{aligned}\frac{1}{2\pi i}\int_{c-i\infty}^{c+i\infty}\frac{\sin\pi(aw-b)}{\sin\pi w}y^w dw &= \frac{y}{2\pi i}\left(\frac{e^{i\pi(a-b)}}{1+e^{i\pi a}y} - \frac{e^{-i\pi(a-b)}}{1+e^{-i\pi a}y}\right)\\ &= \frac{1}{\pi}\left(\frac{y\sin\pi(a-b) - y^2\sin\pi b}{1 + 2y\cos\pi a + y^2}\right)\end{aligned}$$

If $|y| > 1$, then we close the contour over the $\operatorname{Re}(w) < 0$-plane,

$$\begin{aligned}\frac{1}{2\pi i}\int_{c-i\infty}^{c+i\infty}\frac{\sin\pi(aw-b)}{\sin\pi w}y^w dw &= \frac{1}{\pi}\sum_{k=0}^\infty(-1)^k\sin\pi(-ak-b)\,y^{-k}\\ &= \frac{1}{\pi}\left(\frac{y\sin\pi(a-b) - y^2\sin\pi b}{1 + 2y\cos\pi a + y^2}\right)\end{aligned}$$

Both the case $|y| < 1$ and $|y| > 1$ lead to the same result (4.49) and $Q_{a,b}(1) = \frac{1}{2\pi}\left(\frac{\sin\pi(a-b) - \sin\pi b}{1+\cos\pi a}\right) = \frac{1}{2\pi}\frac{\sin\pi\left(\frac{a}{2}-b\right)}{\cos\frac{\pi a}{2}}$.

In summary, for any complex y with $|\arg y| < \pi(1-a)$, any complex b and for $0 < a < 1$, we have proved (4.49).

84. *The function $Q_{a,b}(u)$ for real u.* We deduce some properties of $Q_{a,b}(u)$.

a. From the definition (4.49), we readily deduce the functional equations, for any a, b and y,

$$Q_{a,b}(y) = Q_{a,b}\left(\frac{1}{y}\right) - \frac{\sin\pi b}{\pi}\left(\frac{y^2-1}{1 + 2y\cos\pi a + y^2}\right) \quad (4.53)$$

and

$$Q_{a,b}(y) = -yQ_{a,a-b}\left(\frac{1}{y}\right) \tag{4.54}$$

b. The contour in (4.47) for real $y = u$ can be rewritten as

$$\begin{aligned} Q_{a,b}(u) &= \frac{u^{\frac{a}{2}}}{\pi} \sin\pi\left(\frac{a}{2}-b\right)\int_0^\infty \frac{\cosh\pi at}{\cosh\pi t}\cos(at\log u)\,dt \\ &\quad - \frac{u^{\frac{a}{2}}}{\pi}\cos\pi\left(\frac{a}{2}-b\right)\int_0^\infty \frac{\sinh\pi at}{\cosh\pi t}\sin(at\log u)\,dt \end{aligned}$$

We show that both integrals can be computed. The functional equation (4.53) for $b = 1$ simplifies to $Q_{a,1}(y) = Q_{a,1}\left(\frac{1}{y}\right)$ and shows that

$$\int_0^\infty \frac{\sinh\pi at}{\cosh\pi t}\sin(at\log u)\,dt = \frac{u^a-1}{u^a+1}\tan\left(\frac{\pi a}{2}\right)\int_0^\infty \frac{\cosh\pi at}{\cosh\pi t}\cos(at\log u)\,dt$$

so that

$$\int_0^\infty \frac{\cosh\pi at}{\cosh\pi t}\cos(at\log u)\,dt = \frac{u^{-\frac{a}{2}}\pi\frac{Q_{a,b}(u)}{\sin\pi\left(\frac{a}{2}-b\right)}}{1-\frac{u^a-1}{u^a+1}\cot\pi\left(\frac{a}{2}-b\right)\tan\left(\frac{\pi a}{2}\right)}$$

whose left-hand side is independent of b, so that b can be chosen at will. The simplest choice[6] is $b = \frac{a}{2}$, then

$$\int_0^\infty \frac{\cosh\pi at}{\cosh\pi t}\cos(at\log u)\,dt = \frac{u^{\frac{a}{2}}(1+u^a)\cos\frac{\pi a}{2}}{1+2u^a\cos\pi a + u^{2a}}$$

c. Suppose that we ignore the restriction that the k-sum in

$$Q_{a,b}(t^a) = \frac{1}{\pi}\left(\frac{t^a\sin\pi(a-b) - t^{2a}\sin\pi b}{1+2t^a\cos\pi a + t^{2a}}\right) = \frac{1}{\pi}\sum_{k=1}^\infty \frac{\sin\pi(ak-b)}{(-1)^{k-1}}t^{ak}$$

only converges for $t < 1$ and that we substitute the series formally back in (4.52) and change the order of integration and summation. Then, we find

$$\begin{aligned} E_{a,b}(-z) &= \frac{1}{\pi}\sum_{k=1}^\infty \frac{\sin\pi(ak-b)\,\Gamma(ak-b+1)}{(-z)^k} = -\sum_{k=1}^\infty \frac{1}{(-z)^k\,\Gamma(b-ak)} \\ &= \frac{1}{\Gamma(b)} - E_{-a,b}\left(-\frac{1}{z}\right) \end{aligned}$$

which is precisely equal to (4.11), in spite of the divergence of the series!

[6] It is verified that

$$\frac{\sin\pi(a-b) - u^a\sin\pi b}{\sin\pi\left(\frac{a}{2}-b\right) + \frac{1-u^a}{1+u^a}\cos\pi\left(\frac{a}{2}-b\right)\tan\left(\frac{\pi a}{2}\right)} = \cos\left(\frac{\pi a}{2}\right)(1+u^a)$$

4.4.4 Integral for $E_a(z)$ with $0 < a < 1$

Due to its appearance in various problems, described by fractional calculus, we treat the Mittag-Leffler function $E_a(z) = E_{a,1}(z)$ for $b = 1$ in more depth.

If $b = 1$, then the integral (4.52) simplifies, for and $\text{Re}\left(z^{\frac{1}{a}}\right) > 0$, to

$$E_a(-z) = \frac{\sin \pi a}{\pi} \int_0^\infty \frac{t^{a-1} e^{-z^{\frac{1}{a}} t}}{1 + 2t^a \cos \pi a + t^{2a}} dt \qquad \text{for } 0 < a < 1 \quad (4.55)$$

which is listed in Berberan-Santos (2005b, eq. (34)) for non-negative real $z = x$ and which is deducible from (4.15). As shown in **art.** 68, the companion of (4.55) for $\text{Re}\left(z^{\frac{1}{a}}\right) > 0$ follows from (4.18),

$$E_a(z) = \frac{e^{z^{\frac{1}{a}}}}{a} - \frac{\sin \pi a}{\pi} \int_0^\infty \frac{t^{a-1} e^{-z^{\frac{1}{a}} t}}{t^{2a} - 2t^a \cos(a\pi) + 1} dt \qquad \text{for } 0 < a < 2 \quad (4.56)$$

We present several variations of the integral (4.55).

85. *Integrals deduced from (4.55) by Berberan-Santos for real x.* Berberan-Santos (2005b, eq. (35)) mentions[7] that partial integration of (4.55) results in

$$E_a(-x) = 1 - \frac{1}{2a} + \frac{x^{\frac{1}{a}}}{\pi a} \int_0^\infty \arctan\left(\frac{u^a + \cos(\pi a)}{\sin(\pi a)}\right) e^{-x^{\frac{1}{a}} u} du \qquad (4.57)$$

Indeed, let $y = t^a$ in (4.55), then

$$E_a(-x) = \frac{\sin \pi a}{\pi a} \int_0^\infty \frac{e^{-x^{\frac{1}{a}} y^{\frac{1}{a}}}}{1 + 2y \cos \pi a + y^2} dy$$

With $\frac{1}{1+2y\cos \pi a + y^2} = \frac{1}{2i \sin \pi a}\left(\frac{1}{e^{-i\pi a}+y} - \frac{1}{e^{i\pi a}+y}\right)$, we have

$$E_a(-x) = \frac{1}{2\pi i a} \int_0^\infty \left(\frac{1}{e^{-i\pi a} + y} - \frac{1}{e^{i\pi a} + y}\right) e^{-x^{\frac{1}{a}} y^{\frac{1}{a}}} dy \qquad (4.58)$$

[7] The integral $E_a(-x) = 1 - \frac{1}{2a} + \frac{x^{\frac{1}{a}}}{\pi} \int_0^\infty \arctan\left(\frac{u^a + \cos(\pi a)}{\sin(\pi a)}\right) e^{-x^{\frac{1}{a}} u} du$ in Berberan-Santos (2005b, eq. (35)) misses a factor of $\frac{1}{a}$ before the integral.

as well as

$$\begin{aligned}\int_0^u \frac{dy}{1+2y\cos\pi a+y^2} &= \frac{1}{2i\sin\pi a}\int_0^u\left(\frac{1}{e^{-i\pi a}+y}-\frac{1}{e^{i\pi a}+y}\right)dy \\ &= \frac{1}{2i\sin\pi a}\left(\log\frac{u+e^{-i\pi a}}{e^{-i\pi a}}-\log\frac{u+e^{i\pi a}}{e^{i\pi a}}\right) \\ &= -\frac{i}{2\sin\pi a}\log\frac{1+u\cos\pi a+iu\sin\pi a}{1+u\cos\pi a-iu\sin\pi a} \\ &= \frac{i}{2\sin\pi a}\log\frac{1-i\frac{u\sin\pi a}{1+u\cos\pi a}}{1+i\frac{u\sin\pi a}{1+u\cos\pi a}} \\ &= \frac{1}{\sin\pi a}\arctan\frac{u\sin\pi a}{1+u\cos\pi a}\end{aligned}$$

because $\arctan z=\frac{i}{2}\log\frac{1-iz}{1+iz}$. After substitution of $y=t^a$ for $a>0$, we find

$$\int_0^{u^{\frac{1}{a}}}\frac{t^{a-1}dt}{1+2t^a\cos\pi a+t^{2a}}=\frac{1}{a\sin\pi a}\arctan\frac{u\sin\pi a}{1+u\cos\pi a} \tag{4.59}$$

and similarly,

$$\int_0^{u^{\frac{1}{a}}}\frac{t^{a-1}dt}{1-2t^a\cos\pi a+t^{2a}}=\frac{1}{a\sin\pi a}\arctan\frac{u\sin\pi a}{1-u\cos\pi a} \tag{4.60}$$

Since $\lim_{u\to\infty}\arctan\frac{u\sin\pi a}{1+u\cos\pi a}=\pi a$, (4.59) shows that the integral (4.52) also holds for $z=0$ and equals $E_a(0)=1$.

With this preparation, partial integration of (4.55) yields

$$\begin{aligned}E_a(-x) &= \frac{1}{a\pi}\arctan\left(\frac{\sin\pi a}{\cos\pi a}\right)-\frac{x^{\frac{1}{a}}}{a\pi}\int_0^\infty\arctan\left(\frac{\sin\pi a}{u^a+\cos\pi a}\right)e^{-x^{\frac{1}{a}}u}du \\ &= 1-\frac{x^{\frac{1}{a}}}{a\pi}\int_0^\infty\arctan\left(\frac{\sin\pi a}{u^a+\cos\pi a}\right)e^{-x^{\frac{1}{a}}u}du\end{aligned}$$

After invoking $\arctan x=\frac{\pi}{2}-\arctan\frac{1}{x}$ for $x>0$, we arrive at (4.57).

Further, using the integral for $\arctan z=\int_0^z\frac{dt}{1+t^2}$,

$$1-E_a(-x)=\frac{x^{\frac{1}{a}}}{a\pi}\int_0^\infty du\int_0^{\frac{\sin\pi a}{u^a+\cos\pi a}}dt\frac{e^{-x^{\frac{1}{a}}u}}{1+t^2}$$

and reverse the integrals, provided that $0<a\le\frac{1}{2}$,

$$1-E_a(-x)=\frac{x^{\frac{1}{a}}}{a\pi}\int_0^{\tan\pi a}\frac{\int_0^{\left(\frac{\sin\pi a}{t}-\cos\pi a\right)^{\frac{1}{a}}}e^{-x^{\frac{1}{a}}u}du}{1+t^2}dt$$

Hence, we obtain, for $0<a\le\frac{1}{2}$,

$$E_a(-x)=\frac{1}{a\pi}\int_0^{\tan\pi a}\frac{e^{-\left(x\left(\frac{\sin\pi a}{t}-\cos\pi a\right)\right)^{\frac{1}{a}}}}{1+t^2}dt \tag{4.61}$$

After letting $u^a=\frac{\sin\pi a}{t}-\cos\pi a$ in (4.61), we retrieve (4.55) again.

86. *Integral representations of (4.55).* We transform (4.55) by various substitutions. The alternative form of (4.55), for $\text{Re}\left(z^{\frac{1}{a}}\right) > 0$ and $0 < a < 1$, is

$$E_a(-z) = \frac{\sin \pi a}{\pi} \int_0^\infty \frac{t^{a-1} e^{-z^{\frac{1}{a}} t}}{\sin^2 \pi a + (\cos \pi a + t^a)^2} dt \tag{4.62}$$

We split the integration interval in (4.55) into two parts $[0, 1]$ and $[1, \infty)$, invoke the symmetry (4.53) and obtain

$$E_a(-z) = \frac{\sin \pi a}{\pi} \int_0^1 \frac{u^{a-1}\left(e^{-z^{\frac{1}{a}} u^{-1}} + e^{-z^{\frac{1}{a}} u}\right)}{1 + 2u^a \cos \pi a + u^{2a}} du \tag{4.63}$$

or, alternatively,

$$E_a(-z) = \frac{\sin \pi a}{\pi} \int_1^\infty \frac{u^{a-1}\left(e^{-z^{\frac{1}{a}} u^{-1}} + e^{-z^{\frac{1}{a}} u}\right)}{1 + 2u^a \cos \pi a + u^{2a}} du$$

Substitution $u = \frac{1}{\sin^2 \pi a + (\cos \pi a + t^a)^2}$ in (4.62) results in

$$E_a(-z) = \frac{\sin \pi a}{2\pi a} \int_0^1 \frac{e^{-z^{\frac{1}{a}}\left(\sqrt{\frac{1}{u} - \sin^2 \pi a} - \cos \pi a\right)^{\frac{1}{a}}}}{\sqrt{u}\sqrt{1 - u \sin^2 \pi a}} du \tag{4.64}$$

which is only valid for $0 < a \leq \frac{1}{2}$, where $\cos \pi a \geq 0$.

Substitution of $v^2 = 1 - u \sin^2 \pi a$ in (4.64) leads, for $0 < a < 1$, to

$$E_a(-z) = \frac{1}{\pi a} \int_{\cos \pi a}^1 \frac{e^{-z^{\frac{1}{a}} \sin^{\frac{1}{a}} \pi a \left(\frac{v}{\sqrt{1-v^2}} - \cot \pi a\right)^{\frac{1}{a}}}}{\sqrt{1 - v^2}} dv \tag{4.65}$$

while letting $v = \cos \theta$ transforms (4.65) to

$$E_a(-z) = \frac{1}{\pi a} \int_0^{\pi a} e^{-(z \sin \pi a)^{\frac{1}{a}} (\cot \theta - \cot \pi a)^{\frac{1}{a}}} d\theta \tag{4.66}$$

Substitution of $t = \cot \theta$ in (4.66) yields

$$E_a(-z) = \frac{1}{\pi a} \int_{\cot \pi a}^\infty \frac{e^{-(z \sin \pi a)^{\frac{1}{a}} (t - \cot \pi a)^{\frac{1}{a}}}}{1 + t^2} dt \tag{4.67}$$

which is valid for $0 < a < 1$. The substitution $w = \frac{1}{t}$ in (4.67) leads to Berberan-Santos' representation (4.61), only valid for $0 < a \leq \frac{1}{2}$. Alternatively, the substitution $u = t - \cot \pi a$ in (4.67) brings us back to (4.55).

87. *Alternative form of the Laplace transform (3.10) for $0 < a < 1$ and $b = 1$.* For $\text{Re}(a) > 0$, $\text{Re}(c) > 0$, $\text{Re}(\beta) > 0$, $\text{Re}(s) > 0$ and any x and b, the Laplace transform in (3.10)

$$\int_0^\infty e^{-st} t^{c-1} E_{a,b}\left(-xt^\beta\right) dt = \sum_{k=0}^\infty \frac{\Gamma(c + \beta k)}{\Gamma(b + ak)} \frac{(-1)^k x^k}{s^{c+k\beta}}$$

is available as a series, whose convergence is studied in **art.** 42.

We substitute the integral in (4.66) into the left-hand side Laplace integral, reverse the integrals and we find, for $0 < a < 1$, $b = 1$, $\beta = a$, $x \geq 0$, $\text{Re}(s) \geq 0$ and $\text{Re}(c) > 0$,

$$\int_0^\infty e^{-st} t^{c-1} E_a(-xt^a)\, dt = \frac{\Gamma(c)}{\pi a} \int_0^{\pi a} \frac{d\theta}{\left(s + (x \sin \pi a)^{\frac{1}{a}} (\cot \theta - \cot \pi a)^{\frac{1}{a}}\right)^c} \tag{4.68}$$

The particular case for $s = 0$ in (4.68) becomes a Mellin transform, for all $0 < \text{Re}(c) < a$,

$$\int_0^\infty t^{c-1} E_a(-xt^a)\, dt = \frac{\Gamma(c)}{\pi a (x \sin \pi a)^{\frac{c}{a}}} \int_0^{\pi a} \frac{d\theta}{(\cot \theta - \cot \pi a)^{\frac{c}{a}}} \tag{4.69}$$

which is a variant of the Mellin transform (4.34), which is rewritten as

$$\int_0^\infty t^{ac-1} E_{a,b}(-xt^a)\, dt = \frac{\pi}{x^c \sin \pi c} \frac{1}{\Gamma(b - ac)}$$

valid for $0 < a < 2$, all b and $0 < \text{Re}(c) < 1$.

5
Probability theory and $E_{a,b}(z)$

Section 5.1 introduces the concept of complete monotonicity, which naturally connects to probability theory, via Laplace transformations. In Section 5.2, integrals of the Mittag-Leffler function are associated to both a probability density and a probability generating function, from which we introduce the Mittag-Leffler random variable M_a. Properties such as the sum of i.i.d. Mittag-Leffler random variables and the scaled limit sums, that lead to stable distributions, are deduced in Section 5.3. Section 5.4 links the Weibull and Mittag-Leffler random variables.

5.1 Complete monotonicity

88. *Complete monotonicity of a function.* Widder (1946, Chapter IV) devotes an entire chapter to absolutely and completely monotone functions. The Hausdorff–Bernstein–Widder[1] theorem (Widder, 1946, p. 161, Theorem 12a) states that a necessary and sufficient condition, that a function $\varphi(x)$ on $[0,\infty)$ should be completely monotonic, is that there exists a bounded and non-decreasing function $f(u)$ such that the integral

$$\varphi(x) = \int_0^\infty e^{-xu} df(u)$$

converges for all real $x \geq 0$. In other words, a function $\varphi(x)$ is completely monotonic on $[0,\infty)$ if and only if $\varphi(x)$ is a Laplace transform of a bounded and non-decreasing measure $f(u)$.

Since any probability distribution function $F_X(x) = \Pr[X \leq x]$ of a random variable X is bounded and non-decreasing in the real variable x, complete monotonicity translates (Van Mieghem, 2014, Chapter 2) to

$$\varphi_X(x) = \int_0^\infty e^{-xu} dF_X(u) = E\left[e^{-xX}\right]$$

[1] Widder (1946, p. 144) mentions that Hausdorff, Bernstein and himself have independently proved the theorem.

where $\varphi_X(x)$ is the probability generating function of a *non-negative* random variable X. Clearly, $\varphi_X(0) = \lim_{x\to\infty} F_X(x) - F_X(0) = 1$ for a non-negative random variable. Moreover, if $F_X(x)$ is differentiable for all $x > 0$, then the probability density function $f_X(x) = \frac{dF_X(x)}{dx}$ of the non-negative random variable X is continuous in x and $\varphi_X(x)$ is the Laplace transform of $f_X(x)$. Hence, complete monotonicity of a function $\varphi(x)$ on $[0,\infty)$ implies the existence of a non-negative random variable X with probability generating function $\varphi_X(x) = E\left[e^{-xX}\right]$.

The fact that $(-1)^n \frac{d^n\varphi(x)}{dx^n} = \int_0^\infty x^n e^{-xu} df(u) \geq 0$ for all non-negative integers n is a direct consequence of the Hausdorff–Bernstein–Widder theorem, but the condition $(-1)^n \frac{d^n\varphi(x)}{dx^n} \geq 0$ for all non-negative integers n is also sufficient (Bernstein, 1929, pp. 56-59) and thus an equivalent statement for the complete monotonicity of a function $\varphi(x)$.

89. *Monotonicity of $E_a(-x)$ for $0 < a \leq 1$.* The integral $E_a(-x) = \frac{\sin\pi a}{\pi}\int_0^\infty \frac{t^{a-1}e^{-x^{\frac{1}{a}}t}}{1+2t^a\cos\pi a+t^{2a}}dt$ in (4.55), valid for $0 < a < 1$ and for real $x \geq 0$ (see **art.** 85), possesses a positive integrand, because

$$(1-t^a)^2 < 1 + 2t^a\cos\pi a + t^{2a} = \sin^2\pi a + (\cos\pi a + t^a)^2 < (1+t^a)^2$$

and thus shows that $E_a(-x) > 0$ for $0 < a < 1$. Since $E_1(-x) = e^{-x}$, it holds that $E_a(-x) > 0$ for $0 < a \leq 1$ and that $E_a(-x)$ for $0 < a \leq 1$ is completely monotonic for real $x \geq 0$. Since $E_a(-x)$ is strict decreasing for real x, the complete monotonicity indicates that $E_a(z)$ for $0 < a \leq 1$ has no zeros on the negative real axis.

For $b = a$, the integral (4.52) reduces to

$$x^{1-\frac{1}{a}}E_{a,a}(-x) = \frac{\sin\pi a}{\pi}\int_0^\infty \frac{t^a e^{-x^{\frac{1}{a}}t}}{1+2t^a\cos\pi a + t^{2a}}dt \tag{5.1}$$

illustrating similarly that also $E_{a,a}(-x)$ is complete monotonic for $0 < a \leq 1$ and real $x \geq 0$.

The monotonicity of $E_a(-x)$ for $0 < a \leq 1$ was first conjectured by Feller (1949, Section 7) and later proved by Pollard (1948). Pollard (1948) introduces $\frac{1}{t+z} = \int_0^\infty e^{-s(t+z)}ds$ in Mittag-Leffler's integral (4.26) and obtains a Laplace transform,

$$E_{a,b}(-z) = \int_0^\infty e^{-sz}\left(\frac{1}{2\pi i a}\int_{C_{a\phi}} t^{\frac{1-b}{a}} e^{-st+t^{\frac{1}{a}}}dt\right)ds$$

from which he proves that $E_a(-x)$ is completely monotonic for real $x \geq 0$ and $0 < a \leq 1$, in the sense that $E_a(-x) = \int_0^\infty e^{-xt}dF_a(t)$, where $F_a(t)$ is nondecreasing, bounded and a probability distribution. Pollard (1948) also explicitly determined the non-negative function $F_a'(t)$. However, his proof is not easy and, therefore, omitted, but replaced by our derivation in Section 5.2. Berberan-Santos (2005b) shows that monotonicity, defined for all n by $(-1)^n \frac{d^n}{dx^n}E_a(-x) \geq 0$, follows from his function $g(t) > 0$ in **art.** 50 for $0 < a \leq 1$.

90. *Monotonicity of $E_{a,b}(-x)$ for $0 < a \leq 1$ and $b \geq a$.* Schneider (1996) extended the range of the parameter b by proving that $E_{a,b}(-x) > 0$ for $0 < a \leq 1$ and $b > a$. Schneider's proof is involved, based on Fox functions. Schneider (1996) also derives the corresponding probability measure. Miller and Samko (1997) presented a simple proof, which we include here.

The integral (3.28) with $\lambda = -\frac{z}{x^a}$, $b = a$ and $w = b - a$ becomes

$$E_{a,b}(-z) = \frac{x^{-a}}{\Gamma(b-a)} \int_0^x \left(1 - \frac{u}{x}\right)^{b-a-1} u^{a-1} E_{a,a}\left(-z\left(\frac{u}{x}\right)^a\right) du$$

After substitution $t = \left(\frac{u}{x}\right)^a$, we find, for $b > a > 0$,

$$E_{a,b}(-z) = \frac{1}{a\Gamma(b-a)} \int_0^1 \left(1 - t^{\frac{1}{a}}\right)^{b-a-1} E_{a,a}(-zt)\, dt \qquad (5.2)$$

Art. 89 demonstrates that the function $E_{a,a}(-x)$ is complete monotonic for $0 < a \leq 1$. Since the integrand in (5.2) is non-negative for positive, real z, we conclude that $E_{a,b}(-x) > 0$ and that $E_{a,b}(-x)$ is complete monotonic for $0 < a \leq 1$ and $b \geq a$.

91. *Other complete monotonic functions.* Simon (2015, Theorem 1.2) found that the functions

$$\begin{array}{ll} \frac{c^x}{a} - E_a(x^a) & \text{for } 0 < a \leq 1 \\ E_a(x^a) - \frac{e^x}{a} & \text{for } 1 < a \leq 2 \end{array}$$

are complete monotonic. Simon's proof is long and tedious. His results follow directly from $E_2(x^2) = \cosh(x)$ for $a = 2$ and from the integral (4.18), valid for real $x > 0$,

$$E_a(x^a) = \frac{e^x}{a} - \frac{\sin \pi a}{\pi} \int_0^\infty \frac{t^{a-1} e^{-xt}}{t^{2a} - 2t^a \cos(a\pi) + 1} dt \qquad \text{for } 0 < a < 2$$

because $\sin \pi a \geq 0$ for $0 < a \leq 1$ and $\sin \pi a \leq 0$ for $1 < a < 2$.

Subtracting the formula (4.18) for $E_a(x^a)$ from its derivative with respect to x,

$$\frac{dE_a(x^a)}{dx} = \frac{e^x}{a} + \frac{\sin \pi a}{\pi} \int_0^\infty \frac{t^a e^{-xt}}{t^{2a} - 2t^a \cos(a\pi) + 1} dt \qquad \text{for } 0 < a < 2$$

results in a linear differential equation[2], for real positive x and $0 < a < 2$,

$$\frac{dE_a(x^a)}{dx} - E_a(x^a) = \frac{\sin \pi a}{\pi} \int_0^\infty \frac{t^{a-1}(t+1)e^{-xt}}{t^{2a} - 2t^a \cos(a\pi) + 1} dt \tag{5.3}$$

The right-hand side is positive for $0 < a < 1$ and negative for $1 < a < 2$. In other words, (5.3) and $E_2(x^2) = \cosh(x)$ show that $E_a(x^a) - ax^{a-1}E'_a(x^a)$ is complete monotonic for $1 < a \leq 2$ as well as

$$ax^{a-1}E'_a(x^a) - \frac{e^x}{a} \qquad \text{for } 0 < a \leq 1$$
$$\frac{e^x}{a} - ax^{a-1}E'_a(x^a) \qquad \text{for } 1 < a \leq 2$$

Simon (2015) also extended the range of the parameter b by observing, similarly as in **art**. 90, that the function, whose integral follows from (3.28),

$$\Gamma(b-1)\left(\frac{1}{a}E_{1,b}(x) - E_{a,b}(x^a)\right) = \int_0^1 (1-u)^{b-2}\left(\frac{e^{xu}}{a} - E_a(x^a u^a)\right) du$$

is complete monotonic for $0 < a \leq 1$ and for $b > 1$. A similar reasoning holds for the regime $1 < a \leq 2$. Hence, Simon (2015) found that the extensions

$$E_{1,b}(x) - E_{a,b}(x^a) \qquad \text{for } 0 < a \leq 1 \text{ and } b > 1$$
$$E_{a,b}(x^a) - E_{1,b}(x) \qquad \text{for } 1 < a \leq 2 \text{ and } b > 1$$

are complete monotonic. The derivative version, containing $ax^{a-1}E'_a(x^a)$, can be analogously extended to $b > 1$.

By stochastic arguments relying on Lévy processes, Garrappa *et al.* (2024, Proposition 3) prove, for $1 \leq a \leq 2$ and $1 - a \leq b \leq a$, that $\frac{1}{E_{a,b}(x)}$ is complete monotonic for $x > 0$.

Ferreira (2025) shows, for $0 < a < 1$ and for integers $m > 0$, that $(-1)^{m-1} E_{a,a-m+1}(-x) \geq 0$ and that $E_{a,a-m+1}(-x)$ is bounded for real $x \geq 0$. If $m = 0$, then $E_{a,a+1}(-x)$ was shown to be complete monotonic in **art**. 90 for $b = a + 1$. Ferreira's finding directly follows from the integral (4.51), that simplifies, for $b = a - m + 1$ and for $m > 0$, to

$$(-1)^{m-1} E_{a,a-m+1}(-x) = \frac{\sin \pi a}{\pi} \int_0^\infty \frac{u^{a+m-1}e^{-u}}{(x \sin \pi a)^2 + (x \cos \pi a + u^a)^2} du$$

[2] Since $\frac{dE_a(x^a)}{dx} - E_a(x^a) = e^x \frac{d}{dx}\left(e^{-x}E_a(x^a)\right)$, (5.3) becomes

$$\frac{d}{dx}\left(e^{-x}E_a(x^a)\right) = \frac{\sin \pi a}{\pi} \int_0^\infty \frac{t^{a-1}(t+1)e^{-x(t+1)}}{t^{2a} - 2t^a \cos(a\pi) + 1} dt$$

Integration shows

$$e^{-x}E_a(x^a) - 1 = \frac{\sin \pi a}{\pi}\left(\int_0^\infty \frac{t^{a-1}dt}{t^{2a} - 2t^a \cos(a\pi) + 1} - \int_0^\infty \frac{t^{a-1}e^{-x(t+1)}dt}{t^{2a} - 2t^a \cos(a\pi) + 1}\right)$$

Invoking (4.60) leads to $\int_0^\infty \frac{t^{a-1}dt}{1 - 2t^a \cos \pi a + t^{2a}} = \frac{\pi}{\sin \pi a}\left(\frac{1}{a} - 1\right)$ (see also **art**. 115) and we retrieve (4.18).

5.2 Probability generating *and* distribution function

From the Mittag-Leffler function $E_a(-x)$ with $0 < a < 1$, we construct two probability density functions and show that $E_a(-s^a)$ is both a probability generating function and a probability distribution.

92. *A first probability density function.* After replacing x by s^a in (4.55), the Laplace transform

$$E_a(-s^a) = \frac{\sin \pi a}{\pi} \int_0^\infty \frac{t^{a-1} e^{-st}}{1 + 2t^a \cos \pi a + t^{2a}} dt$$

indicates that

$$f_a(t) = \frac{\sin \pi a}{\pi} \frac{t^{a-1}}{1 + 2t^a \cos \pi a + t^{2a}} \tag{5.4}$$

is a probability density function (pdf) for $t > 0$ and for $0 < a \leq 1$. Indeed, the Laplace transform (B.1) of the non-negative function $f_a(t)$ is (Van Mieghem, 2014) a probability generating function (pgf) $\varphi_X(z) = E\left[e^{-zX}\right]$ of a random variable X, provided $\varphi_X(0) = 1$. Moreover, with $f_a(1) = \frac{1}{2\pi} \tan \frac{\pi a}{2}$, the pdf (5.4) obeys the functional equation – recall that $Q_{a,1}(u) = Q_{a,1}\left(\frac{1}{u}\right)$ in (4.49) –

$$t f_a(t) = \frac{1}{t} f_a\left(\frac{1}{t}\right)$$

The function $f_a(t)$ is plotted for various values of a in Mainardi and Consiglio (2020, Fig. 1) and appears in linear viscoelasticity (Mainardi, 2021). Moreover, $\lim_{a \to 1} f_a(t) = \delta(t-1)$, where $\delta(t)$ is the Dirac impulse.

Its companion, that follows similarly from (5.1) as

$$s^{a-1} E_{a,a}(-s^a) = \frac{\sin \pi a}{\pi} \int_0^\infty \frac{t^a e^{-st}}{1 + 2t^a \cos \pi a + t^{2a}} dt$$

with non-negative function

$$g_a(t) = \frac{\sin \pi a}{\pi} \frac{t^a}{1 + 2t^a \cos \pi a + t^{2a}} = t f_a(t) = g_a\left(\frac{1}{t}\right)$$

is, in contrast to $E_a(-s^a)$, not a probability generating function, because $\lim_{s \to 0} s^{a-1} E_{a,a}(-s^a) = \infty$ for $a < 1$.

93. *The Mittag-Leffler random variable M_a.* The Laplace transform (3.13) with $b = a$ and $x = -1$

$$\int_0^\infty e^{-st} t^{a-1} E_{a,a}(-t^a)\, dt = \frac{1}{s^a + 1} \tag{5.5}$$

is only valid for $|s^a| > 1$. Hence[3], we cannot conclude from $\lim_{s \to 0} \frac{1}{s^a+1} = 1$ for $a > 0$ that (5.5) is a pgf $\varphi_{M_a}(s)$ of a random variable M_a. However, (2.39) taken into account (2.2) shows that

$$-\frac{d}{ds} E_a(-s^a) = s^{a-1} E_{a,a}(-s^a) \tag{5.6}$$

[3] This observation was communicated to me by Rui Ferreira.

Integration of (5.6) yields $\int_p^q t^{a-1} E_{a,a}(-t^a)\, dt = E_a(-p^a) - E_a(-q^a)$. If $p = 0$, then $E_{a,1}(0) = \frac{1}{\Gamma(1)} = 1$, while $\lim_{q\to\infty} E_a(-q^a) = 0$, as follows from the integral (4.55). Hence, it holds that

$$\int_0^\infty t^{a-1} E_{a,a}(-t^a)\, dt = 1$$

illustrating that $t^{a-1} E_{a,a}(-t^a)$ can be regarded as a probability density function and that the integral in (5.5) exists for $\mathrm{Re}(s) \geq 0$ and $a > 0$. Thus, by analytic continuation, the pgf in (5.5)

$$\varphi_{M_a}(s) = E\left[e^{-sM_a}\right] = \frac{1}{s^a + 1} \tag{5.7}$$

is valid for $\mathrm{Re}(s) \geq 0$, with corresponding pdf

$$f_{M_a}(t) = t^{a-1} E_{a,a}(-t^a) \qquad \text{for } 0 < a < 1 \tag{5.8}$$

of a random variable M_a and (5.6) demonstrates that the corresponding Mittag-Leffler distribution for $0 < a < 1$ is

$$F_{M_a}(t) = \Pr[M_a \leq t] = \int_0^t f_{M_a}(u)\, du = 1 - E_a(-t^a) \tag{5.9}$$

The mean $E[M_a] = -\varphi'_{M_a}(0) = \lim_{s\to 0} \frac{as^{a-1}}{(s^a+1)^2} = \infty$. In fact, for $0 < a < 1$, the pgf $\varphi_{M_a}(s)$ in (5.5) is not analytic at $s = 0$, implying that the Taylor series around $s = 0$ does not exist, nor any derivative. Hence, the Mittag-Leffler random variable M_a, defined by the pgf (5.5) and pdf (5.8) for $0 < a < 1$, does not possess any finite moment $E\left[M_a^k\right]$. In the limit $a \to 1$, the Mittag-Leffler random variable M_1 becomes an exponential random variable with mean 1.

94. *Asymptotic behavior of the Mittag-Leffler pdf (5.8).* For large t, the pdf $f_{M_a}(t)$ of the Mittag-Leffler random variable follows from (4.21), for real $x \geq 0$, $0 < a < 1$ and $K < \left\lfloor \frac{1}{a}\left(x^{\frac{1}{a}} + b\right)\right\rfloor$,

$$E_{a,b}(-x) = -\sum_{k=1}^{K} \frac{1}{x^k} \frac{(-1)^k}{\Gamma(b - ka)} + O\left(\frac{1}{z^{K+1}}\right)$$

that becomes with the reflection formula (11.4) of the Gamma function,

$$E_{a,b}(-x) = -\frac{1}{\pi} \sum_{k=1}^{K} \Gamma(1 - b + ka) \sin \pi (b - ka) \frac{(-1)^k}{x^k} + O\left(\frac{1}{z^{K+1}}\right)$$

Thus, for large t, the Mittag-Leffler pdf (5.8) is

$$f_{M_a}(t) = t^{a-1} E_{a,a}(-t^a) = \sum_{k=1}^{K-1} \frac{(-1)^{k-1}\Gamma(1 + ka) \sin \pi (ak)}{\pi t^{ak+1}} + O\left(\frac{1}{t^{aK+1}}\right) \tag{5.10}$$

$$= \frac{\sin \pi a \Gamma(a+1)}{\pi t^{a+1}} \left(1 - \frac{2\cos \pi a}{t^a} \frac{\Gamma(2a+1)}{\Gamma(a+1)} + O\left(\frac{1}{t^{2a}}\right)\right)$$

Differentiation of the asymptotic relation (4.16) yields the first term in (5.10). The same result (5.10) can also be deduced from the integral (4.51) after a tedious computation, invoking the geometric series (1.16), splitting the integration interval into $[0,t]$ and $[t,\infty)$ and bounding the latter integral over $[t,\infty)$ for large t.

5.3 Properties of the Mittag-Leffler random variable

95. *The sum of i.i.d. Mittag-Leffler random variables.* The sum $S_n = \sum_{j=1}^{n} (M_a)_j$ of n i.i.d. Mittag-Leffler random variables $(M_a)_1, \ldots, (M_a)_n$ with the same Mittag-Leffler distribution (5.9), has the pgf (Van Mieghem, 2014, p. 30)

$$\varphi_{S_n}(z) = E\left[e^{-zS_n}\right] = E\left[e^{-z\sum_{j=1}^{n}(M_a)_j}\right] = \varphi_{M_a}^n(z) = (1+z^a)^{-n} \quad (5.11)$$

Inverse Laplace transformation (B.2) returns the pdf of S_n,

$$f_{S_n}(t) = \frac{1}{2\pi i}\int_{c-i\infty}^{c+i\infty} \frac{e^{zt}}{(1+z^a)^n} dz \quad (5.12)$$

where $c > 1$ and $\text{Re}(a) > 0$.

The integral (5.12) can be computed from

$$t^{b-1} E_{a,b}(xt^a) = \frac{1}{2\pi i}\int_{c'-i\infty}^{c'+i\infty} \frac{s^{a-b}e^{st}}{s^a - x} ds$$

in (4.5), valid for $\text{Re}(a) > 0$ and $\text{Re}(b) > 0$ and where $c' > |x|^{\frac{1}{a}}$, by $(n-1)$ times differentiating with respect to x,

$$t^{b-1}\frac{d^{n-1}}{dx^{n-1}} E_{a,b}(xt^a) = \frac{(n-1)!}{2\pi i}\int_{c'-i\infty}^{c'+i\infty} \frac{s^{a-b}e^{st}}{(s^a - x)^n} ds$$

showing that

$$f_{S_n}(t) = \frac{1}{2\pi i}\int_{c-i\infty}^{c+i\infty} \frac{e^{zt}}{(1+z^a)^n} dz = \frac{t^{a-1}}{(n-1)!}\left.\frac{d^{n-1}}{dx^{n-1}} E_{a,a}(xt^a)\right|_{x=-1} \quad (5.13)$$

The chain rule with $z = xt^a$ indicates that $\frac{d}{dx}E_{a,b}(xt^a) = \frac{d}{dz}E_{a,b}(z)\frac{dz}{dx} = t^a \frac{d}{dz}E_{a,b}(z)$ and, thus, $\frac{d^{n-1}}{dx^{n-1}}E_{a,b}(xt^a) = t^{a(n-1)} \left.\frac{d^{n-1}}{dz^{n-1}}E_{a,b}(z)\right|_{z=xt^a}$. Invoking (2.52)

$$(az)^m \frac{d^m}{dz^m} E_{a,b}(z) = \sum_{j=0}^{m} q_j(a,b,m)\, E_{a,b-j}(z)$$

in (5.13) leads to

$$\left.\frac{d^{n-1}}{dx^{n-1}} E_{a,b}(xt^a)\right|_{x=-1} = \frac{1}{(-a)^{n-1}}\sum_{j=0}^{n-1} q_j(a,b,n-1)\, E_{a,b-j}(-t^a)$$

Finally, we find that the pdf of a sum $S_n = \sum_{j=1}^{n} (M_a)_j$ of n i.i.d. Mittag-Leffler random variables equals

$$f_{S_n}(t) = \frac{(-1)^{n-1} t^{a-1}}{(n-1)! a^{n-1}} \sum_{j=0}^{n-1} q_j(a, a, n-1) E_{a,a-j}(-t^a) \tag{5.14}$$

where the polynomial $q_j(a, b, m)$ is specified in (2.50).

Example If $a = 1$, then (5.14) simplifies with $q_j(1, 1, m) = \delta_{jm}$ in (2.67) to

$$f_{S_n}(t) = \frac{(-1)^{n-1}}{(n-1)!} E_{1,1-(n-1)}(-t)$$

Using $E_{1,1-m}(z) = z^m e^z$ for integers $m \geq 0$ in (2.7), leads, for $n \geq 1$, to

$$f_{S_n}(t) = \frac{t^{n-1} e^{-t}}{(n-1)!}$$

which is, indeed, the Erlang probability density function (Van Mieghem, 2014, eq. (3.2.24)) for the sum of n i.i.d. exponential random, all with mean 1.

96. *Limit law of a sum of i.i.d. Mittag-Leffler random variables.* We start from the pgf (5.11) of n i.i.d. Mittag-Leffler random variables and apply a classical scaling argument. If we choose the scaling parameter β in $\varphi_{S_n}(\beta z) = (1 + \beta^a z^a)^{-n}$ equal to $\beta^a = \frac{1}{n}$, then

$$\lim_{n\to\infty} \varphi_{S_n}\left(\frac{z}{n^{\frac{1}{a}}}\right) = \lim_{n\to\infty} E\left[e^{-z\left(n^{-\frac{1}{a}} \sum_{j=1}^{n} (M_a)_j\right)}\right] = \lim_{n\to\infty} \left(1 + \frac{z^a}{n}\right)^{-n} = e^{-z^a}$$

Hence, the random variable $R_a = \lim_{n\to\infty} n^{-\frac{1}{a}} \sum_{j=1}^{n} (M_a)_j$ possesses the pgf

$$\varphi_{R_a}(z) = E\left[e^{-zR_a}\right] = e^{-z^a} \tag{5.15}$$

The random variable R_a belongs to the class of stable[4] distributions. Since the pgf $\varphi_X(z)$ of a continuous, non-negative random variable X is a Laplace transform of its pdf $f_X(t)$, it holds that

$$\int_0^\infty e^{-zt} f_{R_a}(t)\, dt = e^{-z^a} \tag{5.16}$$

If $a \to 0$ and (5.15) tends to a constant independent of z, then R_0 must tend to zero implying that the random variable R_0 does not exist. If $a = 1$, then $R_1 = 1$ and not random. Since $\lim_{z\to 0} e^{-z^a} = 1$, only for $\mathrm{Re}(a) > 0$, the

[4] Let X_1 and X_2 be i.i.d. random variables, similarly distributed as a random variable X. The random variable X is stable if for any constants $a > 0$, $b > 0$, $c > 0$ and d, the random variable $aX_1 + bX_2$ has the same distribution as $cX + d$, denoted as $aX_1 + bX_2 \overset{d}{=} cX + d$. Another definition (Feller, 1971, p. 170) states that X is stable if and only if $\sum_{j=1}^{n} X_j \overset{d}{=} c_n X + d_n$ for any integer $n > 1$ and where the constant $c_n > 0$ and $d_n \in \mathbb{R}$.
Gorenflo and Mainardi (2012) discuss fractional diffusion processes and their relation to Levy stable distributions.

Laplace transform (5.16) shows that $\int_0^\infty f_{R_a}(t)\,dt = \lim_{t\to\infty} \Pr[R_a \le t] = 1$ for $0 < a < 1$ and the stable random variable R_a possesses a probability distribution $F_{R_a}(t) = \Pr[R_a \le t]$ for $0 < a < 1$.

The sum of n i.i.d. random variables $(R_a)_1, (R_a)_2, \ldots, (R_a)_n$ with same distribution $f_{R_a}(t)$ equals $\sum_{j=1}^n (R_a)_j = n^{\frac{1}{a}} R_a$, because

$$E\left[e^{-z\sum_{j=1}^n (R_a)_j}\right] = \varphi_{R_a}^n(z) = e^{-nz^a} = e^{-\left(n^{\frac{1}{a}}z\right)^a} = E\left[e^{-zn^{\frac{1}{a}}R_a}\right]$$

Thus, $\sum_{j=1}^n (R_a)_j = n^{\frac{1}{a}} R_a$ expresses self-similarity: a sum of random variables maintains the same distribution upon scaling, which is an alternative description of a "stable" distribution.

97. *Distribution and pdf of the stable random variable R_a.* Since e^{-z^a} with $a > 0$ in the pgf (5.15) is only analytic for $\mathrm{Re}(z) > 0$, the inverse Laplace transform (B.2) provides us, for $a \neq 1$, with the pdf

$$f_{R_a}(t) = \frac{1}{2\pi i}\int_{c-i\infty}^{c+i\infty} e^{-z^a} e^{zt} dz \qquad \text{for } c > 0 \tag{5.17}$$

The integral (5.17) will be evaluated in different ways.

First, evaluation of (5.17) along the line $z = c + iu$ and using $(c+iu)^a = \left(c^2+u^2\right)^{\frac{a}{2}} e^{ia\arctan\frac{u}{c}}$,

$$f_{R_a}(t) = \frac{e^{tc}}{\pi}\int_0^\infty \frac{\cos\left(tu - \left(c^2+u^2\right)^{\frac{a}{2}}\sin\left(a\arctan\frac{u}{c}\right)\right)}{e^{\left(c^2+u^2\right)^{\frac{a}{2}}\cos\left(a\arctan\frac{u}{c}\right)}} du$$

demonstrates, because $\cos\left(a\arctan\frac{u}{c}\right) > 0$, convergence for finite t and for $0 < a < 1$. Moreover, $|f_{R_a}(t)| < \frac{e^{tc}}{\pi}\int_0^\infty e^{-u^a\cos\frac{\pi a}{2}}du = \frac{e^{tc}}{\pi}\Gamma\left(\frac{1}{a}\right)\left(\cos\frac{\pi a}{2}\right)^{-\frac{1}{a}}$ illustrates that $f_{R_a}(t)$ is bounded for all real $t \ge 0$. If we move the line of integration in (5.17) to $c = 0$, then we must encircle the origin $z = 0$ (e.g. along a circle with radius ε)

$$\begin{aligned} f_{R_a}(t) = {} & \frac{e^{-i\frac{\pi}{2}}}{2\pi i}\int_\infty^\varepsilon e^{-\left(ue^{-i\frac{\pi}{2}}\right)^a} e^{t\left(ue^{-i\frac{\pi}{2}}\right)} du + \frac{\varepsilon}{2\pi}\int_{-\frac{\pi}{2}}^{\frac{\pi}{2}} e^{-\left(\varepsilon e^{i\theta}\right)^a} e^{t\varepsilon e^{i\theta}} e^{i\theta} d\theta \\ & + \frac{e^{i\frac{\pi}{2}}}{2\pi i}\int_\varepsilon^\infty e^{-\left(ue^{i\frac{\pi}{2}}\right)^a} e^{t\left(ue^{i\frac{\pi}{2}}\right)} du \end{aligned}$$

and we find, after simplification,

$$\begin{aligned} f_{R_a}(t) = {} & \frac{1}{\pi}\int_\varepsilon^\infty e^{-u^a\cos\frac{\pi a}{2}} \cos\left(tu - u^a\sin\frac{\pi a}{2}\right) du \\ & + \frac{\varepsilon}{\pi}\int_0^{\frac{\pi}{2}} e^{t\varepsilon\cos\theta - \varepsilon^a\cos a\theta} \cos(t\varepsilon\sin\theta - \varepsilon^a\sin a\theta + \theta) d\theta \end{aligned}$$

For $0 < a < 1$, the last integral vanishes in the limit $\varepsilon \to 0$ and we arrive, for all $t \ge 0$, at the pdf of the stable random variable R_a,

$$f_{R_a}(t) = \frac{1}{\pi}\int_0^\infty e^{-u^a\cos\frac{\pi a}{2}} \cos\left(tu - u^a\sin\frac{\pi a}{2}\right) du \tag{5.18}$$

After integration of (5.18), the distribution $\Pr[R_a \leq t] = \int_0^t f_{R_a}(x)\,dx$ is

$$\Pr[R_a \leq t] = \frac{2}{\pi}\int_0^\infty e^{-u^a \cos\frac{\pi a}{2}} \cos\left(\frac{tu}{2} - u^a \sin\frac{\pi a}{2}\right) \frac{\sin\left(\frac{tu}{2}\right)}{u} du \quad (5.19)$$

and $\lim_{t\to\infty} \Pr[R_a \leq t] = 1$ as shown in **art.** 96.

Second, we deform, as for Hankel's contour explained in **art.** 206, the line of integration in (5.17) parallel with the imaginary axis into two straight lines with angle $\phi > \frac{\pi}{2}$ intersecting at c on the real z-axis. In the resulting integral

$$f_{R_a}(t) = \frac{1}{2\pi i}\int_{c-\infty e^{-i\phi}}^{c+\infty e^{i\phi}} e^{-z^a} e^{zt} dz \qquad c > 0 \text{ and } \phi > \frac{\pi}{2}$$

the function e^{-z^a} can be expanded in a Taylor series and reversal of sum and integral is valid (as opposed to the case for $\phi = \frac{\pi}{2}$), *provided* that $t > 0$,

$$f_{R_a}(t) = \sum_{k=0}^\infty \frac{(-1)^k}{k!}\frac{1}{2\pi i}\int_{c-\infty e^{-i\phi}}^{c+\infty e^{i\phi}} z^{ka} e^{zt} dz \qquad c > 0 \text{ and } \phi > \frac{\pi}{2}$$

Introducing Hankel's integral (15.5), only valid for $t > 0$, yields

$$f_{R_a}(t) = \sum_{k=0}^\infty \frac{(-1)^k}{k!}\frac{t^{-ak-1}}{\Gamma(-ka)} \qquad \text{for } 0 < a < 1 \text{ and } t > 0 \quad (5.20)$$

which is written in terms of the Wright function $W_{\alpha,\beta}(z) = \sum_{k=0}^\infty \frac{1}{\Gamma(\alpha k+\beta)}\frac{z^k}{k!}$ in (3.4), valid for $\alpha > -1$ and all z, as

$$f_{R_a}(t) = \frac{1}{t} W_{-a,0}\left(-\frac{1}{t^a}\right) \qquad \text{for } 0 < a < 1 \text{ and } t > 0 \quad (5.21)$$

The reflection formula (11.4) of the Gamma function results in

$$f_{R_a}(t) = \frac{1}{\pi}\sum_{k=1}^\infty \frac{(-1)^{k-1}}{k!}\frac{\Gamma(ka+1)\sin\pi ka}{t^{ak+1}} \qquad \text{for } 0 < a < 1 \text{ and } t > 0 \quad (5.22)$$

The related asymptotic series (5.10) for large t of the Mittag-Leffler pdf

$$f_{M_a}(t) = \frac{1}{\pi}\sum_{k=1}^K (-1)^{k-1}\frac{\Gamma(ka+1)\sin\pi(ak)}{t^{ak+1}}$$

does not contain the factorial $\frac{1}{k!}$ and diverges if $K \to \infty$. The pdf (5.22) is derived in another, more complicated way by Pollard (1946).

Integration of (5.22) leads, for $0 < a < 1$ and $t > 0$, to the distribution

$$1 - F_{R_a}(t) = \Pr[R_a > t] = \frac{1}{\pi}\sum_{k=1}^\infty \frac{(-1)^{k-1}}{k!}\frac{\Gamma(ka)\sin\pi ka}{t^{ak}}$$

or, in terms of the Wright function,

$$\Pr[R_a > t] = 1 - W_{-a,1}\left(-\frac{1}{t^a}\right) \qquad \text{for } 0 < a < 1 \text{ and } t > 0$$

We can only explore the pdf $f_{R_a}(t)$ at $t = 0$, based on the integral (5.18) and not from the Wright function's integral (4.10), due to the restriction $t > 0$ in (5.20). The k-th derivative of $f_{R_a}(t)$ at $t = 0$ is computed from (5.18), rewritten as $f_{R_a}(t) = \frac{1}{\pi}\operatorname{Re}\left(\int_0^\infty e^{-u^a\left(\cos\frac{\pi a}{2}+i\sin\frac{\pi a}{2}\right)}e^{itu}du\right)$, as

$$\left.\frac{d^k f_{R_a}(t)}{dt^k}\right|_{t=0} = \frac{1}{\pi}\operatorname{Re}\left(i^k\int_0^\infty e^{-u^a\left(\cos\frac{\pi a}{2}+i\sin\frac{\pi a}{2}\right)}u^k du\right)$$

Substituting $x = u^a$, invoking Euler's integral (15.9) with $\operatorname{Re}\left(e^{i\frac{\pi a}{2}}\right) > 0$,

$$\left.\frac{d^k f_{R_a}(t)}{dt^k}\right|_{t=0} = \operatorname{Re}\left(\frac{i^k}{a\pi}\int_0^\infty x^{\frac{k+1}{a}-1}e^{-x\left(e^{i\frac{\pi a}{2}}\right)}dx\right) = \operatorname{Re}\left(\frac{e^{-i\frac{\pi}{2}}}{a\pi}\Gamma\left(\frac{k+1}{a}\right)\right)$$

results in $\left.\frac{d^k f_{R_a}(t)}{dt^k}\right|_{t=0} = 0$ for all $k \geq 0$. In other words, the Taylor series of $f_{R_a}(t)$ around $t = 0$ does not exist and $\lim_{t\to 0} f_{R_a}(t) = f_{R_a}(0) = 0$ for $0 < a < 1$.

5.4 Mittag-Leffler and Weibull random variables

98. *The Mittag-Leffler and Weibull random variable.* We consider the random variable $X = R_a W$, where W is independent of the stable random variable R_a with $0 < a < 1$ and W will be chosen later. The pgf of X is computed by invoking conditional expectations (Van Mieghem, 2014, p. 32),

$$\varphi_X(z) = E\left[e^{-zR_aW}\right] = E_W\left[E_{R_a}\left[e^{-zR_aW}\middle| W\right]\right]$$

With the pgf (5.15) of the stable random variable R_a, the inner conditional expectation is a random variable equal to

$$E_{R_a}\left[e^{-zR_aW}\middle| W\right] = e^{-z^aW^a}$$

so that

$$\varphi_X(z) = E_W\left[e^{-z^aW^a}\right]$$

Let us now define the random variable $Y = W^a$, then the expectation $E_W\left[e^{-z^aW^a}\right]$ becomes

$$\varphi_X(z) = E_Y\left[e^{-z^aY}\right] = \varphi_Y(z^a)$$

If Y is an exponential random variable with mean μ, then the pgf $\varphi_Y(z) = E\left[e^{-zY}\right] = \frac{1}{z+\mu}$. Hence, choosing the mean equal $\mu = 1$, then shows that

$$\varphi_X(z) = \frac{1}{z^a + 1}$$

and the pgf (5.5) demonstrates that $X = M_a$ has a Mittag-Leffler distribution. The random variable $W = Y^{\frac{1}{a}}$ has the distribution $\Pr[W \leq x] =$

$\Pr\left[Y^{\frac{1}{a}} \leq x\right] = \Pr\left[Y \leq x^a\right] = \left(1 - e^{-x^a}\right)$ with the pdf (Van Mieghem, 2014, p. 18)

$$f_W(x) = \frac{d\Pr\left[W \leq x\right]}{dx} = a x^{a-1} e^{-x^a}$$

which illustrates that $W = \widetilde{W}_{a,1}$ is a Weibull random variable (Van Mieghem, 2014, p. 59). The derivation emphasizes the importance of the transformation $g(x) = x^a$, applied to random variables as explained in Van Mieghem (2014, Sec. 2.3.1).

99. *The Weibull random variable.* A Weibull random variable $\widetilde{W}_{a,b}$, with real $a > 0$ and $b > 0$, possesses the probability distribution function[5]

$$F_{\widetilde{W}_{a,b}}(x) = \Pr\left[\widetilde{W}_{a,b} \leq x\right] = 1 - e^{-bx^a} \quad \text{for } x \geq 0 \tag{5.23}$$

and the probability density function

$$f_{\widetilde{W}_{a,b}}(x) = abx^{a-1}e^{-bx^a} \tag{5.24}$$

The tail or survival probability $y = \Pr\left[\widetilde{W}_{a,b} > x\right] = e^{-bx^a}$ obeys the linear differential equation $\frac{dy}{dx} + abx^{a-1}y = 0$. The moments follow from (5.24) as $E\left[\widetilde{W}_{a,b}^c\right] = \int_0^\infty x^c f_{\widetilde{W}_{a,b}}(x)\,dx = b^{-\frac{c}{a}}\Gamma\left(\frac{c}{a} + 1\right)$ for any real $c \geq -a$. Hence, all moments $E\left[\widetilde{W}_{a.b}^k\right]$ for non-negative integer $c = k$ exist. The probability generating function of the Weibull random variable $\widetilde{W}_{a,b}$ is the Laplace transform $\varphi_{\widetilde{W}_{a,b}}(z) = E\left[e^{-z\widetilde{W}_{a,b}}\right] = \int_0^\infty e^{-zx} f_{\widetilde{W}_{a,b}}(x)\,dx$ of pdf in (5.24),

$$\varphi_{\widetilde{W}_{a,b}}(z) = ab\int_0^\infty x^{a-1}e^{-zx-bx^a}\,dx$$

After substitution of $u = bx^a$, we find, for all $a > 0$ and $w = zb^{-\frac{1}{a}}$ with $\mathrm{Re}(w) > 0$, that

$$\varphi_{\widetilde{W}_{a,b}}\left(b^{\frac{1}{a}}w\right) = E\left[e^{-\left(b^{\frac{1}{a}}\widetilde{W}_{a,b}\right)w}\right] = \int_0^\infty e^{-wu^{\frac{1}{a}}-u}\,du \tag{5.25}$$

Hence, the scaled random variable $b^{\frac{1}{a}}\widetilde{W}_{a,b} = W_a$ only depends upon the parameter $a > 0$ and not anymore on the parameter b. We refer for properties and expansions of the pgf $\varphi_{\widetilde{W}_{a,b}}(z)$ to Van Mieghem and Liu (2019, Appendix A).

The Weibull distribution is one of the three possible limit extremal distributions of a sequence of i.i.d. random variables (Berger, 1993, pp. 65-69) and reduces for $a = 1$ to the exponential distribution, just like the Mittag-Leffler distribution (5.9). **Art.** 98 has shown for the parameter $0 < a < 1$ that the Weibull random variable $\widetilde{W}_{a,1} = W_a = \frac{M_a}{R_a}$ is the quotient of the

[5] The tilde-notation $\widetilde{W}_{a,b}$ distinguishes the Weibull random variable from the Wright function $W_{a,b}(z) = \sum_{k=0}^\infty \frac{1}{\Gamma(ak+b)}\frac{z^k}{k!}$ in (3.4).

Mittag-Leffler M_a and stable R_a random variable, whose moments do not exist.

100. *Generalized Mittag-Leffler random variable $M_{a,b}$.* The Weibull random variable $\widetilde{W}_{a,b}$, with real $a > 0$ and real $b > 0$, motivates the search for a two-parameters Mittag-Leffler random variable $M_{a,b}$. **Art.** 90 has shown that $E_{a,b}(-x)$ is complete monotonic for $0 < a \leq 1$ and $b > a$, which allows us to extend the distribution function (5.9) to

$$F_{M_{a,b}}(t) = \Pr[M_{a,b} \leq t] = 1 - \frac{1}{\Gamma(b)} E_{a,b}(-t^a) \tag{5.26}$$

Clearly, if $b = 1$, the distribution function (5.26) simplifies to (5.9) and the Mittag-Leffler random variable becomes $M_{a,1} = M_a$ in **art.** 93.

The corresponding probability density function $f_{M_{a,b}}(t) = \frac{dF_{M_{a,b}}(t)}{dt}$ then equals, with (2.37),

$$f_{M_{a,b}}(t) = \frac{1}{\Gamma(b)} \frac{(b-1) E_{a,b}(-t^a) - E_{a,b-1}(-t^a)}{t}$$

The probability generating function $\varphi_{M_{a,b}}(z) = \int_0^\infty e^{-zt} f_{M_{a,b}}(t)\, dt$ is

$$\varphi_{M_{a,b}}(z) = -\frac{1}{\Gamma(b)} \int_0^\infty e^{-zt} \frac{d}{dt} E_{a,b}(-t^a)\, dt$$

Partial integration yields $\varphi_{M_{a,b}}(z) = 1 - \frac{z}{\Gamma(b)} \int_0^\infty e^{-zt} E_{a,b}(-t^a)\, dt$, illustrating that $\varphi_{M_{a,b}}(0) = 1$. The general Laplace transform (3.10) for $A = a$, $B = 1$ and $x = -1$, converging for $|z| > 1$,

$$\int_0^\infty e^{-zt} E_{a,b}(-t^a)\, dt = \sum_{k=0}^{\infty} \frac{\Gamma(ak+1)}{\Gamma(ak+b)} \frac{(-1)^k}{z^{1+ka}}$$

leads to

$$\begin{aligned} \varphi_{M_{a,b}}(z) &= \frac{1}{\Gamma(b)} \sum_{k=1}^{\infty} \frac{\Gamma(ak+1)}{\Gamma(ak+b)} \frac{(-1)^{k-1}}{z^{ka}} \\ &= \frac{1}{z^a \Gamma(b)} \left(\frac{\Gamma(a+1)}{\Gamma(a+b)} - \frac{\Gamma(2a+1)}{z^a \Gamma(2a+b)} + O\left(\frac{1}{z^{2a}}\right) \right) \end{aligned}$$

A natural question arises whether other two-parameters Mittag-Leffler distributions can be derived besides (5.26) and besides transformation, where the event $\{M_a \leq t^b\}$ for real positive b is equivalent to $\left\{M_a^{\frac{1}{b}} \leq t\right\}$, which creates the random variable $M_a^{\frac{1}{b}}$ that depends upon two parameters with probability distribution $\Pr\left[M_a^{\frac{1}{b}} \leq t\right] = 1 - E_a(-t^{ab})$.

101. *The Weibull random variable as approximation of the Mittag-Leffler random variable.* For $0 < a < 1$, the pdf $f_{W_a}(x)$ in (5.24) of the Weibull random variable approximates the pdf $f_{M_a}(x)$ in (5.8) of the Mittag-Leffler random variable reasonably well for "sufficiently" small x. For small values of x, it holds that $f_{W_a}(x) = ax^{a-1}(1 + O(x^a))$, while $f_{M_a}(x) =$

$\frac{ax^{a-1}}{\Gamma(1+a)}(1 - O(x^a))$ from which we deduce that $f_{M_a}(x) > f_{W_a}(x)$, because $\Gamma(1+a) < 1$ for $0 < a < 1$. For large x, the Weibull pdf $f_{W_a}(x)$ decreases exponentially as $O\left(e^{-x^a}\right)$, whereas the Mittag-Leffler pdf $f_{M_a}(x)$ decreases as $O\left(x^{-a-1}\right)$ by (5.10), explaining the non-existent moments $E\left[M_a^k\right]$ for $k > 0$. Thus, for large x, we deduce that $f_{M_a}(x) > f_{W_a}(x)$. Since the area below any probability distribution equals one, there must be an intermediate regime where $f_{M_a}(x) < f_{W_a}(x)$. Moreover, for $0 < a < 1$, both pdfs are strictly decreasing in x.

In summary, the pdfs $f_{M_a}(x)$ and $f_{W_a}(x)$ intersect in precisely two different point x_1, close to the origin $x = 0$ and x_2, a little further. For larger values of $x > x_2$, the difference $f_{M_a}(x) - f_{W_a}(x)$ increases in x.

The Weibull pdf appears in many empirical measurements as a good fit. Since the Mittag-Leffler functions are the eigenfunctions of fractional *linear* differential equations (Section 1.3), but suffer from infinite moments, the Weibull random variable W_a may serve as a reasonably accurate, physical approximation for small values of M_a. In addition, the Weibull pdf (5.8) is mathematically easier to manipulate and $W_1 = M_1$ (recall $R_1 = 1$) is an exponential random variable, whose probability distribution is key in Markov theory (see e.g. Van Mieghem (2014)).

6
Estimates and bounds

Several types of inequalities are derived. Section 6.1 demonstrates that the Mittag-Leffler function $E_{a,b}(x)$ is log-convex on the entire real x-axis for $0 < a \leq 1$ and $b \geq a$, partially by the property that any completely monotone function on the positive real x-axis is log-convex. The fractional semi-group inequality for the Mittag-Leffler function is studied in Section 6.2, as well as the Garrappa-Popolizio conjecture. Section 6.3 presents inequalities due to Hadamard, who investigated the entire function $\sum_{n=0}^{\infty} \frac{x^n}{(n!)^a}$. For $0 < a \leq 1$ and real $x > 0$, we show the inequality $E_a(x) \geq \sum_{n=0}^{\infty} \frac{x^n}{(n!)^a}$ in (6.23) and $E_a(x^a) > e^x$ in (6.27). The last Section 6.4 presents inequalities based on the Integral Mean Value Theorem.

6.1 Log-convexity of the Mittag-Leffler function

102. *Convex functions.* A real function $f(x)$ on an interval $x \in [\alpha, \beta]$ of the real axis is convex (Hardy *et al.* (1999, Sec. 3.5), Van Mieghem (2014, Sec. 5.2)) if, for any $\alpha \leq u < v \leq \beta$ and real $\lambda \in [0, 1]$, it holds that

$$f(\lambda u + (1-\lambda) v) \leq \lambda f(u) + (1-\lambda) f(v) \tag{6.1}$$

A real function $f(x)$ is concave on an interval I if $-f(x)$ is convex on that interval I. If $f(x)$ is twice differentiable in the interval I, then a necessary and sufficient condition for convexity is that $f''(x) \geq 0$ for each $x \in I$. This theorem is proved in Hardy *et al.* (1999, pp. 76–77).

103. *Log-convexity.* A real function $f(x)$ is log-convex on an interval I of the real axis if $\log f(x)$ is convex on the interval I. The definition (6.1) of convexity of $\log f(x)$ states that $\log f(\lambda u + (1-\lambda) v) \leq \lambda \log f(u) + (1-\lambda) \log f(v)$, which is equivalent to

$$f(\lambda u + (1-\lambda) v) \leq f^{\lambda}(u) f^{1-\lambda}(v) \tag{6.2}$$

Since $-\log x$ is convex on the real positive x-axis, the definition (6.1) implies that $\log(\lambda u + (1-\lambda) v) \geq \lambda \log(u) + (1-\lambda) \log(u)$, which, after exponentiation, becomes

$$u^{\lambda}v^{1-\lambda} \leq \lambda u + (1-\lambda)\,v \tag{6.3}$$

The inequality (6.3) generalizes (see Hardy *et al.* (1999, pp. 76–77)) the arithmetic mean and geometric mean inequality $\sqrt{uv} \leq \frac{u+v}{2}$, which is the special case of (6.3) for $\lambda = \frac{1}{2}$. Applying (6.3) to the right-hand side in the log-convexity inequality (6.2) results in the inequalities

$$f(\lambda u + (1-\lambda)\,v) \leq f^{\lambda}(u)\, f^{1-\lambda}(v) \leq \lambda f(u) + (1-\lambda)\, f(v)$$

Hence, the first and last inequality indicate that every log-convex function is convex. Thus, the concept of log-convexity is *stronger* than that of convexity. However, Webster (1997) points out that log-concavity is *weaker* than concavity. For example, $f(x) = e^{x}$ is log-concave on any interval of the real axis, but is not concave! Similarly, $f(x) = x^2$ is convex, but $\log f(x) = 2\log x$ is concave. The sum of two log-convex functions is also log-convex, which readily follows from the both definitions (6.1) and (6.2).

If $f(x)$ is twice differentiable on the interval $[\alpha, \beta]$, then convexity in **art.** 102 is equivalent to the requirement that $f''(x) \geq 0$ for each $x \in [\alpha, \beta]$. Similarly, log-convexity is satisfied if $\frac{d^2 \log f(x)}{dx^2} \geq 0$ for each $x \in [\alpha, \beta]$. For real functions, $\log f(x)$ is only defined for positive functions, i.e. $f(x) > 0$ for $x \in [\alpha, \beta]$. The inequality $\frac{d^2 \log f(x)}{dx^2} \geq 0$ translates, with $\frac{d^2 \log f(x)}{dx^2} = -\left(\frac{1}{f(x)}\frac{df(x)}{dx}\right)^2 + \frac{1}{f(x)}\frac{d^2 f(x)}{dx^2}$ and $f(x) > 0$, to an equivalent condition for log-convexity

$$f(x)\,\frac{d^2 f(x)}{dx^2} - \left(\frac{df(x)}{dx}\right)^2 \geq 0 \tag{6.4}$$

104. *Log-convexity and complete monotonicity.* Chorfi *et al.* (2022) have shown that any completely monotone function $\varphi(x)$ on $[0, \infty)$ is log-convex. Their elegant proof is a consequence of the definition of log-convexity (**art.** 103) and of complete monotonicity (**art.** 89).

Indeed, the Laplace transform $\varphi(x) = \int_0^{\infty} e^{-xu} df(u)$ indicates that the quadratic polynomial $p_2(y) = \varphi(x)\,y^2 + 2\varphi'(x)\,y + \varphi''(x)$ equals $p_2(y) = \int_0^{\infty}(y-u)^2 e^{-xu} df(u)$, which illustrates, for real y, that $p_2(y) \geq 0$. Hence[1], its discriminant $(\varphi'(x))^2 - \varphi(x)\,\varphi''(x) \leq 0$, because the parabola $p_2(y)$

[1] Courant and Hilbert (1953, p. 2) mention Cauchy's idea to consider the roots of a quadratic polynomial in x,

$$\sum_{k=1}^{n}(a_k x + b_k)^2 = x^2 \sum_{k=1}^{n} a_k^2 + 2x \sum_{k=1}^{n} a_k b_k + \sum_{k=1}^{n} b_k^2 \geq 0$$

whose discriminant $\left(\sum_{k=1}^{n} a_k b_k\right)^2 - \sum_{k=1}^{n} a_k^2 \sum_{k=1}^{n} b_k^2 \leq 0$ must be non-positive, because the parabola lies above the x-axis and only touches the x-axis in one point when the discriminant is zero, which is only possible if $a_k = cb_k$ for all $1 \leq k \leq n$ and $x = -c$. This is one proof of the famous Cauchy-Schwarz inequality; the second proof follows from the Binet-Cauchy determinant formula (Van Mieghem, 2023a, art. 221-222).

does not intersect the y-axis at different points. A non-positive discriminant obeys the log-convexity condition (6.4).

Art. 90 demonstrates that $E_{a,b}(-x)$ is complete monotonous for $0 < a \leq 1$ and $b \geq a$ and real $x > 0$. The property that any completely monotone function on $[0, \infty)$ is log-convex then shows that $E_{a,b}(-x)$ is log-convex for $0 < a \leq 1$ and $b \geq a$ and real $x > 0$.

105. *Log-convexity and Taylor series with positive Taylor coefficients.* Consider the Taylor series $f(z) = \sum_{k=0}^{\infty} f_k(z_0)(z - z_0)^k$ of a complex function $f(z)$ around the point z_0 in the complex plane, that converges for all complex numbers z in a circle $|z - z_0| < R$. If all Taylor coefficients $f_k(z_0)$ are real and non-negative and $z_0 = x_0$ is real, then the quadratic polynomial $q_2(y) = f(x)y^2 + 2f'(x)y + f''(x)$ in real y equals

$$q_2(y) = \sum_{k=0}^{\infty} \left(y^2 f_k(x_0) + 2y(k+1)f_{k+1}(x_0) + (k+2)(k+1)f_{k+2}(x_0)\right)(x - x_0)^k$$

We observe that the quadratic polynomial $q_2(y) \geq 0$ is non-negative for $x - x_0 \geq 0$. By the same argument as in **art.** 104, the discriminant of $q_2(y)$ is non-positive, which is equivalent to the log-convexity (6.4) of the function $f(x)$ in the interval $0 \leq x - x_0 < R$. In other words, we have shown that a function with non-negative real Taylor coefficients $f_k(x_0)$ around a real point x_0 is log-convex in the interval $0 \leq x - x_0 < R$, where R is the radius of convergence of the Taylor series $f(z) = \sum_{k=0}^{\infty} f_k(x_0)(z - x_0)^k$.

Applying this log-convexity property of functions with non-negative Taylor coefficients to the Taylor series $E_{a,b}(z) = \sum_{k=0}^{\infty} \frac{z^k}{\Gamma(b+ak)}$ in (1.1), whose radius of convergence R is infinite for $a > 0$ (as proved in Section 1.4), demonstrates that the Mittag-Leffler function $E_{a,b}(x)$ is log-convex on the positive real x-axis for real $a > 0$ and $b > 0$.

In summary, combining the above result and **art.** 104 demonstrates for $0 < a \leq 1$ and $b \geq a$ that $E_{a,b}(x)$ is log-convex on the entire real x-axis.

6.2 Fractional semi-group inequality

We investigate the difference[2]

$$Y_{a,b}(u, v) = E_{a,b}(u + v) - E_{a,b}(u)\, E_{a,b}(v) \tag{6.5}$$

which reduces, for $a = b = 1$, to the well-known semi-group property of the exponential function as explored in Van Mieghem (2022, Appendix D). Since $Y_{a,b}(u, v) = Y_{a,b}(v, u)$ is symmetric in u and v, there are three possible regimes for real u and v: (i) $\{u > 0, v > 0\}$, (ii) $\{u < 0, v < 0\}$, (iii) $\{u > 0, v < 0\}$. We omit the cases when u and/or v are complex. The aim

---

[2] For real $b > 0$ and $a > 0$, (6.5) shows that $Y_{a,b}(0,0) = \frac{1}{\Gamma(b)}\left(1 - \frac{1}{\Gamma(b)}\right) = \frac{\Gamma(b)-1}{\Gamma^2(b)}$. Thus, $Y_{a,b}(0,0) < 0$ for $1 < b < 2$, else $Y_{a,b}(0,0) > 0$ for $0 < b < 1$ and $b > 2$.

is to deduce the sign of $Y_{a,b}(u,v)$ and to establish "fractional semi-group" inequalities of the type

$$E_{a,b}(u+v) \lesseqgtr E_{a,b}(u)\, E_{a,b}(v)$$

for the three regimes (i), (ii) and (iii) and for particular values of a and b. The fractional semi-group inequality is applied in D'Alessandro and Van Mieghem (2025) to study the dependence of Mittag-Leffler sojourn times in fractional Markov processes.

106. *Taylor series of* $Y_{a,b}(u,v)$. The Taylor series (1.1)

$$E_{a,b}(u+v) = \sum_{k=0}^{\infty} \frac{(u+v)^k}{\Gamma(b+ak)} = \sum_{k=0}^{\infty}\sum_{m=0}^{k} \frac{\binom{k}{m}}{\Gamma(b+ak)} u^{k-m} v^m$$

becomes, after reversing the k- and m-sum,

$$E_{a,b}(u+v) = \sum_{m=0}^{\infty}\sum_{k=m}^{\infty} \frac{\binom{k}{m}}{\Gamma(b+ak)} u^{k-m} v^m = \sum_{m=0}^{\infty}\sum_{k=0}^{\infty} \frac{\binom{k+m}{m}}{\Gamma(b+ak+am)} u^k v^m$$

Hence, the difference function (6.5) equals, for any complex u,v and b, but $\mathrm{Re}(a) > 0$,

$$Y_{a,b}(u,v) = \sum_{m=0}^{\infty}\sum_{k=0}^{\infty} \left(\frac{\binom{k+m}{m}}{\Gamma(b+ak+am)} - \frac{1}{\Gamma(b+ak)\,\Gamma(b+am)} \right) u^k v^m$$

which we rewrite as

$$Y_{a,b}(u,v) = \sum_{m=0}^{\infty}\sum_{k=0}^{\infty} \frac{\left(\frac{\Gamma(1+(k+m))}{\Gamma(1+k)\Gamma(1+m)} \frac{\Gamma(b+ak)\Gamma(b+am)}{\Gamma(b+a(k+m))} - 1 \right)}{\Gamma(b+ak)\,\Gamma(b+am)} u^k v^m \tag{6.6}$$

In summary, the Taylor series of the function $Y_{a,b}(u,v)$ around $u_0 = 0$ and $v_0 = 0$ is

$$Y_{a,b}(u,v) = \sum_{m=0}^{\infty}\sum_{k=0}^{\infty} y_{km}(a,b)\, u^k v^m$$

with Taylor coefficient $y_{km}(a,b) = \frac{1}{k!m!} \left. \frac{d^{k+m} Y_{a,b}(u,v)}{du^k dv^m} \right|_{u=0;v=0}$ equal to

$$y_{km}(a,b) = \frac{\left(\frac{\Gamma(1+(k+m))}{\Gamma(1+k)\Gamma(1+m)} \frac{\Gamma(b+ak)\Gamma(b+am)}{\Gamma(b+a(k+m))} - 1 \right)}{\Gamma(b+ak)\,\Gamma(b+am)} \tag{6.7}$$

We verify that all Taylor coefficients $y_{km}(1,1) = 0$ in (6.7) and that $Y_{1,1}(u,v) = 0$ in the exponential case, where $a = b = 1$. Hence, we recover the important semi-group property of the exponential $e^{u+v} = e^u e^v$ for all complex u and v. Equality $E_{a,1}(u+v) = E_{a,1}(u)\, E_{a,1}(v)$ or $Y_{a,1}(u,v) = 0$ for all complex u and v only occurs if $a = 1$ as follows from Cauchy's Lemma (Berger, 1993, p. 67).

After substitution of $z_1 = ak$ and $z_2 = am$ in (17.9), we obtain

$$\frac{\Gamma(b+ak)\Gamma(b+am)}{\Gamma(b+ak+am)} = \frac{\Gamma(b)}{\prod_{n=0}^{\infty}\left(1+\frac{a^2km}{(n+b)^2+a(k+m)(n+b)}\right)} \tag{6.8}$$

The function $r_n(a) = \frac{a^2km}{(n+b)^2+a(k+m)(n+b)}$ is, for all $n \geq 0$ and $b > 0$, strictly increasing in positive real a, implying that $r_N(a) < r_N(1)$ for $a \in (0,1)$ and $r_N(a) > r_N(1)$ for $a > 1$. Thus, the ratio (6.8) indicates that

$$\begin{aligned} \frac{\Gamma(b+ak)\Gamma(b+am)}{\Gamma(b+ak+am)} &> \frac{\Gamma(b+k)\Gamma(b+m)}{\Gamma(b+k+m)} \quad \text{for } a \in (0,1) \\ \frac{\Gamma(b+ak)\Gamma(b+am)}{\Gamma(b+ak+am)} &< \frac{\Gamma(b+k)\Gamma(b+m)}{\Gamma(b+k+m)} \quad \text{for } a > 1 \end{aligned}$$

These inequalities imply, for $b = 1$, that the Taylor coefficient in (6.7) obeys

$$\begin{aligned} y_{km}(a,1) &> 0 \quad \text{for } a \in (0,1) \\ y_{km}(a,1) &< 0 \quad \text{for } a > 1 \end{aligned}$$

We deduce for $a \in (0,1)$ that $|Y_{a,1}(u,v)| \leq Y_{a,1}(|u|,|v|)$ and that the positive real axis attains the maximum absolute value of the function $Y_{a,1}(u,v)$, while $|Y_{a,1}(u,v)| \geq -Y_{a,1}(|u|,|v|) \geq 0$ holds for $a > 1$, i.e. the positive real axis reaches the minimum absolute value of the function $Y_{a,b}(u,v)$. It also follows for positive real u and v that, if $a \in (0,1)$, then $Y_{a,1}(u,v) > 0$, while if $a > 1$, then $Y_{a,1}(u,v) < 0$.

In conclusion, we have demonstrated, for $b = 1$ and for the first regime (i) $\{u > 0, v > 0\}$, that

$$\begin{aligned} E_a(u+v) - E_a(u)E_a(v) &> 0 \quad \text{if } a \in (0,1) \\ E_a(u+v) - E_a(u)E_a(v) &< 0 \quad \text{if } a > 1 \end{aligned}$$

Art. 107 demonstrates that $Y_{a,1}(u,v) > 0$ for $a \in (0,1)$ holds in regime (ii) $\{u < 0, v < 0\}$. An analytic proof of the existence of a fractional semi-group inequality for regime (iii) is still missing.

107. *Fractional semi-group inequality for $E_a(u)$ with $0 < a < 1$ and $\{u < 0, v < 0\}$.* We demonstrate, for $0 < a < 1$ and regime (ii) $\{u < 0, v < 0\}$ that

$$E_a(u+v) - E_a(u)E_a(v) > 0$$

If $b = 1$ and $0 < a < 1$, then $E_{a,1}(-u)$ is strictly decreasing in real u implying that $\lim_{u\to\infty} E_{a,1}(-u) = 0$ and that the difference $\lim_{u\to\infty} Y_{a,1}(-u,-v) = 0$ for any real, finite v. Moreover, $Y_{a,1}(0,0) = 0$. Hence, at the boundary of the regime (ii) $\{u < 0, v < 0\}$, the difference $Y_{a,1}(u,v)$ in (6.5), which is a continuous function, has two zeros in both u and v.

Suppose that there exists another finite zero $u_0 < 0$ of $Y_{a,1}(u,v)$ inside regime (ii) $\{u < 0, v < 0\}$, then (6.5) indicates that $E_{a,1}(u_0 + v) = E_{a,1}(u_0)E_{a,1}(v)$. The Taylor series around the point u_0

$$E_{a,1}(u_0+v) = E_{a,1}(u_0) + \sum_{k=1}^{\infty} \frac{1}{k!} \left.\frac{d^k E_{a,1}(x)}{dx^k}\right|_{x=u_0} v^k$$

converges for all v, because $E_{a,b}(z)$ is an entire function for $a > 0$. On the other hand, the Taylor series (1.1) leads to

$$E_{a,1}(u_0) E_{a,1}(v) = E_{a,1}(u_0) + E_{a,1}(u_0) \sum_{k=1}^{\infty} \frac{v^k}{\Gamma(ak+1)}$$

Equating corresponding powers in v shows, for $k > 0$, that the equality $E_{a,1}(u_0 + v) = E_{a,1}(u_0) E_{a,1}(v)$ for $v < 0$ implies for all positive integer k that

$$E_{a,1}(u_0) = \frac{\Gamma(ak+1)}{k!} \left. \frac{d^k E_{a,1}(x)}{dx^k} \right|_{x=u_0} \tag{6.9}$$

which only holds if $u_0 = 0$. Indeed, (2.52) and (2.65) show, for $u_0 \neq 0$, that

$$\left. \frac{d^k E_{a,1}(x)}{dx^k} \right|_{x=u_0} = \frac{1}{(au_0)^k} \sum_{j=1}^{k} q_j(a,1,k) E_{a,1-j}(u_0)$$

does not dependent on $E_{a,1}(u_0)$, while the left-hand in (6.9) equals $E_{a,1}(u_0)$. Another argument is that the right-hand side of (6.9) depends upon k and is not constant for $u_0 \neq 0$, for the ratio

$$r_k(u_0) = \frac{\frac{\Gamma(ak+1)}{k!} \left. \frac{d^k E_{a,1}(x)}{dx^k} \right|_{x=u_0}}{\frac{\Gamma(ak+a+1)}{(k+1)!} \left. \frac{d^{k+1} E_{a,1}(x)}{dx^{k+1}} \right|_{x=u_0}}$$

reduces for $k = 1$ to

$$r_1(u_0) = \frac{\Gamma(a+1)}{\Gamma(2a)} u_0 \frac{E_{a,0}(u_0)}{(1-a) E_{a,0}(u_0) + E_{a,-1}(u_0)}$$

and $r_1(u_0)$ is different from $r_2(u_0)$ for $u_0 \neq 0$.

In conclusion, we have shown that there does not exist a finite, non-zero number u_0 for which equality in $E_{a,1}(u_0 + v) = E_{a,1}(u_0) E_{a,1}(v)$ holds. Hence, either $Y_{a,1}(-|u|, -|v|) > 0$ or $Y_{a,1}(-|u|, -|v|) < 0$. The positive sign is found by computing one particular set $(\widetilde{u}, \widetilde{v})$. For example, $\widetilde{u} = \widetilde{v} = -\frac{1}{2}$ gives $Y_{a,1}\left(-\frac{1}{2}, -\frac{1}{2}\right) = E_{a,1}(-1) - E_{a,1}^2\left(-\frac{1}{2}\right) > 0$. Thus, we have shown for $0 < a < 1$ in the regime (ii) $\{u < 0, v < 0\}$ that

$$E_a(u+v) - E_a(u) E_a(v) > 0$$

In addition, Rolle's theorem (Hardy, 2006, p. 263) for real continuous functions indicates that, between two zeros, there exist a real point $w < 0$ for any real number $v < 0$ at which $\left. \frac{\partial Y_{a,1}(u,v)}{\partial u} \right|_{u=w} = 0$, which is, using $\frac{d}{dz} E_{a,1}(z) = \frac{1}{a} E_{a,a}(z)$ from (2.36), equivalent to

$$E_{a,a}(w+v) = E_{a,a}(w) E_{a,1}(v)$$

Since $Y_{a,1}(u,v) > 0$ for both $u < 0$ and $v < 0$, that point w corresponds to a maximum. Numerical computations suggest that there is only one such maximum at w, which seems plausible from the log-convexity of $E_a(x)$ on the real x-axis. A rigorous proof is omitted.

108. *Log-convexity and fractional semi-group inequality.* Since $E_a(x)$ is log-convex (**art.** 105) for real x and $0 < a \leq 1$, the definition (6.2) of log-convexity implies, for $\lambda \in [0, 1]$ and real x and y, that

$$E_a(\lambda x + (1 - \lambda) y) \leq E_a^{\lambda}(x) E_a^{1-\lambda}(y)$$

For real $x, y > 0$, $E_a(-x)$ is strictly decreasing in x so that $E_a(-x - y) < E_a(-\lambda x - (1 - \lambda) y)$. In addition, $0 < E_a(-x) < 1$ for any real $x > 0$ and, thus $E_a^{\lambda}(-x) > E_a(-x)$ for $\lambda \in [0, 1]$. The fractional semi-group inequality in **art.** 107 then leads to the bounds, for positive finite real x and y, $0 < a \leq 1$ and $\lambda \in [0, 1]$:

$$E_a(-x)E_a(-y) < E_a(-x - y) < E_a(-\lambda x - (1 - \lambda) y) \leq E_a^{\lambda}(-x) E_a^{1-\lambda}(-y)$$

Similarly, for real $x, y > 0$, **art.** 106 shows that

$$E_a(\lambda x + (1 - \lambda) y) \leq E_a^{\lambda}(x) E_a^{1-\lambda}(y) < E_a(x) E_a(y) < E_a(x + y)$$

109. *The Garrappa-Popolizio conjecture.* For any complex number z, the Garrappa-Popolizio conjecture[3] claims the truth of the two inequalities

$$|E_{a,b}(z)| \leq E_{a,b}(\operatorname{Re} z) \qquad \text{for } 0 < a \leq 1 \text{ and } a \leq b \tag{6.10}$$

and

$$|E_{a,b}(z)| \geq E_{a,b}(\operatorname{Re} z) \qquad \text{for } a \geq 1 \text{ and } a \geq b \tag{6.11}$$

The inequality (6.11) is true for $a = b = 1$, because $|e^z| = \left|e^{x+iy}\right| = e^x = e^{\operatorname{Re} z}$.

Recently, Garrappa *et al.* (2024) have proved the first inequality (6.10). **Art.** 90 shows that $E_{a,b}(-x)$ is complete monotonic for $x > 0$ and $0 < a \leq 1$ and $a \leq b$. **Art.** 89 then implies that there exist a non-negative real random variable $X_{a,b}$, whose probability generating function satisfies

$$\varphi_{X_{a,b}}(x) = \Gamma(b) E_{a,b}(-x) = E\left[e^{-xX_{a,b}}\right]$$

Since $E_{a,b}(z)$ is an entire function containing a converging Taylor series (1.1) for all finite z, analytic continuation tells us that $\Gamma(b) E_{a,b}(z) = E\left[e^{zX_{a,b}}\right]$ exists[4] for all complex z. The inequality

$$\left|E\left[e^{zX_{a,b}}\right]\right| \leq E\left[\left|e^{zX_{a,b}}\right|\right] = E\left[e^{\operatorname{Re}(z)X_{a,b}}\right]$$

then demonstrates the inequality (6.10) for any complex number z.

For the second inequality (6.11), Garrappa *et al.* (2024) rely on Hadamard's

---

[3] Francesco Mainardi has emailed me this conjecture on 3 December 2023. A few days later on 14 December 2023, Roberto Garrappa has informed me that inequalities (6.10) and (6.11) were verified by a huge number of computations for a wide range of the parameters a and b, but that a proof of (6.10) and (6.11) was still missing.

[4] Equating corresponding powers in z indicates that the moments satisfy $E\left[X_{a,b}^k\right] = \frac{\Gamma(k+1)\Gamma(b)}{\Gamma(ak+b)}$ and are finite, in contrast to those of the Mittag-Leffler random variable M_a in **art.** 93.

factorization theorem. If $a > 1$, then Section 1.4.3 shows for the entire function $E_{a,b}(z)$ that the order $\rho = \frac{1}{a} < 1$, in which case Hadamard's formula (1.19) applies

$$E_{a,b}(z) = \frac{1}{\Gamma(b)} \prod_{n=1}^{\infty} \left(1 - \frac{z}{z_n(a,b)}\right) \tag{6.12}$$

where $z_n(a,b)$ is the n-th largest in absolute value zero of $E_{a,b}(z)$. Taking the absolute value of the product (6.12), $|E_{a,b}(z)| = \frac{1}{|\Gamma(b)|} \prod_{n=1}^{\infty} \left|1 - \frac{z}{z_n(a,b)}\right|$, and observing that

$$\begin{aligned}\left|1 - \frac{z}{z_n(a,b)}\right|^2 &= \left(1 - \operatorname{Re}\left(\frac{z}{z_n(a,b)}\right)\right)^2 + \operatorname{Im}\left(\frac{z}{z_n(a,b)}\right)^2 \\ &\geq \left(1 - \operatorname{Re}\left(\frac{z}{z_n(a,b)}\right)\right)^2\end{aligned}$$

indicates that

$$|E_{a,b}(z)| \geq \frac{1}{|\Gamma(b)|} \prod_{n=1}^{\infty} \left|1 - \operatorname{Re}\left(\frac{z}{z_n(a,b)}\right)\right|$$

Only if all zeros $z_n(a,b)$ are real, then $\operatorname{Re}\left(\frac{z}{z_n(a,b)}\right) = \frac{\operatorname{Re} z}{z_n(a,b)}$ and $|E_{a,b}(z)| \geq |E_{a,b}(\operatorname{Re} z)| \geq E_{a,b}(\operatorname{Re} z)$. Consequently, the second inequality (6.11) holds only if all zeros $z_n(a,b)$ are real. Unfortunately, merely partial knowledge about the zeros of the Mittag-Leffler function $E_{a,b}(z)$ for all a with $\operatorname{Re}(a) > 0$ and all b exists (which is listed as an open problem in Chapter 8). Garrappa *et al.* (2024) prove the second inequality (6.11) partially for $\{a \geq 2 \text{ and } 0 \leq b \leq 2a - 1\}$ and for $\{a \geq 4 \text{ and } 0 \leq b \leq 2a\}$, based on knowledge about the zeros of the Mittag-Leffler function in those parameter regimes.

6.3 Explorations of Hadamard

About 10 years before Mittag-Leffler has introduced his function $E_a(x)$, Hadamard (1893, art. 8, p. 180) investigates, in his study of entire functions, $E_a(x) = \sum_{k=0}^{\infty} \frac{x^k}{\Gamma(ak+1)}$. We give Hadamard's nice argument, starting from $e^{x^{\frac{1}{a}}} = \sum_{k=0}^{\infty} \frac{x^{\frac{k}{a}}}{\Gamma(k+1)}$ and letting $m' = \frac{k}{a}$, which runs over fractions for $a > 1$, resulting in $e^{x^{\frac{1}{a}}} = \sum_{m'=0}^{\infty} \frac{x^{m'}}{\Gamma(m'a+1)}$. Comparing terms with $E_a(x) = \sum_{m=0}^{\infty} \frac{x^m}{\Gamma(am+1)}$, then shows that $E_a(x) < e^{x^{\frac{1}{a}}}$ for $a > 1$. For $a < 1$, Hadamard states that $E_a(x) < \left[\frac{1}{a}\right] x^{\frac{1}{a}} e^{x^{\frac{1}{a}}}$. Hadamard (**art.** 111) derives an exact a-fold integral for the series $\sum_{k=0}^{\infty} \frac{x^k}{(k!)^a}$, from which he deduces that $\sum_{k=0}^{\infty} \frac{x^k}{(k!)^a} < e^{ax^{\frac{1}{a}}}$. **Art.** 112 shows that the bounds, based on the theory of the Mittag-Leffler function, are sharper than Hadamard's estimates.

110. *Hadamard's generalization of Taylor's theorem.* Suppose that $f(z) = \sum_{n=0}^{\infty} f_n z^n$ is convergent for $|z| < R$ and $g(z) = \sum_{n=0}^{\infty} g_n z^n$ is convergent for $|z| < R'$, then Hadamard has shown that

$$F(z) = \sum_{n=0}^{\infty} f_n g_n z^n = \frac{1}{2\pi i} \int_L f(w)\, g\left(\frac{z}{w}\right) \frac{dw}{w} \tag{6.13}$$

is convergent for $|z| < RR'$ and L is a contour, enclosing the origin, on which $|w| < R$ and $|w| > \frac{|z|}{R'}$. Moreover, if $f(z)$ has singularities at $\alpha_1, \alpha_2, \cdots$ and $g(z)$ at $\beta_1, \beta_2, \cdots$, then the singularities of $F(z)$ are to be found among the points $\alpha_m \beta_n$. Hadamard's theorem is demonstrated in Titchmarsh (1964, Section 4.6) as an illustration of analytic continuation.

111. *Hadamard's series* $\sum_{n=0}^{\infty} \frac{1}{(n!)^k} x^n$. After letting $w = x^{\mu} e^{i\theta}$ for positive real x and any real $\mu \in (0,1)$, Hadamard (1893, p. 181) uses in (6.13) a circular contour,

$$\sum_{n=0}^{\infty} f_n g_n x^n = \frac{1}{2\pi} \int_0^{2\pi} f(x^{\mu} e^{i\theta})\, g\left(x^{1-\mu} e^{-i\theta}\right) d\theta \tag{6.14}$$

which he applies to $\sum_{n=0}^{\infty} \frac{1}{(n!)^k} x^n$ for integer k. Hadamard (1893, p. 181) only gives the result (6.16) with some hints, not the derivation, which we give as well as the explicit form of Hadamard's function (6.17) below.

First, choosing $f(z) = g(z) = e^z$ in (6.14) with $\mu = \frac{1}{2}$, then

$$\sum_{n=0}^{\infty} \frac{1}{(n!)^2} x^n = \frac{1}{2\pi} \int_0^{2\pi} e^{x^{\mu} e^{i\theta} + x^{1-\mu} e^{-i\theta}} d\theta = \frac{1}{2\pi} \int_0^{2\pi} e^{2\sqrt{x}\cos\theta} d\theta = I_0\left(2\sqrt{x}\right)$$

where $I_0(z) = \sum_{n=0}^{\infty} \frac{1}{(n!)^2} \left(\left(\frac{z}{2}\right)^2\right)^n = \frac{1}{2\pi} \int_0^{2\pi} e^{z\cos\theta} d\theta$ is the modified Bessel function (Abramowitz and Stegun, 1968, 9.6.12). Second, after choosing $f(z) = e^z$ and $g(z) = h_k(z) = \sum_{n=0}^{\infty} \frac{z^n}{(n!)^k}$ with $h_1(z) = e^z$ in (6.14), we deduce the Hadamard recursion, that holds for real $k \geq 0$ as well,

$$h_k(x) = \frac{1}{2\pi} \int_0^{2\pi} e^{x^{\mu} e^{i\theta}}\, h_{k-1}\left(x^{1-\mu} e^{-i\theta}\right) d\theta \tag{6.15}$$

Iterating (6.15) once yields

$$h_k(x) = \frac{1}{(2\pi)^2} \int_0^{2\pi} d\theta_1 \int_0^{2\pi} d\theta_2 e^{x^{\mu_1} e^{i\theta_1} + x^{(1-\mu_1)\mu_2} e^{-i\mu_2\theta_1 + i\theta_2}}$$
$$\times\, h_{k-2}\left(x^{(1-\mu_1)(1-\mu_2)} e^{-i(1-\mu_2)\theta_1 - i\theta_2}\right)$$

By choosing $\mu_1 = (1-\mu_1)\mu_2$ or $\mu_2 = \frac{\mu_1}{1-\mu_1}$, we have

$$h_k(x) = \int_0^{2\pi} \frac{d\theta_1}{2\pi} \int_0^{2\pi} \frac{d\theta}{2\pi} e^{x^{\mu_1}\left(e^{i\theta_1} + e^{-i\mu_2\theta_1 + i\theta_2}\right)} h_{k-2}\left(x^{1-2\mu_1} e^{-i(1-\mu_2)\theta_1 - i\theta_2}\right)$$

For $k=3$, we thus find with $h_1(z)=e^z$ that

$$\sum_{n=0}^{\infty}\frac{z^n}{(n!)^3}=\int_0^{2\pi}\frac{d\theta_1}{2\pi}\int_0^{2\pi}\frac{d\theta}{2\pi}e^{x^{\mu_1}\left(e^{i\theta_1}+e^{-i\mu_2\theta_1+i\theta_2}\right)+x^{1-2\mu_1}e^{-i(1-\mu_2)\theta_1-i\theta_2}}$$

We choose again equal powers in x. Thus, after equating $\mu_1=1-2\mu_1$, resulting in $\mu_1=\frac{1}{3}$ and $\mu_2=\frac{\mu_1}{1-\mu_1}=\frac{1}{2}$, we arrive at

$$\sum_{n=0}^{\infty}\frac{x^n}{(n!)^3}=\frac{1}{(2\pi)^2}\int_0^{2\pi}d\theta_1\int_0^{2\pi}d\theta_2 e^{x^{\frac{1}{3}}\left(e^{i\theta_1}+2e^{-i\frac{1}{2}\theta_1}\cos\theta_2\right)}$$

A second iteration of (6.15) is

$$\begin{aligned}h_k(x)=&\frac{1}{(2\pi)^3}\int_0^{2\pi}d\theta_1\int_0^{2\pi}d\theta_2\int_0^{2\pi}d\theta_3 e^{x^{\mu_1}\left(e^{i\theta_1}+e^{-i\mu_2\theta_1+i\theta_2}\right)}\\&\times e^{x^{\mu_3(1-\mu_1)(1-\mu_2)}e^{-i\mu_3((1-\mu_2)\theta_1-\theta_2)+i\theta_3}}\\&\times h_{k-3}\left(x^{(1-\mu_1)(1-\mu_2)(1-\mu_3)}e^{-i((1-\mu_2)\theta_1-\theta_2)(1-\mu_3)-i\theta_3}\right)\end{aligned}$$

which we simplify by choosing $\mu_3=\frac{\mu_1}{(1-\mu_1)(1-\mu_2)}=\frac{\mu_2}{1-\mu_2}=\frac{\mu_1}{(1-2\mu_1)}$ as

$$\begin{aligned}h_k(x)=&\int_0^{2\pi}\frac{d\theta_1}{2\pi}\int_0^{2\pi}\frac{d\theta_2}{2\pi}\int_0^{2\pi}\frac{d\theta_3}{2\pi}e^{x^{\mu_1}\left(e^{i\theta_1}+e^{-i\mu_2\theta_1+i\theta_2}+e^{-i\mu_2\theta_1-i\mu_3\theta_2+i\theta_3}\right)}\\&\times h_{k-3}\left(x^{(1-\mu_1)(1-\mu_2)(1-\mu_3)}e^{-i\{(1-\mu_2)(1-\mu_3)\theta_1-\theta_2(1-\mu_3)+\theta_3\}}\right)\end{aligned}$$

For $k=4$, we choose again equal powers in x, implying that the powers of x in $h_{k-3}(z)=e^z$ should be equal to $\mu_1=(1-\mu_1)(1-\mu_2)(1-\mu_3)$. Using $\mu_2=\frac{\mu_1}{1-\mu_1}$, $\mu_3=\frac{\mu_1}{(1-2\mu_1)}$, which results in $\mu_1=\frac{1}{4}$, $\mu_2=\frac{1}{3}$ and $\mu_3=\frac{1}{2}$, we find that

$$\begin{aligned}\sum_{n=0}^{\infty}\frac{x^n}{(n!)^4}=&\int_0^{2\pi}\frac{d\theta_1}{2\pi}\int_0^{2\pi}\frac{d\theta_2}{2\pi}\int_0^{2\pi}\frac{d\theta_3}{2\pi}e^{x^{\frac{1}{4}}\left(e^{i\theta_1}+e^{-i\frac{\theta_1}{3}+i\theta_2}+e^{-i\frac{\theta_1}{3}-i\frac{\theta_2}{2}+i\theta_3}\right)}\\&\times e^{x^{\frac{1}{4}}e^{-i\frac{\theta_1}{3}-\frac{\theta_2}{2}i-i\theta_3}}\end{aligned}$$

From the two last iterations, we can understand Hadamard's claim

$$\sum_{n=0}^{\infty}\frac{x^n}{(n!)^{k+1}}=\frac{1}{(2\pi)^k}\int_0^{2\pi}d\theta_1\int_0^{2\pi}d\theta_2\cdots\int_0^{2\pi}d\theta_k\, e^{x^{\frac{1}{k+1}}F_k(\theta_1,\theta_2,\ldots,\theta_k)}\tag{6.16}$$

where $F_k(\theta_1,\theta_2,\ldots,\theta_k)$ is a finite sum, that obeys after introduction into Hadamard's recursion (6.15) with $\mu=\frac{1}{k}$,

$$F_k(\theta_1,\theta_2,\ldots,\theta_k)=e^{i\theta_k}+e^{-i\frac{\theta_k}{k}}F_{k-1}(\theta_1,\theta_2,\ldots,\theta_{k-1})$$

with $F_1(\theta_1) = 2\cos\theta_1$. Thus, for a few values of k, we find that

$$\begin{aligned}
F_2(\theta_1,\theta_2) &= e^{i\theta_2} + 2e^{-i\frac{\theta_2}{2}}\cos\theta_1 = e^{i\theta_2} + e^{-i\frac{\theta_2}{2}}e^{i\theta_1} + e^{-i\frac{\theta_2}{2}}e^{-i\theta_1} \\
F_3(\theta_1,\theta_2,\theta_3) &= e^{i\theta_3} + e^{-i\frac{\theta_3}{3}}e^{i\theta_2} + e^{-i\frac{\theta_3}{3}}e^{-i\frac{\theta_2}{2}}e^{i\theta_1} + e^{-i\frac{\theta_3}{3}}e^{-i\frac{\theta_2}{2}}e^{-i\theta_1} \\
F_4(\theta_1,\theta_2,\theta_3,\theta_4) &= e^{i\theta_4} + e^{-i\frac{\theta_4}{4}}e^{i\theta_3} + e^{-i\frac{\theta_4}{4}}e^{-i\frac{\theta_3}{3}}e^{i\theta_2} \\
&\quad + e^{-i\frac{\theta_4}{4}}e^{-i\frac{\theta_3}{3}}e^{-i\frac{\theta_2}{2}}e^{i\theta_1} + e^{-i\frac{\theta_4}{4}}e^{-i\frac{\theta_3}{3}}e^{-i\frac{\theta_2}{2}}e^{-i\theta_1}
\end{aligned}$$

which reveals that Hadamard's function $F_k(\theta_1,\theta_2,\ldots,\theta_k)$ can be written as

$$F_k(\theta_1,\theta_2,\ldots,\theta_k) = e^{i\theta_k} + \sum_{j=1}^{k-1} e^{i\theta_j}\prod_{m=j+1}^{k} e^{-i\frac{\theta_m}{m}} + \prod_{m=1}^{k} e^{-i\frac{\theta_m}{m}} \tag{6.17}$$

We readily verify that (6.17) obeys its recursion (6.15). Furthermore, the maximum modulus of $F_k(\theta_1,\theta_2,\ldots,\theta_k)$ follows from (6.17) as

$$|F_k(\theta_1,\theta_2,\ldots,\theta_k)| \le \left|e^{i\theta_k}\right| + \sum_{j=1}^{k-1}\left|e^{-i\left(\theta_j+\sum_{m=j+1}^{k}\frac{\theta_m}{m}\right)}\right| + \left|e^{-i\sum_{m=1}^{k}\frac{\theta_m}{m}}\right| = k+1$$

and we deduce from the k-fold integral (6.16) Hadamard's upper bound

$$\left|\sum_{n=0}^{\infty}\frac{z^n}{(n!)^k}\right| < e^{k|z|^{\frac{1}{k}}} \tag{6.18}$$

A more compact form of Hadamard's function (6.17) is, with $\theta_0 - 0$,

$$F_k(\theta_1,\theta_2,\ldots,\theta_k) = \sum_{j=0}^{k} e^{i\left(\theta_j-\sum_{m=j+1}^{k}\frac{\theta_m}{m}\right)}$$

so that (6.16) possesses the closed form

$$\sum_{n=0}^{\infty}\frac{x^n}{(n!)^{k+1}} = \int_0^{2\pi}\frac{d\theta_1}{2\pi}\int_0^{2\pi}\frac{d\theta_2}{2\pi}\cdots\int_0^{2\pi}\frac{d\theta_k}{2\pi}\, e^{x^{\frac{1}{k+1}}\sum_{j=0}^{k} e^{i\left(\theta_j-\sum_{m=j+1}^{k}\frac{\theta_m}{m}\right)}}$$

112. *Hadamard's series* $\sum_{k=0}^{\infty}\frac{z^k}{(k!)^a}$ *and* $E_a(z)$. Bounds of the Gamma function for $ak+b>0$ follow after exponentiation from (13.14) in Chapter 13,

$$\begin{aligned}
\sqrt{2\pi}\,(ak+b)^{ak+b-\frac{1}{2}}\,e^{-(ak+b)} &\le \Gamma(ak+b) \\
&\le \sqrt{2\pi}\,(ak+b)^{ak+b-\frac{1}{2}}\,e^{-(ak+b)}e^{\frac{1}{12(ak+b)}}
\end{aligned}$$

As $\frac{(ak+b)^{ak+b-\frac{1}{2}}}{e^{(ak+b)}} = a^{a\left(k+\frac{b}{a}\right)-\frac{1}{2}}\left(k+\frac{b}{a}\right)^{\frac{a-1}{2}}\left(\left(k+\frac{b}{a}\right)^{\left(k+\frac{b}{a}-\frac{1}{2}\right)}e^{-\left(k+\frac{b}{a}\right)}\right)^a$ and again using (13.14),

$$\sqrt{2\pi}\,(ak+b)^{ak+b-\frac{1}{2}}\,e^{-(ak+b)} \le \sqrt{2\pi}a^{a\left(k+\frac{b}{a}\right)-\frac{1}{2}}\left(k+\frac{b}{a}\right)^{\frac{a-1}{2}}\left(\frac{\Gamma\left(k+\frac{b}{a}\right)}{\sqrt{2\pi}}\right)^a$$

and replacing the inequality by an order estimate because $e^{\frac{1}{12(ak+b)}} \approx 1$ for large $ak+b$, we have

$$\Gamma(ak+b) = \frac{\left(\sqrt{2\pi}\right)^{1-a} a^{b-\frac{1}{2}}}{\sqrt{k+\frac{b}{a}}} \left(a^k \sqrt{k+\frac{b}{a}}\Gamma\left(k+\frac{b}{a}\right)\right)^a \left(1+O\left(\frac{1}{ak+b}\right)\right)$$

If $b=1$, then for $ak+1>0$, the above reduces to

$$\begin{aligned}\Gamma(ak+1) &\approx \frac{\sqrt{a}\left(\sqrt{2\pi}\right)^{1-a}}{\sqrt{k+\frac{1}{a}}} \left(a^k \sqrt{k+\frac{1}{a}}\Gamma\left(k+\frac{1}{a}\right)\right)^a \\ &< \frac{\sqrt{a}\left(\sqrt{2\pi}\right)^{1-a}}{\sqrt{k}} \left(a^k \Gamma\left(k+\frac{1}{a}+1\right)\right)^a\end{aligned}$$

For sufficiently large a, we approximate as

$$\Gamma(ak+1) \approx \frac{\sqrt{a}\left(\sqrt{2\pi}\right)^{1-a}}{\sqrt{k}} \left(a^k \Gamma(k+1)\right)^a \tag{6.19}$$

After introducing (6.19) in the definition (1.1) of the Mittag-Leffler function for real, positive x, we approximately obtain

$$E_a(x) = \sum_{k=0}^{\infty} \frac{x^k}{\Gamma(ak+1)} \approx \frac{\left(\sqrt{2\pi}\right)^{a-1}}{\sqrt{a}} \sum_{k=0}^{\infty} \frac{\sqrt{k}\left(\frac{x}{a^a}\right)^k}{(k!)^a} \tag{6.20}$$

Gauss's multiplication formula (11.12) in **art.** 151 indicates for integer values of $z=k$ that

$$\Gamma(nk+1) = \sqrt{n}\left(\sqrt{2\pi}\right)^{1-n} n^{nk} \prod_{j=1}^{n} \Gamma\left(k+\frac{j}{n}\right)$$

For integers $k \geq 2$ and $n>1$, it holds that $\Gamma(k) < \Gamma\left(k+\frac{j}{n}\right) < \Gamma(k+1)$ for $j<n$ and we find the bounds

$$\sqrt{n}\left(\sqrt{2\pi}\right)^{1-n} \left(n^k\Gamma(k)\right)^n < \Gamma(nk+1) < \sqrt{n}\left(\sqrt{2\pi}\right)^{1-n} \left(n^k\Gamma(k+1)\right)^n$$

Although the bounds have been derived for integer $n>1$, the bounds also hold for real $a>1$. Hence, Gauss's multiplication formula produces the bounds, for $a>1$,

$$\frac{\left(\sqrt{2\pi}\right)^{a-1}}{\sqrt{a}} \sum_{k=2}^{\infty} \frac{\left(\frac{x}{a^a}\right)^k}{(k!)^a} < E_a(x) - 1 - \frac{x}{\Gamma(a+1)} < \frac{\left(\sqrt{2\pi}\right)^{a-1}}{\sqrt{a}} \frac{x}{a} \sum_{k=1}^{\infty} \frac{\left(\frac{x}{a^a}\right)^k}{(k!)^a} \tag{6.21}$$

Bieberbach's second bound (4.29) in **art.** 75 shows that $E_a(x) < \frac{1}{a}e^{x^{\frac{1}{a}}}$. Combined with (6.21) leads, for $a>1$ with equality for $a=1$ and real $x \geq 0$, to

$$\sum_{k=2}^{\infty} \frac{x^k}{(k!)^a} < \frac{\left(\sqrt{2\pi}\right)^{1-a}}{\sqrt{a}} e^{ax^{\frac{1}{a}}} \tag{6.22}$$

The bound (6.22) has the same exponential factor as Hadamard's bound (6.18), but is sharper than (6.18).

113. *A lower bound for the Mittag-Leffler function $E_{a,b}(x)$ for $a \in (0,1)$ and positive real x.* The log-convexity of the Gamma function (Section 10.4) implies, for $\lambda \in [0,1]$ and positive real x and y, that

$$\Gamma(\lambda x + (1-\lambda) y) \leq \Gamma^{\lambda}(x) \Gamma^{1-\lambda}(y)$$

Hence, for $a \in (0,1)$, we obtain

$$\Gamma(b+ak) = \Gamma(a(k+b) + (1-a) b) \leq \Gamma^{a}(k+b) \Gamma^{1-a}(b)$$

The Taylor series (1.1) for positive real numbers x and b becomes, for $0 < a < 1$,

$$E_{a,b}(x) = \sum_{k=0}^{\infty} \frac{x^k}{\Gamma(b+ak)} \geq \frac{1}{\Gamma^{1-a}(b)} \sum_{k=0}^{\infty} \frac{x^k}{\Gamma^{a}(k+b)}$$

For $b = 1$ and $0 < a \leq 1$, we complement Hadamard's explorations in **art.** 111 with the lower bound for $x > 0$,

$$E_a(x) \geq \sum_{k=0}^{\infty} \frac{x^k}{(k!)^a} \tag{6.23}$$

For $0 < a \leq 1$, it holds that $\frac{1}{(k!)^a} \geq \frac{1}{k!}$ for integer $k \geq 0$ and the lower bound (6.23) leads to the weaker lower bound $E_a(x) \geq e^x$. In fact, a sharper lower bound for $0 < a < 1$ and real $x > 0$ exists as shown in **art.** 115.

6.4 The Integral Mean Value Theorem

114. *Integral Mean Value Theorem for $E_a(-x)$.* The Integral Mean Value Theorem (Hardy, 2006, p. 322) states: For a real continuous function $g(x)$ on the interval $[\alpha, \beta]$ and a function $p(x)$, which is positive and integrable on the interval $[\alpha, \beta]$, there exist a real number $\varphi \in [\alpha, \beta]$ such that

$$\int_{\alpha}^{\beta} p(x) g(x) dx = g(\varphi) \int_{\alpha}^{\beta} p(x) dx \tag{6.24}$$

If the integrals at both side of (6.24) exist for an infinite interval $[\alpha, \beta]$, then the Integral Mean Value Theorem also applies.

We apply (6.24) to the integral (4.52) with real, positive $z = x$ and $a \in (0,1)$. First, we choose as positive function $p(t) = t^{a-1} e^{-x^{\frac{1}{a}} t}$ and as continuous function $g(t) = \frac{1}{1+2t^a \cos \pi a + t^{2a}}$ with maximum $g_{\max} = g(0) = 1$ and minimum $g_{\min} = g(\infty) = 0$. For a real number $\varphi \in [0, \infty)$ and real $x \geq 0$, (6.24) indicates that

$$\begin{aligned} E_a(-x) &= \frac{\sin \pi a}{\pi} \frac{1}{1 + 2\varphi^a \cos \pi a + \varphi^{2a}} \int_0^{\infty} t^{a-1} e^{-x^{\frac{1}{a}} t} dt \\ &= \frac{\frac{\sin \pi a}{\pi}}{1 + 2\varphi^a \cos \pi a + \varphi^{2a}} \frac{\Gamma(a)}{x} \end{aligned}$$

where $1 + 2\varphi^a \cos \pi a + \varphi^{2a} = (\cos \pi a + \varphi)^2 + \sin^2 \pi a > 0$ for $0 < a < 1$. For $\varphi = 0$, we retrieve the asymptotic form (4.16) for large x.

Applying the Integral Mean Value Theorem (6.24) to the integral (4.63) leads, for a real number $\psi_{a,x} \in [0, 1]$, that depends on $a \in (0, 1)$ and real $x \geq 0$, to

$$E_a(-x) = \frac{\sin \pi a}{\pi} \left(e^{-\frac{x^{\frac{1}{a}}}{\psi_{a,x}}} + e^{-x^{\frac{1}{a}} \psi_{a,x}} \right) \int_0^1 \frac{u^{a-1}}{1 + 2u^a \cos \pi a + u^{2a}} du$$

Invoking (4.59) indicates that $\int_0^1 \frac{u^{a-1}}{1+2u^a \cos \pi a + u^{2a}} du = \frac{\pi}{2 \sin \pi a}$, resulting in[5]

$$E_a(-x) = \frac{1}{2} \left(e^{-\frac{x^{\frac{1}{a}}}{\psi_{a,x}}} + e^{-x^{\frac{1}{a}} \psi_{a,x}} \right) \tag{6.25}$$

We must exclude the essential singularity at $\psi_{a,x} = 0$, because the right-hand side of (6.25) equals $\frac{1}{2}$ for all $x \geq 0$, while the maximum is $E_a(0) = 1$. If $\psi_{a,x} = 1$, then (6.25) becomes $E_a(-x) = e^{-x^{\frac{1}{a}}}$, which only holds for $a = 1$.

In summary, for each real $x > 0$ and $0 < a < 1$, there exists real, positive numbers $\varphi \in [0, \infty)$ and $\psi_{a,x} \in (0, 1)$, satisfying

$$E_a(-x) = \frac{\frac{\sin \pi a}{\pi} \Gamma(a)}{1 + 2\varphi^a \cos \pi a + \varphi^{2a}} \frac{1}{x} \text{ and } E_a(-x) = \frac{1}{2} \left(e^{-\frac{x^{\frac{1}{a}}}{\psi_{a,x}}} + e^{-x^{\frac{1}{a}} \psi_{a,x}} \right) \tag{6.26}$$

115. *Another lower bound for the Mittag-Leffler function $E_{a,b}(x)$ for $a \in (0,1)$ and positive real x.* As mentioned in the beginning of Section 6.3, Hadamard (1893, p. 180) has shown the upper bound $E_a(x^a) < e^x$ for $a > 1$ and for real $x > 0$. Here, we prove the sharper lower bound than $E_a(x) \geq e^x$ in **art.** 113 for $0 < a < 1$ and real $x > 0$,

$$\frac{e^x}{a} > E_a(x^a) > e^x \tag{6.27}$$

The upper bound in (6.27) follows directly from complete monotonicity in **art.** 91. The lower bound[6] in (6.27) follows from the differential equation (5.3). Indeed, for $0 < a < 1$, (5.3) becomes a differential inequality $\frac{dE_a(x^a)}{dx} - E_a(x^a) > 0$ is rewritten as $\frac{d}{dx} \log E_a(x^a) > 1$ for real $x \geq 0$. Integrated from 0 to x, using $E_a(0) = 1$, results in $E_a(x^a) > e^x$.

Another approach considers the difference, for real $x > 0$ and $0 < a < 2$, of the integral (4.18) and the exponential function,

$$E_a(x^a) - e^x = \left(\frac{1}{a} - 1 \right) e^x - \frac{\sin \pi a}{\pi} \int_0^\infty \frac{t^{a-1} e^{-xt}}{t^{2a} - 2t^a \cos(a\pi) + 1} dt$$

[5] We mention for $a = 2$ that $E_2(z) = \cosh(\sqrt{z}) = \frac{1}{2} \left(e^{z^{\frac{1}{2}}} + e^{-z^{\frac{1}{2}}} \right)$.

[6] The idea is due to Rui Ferreira (private communication, August 2025).

Similarly to (4.63), we split the integration interval into two parts $[0,1]$ and $[1,\infty)$, invoke the symmetry (4.53) and obtain

$$\int_0^\infty \frac{t^{a-1}e^{-xt}}{t^{2a}-2t^a\cos(a\pi)+1}dt = \int_0^1 \frac{t^{a-1}\left(e^{-xt}+e^{-\frac{x}{t}}\right)}{t^{2a}-2t^a\cos(a\pi)+1}dt$$

The Integral Mean Value Theorem (**art**. 114) then states that there exist a real number $\psi_{a,x}\in(0,1)$, for each $a\in(0,2)$ and $x\geq 0$, so that

$$\int_0^1 \frac{t^{a-1}\left(e^{-xt}+e^{-\frac{x}{t}}\right)dt}{t^{2a}-2t^a\cos(a\pi)+1} = \left(e^{-x\psi_a}+e^{-\frac{x}{\psi_a}}\right)\int_0^1 \frac{t^{a-1}dt}{t^{2a}-2t^a\cos(a\pi)+1}$$

Using (4.60), $\int_0^1 \frac{u^{a-1}du}{1-2u^a\cos\pi a+u^{2a}} - \int_1^\infty \frac{u^{a-1}du}{1-2u^a\cos\pi a+u^{2a}} = \frac{1}{2}\frac{\pi}{\sin\pi a}\left(\frac{1}{a}-1\right)$, we obtain

$$\int_0^\infty \frac{t^{a-1}e^{-xt}}{t^{2a}-2t^a\cos(a\pi)+1}dt = \left(e^{-x\psi_{a,x}}+e^{-\frac{x}{\psi_{a,x}}}\right)\frac{1}{2}\frac{\pi}{\sin\pi a}\left(\frac{1}{a}-1\right)$$

Finally, for $0<a<2$, we arrive, for $\psi_{a,x}\in(0,1]$ and $x\geq 0$, at

$$E_a(x^a)-e^x = \left(\frac{1}{a}-1\right)\left\{e^x-\left(\frac{e^{-x\psi_{a,x}}+e^{-\frac{x}{\psi_{a,x}}}}{2}\right)\right\} \tag{6.28}$$

Formula (6.28) also holds for $a=2$, where $E_2\left(x^2\right)=\cosh x$ and $\psi_{2,x}=1$. Hence, (6.28) proves, for $0<a\leq 1$, the lower bound (6.27) for real $x>0$ and, for $1<a\leq 2$, Hadamard's upper bound $E_a(x^a)<e^x$. Alternatively, since $\psi_{a,x}\in(0,1]$,

$$E_a(x^a) = \frac{e^x}{a} - \left(\frac{1}{a}-1\right)\left(\frac{e^{-x\psi_{a,x}}+e^{-\frac{x}{\psi_{a,x}}}}{2}\right)$$

shows that the last term is exponentially small in real $x>0$ (see e.g. **art**. 71). Comparison with (3.52) indicates that the remainder (3.51) for $b=1$ equals, for $0<a\leq 2$,

$$\int_0^\infty \left(\frac{\lfloor u\rfloor}{a}-\left\lfloor\frac{u}{a}\right\rfloor\right)\frac{d}{du}\left(\frac{x^u}{\Gamma(u+1)}\right)du = \left(\frac{1}{a}-1\right)\left(1-\frac{e^{-x\psi_{a,x}}+e^{-\frac{x}{\psi_{a,x}}}}{2}\right)$$

7
Miscellanea

The penultimate Chapter 7 in Part I on the Mittag-Leffler function collects several different results, that I found interesting enough to include. Taylor series play a crucial role in the theory of functions and Section 7.1 adds two types of Taylor series for $E_{a,b}(z)$, (7.5) and its companion (7.7) and another type in (7.8). Both types are based on the Taylor series of the entire function $\frac{1}{\Gamma(z)}$, studied in Section 15.2 in Part II. The short Section 7.2 presents Möbius inversion pairs for the Mittag-Leffler function, that usually occur in Dirichlet theory and number theory. Section 7.3 defines Apelblat series and gives examples, that earlier have appeared in Apelblat (2020). Inspired by Gauss's deductions of the Euler-Gauss product, which can be written as a limit, Section 7.4 expresses the Mittag-Leffler function as a limit.

While most of the Sections in Chapter 7 can be regarded as curiosities, the last Section 7.5 on the integral $I_{a,b}(z) = \int_0^\infty \frac{z^u}{\Gamma(b+au)}\,du$ in (1.2) contains important properties, that are also shared in some way, exact or approximate, with its discrete counter part, the Mittag-Leffler function $E_{a,b}(z) = \sum_{k=0}^\infty \frac{z^k}{\Gamma(b+ak)}$.

7.1 Other types of Taylor expansions for $E_{a,b}(z)$

116. *A Taylor series approach with Fermi-Dirac integrals.* We introduce the Taylor series of the entire function $\frac{1}{\Gamma(b+ak)}$ around q into the definition (1.1) of $E_{a,b}(z) = \sum_{k=0}^\infty \frac{z^k}{\Gamma(b+ak)}$,

$$E_{a,b}(z) = \frac{1}{\Gamma(b)} + \sum_{k=1}^\infty z^k \sum_{m=0}^\infty \frac{1}{m!} \frac{d^m}{du^m} \frac{1}{\Gamma(u)}\bigg|_{u=q} (ak+b-q)^m$$

The reversal in the k- and m-sum,

$$E_{a,b}(z) = \frac{1}{\Gamma(b)} + \frac{\frac{z}{1-z}}{\Gamma(q)} + \sum_{m=1}^\infty \frac{1}{m!} \frac{d^m}{du^m} \frac{1}{\Gamma(u)}\bigg|_{u=q} a^m \sum_{k=1}^\infty z^k \left(k + \frac{b-q}{a}\right)^m$$

leads to a confinement of $|z| < 1$. We will now choose $q = b$ and evaluate the series $\sum_{k=1}^{\infty} k^m z^k$, that converges for $|z| < 1$.

The Fermi-Dirac integral[1] of order p is defined (Dingle, 1956) as

$$F_p(y) = \frac{1}{\Gamma(p+1)} \int_0^{\infty} \frac{x^p}{1+e^{x-y}}\, dx \tag{7.1}$$

The value of the zero argument in y is immediately written in terms of the Eta function,

$$F_p(0) = \eta(p+1) \tag{7.2}$$

where the Eta function $\eta(s)$ is related to the Riemann Zeta-function $\zeta(s)$ as

$$\eta(s) = (1 - 2^{1-s})\zeta(s) \tag{7.3}$$

After expanding $\frac{1}{1+e^{x-y}} = \frac{e^{-x+y}}{1+e^{-x+y}} = \sum_{k=1}^{\infty}(-1)^{k-1}e^{-k(x-y)}$ for $\operatorname{Re}(y) < 0$ in (7.1) and invoking Euler's integral (10.2) of the Gamma function, the Dirichlet series for all complex p is deduced as

$$F_p(y) = \sum_{k=1}^{\infty} (-1)^{k-1} \frac{(e^y)^k}{k^{p+1}} \tag{7.4}$$

For example, $F_{-1}(y) = \frac{1}{1+e^{-y}}$ and $F_0(y) = \log(1+e^y) = -\log[F_{-1}(-y)]$. For $p = -m-1$, (7.4) reduces to $\sum_{k=1}^{\infty} k^m (-z)^k = -F_{-m-1}(\log z)$ for $|z| < 1$ and we find

$$E_{a,b}(-z) = \frac{1}{\Gamma(b)} - \sum_{m=0}^{\infty} \frac{1}{m!} \frac{d^m}{du^m} \frac{1}{\Gamma(u)}\bigg|_{u=b} F_{-m-1}(\log z) a^m \tag{7.5}$$

For integer negative order $p = -k$ and $k > 0$, we can show (Van Mieghem, 2026) that

$$F_{-k}(y) = \frac{d^{k-1}}{dy^{k-1}} \left(\frac{1}{1+e^{-y}}\right) = \sum_{m=1}^{k} (m-1)!(-1)^{m-1} \mathcal{S}_k^{(m)} \left(\frac{1}{1+e^{-y}}\right)^m$$

where $\mathcal{S}_k^{(m)}$ is the Stirling number of the second kind (Appendix A). Since $\frac{1}{1+e^{-y}} = 1 - \frac{1}{1+e^{y}}$, which is equivalent to $F_{-1}(y) = 1 - F_{-1}(-y)$, the k-th derivative shows that, for $k > 1$,

$$F_{-k}(y) = (-1)^k F_{-k}(-y) \tag{7.6}$$

Thus, we extend (7.5) for $|z| < 1$ to,

$$E_{a,b}(-z) = \sum_{m=0}^{\infty} \frac{1}{m!} \frac{d^m}{du^m} \frac{1}{\Gamma(u)}\bigg|_{u=b} F_{-m-1}(-\log z)(-a)^m \tag{7.7}$$

[1] The Fermi-Dirac integral $F_p(y)$ plays an important role in solid-state physics (see e.g. Blakemore (1962), Madelung (1981), Ashcroft and Mermin (1981) and Kittel (1986)).

Stretching the convergence constraint in (7.7) to $z = 1$ and using (7.2) results in

$$E_{a,b}(-1) = \sum_{m=0}^{\infty} \frac{1}{m!} \frac{d^m}{du^m} \frac{1}{\Gamma(u)}\bigg|_{u=b} \eta(-m)(-a)^m$$

Further by (7.3), $\zeta(-n) = \frac{(-1)^n}{n+1} B_{n+1}$ and $\zeta(-2n) = 0$ for $n > 0$, it holds that $\eta(-m) = (1 - 2^{1+m})\frac{(-1)^m}{m+1} B_{m+1}$. Taking into account (**art**. 242) that the odd Bernoulli numbers $B_{2m+1} = 0$ for $m > 0$, we find

$$E_{a,b}(-1) = -\frac{1}{2\Gamma(b)} - \sum_{k=1}^{\infty} \frac{B_{2k}}{(2k)!} \frac{d^{2k-1}}{dz^{2k-1}} \frac{1}{\Gamma(z)}\bigg|_{z=b} (2^{2k} - 1)a^{2k-1}$$

which converges fast for small a. Since $E_{a,b}(-1)$ in (1.1) is an alternating series with decreasing coefficients for $a > 0$ and $b > 1.462$, it holds that $\frac{1}{\Gamma(b)} < E_{a,b}(-1) < \frac{1}{\Gamma(a+b)}$.

The major interest of the expansions (7.5) and (7.7) lies in its fast convergence for small a, whereas the definition (1.1) is converging slower for small a. Substituting (7.7) for $|z| < 1$ into the cyclotomic formula (2.15)

$$E_{a,b}(z) = \frac{1}{n} \sum_{r=0}^{n-1} E_{\frac{a}{n},b}\left(z^{\frac{1}{n}} e^{i\frac{2\pi r}{n}}\right)$$

and choosing $n = \lfloor a \rfloor + 1$, where $\lfloor a \rfloor$ is the largest integer smaller than or equal to a, yields, for $|z| < 1$ and any real a, a series that faster converges than (7.7) for large a,

$$E_{a,b}(z) = \sum_{m=0}^{\infty} \frac{\frac{d^m}{du^m}\frac{1}{\Gamma(u)}\Big|_{u=b}}{m!} \frac{\sum_{r=0}^{\lfloor a \rfloor} F_{-m-1}\left(-\frac{\log z}{\lfloor a \rfloor+1} - i\pi\left(\frac{2r}{\lfloor a \rfloor+1} + 1\right)\right)}{(\lfloor a \rfloor + 1)\left(-\frac{\lfloor a \rfloor+1}{a}\right)^m}$$

117. *A Taylor series approach based on the inverse of the Gamma function.* We present a related approach as in **art**. 116 based on the modified Taylor series (15.20)

$$\frac{1}{\Gamma(z)} = e^{-pz} q^{1-z} \sum_{k=0}^{\infty} \widetilde{c}_k(p,q)\, z^k$$

tunable in the complex parameter p and q, with $\operatorname{Re}(q) > 0$, with Taylor coefficient (15.21)

$$\widetilde{c}_k(p,q) = \frac{1}{2\pi i} \frac{1}{k!} \int_0^{\infty} e^{-qt} \left\{(p - \log t + i\pi)^k - (p - \log t - i\pi)^k\right\} dt$$

derived in **art**. 213. After substitution of (15.20) into the Taylor series (1.1) of the Mittag-Leffler function,

$$E_{a,b}(z) = \sum_{n=0}^{\infty} \frac{z^n}{\Gamma(an+b)} = q^{1-b} \sum_{n=0}^{\infty} \left(\frac{z}{q^a}\right)^n \sum_{k=0}^{\infty} \widetilde{c}_k(p,q)\, e^{-p(an+b)} (an+b)^k$$

$$= q^{1-b} \sum_{n=0}^{\infty} \left(\frac{z}{q^a}\right)^n \sum_{k=0}^{\infty} \widetilde{c}_k(p,q)(-1)^k \left.\frac{d^k e^{-t(an+b)}}{dt^k}\right|_{t=p}$$

and formally reversing the n- and k-summations,

$$E_{a,b}(z) = q^{1-b} \sum_{k=0}^{\infty} \widetilde{c}_k(p,q)(-1)^k \frac{d^k}{dt^k} \left.\left(\sum_{n=0}^{\infty} \left(\frac{z}{q^a}\right)^n e^{-t(an+b)} \right)\right|_{t=p}$$

we obtain

$$E_{a,b}(z) = q^{1-b} \sum_{k=0}^{\infty} \widetilde{c}_k(p,q)(-1)^k \frac{d^k}{dt^k} \left.\left(\frac{e^{-tb}}{1 - \frac{z}{q^a} e^{-ta}} \right)\right|_{t=p} \tag{7.8}$$

where the convergence of the geometric series (1.15) requires that $\left|\frac{z}{q^a} e^{-pa}\right| < 1$, thus $\operatorname{Re} p > \frac{\log|z|}{a} - \log|q|$. The k-th derivative in (7.8) is computed exactly in **art.** 123.

After introducing the Taylor coefficient (15.21) and again assuming that a reversal of summation and integration is allowed, (7.8) becomes

$$E_{a,b}(z) = \frac{q^{1-b}}{2\pi i} \int_0^{\infty} e^{-qt} \{A(\log t - i\pi) - A(\log t + i\pi)\}\, dt$$

where the Taylor series $A(\log t \pm i\pi)$ around $t_0 = p$ is

$$\sum_{k=0}^{\infty} \frac{1}{k!} \frac{d^k}{dt^k} \left.\left(\frac{e^{-tb}}{1 - \frac{z}{q^a} e^{-ta}} \right)\right|_{t=p} (\log t - p \pm i\pi)^k = \frac{e^{-(\log t \pm i\pi)b}}{1 - \frac{z}{q^a} e^{-(\log t \pm i\pi)a}}$$

which leads to

$$E_{a,b}(z) = \frac{q^{1-b}}{2\pi i} \int_0^{\infty} t^{-b} e^{-qt} \left\{ \frac{e^{i\pi b}}{1 - \frac{z}{q^a} t^{-a} e^{i\pi a}} - \frac{e^{-i\pi b}}{1 - \frac{z}{q^a} t^{-a} e^{-i\pi a}} \right\} dt$$

Since $\frac{e^{bi\pi}}{1 - Zt^{-a} e^{ai\pi}} - \frac{e^{-bi\pi}}{1 - Zt^{-a} e^{-ai\pi}} = 2i \left(\frac{\sin \pi b + Zt^{-a} \sin \pi(a-b)}{1 - 2Zt^{-a} \cos a\pi + Z^2 t^{-2a}} \right)$, we arrive at,

$$E_{a,b}(z) = \frac{q^{1-b}}{\pi} \int_0^{\infty} t^{-b} e^{-qt} \left(\frac{\sin \pi b + \frac{z}{(qt)^a} \sin \pi (a-b)}{1 - 2\frac{z}{(tq)^a} \cos a\pi + \frac{z^2}{(tq)^{2a}}} \right) dt$$

$$= \frac{1}{\pi} \int_0^{\infty} u^{a-b} e^{-u} \left(\frac{u^a \sin \pi b + z \sin \pi (a-b)}{u^{2a} - 2zu^a \cos a\pi + z^2} \right) du$$

which equals (4.50). However, as shown in **art.** 68, (4.50) is only correct if $0 < a \leq 1$ and $a - b > -1$. In other words, the reversal of operators has limited the scope of the parameters $a \in (0, 1]$ and $b < a + 1$. Also, the parameter q does not seem to play a role and we could have started from the one parameter modified Taylor series (15.18).

7.2 The Mittag-Leffler function and number theory

118. *Möbius inversion.* The first Möbius inversion pair is

$$g(x) = \sum_{n=1}^{\infty} f(nx) \qquad \Longleftrightarrow \qquad f(x) = \sum_{n=1}^{\infty} \mu(n) g(nx) \tag{7.9}$$

where $\mu(n)$ is the Möbius function. The Möbius function satisfies the functional equation

$$\sum_{k|n} \mu(k) = \delta_{n,1} \tag{7.10}$$

and shows that $\mu(1) = 1$. Since $k = 1$ is always a divisor, we obtain from (7.10) the recursion

$$\mu(m) = -1 - \sum_{k|m;1<k<m} \mu(k)$$

from which $\mu(2) = -1$, $\mu(3) = -1$, $\mu(4) = 0$, $\mu(5) = -1$, $\mu(6) = 1$ and so on. The Möbius function is an important function in number theory and in the theory of the Riemann Zeta-function. Hardy and Wright (2008) show that $\mu(n) = (-1)^k$ if the prime factorization of n contains k different primes, else $\mu(n) = 0$. Hence, $|\mu(n)| = 1$.

Let $f(x) = \frac{z^x}{\Gamma(b+x)}$ in (7.9), then

$$g(x) = \sum_{n=1}^{\infty} \frac{z^{nx}}{\Gamma(b+nx)} - E_{x,b}(z^x) - \frac{1}{\Gamma(b)}$$

and Möbius inversion $f(x) - \sum_{n=1}^{\infty} \mu(n) g(nx)$ in (7.9) yields, for $x \neq 0$,

$$\frac{z^x}{\Gamma(b+x)} = \sum_{n=1}^{\infty} \mu(n) \left(E_{nx,b}(z^{nx}) - \frac{1}{\Gamma(b)} \right) \tag{7.11}$$

With $y = z^x$, (7.11) simplifies to $\frac{y}{\Gamma(b+x)} = \sum_{n=1}^{\infty} \mu(n) \left(E_{nx,b}(y^n) - \frac{1}{\Gamma(b)} \right)$.

Invoking $E_{2a,b}(z^2) = \frac{E_{a,b}(z)+E_{a,b}(-z)}{2}$ in **art.** 4 in (7.11)

$$\begin{aligned}
\frac{z^x}{\Gamma(b+x)} &= \sum_{n=1}^{\infty} \mu(n) \left(2E_{2nx,b}(z^{2nx}) - E_{nx,b}(-z^{nx}) - \frac{1}{\Gamma(b)} \right) \\
&= 2\sum_{n=1}^{\infty} \mu(n) \left(E_{2nx,b}(z^{2nx}) - \frac{1}{\Gamma(b)} \right) \\
&\quad - \sum_{n=1}^{\infty} \mu(n) \left(E_{nx,b}(-z^{nx}) - \frac{1}{\Gamma(b)} \right)
\end{aligned}$$

and using (7.11) leads to

$$\sum_{n=1}^{\infty} \mu(n) \left(E_{nx,b}(-z^{nx}) - \frac{1}{\Gamma(b)} \right) = \frac{2z^{2x}}{\Gamma(b+2x)} - \frac{z^x}{\Gamma(b+x)} \tag{7.12}$$

which is an instance of the general Möbius function identity, proved in Van Mieghem (2026),

$$\sum_{j=1}^{n}\mu(j)\sum_{q=1}^{\left\lfloor \frac{n}{j}\right\rfloor}(-1)^{q}f(qj)=2f(2)-f(1) \tag{7.13}$$

The Möbius function identity (7.13) holds for any function f and any $n>1$.

119. *Mertens function.* Applying Abel summation to (7.9) using the Mertens[2] function $\gamma_{-1}(k)=\sum_{l=1}^{k}\mu(l)$, we obtain

$$f(x)=\sum_{k=1}^{\infty}\gamma_{-1}(k)\ (g(kx)-g((k+1)x))+\lim_{N\to\infty}g(Nx)\gamma_{-1}(N)$$

Since $\lim_{N\to\infty}g(Nx)=\lim_{N\to\infty}\sum_{n=1}^{\infty}\frac{z^{nNx}}{\Gamma(b+nNx)}=0$ for real $x>0$ and all z, we find

$$\frac{z^{x}}{\Gamma(b+x)}=\sum_{k=1}^{\infty}\gamma_{-1}(k)\ \left(E_{kx,b}\left(z^{kx}\right)-E_{kx+x,b}\left(z^{kx+x}\right)\right) \tag{7.14}$$

With $\int_{kx}^{(k+1)x}\frac{d}{da}E_{a,b}\left(z^{a}\right)da=E_{kx+x,b}\left(z^{kx+x}\right)-E_{kx,b}\left(z^{kx}\right)$, the corresponding integral representation of (7.14) is

$$\begin{aligned}\frac{z^{x}}{\Gamma(b+x)}&=-\sum_{k=1}^{\infty}\gamma_{-1}(k)\int_{kx}^{(k+1)x}\frac{d}{da}E_{a,b}\left(z^{a}\right)da\\&=-\sum_{k=1}^{\infty}\int_{kx}^{(k+1)x}\gamma_{-1}\left(\left\lfloor\frac{a}{x}\right\rfloor\right)\frac{d}{da}E_{a,b}\left(z^{a}\right)da\end{aligned}$$

and

$$\frac{z^{x}}{\Gamma(b+x)}=-\int_{x}^{\infty}\gamma_{-1}\left(\left\lfloor\frac{a}{x}\right\rfloor\right)\frac{d}{da}E_{a,b}\left(z^{a}\right)da \qquad \text{for } x>0 \tag{7.15}$$

Similarly for (7.12), it holds that

$$\frac{2z^{2x}}{\Gamma(b+2x)}-\frac{z^{x}}{\Gamma(b+x)}=-\int_{x}^{\infty}\gamma_{-1}\left(\left\lfloor\frac{a}{x}\right\rfloor\right)\frac{d}{da}E_{a,b}\left(-z^{a}\right)da \qquad \text{for } x>0 \tag{7.16}$$

Although (7.15) and (7.16) are remarkable, further progress to estimate the order of $\gamma_{-1}([x])$ requires a study of $\frac{d}{da}E_{a,b}\left(\lambda z^{a}\right)$.

7.3 Apelblat series

120. *Apelblat series.* Inspired by series in Apelblat (2020), we call

$$g_{a,b}(t)=\sum_{k=0}^{\infty}f_{k}t^{b+k-1}E_{a,b+k}\left(xt^{a}\right) \tag{7.17}$$

[2] A sufficient condition to prove the Riemann Hypothesis is to demonstate that the Mertens function behaves as $\gamma_{-1}(x)=O\left(x^{\frac{1}{2}+\varepsilon}\right)$ for large x.

an Apelblat series, where the Taylor series of the function $f(z) = \sum_{k=0}^{\infty} f_k z^k$ around $z_0 = 0$ converges for $|z| \leq R$. Evidently, if $f_k = 0$ for $k > n$, then $f(z)$ is a polynomial of order n in z and $f(z)$ is an entire function. Earlier in Section 2.7, we have already encountered Apelblat series in (2.73) and (2.75) for $\int_0^z E_{a,b}(u)\,du$.

We generalize the method of Apelblat (2020). We take the Laplace transform $\mathcal{L}[.]$ of both sides of (7.17) and use (3.13),

$$\mathcal{L}[g_{a,b}(t)] = \sum_{k=0}^{\infty} f_k \mathcal{L}\left[t^{b+k-1} E_{a,b+k}(xt^a)\right] = \frac{s^{a-b+\beta}}{s^a - x} \sum_{k=0}^{\infty} f_k s^{-k-\beta}$$

and obtain a product of Laplace transformed functions

$$\mathcal{L}[g_{a,b}(t)] = \mathcal{L}\left[t^{b-\beta-1} E_{a,b-\beta}(xt^a)\right] \frac{1}{s^\beta} f\left(\frac{1}{s}\right) \tag{7.18}$$

The convolution theorem for the Laplace transform suggests us to find the inverse Laplace transform $\mathcal{L}^{-1}\left[\frac{1}{s^\beta} f\left(\frac{1}{s}\right)\right]$. Apelblat (2020) observes and demonstrates that elegant series follow if a closed form for $\mathcal{L}^{-1}\left[\frac{1}{s^\beta} f\left(\frac{1}{s}\right)\right]$ exist, else we can proceed with (B.2) and Hankel's integral (15.5)

$$\mathcal{L}^{-1}\left[\frac{1}{s^\beta} f\left(\frac{1}{s}\right)\right] = \frac{1}{2\pi i}\int_{c-i\infty}^{c+i\infty} \frac{1}{s^\beta} f\left(\frac{1}{s}\right) e^{st} ds = \frac{1}{2\pi i}\sum_{k=0}^{\infty} f_k \int_{c-i\infty}^{c+i\infty} \frac{e^{st} ds}{s^{k+\beta}}$$
$$= t^{\beta-1} \sum_{k=0}^{\infty} f_k \frac{t^k}{\Gamma(\beta + k)}$$

After taking the inverse Laplace transform of both sides in (7.18), we formally arrive at the Apelblat series (7.17), for the free parameter $\mathrm{Re}(\beta) > 0$,

$$g_{a,b}(t) = \int_0^t (t-u)^{b-\beta-1} E_{a,b-\beta}\left(x(t-u)^a\right) u^{\beta-1} \sum_{k=0}^{\infty} \frac{f_k u^k}{\Gamma(\beta+k)} du \tag{7.19}$$

which directly follows from (3.28) for $w = \beta + k$ and $b \to b - \beta$. Hence, the property (7.19) of the Apelblat series is a consequence of the generalized integration property in **art.** 47.

121. *Examples of Apelblat series.* We include several instances of the general Apelblat series (7.19).

Example The Taylor series of $(1+cz)^p = \sum_{k=0}^{\infty} \binom{p}{k} c^k z^k$ converges for all complex p provided $|z| < \frac{1}{c}$. The right-hand series in (7.19) becomes with $f_k = \binom{p}{k} c^k = \frac{\Gamma(p+1)}{k!\Gamma(p-k+1)} c^k = \frac{\Gamma(-p+k)}{k!\Gamma(-p)} (-c)^k$

$$\sum_{k=0}^{\infty} f_k \frac{u^k}{\Gamma(\beta+k)} = \frac{1}{\Gamma(-p)} \sum_{k=0}^{\infty} \frac{\Gamma(-p+k)}{k!\Gamma(\beta+k)} (-cu)^k$$

which reduces if we choose $\beta = -p$ to $\sum_{k=0}^{\infty} f_k \frac{u^k}{\Gamma(\beta+k)} = \frac{1}{\Gamma(-p)} e^{-cu}$. The

Apelblat series (7.19) then becomes, for $q=-p$ and $0<\mathrm{Re}\,(q)<\mathrm{Re}\,(b)$,

$$\sum_{k=0}^{\infty}\binom{-q}{k}c^k t^{b+k-1}E_{a,b+k}\left(xt^a\right)=\int_0^t\frac{u^{b-q-1}\left(t-u\right)^{q-1}}{e^{c(t-u)}}\frac{E_{a,b-q}\left(xu^a\right)}{\Gamma\left(q\right)}du$$

For $q=1$, $b=2$, $c=x=1$, we retrieve the series Apelblat (2020, eq. (64)) and for $q=2$, $b=2$, $c=x=1$, the series Apelblat (2020, eq.(68)).

Example The Taylor series $\left(\frac{z}{2}\right)^{-p}J_p\left(z\right)=\sum_{k=0}^{\infty}\frac{(-1)^k}{k!\Gamma(p+1+k)}\left(\frac{z}{2}\right)^{2k}$ in Abramowitz and Stegun (1968, 9.1.10) of the Bessel function $J_p\left(z\right)$ has only even Taylor coefficients $j_{2k}=\frac{\left(-\frac{1}{4}\right)^k}{k!\Gamma(p+1+k)}$ and the odd $j_{2k+1}=0$. If $f_{2k+1}=0$, then

$$\sum_{k=0}^{\infty}f_k\frac{u^k}{\Gamma\left(\beta+k\right)}=\sum_{k=0}^{\infty}f_{2k}\frac{u^{2k}}{\Gamma\left(\beta+2k\right)}=\frac{\sqrt{\pi}}{2^{\beta-1}}\sum_{k=0}^{\infty}f_{2k}\frac{\left(\frac{u}{2}\right)^{2k}}{\Gamma\left(\frac{\beta}{2}+k\right)\Gamma\left(\frac{\beta+1}{2}+k\right)}$$

where the duplication formula (11.9) of Gamma function is used. If $f_{2k}=\binom{p}{k}c^{2k}=\frac{\Gamma(-p+k)}{k!\Gamma(-p)}\left(-c^2\right)^k$ then

$$\sum_{k=0}^{\infty}f_k\frac{u^k}{\Gamma\left(\beta+k\right)}=\frac{\sqrt{\pi}}{2^{\beta-1}\Gamma\left(-p\right)}\sum_{k=0}^{\infty}\frac{\Gamma\left(-p+k\right)\left(-1\right)^k\left(\frac{cu}{2}\right)^{2k}}{k!\Gamma\left(\frac{\beta}{2}+k\right)\Gamma\left(\frac{\beta+1}{2}+k\right)}$$

Choosing $-p=\frac{\beta}{2}$ and the Taylor series of the Bessel function,

$$\sum_{k=0}^{\infty}f_k\frac{u^k}{\Gamma\left(\beta+k\right)}=\frac{\sqrt{\pi}\left(\frac{cu}{2}\right)^{-\frac{\beta-1}{2}}}{2^{\beta-1}\Gamma\left(\frac{\beta}{2}\right)}J_{\frac{\beta-1}{2}}\left(cu\right)$$

yields the Apelblat series (7.19)

$$\begin{aligned}A_1&=\sum_{k=0}^{\infty}\binom{-\frac{\beta}{2}}{k}c^{2k}t^{b+2k-1}E_{a,b+2k}\left(xt^a\right)\\&=\frac{\sqrt{\pi}\left(\frac{2}{c}\right)^{\frac{\beta-1}{2}}}{2^{\beta-1}\Gamma\left(\frac{\beta}{2}\right)}\int_0^t\left(t-u\right)^{b-\beta-1}E_{a,b-\beta}\left(x\left(t-u\right)^a\right)u^{\frac{\beta-1}{2}}J_{\frac{\beta-1}{2}}\left(cu\right)du\end{aligned}$$

Choosing $-p=\frac{\beta+1}{2}$ yields

$$\sum_{k=0}^{\infty}f_k\frac{u^k}{\Gamma\left(\beta+k\right)}=\frac{\sqrt{\pi}\left(\frac{cu}{2}\right)^{-\frac{\beta-2}{2}}}{2^{\beta-1}\Gamma\left(\frac{\beta+1}{2}\right)}J_{\frac{\beta-2}{2}}\left(cu\right)$$

and the Apelblat series (7.19) then becomes

$$\begin{aligned}A_2&=\sum_{k=0}^{\infty}\binom{-\frac{\beta+1}{2}}{k}c^{2k}t^{b+2k-1}E_{a,b+2k}\left(xt^a\right)\\&=\frac{\sqrt{\pi}\left(\frac{c}{2}\right)^{1-\frac{\beta}{2}}}{2^{\beta-1}\Gamma\left(\frac{\beta+1}{2}\right)}\int_0^t\left(t-u\right)^{b-\beta-1}E_{a,b-\beta}\left(x\left(t-u\right)^a\right)u^{\frac{\beta}{2}}J_{\frac{\beta}{2}-1}(cu)\,du\end{aligned}$$

For $\beta = 0$, $b = 1$, $c = x = 1$, we retrieve the series (Apelblat, 2020, eq. (74)),

$$\sum_{k=0}^{\infty} \binom{-\frac{1}{2}}{k} t^{2k} E_{a,1+2k}\left(t^{a}\right) = -\int_{0}^{t} E_{a}\left(u^{a}\right) J_{1}\left(t-u\right) du$$

Example The hypergeometric function (1.17) with Taylor series around the origin,

$$F\left(p,q;r;z\right) = \frac{\Gamma\left(r\right)}{\Gamma\left(p\right)\Gamma\left(q\right)} \sum_{k=0}^{\infty} \frac{\Gamma\left(p+k\right)\Gamma\left(q+k\right)}{\Gamma\left(r+k\right)k!} z^{k}$$

leads to the Apelblat series (7.19)

$$\sum_{k=0}^{\infty} f_{k} \frac{c^{k}u^{k}}{\Gamma\left(\beta+k\right)} = \frac{\Gamma\left(r\right)}{\Gamma\left(p\right)\Gamma\left(q\right)} \sum_{k=0}^{\infty} \frac{\Gamma\left(p+k\right)\Gamma\left(q+k\right)}{\Gamma\left(r+k\right)\Gamma\left(\beta+k\right)} \frac{c^{k}u^{k}}{k!}$$

If we choose q equal to β, then

$$\sum_{k=0}^{\infty} f_{k} \frac{c^{k}u^{k}}{\Gamma\left(\beta+k\right)} = \frac{\Gamma\left(r\right)}{\Gamma\left(p\right)\Gamma\left(\beta\right)} \sum_{k=0}^{\infty} \frac{\Gamma\left(p+k\right)}{\Gamma\left(r+k\right)} \frac{\left(cu\right)^{k}}{k!} = \frac{1}{\Gamma\left(\beta\right)} M\left(p,r,cu\right)$$

where $M\left(a,b,z\right) = \frac{\Gamma(b)}{\Gamma(a)} \sum_{k=0}^{\infty} \frac{\Gamma(a+k)}{\Gamma(b+k)} \frac{u^{k}}{k!}$ in (2.14) is Kummer's confluent hypergeometric function. The Apelblat series (7.19) thus becomes

$$\begin{aligned} A_{3} &= \frac{\Gamma\left(r\right)}{\Gamma\left(p\right)} \sum_{k=0}^{\infty} \frac{\Gamma\left(p+k\right)\Gamma\left(\beta+k\right)}{\Gamma\left(r+k\right)k!} c^{k} t^{b+k-1} E_{a,b+k}\left(xt^{a}\right) \\ &= \int_{0}^{t} \left(t-u\right)^{b-\beta-1} E_{a,b-\beta}\left(x\left(t-u\right)^{a}\right) u^{\beta-1} M(p,r,cu) du \end{aligned}$$

In order to use the property of the Kummer function (Abramowitz and Stegun, 1968, 13.3.2),

$$\lim_{a\to\infty} M\left(a,b,-\frac{z}{a}\right) = \Gamma\left(b\right) z^{\frac{1-b}{2}} J_{b-1}\left(2\sqrt{z}\right)$$

we first choose $c = -\frac{1}{p}$ and then take the limit $p \to \infty$ of both sides, with $\lim_{p\to\infty} \frac{\Gamma(p+k)}{\Gamma(p)p^{k}} = 1$ in **art.** 198,

$$\begin{aligned} A_{4} &= \sum_{k=0}^{\infty} \frac{\Gamma\left(\beta+k\right)\left(-1\right)^{k}}{\Gamma\left(r+k\right)k!} t^{b+k-1} E_{a,b+k}\left(xt^{a}\right) \\ &= \int_{0}^{t} \left(t-u\right)^{b-\beta-1} E_{a,b-\beta}\left(x\left(t-u\right)^{a}\right) u^{\beta+\frac{1-r}{2}-1} J_{r-1}\left(2\sqrt{u}\right) du \end{aligned}$$

After letting $\beta = r$ and then $r = b-1$, we obtain

$$\sum_{k=0}^{\infty} \left(-1\right)^{k} \frac{t^{b+k-1}}{k!} E_{a,b+k}\left(xt^{a}\right) = \int_{0}^{t} E_{a}\left(x\left(t-u\right)^{a}\right) u^{\frac{b-2}{2}} J_{b-2}\left(2\sqrt{u}\right) du$$

which simplified for $b=2$ to

$$\sum_{k=1}^{\infty}\frac{(-1)^{k-1}t^k}{(k-1)!}E_{a,k+1}\left(xt^a\right)=\int_0^t E_a\left(xu^a\right)J_0\left(2\sqrt{t-u}\right)du$$

7.4 The Mittag-Leffler function expressed as a limit

122. *First limit for $E_{a,b}(z)$.* The Gauss product $\Pi(n,z)$ in (10.3) indicates that

$$\frac{1}{\Pi(n,z)}=(1+z)\left(1+\frac{z}{2}\right)\left(1+\frac{z}{3}\right)\dots\left(1+\frac{z}{n}\right)e^{-z\log n}$$

and defines the Gamma function $\Gamma(z+1)=\lim_{n\to\infty}\Pi(n,z)$ as a limit. Thus, $\Gamma(z)=\frac{1}{z}\lim_{n\to\infty}\Pi(n,z)$ and

$$\begin{aligned}E_{a,b}(z)&=\sum_{k=0}^{\infty}\frac{z^k}{\Gamma(b+ak)}=\sum_{k=0}^{\infty}\lim_{n\to\infty}\frac{(b+ak)\,z^k}{\Pi(n,b+ak)}\\&=\sum_{k=0}^{\infty}\lim_{n\to\infty}(b+ak)\,z^k\prod_{j=1}^{n}\left(1+\frac{b+ak}{j}\right)e^{-(b+ak)\log n}\end{aligned}$$

With the generating function (A.5) of the Stirling numbers of the first kind for $k>0$,

$$\prod_{j=1}^{n}\left(1+\frac{z}{j}\right)=\frac{1}{n!}\prod_{j=1}^{n}(z+j)=\frac{1}{n!}\sum_{j=0}^{n}S_{n+1}^{(j+1)}(-1)^{n-j}z^j$$

we obtain

$$E_{a,b}(z)=\sum_{k=0}^{\infty}\lim_{n\to\infty}\frac{1}{n!}\sum_{j=0}^{n}S_{n+1}^{(j+1)}(-1)^{n-j}z^k(b+ak)^{j+1}e^{-(b+ak)\log n}$$

and arrive at the first limit

$$E_{a,b}(z)=\lim_{n\to\infty}\frac{1}{n!}\sum_{j=0}^{n}S_{n+1}^{(j+1)}(-1)^{n-j}\left(\sum_{k=0}^{\infty}\frac{z^k(b+ak)^{j+1}}{n^{b+ak}}\right)\qquad(7.20)$$

123. We will derive a finite series for the infinite series in brackets in (7.20), which we define as

$$T_j(z;n)=\sum_{k=0}^{\infty}\frac{z^k(b+ak)^j}{n^{b+ak}}\qquad(7.21)$$

Since n in (7.20) grows unboundedly, we may assume that $\left|\frac{z}{n^a}\right|<1$ and we find

$$T_0(z;n)=\sum_{k=0}^{\infty}\frac{z^k}{n^{b+ak}}=\frac{1}{n^b}\sum_{k=0}^{\infty}\left(\frac{z}{n^a}\right)^k=\frac{1}{n^b}\frac{1}{1-\frac{z}{n^a}}=\frac{n^{a-b}}{n^a-z}$$

Let $n=e^x$, then $T_j(z;e^x)=\sum_{k=0}^{\infty}z^k(b+ak)^j e^{-(b+ak)x}$ and $T_0(z;e^x)=$

$\frac{e^{(a-b)x}}{e^{ax}-z} = \frac{e^{-bx}}{1-e^{\log z - ax}}$. Differentiation shows that $\frac{d}{dx}T_{j-1}(z;e^x) = -T_j(z;e^x)$ and iteration leads us to

$$T_j(z;e^x) = (-1)^j \frac{d^j}{dx^j} T_0(z;e^x) = (-1)^j \frac{d^j}{dx^j}\left(\frac{e^{(a-b)x}}{e^{ax}-z}\right)$$

Invoking Leibniz' rule,

$$(-1)^j T_j(z;e^x) = \sum_{m=0}^{j} \binom{j}{m} \frac{d^{j-m}}{dx^{j-m}} e^{(a-b)x} \frac{d^m}{dx^m}\left(\frac{1}{e^{ax}-z}\right) \tag{7.22}$$

Now, $\frac{d^{j-m}}{dx^{j-m}} e^{(a-b)x} = (a-b)^{j-m} e^{(a-b)x}$ and

$$\frac{d^m}{dx^m}\left(\frac{1}{e^{ax}-z}\right) = \frac{d^m}{dy^m}\left(\frac{1}{e^y-z}\right)\frac{d^m y}{dx^m} = a^m \frac{d^m}{dy^m}\left(\frac{1}{e^y-z}\right)$$

Provided that $b \neq 0$ and recalling that $\mathcal{S}_m^{(0)} = \delta_{m0}$ for the Stirling numbers of the second kind $\mathcal{S}_m^{(k)}$, the Taylor series, derived in Van Mieghem (2026),

$$\frac{1}{e^{z+b}-1} = \sum_{m=0}^{\infty}\left(\sum_{k=0}^{m} \frac{(-1)^k k! e^{kb}}{(e^b-1)^{k+1}} \mathcal{S}_m^{(k)}\right)\frac{z^m}{m!} \tag{7.23}$$

indicates that

$$\left.\frac{d^m}{dz^m}\left(\frac{1}{e^{z+b}-1}\right)\right|_{z=0} = \sum_{k=0}^{m} \frac{(-1)^k k! e^{kb}}{(e^b-1)^{k+1}} \mathcal{S}_m^{(k)} = \left.\sum_{k=0}^{m} \frac{(-1)^k k! e^{k(b+z)}}{(e^{b+z}-1)^{k+1}} \mathcal{S}_m^{(k)}\right|_{z=0}$$

Thus, $\frac{d^m}{dz^m}\left(\frac{1}{e^{z+b}-1}\right) = \sum_{k=0}^{m} \frac{(-1)^k k! e^{k(b+z)}}{(e^{b+z}-1)^{k+1}} \mathcal{S}_m^{(k)}$ and with $\frac{1}{e^y-z} = \frac{1}{z}\frac{1}{e^{y-\log z}-1}$, we find

$$\frac{d^m}{dy^m}\left(\frac{1}{e^y-z}\right) = \frac{1}{z}\frac{d^m}{dy^m}\left(\frac{1}{e^{y-\log z}-1}\right) = \frac{1}{z}\sum_{k=1}^{m} \frac{(-1)^k k! e^{k(y-\log z)}}{(e^{y-\log z}-1)^{k+1}} \mathcal{S}_m^{(k)}$$

and

$$\frac{d^m}{dx^m}\left(\frac{1}{e^{ax}-z}\right) = a^m \frac{d^m}{dy^m}\left(\frac{1}{e^y-z}\right) = \frac{a^m}{z}\sum_{k=1}^{m} \frac{(-1)^k k! e^{k(ax-\log z)}}{(e^{ax-\log z}-1)^{k+1}} \mathcal{S}_m^{(k)}$$
$$= a^m \sum_{k=1}^{m} \frac{(-1)^k k! e^{kax}}{(e^{ax}-z)^{k+1}} \mathcal{S}_m^{(k)}$$

In summary, provided that $e^x > |z|$ and combining all in (7.22) results in

$$T_j(z;e^x) = \sum_{k=0}^{\infty} z^k (b+ak)^j e^{-(b+ak)x} \tag{7.24}$$
$$= (b-a)^j e^{(a-b)x} \sum_{m=0}^{j} \binom{j}{m}\left(\frac{a}{a-b}\right)^m \sum_{k=0}^{m} \frac{(-1)^k k! e^{kax}}{(e^{ax}-z)^{k+1}} \mathcal{S}_m^{(k)}$$

Example Let $ze^{-ax} = y$ and $|y| < 1$, then $T_j(z;e^x)$ in (7.24) is rewritten, after reversal of m- and k-sum, as a Taylor series in y around $y_0 = 0$,

$$\sum_{k=0}^{\infty} (b+ak)^j \, y^k = (b-a)^j \sum_{k=0}^{j} \frac{(-1)^k k!}{(1-y)^{k+1}} \sum_{m=k}^{j} \binom{j}{m} \left(\frac{a}{a-b}\right)^m \mathcal{S}_m^{(k)} \tag{7.25}$$

If $b \neq a$, then (7.25) only depends on one parameter $c = \frac{a}{b-a}$,

$$\sum_{k=1}^{\infty} (1+ck)^j \, y^{k-1} = \sum_{k=0}^{j} \frac{(-1)^k k!}{(1-y)^{k+1}} \sum_{m=k}^{j} \binom{j}{m} (-c)^m \, \mathcal{S}_m^{(k)}$$

For $b = 0$ and any $a > 0$, (7.25) reduces to

$$\sum_{k=0}^{\infty} k^j y^k = (-1)^j \sum_{k=0}^{j} \frac{(-1)^k k!}{(1-y)^{k+1}} \sum_{m=k}^{j} \binom{j}{m} \mathcal{S}_m^{(k)}$$

while the instance of (7.25) with $a = b = 1$ is simpler and also follows from (7.23),

$$\sum_{k=1}^{\infty} k^j y^{k-1} = \sum_{k=0}^{\infty} (k+1)^j \, y^k = (-1)^j \sum_{k=0}^{j} \frac{(-1)^k k! \mathcal{S}_j^{(k)}}{(1-y)^{k+1}} \tag{7.26}$$

124. *Second limit for $E_{a,b}(z)$.* We transform $n = e^x$ back in (7.24),

$$\begin{aligned} T_j(z;n) &= \sum_{k=0}^{\infty} z^k \frac{(b+ak)^j}{n^{(b+ak)}} \\ &= (b-a)^j \, n^{(a-b)} \left(\sum_{m=0}^{j} \binom{j}{m} \left(\frac{a}{a-b}\right)^m \sum_{k=0}^{m} \frac{(-1)^k k! n^{ka}}{(n^a - z)^{k+1}} \mathcal{S}_m^{(k)} \right) \end{aligned}$$

Substitution into the first limit (7.20) and using $S_m^{(0)} = \delta_{0m}$ yield the second limit

$$E_{a,b}(z) = \lim_{n\to\infty} \frac{(-1)^{n-1} \, n^{(a-b)}}{n!} U_n(z) \tag{7.27}$$

where the triple sum is

$$U_n(z) = \sum_{j=0}^{n+1} S_{n+1}^{(j)} (a-b)^j \sum_{m=0}^{j} \binom{j}{m} \left(\frac{a}{a-b}\right)^m \sum_{k=0}^{m} \frac{(-1)^k k! n^{ka}}{(n^a - z)^{k+1}} S_m^{(k)} \tag{7.28}$$

The second limit (7.27) is related to (7.8). Moreover, the derivative $\frac{d^m}{dz^m} E_{a,b}(z)$ directly follows from (7.27) and comparison with (2.52) may shed further light on the polynomials $q_j(a,b,m)$ in Section 2.6.

125. *The triple sum $U_n(z)$ in (7.28).* We concentrate on the finite triple sum $U_n(z)$ in (7.28) and will present several alternatives for $U_n(z)$ and thus slightly different limits of (7.27).

We reverse the m-sum and k-sum in (7.28),

$$U_n(z) = \sum_{j=0}^{n+1} S_{n+1}^{(j)} (a-b)^j \sum_{k=0}^{j} \frac{(-1)^k k! n^{ka}}{(n^a - z)^{k+1}} \sum_{m=k}^{j} \binom{j}{m} \left(\frac{a}{a-b}\right)^m \mathcal{S}_m^{(k)}$$

Reversing additionally the j-sum and k-sum,

$$U_n(z) = \sum_{k=0}^{n+1} (-1)^k k! \frac{\left(\frac{n^a}{n^a - z}\right)^k}{n^a - z} \sum_{j=k}^{n+1} S_{n+1}^{(j)} (a-b)^j \sum_{m=k}^{j} \binom{j}{m} \left(\frac{a}{a-b}\right)^m \mathcal{S}_m^{(k)}$$

Finally, we reverse the j- and m-sum,

$$U_n(z) = \sum_{k=0}^{n+1} (-1)^k k! \frac{\left(\frac{n^a}{n^a - z}\right)^k}{n^a - z} \sum_{m=k}^{n+1} \left(\frac{a}{a-b}\right)^m \mathcal{S}_m^{(k)} \sum_{j=m}^{n+1} \binom{j}{m} S_{n+1}^{(j)} (a-b)^j$$

The sum $Q = \sum_{j=m}^{n+1} \binom{j}{m} S_{n+1}^{(j)} (a-b)^j = \frac{1}{m!} \sum_{j=m}^{n+1} \frac{j!}{(j-m)!} S_{n+1}^{(j)} (a-b)^j$ is computed from the generating function $\frac{\Gamma(x+1)}{\Gamma(x+1-m)} = \sum_{k=0}^{m} S_m^{(k)} x^k$ in (A.5) as $\frac{d^n}{dx^n} \frac{\Gamma(x+1)}{\Gamma(x+1-m)} = \sum_{k=0}^{m} S_m^{(k)} \frac{d^n}{dx^n} x^k = \sum_{j=0}^{m} S_m^{(j)} \frac{j!}{(j-n)!} x^{j-n}$. Hence,

$$\begin{aligned} Q &= \frac{1}{m!} \sum_{j=m}^{n+1} \frac{j!}{(j-m)!} S_{n+1}^{(j)} (a-b)^j = \frac{(a-b)^m}{m!} \sum_{j=m}^{n+1} \frac{j! S_{n+1}^{(j)}}{(j-m)!} (a-b)^{j-m} \\ &= \frac{(a-b)^m}{m!} \frac{d^m}{dx^m} \frac{\Gamma(x+1)}{\Gamma(x-n)} \bigg|_{x=a-b} = \frac{(a-b)^m}{m!} \frac{d^m}{dx^m} \prod_{k=0}^{n} (x-k) \bigg|_{x=a-b} \end{aligned}$$

We return to the triple sum,

$$U_n(z) = \sum_{k=0}^{n+1} (-1)^k k! \frac{\left(\frac{n^a}{n^a - z}\right)^k}{n^a - z} \sum_{m=k}^{n+1} a^m \frac{\mathcal{S}_m^{(k)}}{m!} \frac{d^m}{dx^m} \left(\frac{\Gamma(x+1)}{\Gamma(x-n)}\right)\bigg|_{x=a-b}$$

Reversing the k- and m-sum yields

$$U_n(z) = \frac{1}{n^a} \sum_{m=0}^{n+1} \frac{a^m}{m!} \frac{d^m}{dx^m} \left(\frac{\Gamma(x+1)}{\Gamma(x-n)}\right)\bigg|_{x=a-b} \sum_{k=0}^{m} \frac{(-1)^k k! \mathcal{S}_m^{(k)}}{(1 - n^{-a} z)^{k+1}} \qquad (7.29)$$

126. *From the limit (7.27) back to the Taylor series (1.1).* From the variants of $U_n(z)$ in **art**. 124, we now verify the correctness of (7.27). With $y = n^{-a} z$ in (7.26), thus $\sum_{k=0}^{m} \frac{(-1)^k k! \mathcal{S}_m^{(k)}}{(1-n^{-a}z)^{k+1}} = (-1)^m \sum_{k=1}^{\infty} k^m (n^{-a} z)^{k-1}$,

(7.29) becomes

$$
\begin{aligned}
U_n(z) &= \frac{1}{n^a}\sum_{m=0}^{n+1}\frac{a^m}{m!}\frac{d^m}{dx^m}\left(\frac{\Gamma(x+1)}{\Gamma(x-n)}\right)\bigg|_{x=a-b}(-1)^m\sum_{k=1}^{\infty}k^m\left(n^{-a}z\right)^{k-1} \\
&= \frac{1}{z}\sum_{m=0}^{n+1}\frac{(-a)^m}{m!}\frac{d^m}{dx^m}\left(\frac{\Gamma(x+1)}{\Gamma(x-n)}\right)\bigg|_{x=a-b}\sum_{k=1}^{\infty}\frac{k^m z^k}{n^{ak}} \\
&= \frac{1}{z}\sum_{k=1}^{\infty}\left(\frac{z}{n^a}\right)^k\sum_{m=0}^{n+1}\frac{(-ak)^m}{m!}\frac{d^m}{dx^m}\left(\frac{\Gamma(x+1)}{\Gamma(x-n)}\right)\bigg|_{x=a-b}
\end{aligned}
$$

The Taylor series (1.12) indicates that

$$
\sum_{m=0}^{\infty}\frac{(-ak)^m}{m!}\frac{d^m}{dx^m}\left(\frac{\Gamma(x+1)}{\Gamma(x-n)}\right)\bigg|_{x=a-b} = \frac{\Gamma(-a(k-1)-b+1)}{\Gamma(-a(k-1)-b-n)}
$$

If n is large and with $E_{a,b}(z) = \lim_{n\to\infty}\frac{(-1)^{n-1}n^{(a-b)}}{n!}U_n(z)$ in (7.27), we obtain

$$
E_{a,b}(z) = \lim_{n\to\infty}\frac{(-1)^{n-1}n^{(a-b)}}{n!}\frac{1}{z}\sum_{k=1}^{\infty}\left(\frac{z}{n^a}\right)^k\frac{\Gamma(-a(k-1)-b+1)}{\Gamma(-a(k-1)-b-n)}
$$

The reflection formula (11.4) yields $\frac{\Gamma(-a(k-1)-b+1)}{\Gamma(-a(k-1)-b-n)} = \frac{\Gamma(1+a(k-1)+b+n)}{(-1)^{n-1}\Gamma(a(k-1)+b)}$ and

$$
\begin{aligned}
E_{a,b}(z) &= \lim_{n\to\infty}\frac{n^{(a-b)}}{z}\sum_{k=1}^{\infty}\left(\frac{z}{n^a}\right)^k\frac{\Gamma(1+a(k-1)+b+n)}{n!\Gamma(a(k-1)+b)} \\
&= \sum_{k=0}^{\infty}\frac{z^k}{\Gamma(ak+b)}\lim_{n\to\infty}\frac{n^{(a-b)}\Gamma(n+1+ak+b)}{n^{a(k+1)}\Gamma(n+1)}
\end{aligned}
$$

Finally with (14.12), $\lim_{n\to\infty}\frac{n^{(a-b)}\Gamma(n+1+ak+b)}{n^{a(k+1)}\Gamma(n+1)} = \lim_{n\to\infty}\frac{n^{(a-b)}}{n^{a(k+1)}}n^{ak+b} = 1$ and we arrive again at the Taylor series (1.1) of the Mittag-Leffler function $E_{a,b}(z)$.

7.5 The integral $I_{a,b}(z)$

We will study the integral $I_{a,b}(z) = \int_0^\infty \frac{z^u}{\Gamma(b+au)}du$ in (1.2), which is the "continuous-sum" variant of the Mittag-Leffler function $E_{a,b}(z) = \sum_{k=0}^{\infty}\frac{z^k}{\Gamma(b+ak)}$ and which we have already encountered in Section 3.6 on Euler-Maclaurin summation.

127. *Complex argument* z. Since $z^u = e^{u\log z}$ is a multi-valued function with a branch cut at the negative z-axis, the integral $I_{a,b}(z)$ is also a multi-valued function, which does not exist for negative real z. For $z = re^{i\theta}$, the integral in (1.2),

$$
I_{a,b}\left(re^{i\theta}\right) = \int_0^\infty \frac{r^u e^{i\theta u}}{\Gamma(b+au)}du \tag{7.30}
$$

is split up in a real and imaginary part, assuming that a and b are real, as

$$I_{a,b}\left(re^{i\theta}\right) = \int_0^\infty \frac{r^u \cos\theta u}{\Gamma(b+au)}\,du + i\int_0^\infty \frac{r^u \sin\theta u}{\Gamma(b+au)}\,du \tag{7.31}$$

which illustrates that $I_{a,b}\left(re^{i\theta}\right)$ is only real if $\theta = 0$, i.e. only when z is a real non-negative number. In contrast, the Mittag-Leffler function $E_{a,b}(z)$ in (2.1)

$$E_{a,b}\left(re^{i\theta}\right) = \sum_{k=0}^{\infty} \frac{r^k \cos k\theta}{\Gamma(b+ak)} + i\sum_{k=0}^{\infty} \frac{r^k \sin k\theta}{\Gamma(b+ak)}$$

demonstrates that $E_{a,b}\left(re^{i\theta}\right)$ is real along the entire real axis, because $\sin\theta = 0$ for $\theta = m\pi$ for $m \in \mathbb{Z}$. The point $z = 0$ is a singularity as shown in **art**. 135 below. For real a and b, $I_{a,b}(x)$ is real on the real axis x and the reflection principle (Titchmarsh, 1964, p. 155) states that complex conjugate $I_{a,b}^*(z) = I_{a,b}(z^*)$ for $\operatorname{Re}(z) > 0$.

Both integrals in (7.31) decrease in θ for positive a and b and demonstrate that $\left|I_{a,b}\left(re^{i\theta}\right)\right|$ decreases with $\theta \geq 0$. Hence, we find that $\left|I_{a,b}\left(re^{i\theta}\right)\right| \leq I_{a,b}(r)$ and that $\lim_{\theta\to\infty} I_{a,b}\left(re^{i\theta}\right) = 0$.

After substitution of $t = \theta u$ for $\theta \neq 0$ in (7.30),

$$I_{a,b}\left(re^{i\theta}\right) = \frac{1}{\theta}\int_0^\infty \frac{\left(r^{\frac{1}{\theta}}\right)^t e^{it}}{\Gamma\left(b + \frac{a}{\theta}t\right)}\,dt$$

shows, after comparison with (7.30), that

$$I_{a,b}\left(re^{i\theta}\right) = \frac{1}{\theta} I_{\frac{a}{\theta},b}\left(r^{\frac{1}{\theta}} e^{i}\right)$$

Thus, the integral $I_{a,b}\left(re^{i\theta}\right)$ at any complex number $z = re^{i\theta}$ with for $\theta \neq 0$ can be mapped to an evaluation along the straight line in the complex z-plane with angle equal to 1 radius.

In summary, $I_{a,b}(z)$ is a multi-valued function, which necessitates, just as for the logarithm, to limit the argument θ of z to the usual range $[0, 2\pi]$ or $[-\pi, \pi]$. In the sequel, we implicitly assume that z is not a negative real number and that $I_{a.b}(z)$ is limited to its principle branch or Riemann sheet.

128. *Functional relations of the integral* $I_{a,b}(z)$. In contrast to the Mittag-Leffler function $E_{a,b}(z)$, we can scale the integral $I_{a,b}(z)$ easily by considering various real substitutions in (1.2) (see also **art**. 58). We consider a linear transformation $b + au = \beta + \alpha v$ with $a > 0$ and $\alpha > 0$. Thus, let $u = \frac{\beta - b}{a} + \frac{\alpha}{a}v$ in (1.2), then

$$\begin{aligned} I_{a,b}(z) &= \int_0^\infty \frac{z^u}{\Gamma(b+au)}\,du = \frac{\alpha}{a} z^{\frac{\beta-b}{a}} \int_{\frac{b-\beta}{\alpha}}^\infty \frac{z^{\frac{\alpha}{a}v}}{\Gamma(\beta+\alpha v)}\,dv \\ &= \frac{\alpha}{a} z^{\frac{\beta-b}{a}} \left(\int_0^\infty \frac{\left(z^{\frac{\alpha}{a}}\right)^v}{\Gamma(\beta+\alpha v)}\,dv - \int_0^{\frac{b-\beta}{\alpha}} \frac{z^{\frac{\alpha}{a}v}}{\Gamma(\beta+\alpha v)}\,dv \right) \end{aligned}$$

Using the definition (1.2) of $I_{a,b}(z)$ leads to the linear transformation law

$$I_{a,b}(z) = \frac{\alpha}{a} z^{\frac{\beta-b}{a}} \left(I_{\alpha,\beta}\left(z^{\frac{\alpha}{a}}\right) - \int_0^{\frac{b-\beta}{\alpha}} \frac{z^{\frac{\alpha}{a}v}}{\Gamma(\beta+\alpha v)} dv \right) \tag{7.32}$$

Example If $\beta = b$, then the linear transformation law (7.32) simplifies to

$$I_{a,b}(z) = \frac{\alpha}{a} I_{\alpha,b}\left(z^{\frac{\alpha}{a}}\right)$$

In particular, for $a > 0$ and $\alpha = \frac{1}{a}$, the map

$$I_{\frac{1}{a},b}(z) = a^2 I_{a,b}\left(z^{a^2}\right) \tag{7.33}$$

is the exact companion of the approximate $E_{\frac{1}{a},b}(z) \approx a^2 E_{a,b}\left(z^{a^2}\right)$ between $E_{\frac{1}{a},b}(z)$ and $E_{a,b}(z)$, based on Bieberbach's deductions in **art.** 75.

129. *Properties deduced from the linear transformation law (7.32).* We consider some particular cases for the auxiliary parameters α and β in the linear transformation law (7.32).

Property Differentiating $z^{-\frac{\beta-b}{a}} I_{a,b}(z) = \frac{\alpha}{a} \int_{\frac{b-\beta}{\alpha}}^{\infty} \frac{z^{\frac{\alpha}{a}v}}{\Gamma(\beta+\alpha v)} dv$ with respect to b yields

$$\frac{\partial}{\partial b}\left(z^{\frac{b-\beta}{a}} I_{a,b}(z)\right) = -\frac{1}{a} \frac{z^{\frac{b-\beta}{a}}}{\Gamma(b)}$$

which is independent of α and which leads to the linear differential equation in b,

$$\frac{\partial I_{a,b}(z)}{\partial b} + \frac{\log z}{a} I_{a,b}(z) = -\frac{1}{a\Gamma(b)}$$

Property If $\alpha = 1$ and $\beta = 0$, then the linear transformation law (7.32) becomes

$$I_{a,b}(z) = \frac{z^{-\frac{b}{a}}}{a} \int_b^{\infty} \frac{\left(z^{\frac{1}{a}}\right)^v}{\Gamma(v)} dv \tag{7.34}$$

which, similarly to the Mittag-Leffler function $E_{a,b}(z)$ and exhibited in the differential recursion (2.39) in **art.** 16, illustrates the appearance of the "natural" $w = z^a$ map, suggesting to consider $I_{a,b}(z^a)$ rather than $I_{a,b}(z)$. Indeed, the integral in (7.34) further looses the parameter a,

$$I_{a,b}(z^a) = \int_0^{\infty} \frac{z^{au} du}{\Gamma(b+au)} = \frac{1}{a} \int_0^{\infty} \frac{z^t dt}{\Gamma(b+t)} = \frac{z^{-b}}{a} \int_b^{\infty} \frac{e^{v \log z}}{\Gamma(v)} dv$$

from which we deduce that

$$\frac{\partial}{\partial a}\left(a I_{a,b}(z^a)\right) = 0 \tag{7.35}$$

The differential equation (7.35) is equivalent, for any positive real a_1 and a_2, to

$$a_1 I_{a_1,b}\left(z^{a_1}\right) = a_2 I_{a_2,b}\left(z^{a_2}\right)$$

In fact, (7.33) is an instance of (7.35). The complex integral $aI_{a,b}\left(z^a\right) = \frac{z^{1-b}}{2\pi i}\int_{c-i\infty}^{c+i\infty}\frac{e^{(zs-b\log s)}}{\log s}ds$ in (7.38) below also confirms (7.35).

Property If $b > \beta > 0$ and $z = x$ is real and positive, then the integral in (7.32) is non-negative and

$$I_{a,b}\left(x\right) < \frac{\alpha}{a}x^{\frac{\beta-b}{a}}I_{\alpha,\beta}\left(x^{\frac{\alpha}{a}}\right)$$

which bears resemblance to the inequality $E_{m,b}\left(x\right) < \frac{x^{\frac{1-b}{m}}}{m}e^{x^{\frac{1}{m}}}$ in **art.** 17.

130. *Differential recursion.* Differentiating the integral $I_{a,b}\left(z\right)$ in (1.2) and using the functional equation (10.1) of the Gamma function,

$$\frac{d\left\{z^{b-1}I_{a,b}\left(z^a\right)\right\}}{dz} = \int_0^\infty \frac{(au+b-1)\,z^{au+b-2}}{\Gamma\left(b+au\right)}du = \int_0^\infty \frac{z^{au+b-2}}{\Gamma\left(b-1+au\right)}du$$

leads to the differential recursion equation

$$\frac{d}{dz}\left\{z^{b-1}I_{a,b}\left(z^a\right)\right\} = z^{b-2}I_{a,b-1}\left(z^a\right) \tag{7.36}$$

which is precisely the same as for the Mittag-Leffler function $E_{a,b}\left(z^a\right)$ in (2.39).

131. *Integral Mean Value Theorem for $I_{a,b}\left(x\right)$.* The Integral Mean Value Theorem (**art.** 114) shows, for positive real a, b and x, that there exists a real number $\varphi > 0$ such that

$$I_{a,b}\left(x\right) = \int_0^\infty \frac{x^u}{\Gamma\left(b+au\right)}du = \frac{1}{\Gamma\left(b+a\varphi\right)}\int_0^\infty e^{u\log x}du$$

The last integral only exists for $0 \leq x < 1$. Hence, for $a > 0, b > 0$, $0 \leq x < 1$ and $\varphi \geq 0$, it holds that

$$I_{a,b}\left(x\right) = \frac{1}{\Gamma\left(b+a\varphi\right)}\frac{1}{\left|\log x\right|} \leq \frac{1.12918}{\left|\log x\right|}$$

where the inequality follows from the maximum value (**art.** 154) of $\frac{1}{\Gamma(u)}$ for $u \geq 0$. Hence, approaching the essential singularity at $z = 0$ along the positive real axis gives $\lim_{x\to 0} I_{a,b}\left(x\right) = 0$.

132. *A complex integral representation for $I_{a,b}\left(z\right)$.* We start by concentrating on the integral

$$z^{b-1}I_{a,b}\left(z^a\right) = \int_0^\infty \frac{z^{b-1+au}}{\Gamma\left(b+au\right)}du$$

whose Laplace transform is

$$L_{a,b}\left(s\right) = \int_0^\infty \left\{z^{b-1}I_{a,b}\left(z^a\right)\right\}e^{-zs}dz = \int_0^\infty e^{-zs}\int_0^\infty \frac{z^{b-1+au}}{\Gamma\left(b+au\right)}du\,dz$$

The reversal of the integrals is justified by absolute convergence,

$$L_{a,b}(s) = \int_0^\infty \frac{du}{\Gamma(b+au)} \int_0^\infty z^{b+au-1} e^{-zs} dz = \int_0^\infty \frac{du}{\Gamma(b+au)} \frac{\Gamma(b+au)}{s^{b+au}}$$
$$= s^{-b} \int_0^\infty e^{-au \log s} du = \frac{s^{-b}}{a \log s}$$

The inverse Laplace transform (B.2) returns a complex integral,

$$z^{b-1} I_{a,b}(z^a) = \frac{1}{2\pi i a} \int_{c-i\infty}^{c+i\infty} \frac{e^{zs}}{s^b \log s} ds \tag{7.37}$$

where $c > 1$ because $L_{a,b}(s)$ is only analytic for $\operatorname{Re}(s) > 1$ due to the pole at $s = 1$. After replacing z^a by z in (7.37) and combining with the definition (1.2), we obtain

$$I_{a,b}(z) = \int_0^\infty \frac{e^{u \log z}}{\Gamma(b+au)} du = \frac{z^{\frac{1-b}{a}}}{2\pi i a} \int_{c-i\infty}^{c+i\infty} \frac{e^{z^{\frac{1}{a}} s}}{s^b \log s} ds \qquad c > 1 \tag{7.38}$$

Another derivation of (7.38) follows directly from Hankel's contour (15.1) as

$$I_{a,b}(z) = \frac{1}{2\pi i} \int_C w^{-b} e^w dw \int_0^\infty e^{-u(a \log w - \log z)} du$$

Only if $\operatorname{Re}(a \log w - \log z) > 0$, which is equivalent to $\operatorname{Re}\left(\log \frac{w^a}{z}\right) = \log\left|\frac{w^a}{z}\right| > 0$ and $\left|\frac{w^a}{z}\right| > 1$, then

$$I_{a,b}(z) = \frac{1}{2\pi i} \int_C \frac{w^{-b} e^w}{a \log w - \log z} dw \qquad \text{with } |w| > |z|^{\frac{1}{a}} \tag{7.39}$$

where the constraint $|w| > |z|^{\frac{1}{a}}$ requires to deform the contour C (as explained in **art.** 74). A formal substitution $w = z^{\frac{1}{a}} s$ transforms, except for the Hankel contour C, (7.39) into (7.38). The Hankel contour is less restricted than the line parallel with the imaginary axis (see **art.** 206).

133. *Evaluation of the complex integral (7.38) for* $I_{a,b}(z)$. We distinguish between two regimes: $\operatorname{Re}\left(z^{\frac{1}{a}}\right) < 0$ and $\operatorname{Re}\left(z^{\frac{1}{a}}\right) > 0$.

a. For $\operatorname{Re}\left(z^{\frac{1}{a}}\right) < 0$, the contour in (7.38) can be closed over the positive $\operatorname{Re}(s)$-plane, in which the integrand is analytic and[3] $I_{a,b}(z) = 0$. Let

[3] Hence, for $\operatorname{Re}\left(z^{\frac{1}{a}}\right) < 0$, it holds that

$$0 = \int_0^\infty \frac{e^{u \log r} e^{iu\theta}}{\Gamma(b+au)} du = \int_b^\infty \frac{e^{w \frac{\log r}{a}} \left\{\cos\left(\frac{\theta}{a}(w-b)\right) + i \sin\left(\frac{\theta}{a}(w-b)\right)\right\}}{a r^{\frac{b}{a}} \Gamma(w)} dw$$

implying that

$$\int_b^\infty \frac{e^{w \frac{\log r}{a}} \cos\left(\frac{\theta}{a}(w-b)\right)}{\Gamma(w)} dw = \int_b^\infty \frac{e^{w \frac{\log r}{a}} \sin\left(\frac{\theta}{a}(w-b)\right)}{\Gamma(w)} dw = 0$$

$z = re^{i\theta}$ with $\theta = \arg z \in [-\pi, \pi]$ and recalling that $a > 0$, then $\mathrm{Re}\left(z^{\frac{1}{a}}\right) = r^{\frac{1}{a}} \cos\frac{\theta}{a}$ and $\mathrm{Re}\left(z^{\frac{1}{a}}\right) < 0$ requires that $\cos\frac{\theta}{a} < 0$, which is equivalent to $\frac{\pi}{2} < \frac{\pm\theta}{a} < \pi$ or $\frac{\pi a}{2} < |\arg z| < \pi a$. The latter condition, combined with $0 < |\arg z| < \pi$ is only possible if $0 < a < 2$. If $1 \leq a < 2$, then the combined condition means that $\frac{\pi}{2} < |\arg z| < \pi$ or that $\mathrm{Re}(z) < 0$. Only if $0 < a < \frac{1}{2}$, then the combined condition means that $\frac{\pi}{2} < |\arg z| < \frac{\pi a}{2}$ or that $\mathrm{Re}(z) > 0$, while for $\frac{1}{2} \leq a \leq 1$, $\mathrm{Re}(z)$ can be either sign.

b. For $\mathrm{Re}\left(z^{\frac{1}{a}}\right) > 0$, we close the contour in (7.38) over the negative $\mathrm{Re}(s)$-plane around the branch cut of $s^b \ln(s)$, which is the negative real axis. Thus, we consider a Hankel-type contour C that consists of the line at $c > 1$, the quarter of a circle with infinite radius from $\frac{\pi}{2}$ to $\pi - \varepsilon$, the line segment above the real negative axis from minus infinity to $s = 0$, the circle around the origin $s = 0$ from $\pi - \varepsilon$ back to $-\pi - \varepsilon$ with radius δ, the line segment below the real negative axis from $s = 0$ towards minus infinity, the quarter circle with infinite radius back to close the contour C. This contour encloses the pole at $s = 1$, whose residue is $\lim_{s\to 1} \frac{e^{z^{\frac{1}{a}} s}(s-1)}{s^b \ln s} = e^{z^{\frac{1}{a}}}$. Cauchy's Residue Theorem (Titchmarsh, 1964) results in

$$\frac{1}{2\pi i} \int_C \frac{e^{z^{\frac{1}{a}} s}}{s^b \ln s} ds = e^{z^{\frac{1}{a}}}$$

while the evaluation of the contour C yields

$$\frac{1}{2\pi i} \int_C \frac{e^{z^{\frac{1}{a}} s}}{s^b \ln s} ds = a z^{\frac{b-1}{a}} I_{a,b}(z) + \frac{1}{2\pi i} \int_\infty^0 \frac{e^{-z^{\frac{1}{a}} x} d\left(e^{i(\pi-\varepsilon)} x\right)}{x^b e^{ib(\pi-\varepsilon)} \ln\left(x e^{i(\pi-\varepsilon)}\right)}$$
$$+ \frac{1}{2\pi i} \int_0^\infty \frac{e^{-z^{\frac{1}{a}} x} d\left(e^{i(-\pi-\varepsilon)} x\right)}{x^b e^{ib(-\pi-\varepsilon)} \ln\left(x e^{i(-\pi-\varepsilon)}\right)}$$

since the parts of C along the circles vanish for $\mathrm{Re}(z) > 0$, but for $\delta \to 0$ only provided $\mathrm{Re}(b) \leq 1$. Hence, we obtain

$$a z^{\frac{b-1}{a}} I_{a,b}(z) = e^{z^{\frac{1}{a}}} + \frac{1}{2\pi i} \int_0^\infty \frac{e^{-z^{\frac{1}{a}} x}}{x^b} \left(\frac{e^{ib\pi}}{\ln x - i\pi} - \frac{e^{-ib\pi}}{\ln x + i\pi}\right) dx$$

In summary, with $\frac{e^{ib\pi}}{\ln x - i\pi} - \frac{e^{-ib\pi}}{\ln x + i\pi} = 2\pi i \frac{\frac{\sin b\pi}{\pi} \ln x + \cos b\pi}{\left(\pi^2 + (\ln x)^2\right)}$ and the definition (1.2) of $I_{a,b}(z)$, we arrive, for $\mathrm{Re}(b) \leq 1$ and $\mathrm{Re}\left(z^{\frac{1}{a}}\right) > 0$, at

$$\int_0^\infty \frac{z^u}{\Gamma(b + au)} du = \frac{z^{\frac{1-b}{a}}}{a} \left\{ e^{z^{\frac{1}{a}}} + \int_0^\infty \frac{e^{-z^{\frac{1}{a}} x}}{x^b} \left(\frac{\frac{\sin b\pi}{\pi} \ln x + \cos b\pi}{\pi^2 + (\ln x)^2} \right) dx \right\} \tag{7.40}$$

while, for $\mathrm{Re}\left(z^{\frac{1}{a}}\right) < 0$, it holds that $I_{a,b}(z) = 0$.

I found later in Hardy (1978, p. 196, eq. (11.11.1)) that Ramanujan has proved (7.40) differently in the form, for real $y \geq 0$ and $\xi \geq 0$,

$$\int_{-\xi}^{\infty} \frac{y^u du}{\Gamma(1+u)} + \int_0^{\infty} x^{\xi-1} e^{-yx} \left(\frac{\cos \pi\xi - \frac{\sin \pi\xi}{\pi} \ln x}{\pi^2 + (\ln x)^2} \right) dx = e^y$$

As Hardy wrote, Ramanujan's proof is very ingenious. If the left-hand side is denoted by $p(y, \xi)$, then $\frac{\partial p}{\partial \xi} = 0$, implying that $p(y, \xi)$ is a function $P(y)$ of y only. Further, Ramanujan finds by differentiating $p(y, \xi)$ with respect to y that $\frac{dP(y)}{dy} = P(y)$. Hence, $P(y) = Ce^y$, where the constant C is determined by letting $y = \xi = 0$.

134. *Deductions from the integral (7.40) for $I_{a,b}(z)$.* The analysis that led to (7.40) is analogous to that in **art.** 67 and to the Euler-Maclaurin transformation (3.50) of $E_{a,b}(z)$ in **art.** 58. Expression (7.40) is also related to Bierberbach's integral in **art.** 75, written as

$$E_{a,b}(z) = \frac{z^{\frac{1-b}{a}}}{a} \left\{ e^{z^{\frac{1}{a}}} + \frac{1}{2\pi i} \frac{1}{z} \int_{C'_{a\phi}} \frac{\left(\frac{t}{z}\right)^{\frac{1-b}{a}} e^{t^{\frac{1}{a}}}}{\left(\frac{t}{z}\right) - 1} dt \right\} \tag{7.41}$$

For $b = 1$, (7.40) simplifies to

$$I_{a,1}(z) = \frac{1}{a} \left\{ e^{z^{\frac{1}{a}}} - \int_0^{\infty} \frac{e^{-z^{\frac{1}{a}} x}}{x\left(\pi^2 + (\ln x)^2\right)} dx \right\}$$

which is the companion of (4.56) for $E_a(z)$. Since $\int_0^{\infty} \frac{e^{-\lambda x}}{x\left(\pi^2 + (\ln x)^2\right)} dx$ is increasing[4] in $\text{Re}(\lambda) > 0$ from 0 to 1, the following lower and upper bound for $\text{Re}\left(z^{\frac{1}{a}}\right) > 0$ hold,

$$\frac{1}{a} e^{z^{\frac{1}{a}}} - \frac{1}{a} < I_{a,1}(z) < \frac{1}{a} e^{z^{\frac{1}{a}}}$$

With $\alpha = \beta = 1$ in the linear map (7.32) and $I_{1,1}\left(z^{\frac{1}{a}}\right)$ from (7.40), we find for $\text{Re}\left(z^{\frac{1}{a}}\right) > 0$,

$$I_{a,b}(z) = \frac{z^{\frac{1-b}{a}}}{a} \left(e^{z^{\frac{1}{a}}} - \int_0^{\infty} \frac{e^{-z^{\frac{1}{a}} x}}{x\left(\pi^2 + (\ln x)^2\right)} dx - \int_0^{b-1} \frac{\left(z^{\frac{1}{a}}\right)^x}{x\Gamma(v)} dx \right)$$

Comparison with (7.40) indicates, for $b \leq 1$ and $\text{Re}(z) > 0$, that

$$\int_0^{\infty} \frac{e^{-zx}}{x^b} \left(\frac{\frac{\sin b\pi}{\pi} \ln x + \cos b\pi}{\pi^2 + (\ln x)^2} \right) dx = \int_0^{\infty} \frac{-e^{-zx}}{\pi^2 + (\ln x)^2} \frac{dx}{x} - \int_0^{b-1} \frac{z^v dv}{\Gamma(v+1)}$$

[4] Indeed (see Van Mieghem (2014, p.73)),

$$\int_0^{\infty} \frac{e^{-\lambda x}}{x\left(\pi^2 + (\ln x)^2\right)} dx = \int_{-\infty}^{\infty} \frac{e^{-\lambda e^t}}{\pi^2 + t^2} dt \leq \int_{-\infty}^{\infty} \frac{dt}{\pi^2 + t^2} = 1$$

135. *A series for $I_{a,b}(z)$.* The Taylor series of $I_{a,b}(z)$ around $z = w$ equals

$$I_{a,b}(z) = \sum_{j=0}^{\infty} \frac{1}{j!} \left. \frac{d^j I_{a,b}(z)}{dz^j} \right|_{z=w} (z-w)^j$$

For real $a > 0$, the derivative $\left. \frac{d^j I_{a,b}(z)}{dz^j} \right|_{z=w}$ equals

$$\int_0^{\infty} \frac{\left. \frac{d^j}{dz^j} z^u \right|_{z=w}}{\Gamma(b+au)} du = \frac{1}{w^j} \int_0^{\infty} \frac{u(u-1)\cdots(u-j+1)\, e^{u \log w}}{\Gamma(b+au)} du$$

and converges for all finite j, except for real $w < 0$. Since the integrand $\frac{uw^{u-1}}{\Gamma(b+au)} > 0$ for $u, a, b > 0$, implying that $\left. \frac{dI_{a,b}(z)}{dz} \right|_{z=x} > 0$ for positive real x and a, b, the function $I_{a,b}(x)$ increases strictly for all $x > 0$.

In contrast to $E_{a,b}(z)$, the function $I_{a,b}(z)$ is not entire and has an essential singularity at $z = 0$, where none of the derivatives exists. Despite the fact that $I_{a,b}(z)$ is not an entire function in z, $I_{a,b}(z)$ shares the same differential recursion (7.36) as the Mittag-Leffler function. Thus, an analogous method as in **art.** 18 leads to the same closed form for $\left. \frac{d^j I_{a,b}(z)}{dz^j} \right|_{z=w}$ as (2.48) and (2.52), with formally $E_{a,b}$ replaced by $I_{a,b}$. The observation holds for a general class (see **art.** 16), which enlarges the role of the polynomial $q_j(a, b, m)$ in Section 2.6.

We expand the integrand of the integral $I_{a,b}(z)$, defined in (1.2), in a Taylor series around b,

$$\frac{1}{\Gamma(b+au)} = \sum_{j=0}^{\infty} \frac{1}{j!} \left. \frac{d^j}{dy^j} \frac{1}{\Gamma(y)} \right|_{y=b} (au)^j$$

which converges for all u and b, and obtain, assuming that reversal of the sum and integral is allowed,

$$I_{a,b}(z) = \int_0^{\infty} \frac{z^u}{\Gamma(b+au)} du = \sum_{j=0}^{\infty} \frac{a^j}{j!} \left. \frac{d^j}{dy^j} \frac{1}{\Gamma(y)} \right|_{y=b} \int_0^{\infty} u^j e^{-u(-\log z)} du$$

Only if $\operatorname{Re}(\log z) < 0$, thus $0 \le |z| < 1$, then we arrive at

$$I_{a,b}(z) = -\frac{1}{\log z} \sum_{j=0}^{\infty} \left. \frac{d^j}{dy^j} \frac{1}{\Gamma(y)} \right|_{y=b} \left(-\frac{a}{\log z} \right)^j \tag{7.42}$$

However, the reversal is only allowed after limiting the range in $|z|$ further. The series (7.42) only converges if $\left| \frac{a}{\log z} \right|$ is sufficiently small. In particular, for $b = 1$, our asymptotic formula (15.36) indicates that $\left| \frac{a}{\log z} \right| < \frac{1}{\sqrt{\pi^2+\gamma^2}}$, which is numerically verified: for $z = \frac{1}{30}$, $a = b = 1$, then $K = 20$ terms in (7.42) give four digits accurate.

The series (7.42) indicates (as in **art.** 131) that $I_{a,b}(0) = 0$.

8
Epilogue

Driven by the surge in fractional analysis Mainardi (2021), the Mittag-Leffler function $E_{a,b}(z)$, called by Mainardi (2020) the "Queen Function of the Fractional Calculus", gains increasing interest. I will end this work by enumerating some open problems.

Although partial results exist, the determination of the zeros of $E_{a,b}(z)$ in the complex plane still stands on the agenda. Indeed, Weierstrass's entire function theory (Titchmarsh, 1964) and Hadamard factorization theorem (1.18) show that any entire function can be represented as a product form that contains all the zeros and such general product form for $E_{a,b}(z)$ has not been found yet. A general Lagrange series for $G\left(f^{-1}(z)\right)$, thus a function $G(z)$ of the inverse function $f^{-1}(z)$ of the function $f(z)$, in terms of characteristic coefficients (**art.** 18), is available (see e.g. Van Mieghem (2023a, art. 342), Van Mieghem (1996, Appendix A)). Thus, the zero most close to a point z_0 in the complex plane can be approximated accurately, if the Taylor coefficients of $f(z)$ around z_0 are known and easily computable.

Related to this root-locus problem is the study of the inverse function, which is the solution $z = E_{a,b}^{-1}(w)$ of $w = E_{a,b}(z)$. The inverse function $f^{-1}(z)$ satisfies both $f\left(f^{-1}(z)\right) = z = f^{-1}(f(z))$ and the derivative equation $\left.\frac{df(u)}{du}\right|_{u=f^{-1}(z)} \frac{df^{-1}(z)}{dz} = 1$. Hence, the derivative of the inverse Mittag-Leffler function becomes, with $\frac{d}{dz}E_{a,b}(z) = \frac{E_{a,b-1}(z)-(b-1)E_{a,b}(z)}{az}$ in (2.37),

$$\frac{d}{dz}E_{a,b}^{-1}(z) = \frac{1}{\left.\frac{d}{dw}E_{a,b}(w)\right|_{w=E_{a,b}^{-1}(z)}} = \frac{aE_{a,b}^{-1}(z)}{E_{a,b-1}\left(E_{a,b}^{-1}(z)\right) - (b-1)z}$$

from which the logarithmic derivative of the inverse Mittag-Leffler function satisfies

$$\frac{d}{dz}\log E_{a,b}^{-1}(z) = \frac{a}{E_{a,b-1}\left(E_{a,b}^{-1}(z)\right) - (b-1)z} \tag{8.1}$$

If $b = 1$, then (8.1) simplifies with $E_{a,0}(z) = zE_{a,a}(z)$ in (2.2) to

$$\frac{d}{dz}E_a^{-1}(z) = \frac{a}{E_{a,a}\left(E_a^{-1}(z)\right)}$$

At any rate, an analytic evaluation of $E_{a,b-1}\left(E_{a,b}^{-1}(z)\right)$ seems crucial to proceed with (8.1). In other words, can we express $E_{a,b-1}(z)$ solely in terms of $E_{a,b}(z)$? Mainly numerical computations of the inverse Mittag-Leffler function exist (see e.g. Hilfer and Seybold (2006)), but few analytic results. Zeng and Chen (2015) concentrate on a Padé approximation for $E_{a,b}(-x)$ for real $x \geq 0$ with $0 < a \leq 1$ and $b > a$, which is complete monotonic (**art.** 90),

$$\Gamma(b-a)\, xE_{a,b}(-x) \approx \frac{p_0 + p_1 x + \cdots + p_n x^n}{q_0 + q_1 x + \cdots + q_n x^n} \tag{8.2}$$

where the problem consists of finding the coefficients p_j and q_j for $0 \leq j \leq n$ of the polynomials of degree n so that (8.2) has correct expansions at $x = 0$ and at $x \to \infty$. They confine to the case $n = 2$ and obtain fairly accurate approximations of the type $y = E_{a,b}(-x) \approx \frac{r_0 + r_1 x + r_2 x^2}{t_0 + t_1 x + t_2 x^2}$, that, in addition, allow analytic inversion $x \approx -E_{a,b}^{-1}(y)$ for $0 < a < 1$ and $b > a$.

Expansions of a function into a set of independent functions frequently appears in the theory of functions, for example, in Fourier analysis. Another example appears in the theory of Bessel functions (Watson, 1995, Chapter XVI). **Art.** 31 shows that, for non-rational $a > 0$ and non-integer b, the set of m functions $E_{a,b-1}(z), E_{a,b-2}(z), \ldots, E_{a,b-m}(z)$ is linearly independent, implying that any "sufficiently well-behaved" function $f(z)$ can be expressed as

$$f(z) = \sum_{j=0}^{\infty} c_j(f)\, E_{a,b-j}(z)$$

for a particular set of coefficients $\{c_j(f)\}_{j\geq 0}$ that are independent of z. How to find the coefficient $c_j(f)$ for a given function $f(z)$? What is the closed form (Van Mieghem, 2023a, art. 350-351) of the scalar product

$$\left(E_{a,b-j}(z), E_{a,b-k}(z)\right) = \int_{\alpha}^{\beta} E_{a,b-j}(z)\, E_{a,b-k}(z)\, dW(z)$$

where $W(z)$ is non-negative weight function over an interval $[\alpha, \beta]$? How to construct an orthogonal set from the set $\{E_{a,b-j}\}_{j\geq 0}$?

The set of linearly independent functions $\{E_{a,b-j}(z)\}_{j\geq 1}$ can be extended. Indeed, properties of the Mittag-Leffler function in Chapter 2, deduced from the general differential recursion (2.41) in b in **art.** 16, also hold for a general class of functions, such as $f_{a,b}(z) = \sum_{k=0}^{\infty} \frac{f_k z^k}{\Gamma(b+ak)}$ and associated integral $\mathcal{F}_{a,b}(z) = \int_0^{\infty} \frac{f(u) z^u}{\Gamma(b+au)} du$. Those functions feature the same form (2.53) of Taylor series around an arbitrary point z_0 in terms of the polynomials $q_j(a, b, m)$ (**art.** 20), the non-existence of linear differential equations for non-rational parameter a (Section 2.8) and a Taylor series as (2.93) for their logarithm $\log f_{a,b}(z)$. Hence, the general differential recursion (2.41) in b may lead to an "algebra" of common properties of all functions that obey (2.41).

Other open problems in the theory of Mittag-Leffler functions are parts of the fractional semi-group inequality (**art**. 106) and of the Garrappa-Popolizio conjecture (**art**. 109). Further, are there explicit, closed forms of the logarithm of the Mittag-Leffler function $E_{a,b}(z)$ or do there exist product forms, rather than series expansions?

I would like to end the first part on the Mittag-Leffler function with the conviction that there are still beautiful, mathematical "Mittag-Leffler treasures" hidden in the complex plane.

Part II

The Gamma function

9
Introduction

Besides the basic functions (see e.g. Jordan (1909, Chapitre II, Sections 3-5), Hardy (2006, Chapters IX and X)) like the polynomial, fractions and powers of polynomials, the exponential, logarithm, circular or trigonometric functions (as sinus, cosinus, tangens, etc.), the Gamma function is the next important complex function. Nearly all books on function theory (see e.g. Jordan (1909); Erdélyi *et al.* (1953b); Sansone and Gerretsen (1960); Titchmarsh (1964); Whittaker and Watson (1996); Markushevich (1985); Evgrafov (2019); Bourbaki (2004)) treat the Gamma function. I recommend the historical account on the Gamma function, from the point of view of an interpolation problem, written by Davis (1959), for which he received the Chauvenet Prize in 1963. Section 9.4 gives a brief flavor of how mathematics reforms into a rigorous building of beautiful logic and formulas, inspired by the brilliant ideas of our great genii.

In his beautiful book on the Gamma function, Nielsen (1956) cites the historic achievements and reviews most contributions before 1906. Nielsen (1956) starts his book with the functional equation $\Gamma(z+1) = z\Gamma(z)$ in (10.1) of the Gamma function and immediately remarks that any solution can be multiplied by a periodic function $\omega(z) = \omega(z+1)$ with period 1. A similar approach was followed by Jensen and Gronwall (1916), whose great paper offers a nice treatment of the Gamma function, without using integrals! Artin (1964) also starts from the functional equation (10.1), but adds convexity. We briefly discuss Artin's approach in Section 10.4. Next, Nielsen (1956) concentrates on the digamma function, which is the logarithmic derivative $\psi(z) = \frac{d}{dz}\log\Gamma(z)$ of the Gamma function and which we treat in Chapter 12. Nielsen then applies Weierstrass's factorization theory for entire functions (Titchmarsh, 1964) and deduces Weierstrass's product (11.1), $\frac{1}{\Gamma(z+1)} = e^{\gamma z}\prod_{n=1}^{\infty}\left(1+\frac{z}{n}\right)e^{-\frac{z}{n}}$, which we will derive from the Euler-Gauss product (10.10)

$$\Gamma(z+1) = \prod_{n=1}^{\infty}\left(1+\frac{1}{n}\right)^z\left(1+\frac{z}{n}\right)^{-1}$$

in **art.** 148. Weierstrass created his magnificent theory for entire functions, a pearl of complex function theory, inspired by the Euler-Gauss product (10.10) and Gauss's remark on factorization in Gauss (1813, p. 146).

We approach the theory of the Gamma function $\Gamma(z)$ by concentrating on the Euler-Gauss product (10.10), valid for all complex z, rather than starting from Euler's integral $\Gamma(z) = \int_0^\infty e^{-t} t^{z-1} dt$ in (10.2), which is only valid for $\operatorname{Re} z > 0$. Jordan (1913, Chapitre III, Section 2), on the other hand, derives (a part of) the theory from Euler's integral (10.2) without mentioning the digamma function. In particular (**art.** 165), Jordan (1913, art. 201) computes the integral $\Gamma'(z) = \int_0^\infty e^{-t} t^{z-1} \log t \, dt$ by substituting $\log t = \int_0^\infty \frac{e^{-u} - e^{-tu}}{u} du$ (as in **art.** 164), shows that reversal of integration is allowed and differently arrives at the integral (12.54) for $\log \Gamma(z)$ in **art.** 171. By using the generating function (A.13) of the Bernoulli numbers, Jordan (1913, art. 215) finds Stirling's asymptotic formula (14.1).

9.1 Outline of Part II on the Gamma function

This second Part on the Gamma function starts in Chapter 10 with the question how to define the Gamma function $\Gamma(z)$ for complex z as a generalization of the factorial $n! = 1.2.3 \ldots n$. The general definition of the Gamma function is the Euler-Gauss product (10.10) that holds for all complex z, although Euler's integral $\Gamma(z) = \int_0^\infty e^{-t} t^{z-1} dt$ in (10.2) is easier and more used, but only holds for $\operatorname{Re}(z) > 0$. The Euler-Gauss product (10.10) was inspired by Euler's infinite product (9.9). However, Euler did not explain how he came to his invention as explained Section 9.4. Several different viewpoints of Rademacher (**art.** 143), Jensen (**art.** 144), Laugwitz and Rodewald (**art.** 145) and Artin (**art.** 146) are discussed.

Chapter 11 deduces the most important properties of the Gamma function $\Gamma(z)$ from the Euler-Gauss product (10.10): the Weierstrass product (**art.** 148), the reflection formula $\Gamma(z)\Gamma(1-z) = \frac{\pi}{\sin \pi z}$ in (11.4) in **art.** 149, explicit expressions of the absolute value of $|\Gamma(z)|$ at certain values of z in **art.** 150, Gauss's multiplication formula (11.8) in **art.** 151 and products of Gamma functions (**art.** 153).

The Euler-Gauss product (10.10) is also the natural route to study the digamma function $\psi(z) = \frac{d}{dz} \log \Gamma(z)$ in Chapter 12. In fact, the digamma function $\psi(z+1) = -\gamma + \sum_{n=1}^\infty \left(\frac{1}{n} - \frac{1}{z+n} \right)$ in (12.3) is mathematically a little easier than the Gamma function. The digamma function $\psi(z)$ is related to the Hurwitz Zeta-function $\zeta(s,z)$ in **art.** 156, while Lerch's elegant formula $\log \Gamma(z) = \zeta'(0,z) - \zeta'(0)$ in (12.16) outlines its relation to $\log \Gamma(z)$. Section 12.2 derives Gauss's remarkable closed form of digamma function (12.23) for fractions $z = \frac{j}{m} \in \mathbb{Q}$. We generalize Gauss's approach with an exact closed form (12.32) of the generating function $\sum_{j=1}^m \psi\left(\frac{j}{m}\right) z^{j-1}$ over the fractions in **art.** 162. Section 12.3 of Chapter 12 derives several integrals for $\log \Gamma(z)$, that open the way towards Binet's function $\mu(z) = \log \Gamma(z) - \left(z - \frac{1}{2}\right) \log z + z - \frac{1}{2} \log(2\pi)$ in (13.1).

Section 12.4 on interpolation presents the binomial or Newton series for the digamma function $\psi(z)$ in **art.** 170 and for $\log\Gamma(z)$ in **art.** 172.

Chapter 13 studies Binet's converging factorial series for the Binet function $\mu(z)$. Binet (1839) has derived two different factorial series (13.4) and (13.30), whereas Gilbert (and we independently) show in Theorem 5 of Chapter 13 that there exists infinitely many converging factorial series for the Binet function $\mu(z)$. Although the Binet-types of factorial series converge as opposed to the Stirling asymptotic expansion (14.1), they are far less known and used than Stirling's asymptotic expansion (Section 14.1).

Chapter 14 further consider the behavior of both $\log\Gamma(z)$ and $\Gamma(z)$ for large values z. In **art.** 198, we derive the asymptotic expansion of $\frac{\Gamma(a+z)}{\Gamma(b+z)}$ in (14.12). **Art.** 200 computes the Stirling variant for $\Gamma(z+1) \simeq \sqrt{2\pi z}\left(\frac{z}{e}\right)^z\left(1+\sum_{m=1}^{K}\frac{\widetilde{c}_m}{z^m}\right)$ up to arbitrary K in (14.17) and list its coefficients $\widetilde{c}_m$ exactly up to $m = 15$ terms. We end Chapter 14 with the beautiful, converging counter part of Stirling's expansion (14.17), the factorial series (14.26) of Cornelius Lanczos.

Chapter 15 investigates the entire function $\frac{1}{\Gamma(z)}$. The Euler-Gauss product (10.10) shows that $\frac{1}{\Gamma(z)}$ is an entire function in the complex plane, with zeros at the non-positive integers $z = -k$ with $k \in \mathbb{N}$. First in Section 15.1, we derive Hankel's complex integral $\frac{1}{\Gamma(z)} = \frac{1}{2\pi i}\int_C w^{-z}e^w dw$ in (15.1), with the characteristic counter C around the negative real w-axis, where w^z has a branch cut and where w^z is thus not analytic. Hankel's contour integral generalizes Euler's integral $\Gamma(z) = \int_0^\infty e^{-t}t^{z-1}dt$ to all complex z. Hankel's contour integral (15.1) plays a crucial role in the theory of the Mittag-Leffler as displayed in Chapter 4. Section 15.2 studies different ways to deduce the Taylor coefficients c_k around $z_0 = 0$ of the entire function $\frac{1}{\Gamma(z)} = \sum_{k=0}^{\infty} c_k z^k$. Based on our characteristic coefficients in Appendix A, **art.** 217 and 218 present two new closed-form expression (15.31) and (15.35) of the Taylor coefficients c_k and an elegant asymptotic formula (15.36) of c_k for large k, derived from (15.35).

Chapter 16 concentrates on $\gamma^*(a,z)$, the entire incomplete Gamma function, and the related integral $\Gamma(a,z) = \int_z^\infty e^{-x}x^{a-1}dx$, which is a rather straightforward variant of Euler's integral. The other motivation for its inclusion is the special case $E_{1,b}(z) = e^z\gamma^*(b-1,z)$ in (2.11) of the Mittag-Leffler function $E_{a,b}(z)$ for the parameter $a = 1$ in **art.** 5, further investigated in Section 16.4, and the general form (2.35) of $E_{\frac{m}{n},b}(z)$ for fractional $a = \frac{m}{n}$ in terms of $\gamma^*(b,z)$. The entire function $\gamma^*(a,z)$ possesses two series (16.7) and (16.18)

$$\gamma^*(a,z) = \frac{1}{\Gamma(a)}\sum_{k=0}^{\infty}\frac{(-1)^k z^k}{k!\,(a+k)} = e^{-z}\sum_{k=0}^{\infty}\frac{z^k}{\Gamma(a+1+k)}$$

converging for all complex z. We derive recursions for $\Gamma(a,z)$ in **art.** 221, for $\gamma^*(a,z)$ in **art.** 222 and asymptotic expansions in Section 16.3. Chapter

16 ends with Laplace and Mellin transforms of $\Gamma(a,z)$ and $\gamma^*(a,z)$ in Section 16.5, that are applied to the Mittag-Leffler function in **art.** 45.

Part II on the Gamma function concludes with Chapter 17 on the Beta function, defined in (10.11) by both the fraction of Gamma functions and its integral, for $\operatorname{Re}(p) > 0$ and $\operatorname{Re}(q) > 0$,

$$B(p,q) = \frac{\Gamma(p)\Gamma(q)}{\Gamma(p+q)} = \int_0^1 u^{p-1}(1-u)^{q-1}\,du$$

Historically, the Beta function was well studied before the Gamma function and treated in the comprehensive treatise of Binet (1839), that reviews the theory, including nearly all formulae, on the Beta function $B(p,q)$ known at that time. In **art.** 233, we show the intimate connection of the Beta function with Newton's binomial series (17.1). Newton's series (17.1) and the geometric series (1.15) are the most basic, finite sum formulae in calculus. Some transformations of the Beta integral are derived in **art.** 237. Chapter 17 terminates with the classical proof in **art.** 238 of $B(p,q) = \frac{\Gamma(p)\Gamma(q)}{\Gamma(p+q)}$, derived from Euler's integral $\Gamma(z) = \int_0^\infty e^{-t}t^{z-1}dt$ and with Nanjundiah's proof (**art.** 239).

Due to the long history of the Gamma function, we have confined ourselves even more than in Part I on the Mittag-Leffler function. The aim was to present the complex theory of the Gamma function from its start and to show the enrichment of the theory over the years by many scientists. We did not include generalizations such as q-extensions (Askey, 1980) and integrals or sums over products of Gamma functions. We have omitted the Gamma distribution function in probability theory, numerical aspects, continued fractions for the Gamma function and considerations about the inverse function $\Gamma^{-1}(z)$, that obeys $\Gamma^{-1}(\Gamma(z)) = \Gamma\left(\Gamma^{-1}(z)\right) = z$, but refer to Borwein and Corless (2018).

9.2 The harmonic series and Euler's constant

Before concentrating on the Gamma function, we briefly discuss the harmonic series and an important constant $\gamma \simeq 0.577216$ due to Euler.

136. *The harmonic series.* The harmonic series is defined as

$$H_N = \sum_{k=1}^{N} \frac{1}{k} = 1 + \frac{1}{2} + \frac{1}{3} + \cdots + \frac{1}{N} \tag{9.1}$$

The harmonic series plays a crucial role in the study of the convergence of series and integrals.

Historically, the divergence of the harmonic series H_N has been elegantly demonstrated by Nicole Oresme[1], without resorting to calculus, as follows,

$$\begin{aligned} H_N &= 1 + \frac{1}{2} + \underbrace{\frac{1}{3} + \frac{1}{4}} + \underbrace{\frac{1}{5} + \frac{1}{6} + \frac{1}{7} + \frac{1}{8}} + \cdots + \frac{1}{N} \\ &> 1 + \frac{1}{2} + \underbrace{\frac{1}{4} + \frac{1}{4}} + \underbrace{\frac{1}{8} + \frac{1}{8} + \frac{1}{8} + \frac{1}{8}} + \cdots + \frac{1}{2^{\lfloor \log_2 N \rfloor}} \\ &= 1 + \frac{1}{2} + \quad \frac{1}{2} \quad + \qquad \frac{1}{2} \qquad + \cdots + \frac{1}{2} \end{aligned}$$

resulting in

$$H_N > 1 + \frac{\lfloor \log_2 N \rfloor}{2}$$

where $2^{\lfloor \log_2 N \rfloor} \leq N$ and equality only holds if N is a power of 2. Clearly, if $N \to \infty$ so does $\log_2 N$ and the logarithmic divergence of the harmonic series H_N is established. Since $H_N > 1 + \frac{1}{2} \left\lfloor \frac{\log N}{\log 2} \right\rfloor > \frac{\log N}{2 \log 2}$, Oresme's bound for large N is $H_N \gtrsim \frac{\log N}{2 \log 2}$, while the correct order is $H_N \sim \log N$.

Euler (1755, Pars posterior, Chapt. 6, art. 142-144, pp. 442-446) computes the harmonic series (9.1) up to an astonishing $N = 10^6$ terms, based on his variant (3.41) of the Euler-Maclaurin summation formula in **art.** 53 for $f(t) = \frac{1}{t}$ as

$$\sum_{k=1}^{N} \frac{1}{k} = \log N + \frac{1}{2N} - \frac{1}{12} \frac{1}{N^2} - \sum_{n=2}^{M} \frac{B_{2n}}{2n} \frac{1}{N^{2n}} + C \tag{9.2}$$

where the constant $C = 1 + \sum_{n=1}^{M} \frac{B_n}{n} = \gamma$ (**art.** 137). Euler (1755, Pars posterior, Chapt. 6, art. 143, p. 444) mentions that the series $\sum_{n=2}^{M} \frac{B_{2n}}{2n} \frac{1}{N^{2n}}$ diverges for $M \to \infty$ and that the constant C cannot be determined, but he suggests to compute the constant C for $N = 10$ in (9.2). He computes both the left-hand side $\sum_{k=1}^{10} \frac{1}{k} = 2.928968253968253968$ up to 18 digits correct, $\log 10$ at the right-hand side via fast convergent series for the logarithm of rational numbers (deduced in the previous parts of his book and in Euler (1748, p. 91)), the right-hand side Bernoulli number series $\sum_{n=1}^{M} \frac{B_n}{n} \frac{1}{10^n}$ for $M = 18$, resulting in a determination of $\gamma = 0.5772156649015325$ up to 16 digits correct!

Euler thus found from (9.2) the stronger result than Oresme that $\gamma = \lim_{N \to \infty} (H_N - \log N)$. A fast converging series for the harmonic series H_N for finite N is given in (13.61).

137. *Euler's constant.* Euler's constant[2] is defined (Abramowitz and Ste-

[1] Nicole Oresme (ca. 1325-1382) was one of the most erudite persons of his era and often compared to Leonardo Da Vinci (1452-1519). He was bishop of Lisieux in France. Absolute infinity was associated with God, also contemplated by Augustinus, and was a medieval theme. The fact that the terms in the harmonic series H_N decrease to zero, i.e. $\lim_{N \to \infty} \frac{1}{N} = 0$, and yet that the sum $\lim_{N \to \infty} H_N$ diverges was considered a mystic property of the numbers in which God must have a hand to play.

[2] Gauss (1813, 154, footnote) gives the Euler-Mascheroni constant γ up to 40 decimals accurate. Gauss provided the method (i.e. the power series of the digamma function

gun, 1968, 6.1.3) as

$$\gamma = \lim_{k\to\infty}\left(\sum_{n=1}^{k}\frac{1}{n} - \log k\right) = 0.57721\ldots \tag{9.3}$$

Whittaker and Watson (1996, p. 235) elegantly demonstrate that the limit in (9.3) exists. They define

$$u_n = \int_0^1 \frac{t}{n\,(n+t)}dt = \frac{1}{n} - \log\frac{n+1}{n}$$

With $\sum_{n=1}^{k-1}\log\frac{n+1}{n} = \sum_{n=1}^{k-1}\log(n+1) - \sum_{n=1}^{k-1}\log n = \log(k)$ and

$$\begin{aligned}\sum_{n=1}^{k}\frac{1}{n} - \log k &= \sum_{n=1}^{k}\left(\frac{1}{n} - \log\frac{n+1}{n}\right) + \log\frac{k+1}{k} \\ &= \sum_{n=1}^{k} u_n + \log\left(1+\frac{1}{k}\right)\end{aligned} \tag{9.4}$$

an alternative representation of Euler's constant (9.3) is obtained as $\gamma = \sum_{n=1}^{\infty} u_n$. Since $0 < u_n = \int_0^1 \frac{t}{n(n+t)}dt < \int_0^1 \frac{1}{n^2}dt = \frac{1}{n^2}$ and $\sum_{n=1}^{\infty}\frac{1}{n^2} = \zeta(2) = \frac{\pi^2}{6} = 1.64493$, it holds that $0 < \gamma < \frac{\pi^2}{6}$.

The bounds can be sharpened by an instance of $\frac{1}{n+1} < \int_n^{n+1}\frac{dx}{x} = \log\frac{n+1}{n} < \frac{1}{n}$ in (9.7) below, valid for any real $n > 0$. Indeed, since $\frac{1}{n} - \log\frac{n+1}{n} > 0$, the identity (9.4) provides the lower bound

$$\sum_{n=1}^{k}\frac{1}{n} - \log k = 1 - \log 2 + \sum_{n=2}^{k}\left(\frac{1}{n} - \log\frac{n+1}{n}\right) + \log\frac{k+1}{k} > 1 - \log 2 \simeq 0.307$$

Similarly, rewriting the identity (9.4) and using $\log\frac{n+1}{n} - \frac{1}{n+1} > 0$ gives us the upper bound

$$\begin{aligned}\sum_{n=1}^{k}\frac{1}{n} - \log k &= 1 + \sum_{n=2}^{k}\frac{1}{n} - \sum_{n=1}^{k-1}\log\frac{n+1}{n} = 1 - \sum_{n=1}^{k-1}\left(\log\frac{n+1}{n} - \frac{1}{n+1}\right) \\ &< 1 - \left(\log 2 - \frac{1}{2}\right) \simeq 0.807\end{aligned}$$

In summary, we find $1 - \log 2 < \gamma < \frac{3}{2} - \log 2$. Sharper bounds follows from Gauss's integral (12.38) and Poisson's integral (13.62).

9.3 The Riemann Zeta-function

138. *The Riemann Zeta-function and Harmonic Series.* The Riemann

$\psi(z)$) and Fredericus Bernhardus Gothofredus Nicolai has performed the computation. Whittaker and Watson (1996, p. 235) mention that J. C. Adams computed γ up to 260 decimals accurate.

Zeta-function, defined by the Dirichlet series in the complex number z,

$$\zeta(z) = \lim_{N\to\infty} \sum_{n=1}^{N} \frac{1}{n^z} = 1 + \frac{1}{2^z} + \frac{1}{3^z} + \cdots \tag{9.5}$$

is the direct generalization in z of the harmonic series H_N in (9.1), where $z = 1$. A key result is the Euler product, beautifully deduced by Euler (1748, Caput XV, art. 283) and valid for $\mathrm{Re}(z) > 1$,

$$\zeta(z) = \prod_{p:\text{ all primes}} \left(1 - \frac{1}{p^z}\right)^{-1} = \frac{1}{\left(1-\frac{1}{2^z}\right)\left(1-\frac{1}{3^z}\right)\cdots\left(1-\frac{1}{p^z}\right)\cdots} \tag{9.6}$$

which is a generating function of the prime numbers $2, 3, 5, 7, 11, 13, 17, \ldots$

The appearance of the prime numbers in (9.6) is one of the most fascinating properties of the Riemann Zeta-function $\zeta(z)$. The theory of the Riemann Zeta-function in the complex z-plane is rich and surprisingly complicated as illustrated by the book of Titchmarsh and Heath-Brown (1986). At the same time, the theory is still incomplete as witnessed by the famous Riemann hypothesis (Riemann, 1860), claiming that all non-trivial zeros of $\zeta(z)$ lie on the line $\mathrm{Re}(z) = \frac{1}{2}$, which is at present the most challenging open mathematical problem.

The right-hand side Dirichlet series in (9.5) converges for $\mathrm{Re}(z) > 1$. Since $n^z = n^\sigma n^{iT}$ and $|n^z| \le n^\sigma$ for $z = \sigma + iT$, it follows directly that

$$|\zeta(\sigma + iT)| \le \sum_{n=1}^{\infty} \frac{1}{n^\sigma} = \zeta(\sigma)$$

illustrating that $\zeta(z)$ for real $z = \sigma$ is maximal for any complex number z with $\mathrm{Re}(z) = \sigma > 1$. In other words, if the Dirichlet series converges for real σ, then the Dirichlet series converges absolutely for any $z = \sigma + iT$ with $\mathrm{Re}(z) = \sigma$. We determine the values of $\zeta(\sigma)$ in relation to convergence.

If a function $f(x)$ is positive and increasing for all $x \in [a, b]$, the integral is bounded by the lower and upper Riemann sum (Jordan, 1909, art. 307),

$$f(x) < \int_x^{x+1} f(u)\, du < f(x+1)$$

such that for integers $n, N \in [a, b]$, we obtain the inequalities

$$\sum_{k=n}^{N} f(k) < \int_n^{N+1} f(u)\, du < \sum_{k=n+1}^{N+1} f(k) \tag{9.7}$$

with the opposite inequality signs if $f(x)$ is positive and decreasing.

We apply the inequalities (9.7) to the Riemann Zeta-function $\zeta(\sigma)$, where $f(x) = x^{-\sigma}$ is decreasing for $\sigma > 0$. With $\int_n^{N+1} \frac{dx}{x} = \log\left(\frac{N+1}{n}\right)$, but $\int_n^{N+1} x^{-\sigma} dx = \frac{(N+1)^{1-\sigma} - n^{1-\sigma}}{1-\sigma}$ for $\sigma \neq 1$, (9.7) becomes

$$\sum_{k=1}^{N} \frac{1}{k^\sigma} > \frac{(N+1)^{1-\sigma} - 1}{1-\sigma} > \sum_{k=2}^{N+1} \frac{1}{k^\sigma} \tag{9.8}$$

The first inequality illustrates that the Dirichlet series $\zeta(\sigma) = \sum_{k=1}^{\infty} \frac{1}{k^\sigma}$ diverges for $\sigma < 1$. For $\sigma > 1$, the second inequality demonstrates convergence of $\zeta(\sigma) = 1 + \sum_{k=2}^{\infty} \frac{1}{k^\sigma} < 1 + \frac{1}{\sigma-1} = \frac{\sigma}{\sigma-1}$. For $\sigma = 1$, the first inequality reduces to $H_N = \sum_{k=1}^{N} \frac{1}{k} > \log(N+1)$ and shows that the harmonic series H_N diverges logarithmically in N, as already established differently in **art.** 136.

In summary, the Dirichlet series $\zeta(z) = \sum_{n=1}^{\infty} \frac{1}{n^z}$ for $z = \sigma + iT$ converges absolutely for $\text{Re}(z) = \sigma > 1$ and diverges[3] for $\text{Re}(z) < 1$.

9.4 Euler's product and Euler's integral for the Gamma function

Davis (1959) situated the birth of the Gamma function with a correspondence between the 22 years-young genius Euler (1707-1783) and the savant Christian Goldbach (1690-1764), in which Euler informed Goldback that the interpolation problem for the factorial $n!$ is not algebraically solvable. In fact, the title of the paper of Euler (1738) is translated as *On transcendental progressions whose general terms cannot be expressed algebraically.* Nielsen (1956) shows that the solution to the "interpolation problem" of finding a function f that passes through the set $(1,1), (2,2), (3,6), (4,24), \ldots, (n, n!)$ is not unique. In the early decades after the invention of calculus by Newton and Leibniz, interpolation theory consisted of finding a closed-form analytic expression $f(z)$ that produces the values, i.e. $f(j) = a_j$ at integer $z = j \in [1, n]$, in a given sequence $(1, a_1), (2, a_2), \ldots, (n, a_n)$, where a_j with $1 \leq j \leq n$ is a real number.

Euler (1738, art. 1) started with the formula, which he discovered in 1729,

$$\frac{1.2^n}{1+n}\frac{2^{1-n}.3^n}{2+n}\frac{3^{1-n}.4^n}{3+n}\frac{4^{1-n}.5^n}{4+n}\ldots = n! \tag{9.9}$$

which Davis (1959) writes as

$$\left[\left(\frac{2}{1}\right)^n \frac{1}{1+n}\right]\left[\left(\frac{3}{2}\right)^n \frac{2}{2+n}\right]\left[\left(\frac{4}{3}\right)^n \frac{3}{3+n}\right]\cdots = n!$$

More generally, (9.9) in terms of the complex number z instead of n is rewritten as

$$\Gamma(z+1) = \prod_{k=1}^{\infty} \frac{(k+1)^z}{k^{z-1}(k+z)} \tag{9.10}$$

Davis (1959) ironically referred to the sudden and unmotivated placement as Euler's product (9.10) as a *deus ex machina*, quite popular in the theaters of the Old Greeks and Romans. Perhaps, the most important contribution

[3] The behavior of $\zeta(z)$ along the line $\text{Re}(z) = 1$, except for the simple pole at $z = 1$, is more complicated and was established first in 1896 by de la Vallée Poussin and, independently, by Hadamard (see e.g. Edwards (1974, pp. 91-95), Landau (1974)) in their proof of the Prime-number Theorem (Titchmarsh and Heath-Brown, 1986, Chapter III), i.e. the number of primes $\pi(x)$ not exceeding x is $\pi(x) \sim \frac{x}{\log x}$ for large x.

of Euler's paper was the Euler product (9.9) posed on line 8 of the first page of his 22 pages long paper, because (9.9) has motivated Gauss (1813) to deduce and study that general Euler-Gauss product (9.10), which equals (10.10) below and on which nearly the entire theory of the Gamma function is based! A similar way of posing big results, without telling how they were found, is also characteristic in Gauss's work (**art.** 142).

Euler verifies (9.9) by introducing $n = 0, 1, 2, 3$, e.g., for $n = 3$,

$$\frac{2.2.2}{1.1.4}\frac{3.3.3}{2.2.5}\frac{4.4.4}{3.3.6}\frac{5.5.5}{4.4.7}\cdots = \frac{2}{4}\frac{3}{5}\frac{4}{6}\frac{5}{7}\cdots = 2.3\frac{1}{4}\frac{1}{5}\frac{4}{6}\frac{5}{7}\cdots = 6$$

Euler also computes his product (9.9) for $n = \frac{1}{2}$,

$$\sqrt{\frac{2.4}{3.3}\frac{4.6}{5.5}\frac{6.8}{7.7}\frac{8.10}{9.9}\cdots} = \sqrt{\frac{1}{2}\left(\frac{2.2}{1.3}\right)\left(\frac{4.4}{3.5}\right)\left(\frac{6.6}{5.7}\right)\left(\frac{8.8}{7.9}\right)\cdots}$$

recognizes Wallis' product[4], deduced by interpolation of $\int_0^\pi \sin^k x \, dx$ and published in 1656,

$$\frac{\pi}{2} = \prod_{k=1}^{\infty} \frac{4k^2}{4k^2 - 1} = \left(\frac{2.2}{1.3}\right)\left(\frac{4.4}{3.5}\right)\left(\frac{6.6}{5.7}\right)\left(\frac{8.8}{7.9}\right)\cdots \tag{9.11}$$

and finds that $\left(\frac{1}{2}\right)! = \Gamma\left(\frac{3}{2}\right) = \frac{\sqrt{\pi}}{2}$. Euler (1738, art. 3) realizes that the Euler product (9.9) for integer n returns an integer, but for non-integers as $z = \frac{1}{2}$, suddenly π pops up, which is not algebraically expressible[5], related to circles and their area, and the latter implies integrals (Euler, 1738, art. 4-5). Hence, Euler (1738, art. 6) concludes that his product (9.9) might be expressed as an integral and proposed to start (Euler, 1738, art. 8) with the Beta-type integral $\int_0^z x^a (1-x)^n \, dx$, which was earlier investigated by Newton, Wallis and Stirling, as mentioned by Davis (1959). Actually, Euler (1738, art. 9) considers further the case $z = 1$ and finds, from Newton's binomium (**art.** 233), the Beta-function's fundamental relation $B(p, q) = \frac{\Gamma(p)\Gamma(q)}{\Gamma(p+q)}$,

$$\int_0^1 x^a (1-x)^n \, dx = \frac{1.2.\ldots.n}{(a+1)(a+2)\cdots(a+n+1)}$$

Euler (1738, art. 9) replaces $a = \frac{f}{g}$,

$$\int_0^1 x^{\frac{f}{g}} (1-x)^n \, dx = \frac{g^{n+1}}{f + (n+1)g} \frac{1.2.\ldots.n}{(f+1.g)(f+2.g)\cdots(f+n.g)}$$

and observes from

$$\frac{f + (n+1)g}{g^{n+1}} \int_0^1 x^{\frac{f}{g}} (1-x)^n \, dx = \frac{n!}{(f+1.g)(f+2.g)\cdots(f+n.g)}$$

[4] Wallis' product (9.11) follows from the infinite product (11.2) of $\sin(\pi z)$ for $z = \frac{1}{2}$ as $\frac{\pi}{2} = \prod_{k=1}^{\infty} \left(1 - \frac{1}{2^2 k^2}\right)^{-1} = \prod_{k=1}^{\infty} \frac{2k}{2k-1}\frac{2k}{2k+1}$.

[5] It is not a finite sum nor product of ratios, powers or roots, which explains the title of Euler's article.

that, for $f = 1$ and $g = 0$, the right-hand side equals $n!$ The sequel consists of evaluating the left-hand side integral in the limit $g \to 0$ with $f = 1$. Davis (1959) outlines each step in Euler's ingenious, but less rigorous manipulations that led him to

$$n! = \int_0^1 (-\log x)^n \, dx$$

which transforms, after substitution $x = e^{-t}$, to Euler's basic integral $\Gamma(z) = \int_0^\infty e^{-t} t^{z-1} dt$ in (10.2) in **art.** 140.

Only much later, Euler (1755, Pars posterior, Chapt. 17) explains his proposition (9.9) by giving the general procedure to find intermediate terms of a sequence by interpolation. More generally, Euler (1755, Pars posterior, Chapt. 17) considers several types of general sequences $(1, a_1), (2, a_2), \ldots, (n, a_n)$, and, in particular (Euler, 1755, Pars posterior, Chapt. 17, art. 401, p. 832), the sequence, where $a_k = \prod_{j=0}^{k-1} (\alpha + j\beta)$, for which his procedure gives the value Σ_ω at index ω as

$$\Sigma_\omega = \alpha^\omega \frac{\alpha^{1-\omega} (\alpha+\beta)^\omega}{\alpha + \omega\beta} \frac{(\alpha+\beta)^{1-\omega} (\alpha+2\beta)^\omega}{\alpha+\beta+\omega\beta} \frac{(\alpha+2\beta)^{1-\omega} (\alpha+3\beta)^\omega}{\alpha+2\beta+\omega\beta} \cdots$$

which reduces to (9.9) for $\alpha = \beta = 1$. For that case $\alpha = \beta = 1$, Euler (1755, Pars posterior, Chapt. 17, art. 402, exemplum 1, p. 835) computes again the value $\Sigma_{\frac{1}{2}} = \sqrt{\pi}$ at index $\omega = \frac{1}{2}$.

10
Defining the Gamma function

Chapter 10 concentrates on the problem how to define the Gamma function $\Gamma(z)$ for complex numbers z. As mentioned in Section 9.4, the search for an interpolating function of the factorial $n!$ at the dawn of calculus was the initial problem, that gave birth to the beautiful theory of the Gamma function. Euler (1738) already pointed out that the problem was not easy, but gave the essential boost by (9.9) to the theory. Section 10.1 starts with essential building blocks, such as the functional equation (10.1) of the Gamma function, derived from the factorial, and Euler's integral (10.2), the most prominent integral in the Gamma function's theory. Section 10.2 treats Gauss's work. It was Gauss, who has derived the first rigorous analysis of the Gamma function in his magnificent paper (Gauss, 1813) on the hypergeometric series. Section 10.3 presents the other approaches of Rademacher (**art.** 143), Jensen (**art.** 144), Laugwitz and Rodewald (**art.** 145), followed in Section 10.4 by the more function theoretical characterization of Artin (**art.** 146) and Wielandt (**art.** 147).

10.1 The functional equation of the Gamma function

139. *Functional equation of the Gamma function.* The Gamma function $\Gamma(z)$ is an extension of the factorial $n! = 1.2.3\ldots n$ in the integers $n \geq 1$ to complex numbers z. The factorial obeys

$$n! = n(n-1)!$$

which directly generalizes to the functional equation $\Pi(z) = z\Pi(z-1)$ with $n! = \Pi(n)$ and $\Pi(0) = 1$, in the notation of Gauss in his truly impressive manuscript (Gauss, 1813, p. 146). Later in 1814, Legendre defined the Gamma function by its current notation $\Gamma(z) = \Pi(z-1)$, with $\Gamma(1) = 1$, and the functional equation $\Pi(z) = z\Pi(z-1)$ translates to the Gamma function $\Gamma(z)$ as

$$\Gamma(z+1) = z\Gamma(z) \tag{10.1}$$

The first step in the theory of the Gamma function consists of finding a solution of the functional equation (10.1).

140. *Euler's integral for the Gamma function.* Later scholars, in particular Gauss, have studied, polished and generalized Euler's ideas and ingenuity (see Section 9.4), that resulted in the famous and established integral

$$\Gamma(z) = \int_0^\infty e^{-t} t^{z-1} dt \qquad \text{for } \operatorname{Re}(z) > 0 \tag{10.2}$$

Partial integration of (10.2) shows that Euler's integral (10.2) obeys the functional equation (10.1) and $\Gamma(1) = \int_0^\infty e^{-t} dt = 1$. Moreover with $\sigma = \operatorname{Re}(z)$, Euler's integral shows that $|\Gamma(z)| \leq \int_0^\infty e^{-t} t^{\sigma-1} dt = \Gamma(\sigma)$, illustrating that the maximum absolute value of any complex z with $\operatorname{Re}(z) = \sigma$ is attained at the real axis and $\Gamma(\sigma)$ is always positive for $\sigma > 0$, because the integrand $e^{-t} t^{\sigma-1}$ is never negative for $t \geq 0$. The functional equation (10.1) indicates that $\Gamma(\sigma) = \frac{\Gamma(\sigma+1)}{\sigma} = O\left(\frac{1}{\sigma}\right)$ as $\sigma \to 0$ and that $\Gamma(z)$ has a pole at $z = 0$, thus explaining the validity range $\operatorname{Re}(z) > 0$ in Euler's integral (10.2). Indeed, writing Euler's integral (10.2) as

$$\Gamma(\sigma) = \int_0^1 e^{-t} t^{\sigma-1} dt + \int_1^\infty e^{-t} t^{\sigma-1} dt$$

and observing that the last integral exists for any real σ, the convergence limitation of Euler's integral (10.2) lies in the first, positive integral $\int_0^1 e^{-t} t^{\sigma-1} dt > e^{-1} \int_0^1 t^{\sigma-1} dt = \frac{1 - \lim_{t \to 0} t^\sigma}{e\sigma}$, which exists only if $\sigma > 0$. Since Euler's integral is only valid for $\operatorname{Re}(z) > 0$, other ingenious methods have been devised that are valid for all complex numbers z.

In spite of the limitation $\operatorname{Re}(z) > 0$, Euler's integral (10.2) is often the main entrance to the exploration of the Gamma function as e.g. in Jordan (1913, Chapitre III, Section 2). Euler's integral (10.2) is also known (Jordan, 1913, art. 207) as the *Eulerean integral of the second kind*, while the Beta-integral (10.11) is the *Eulerean integral of the first kind.*

10.2 Gauss's approach

141. *Gauss's infinite product for the Gamma function.* Iterating n times the functional equation (10.1) gives $\Gamma(z) = \frac{\Gamma(z+n)}{z(z+1)\ldots(z+n-1)}$, but purely iterating (10.1) for non-integer values is not successful. Therefore and inspired by Euler's product (9.9), Gauss (1813, p. 144) proposes to consider the more general form

$$\Pi(k, z) = \frac{\Pi(k)\,\Pi(z)}{\Pi(k+z)} k^z = \frac{1}{(z+1)} \frac{2}{(z+2)} \frac{3}{(z+3)} \cdots \frac{k}{(z+k)} k^z \tag{10.3}$$

which satisfies

$$\Pi(k, z+1) = \Pi(k, z) \frac{z+1}{1 + \frac{z+1}{k}} \tag{10.4}$$

as well as

$$\Pi(k+1,z) = \Pi(k,z)\frac{\left(1+\frac{1}{k}\right)^{z+1}}{1+\frac{z+1}{k}} = \Pi(k,z)\frac{(k+1)^{z+1}}{k^z(k+z+1)} \tag{10.5}$$

Iterating (10.5) in the integer k, with $\Pi(1,z) = \frac{1}{z+1}$, results in

$$\Pi(k,z) = \frac{1}{z+1}\frac{2^{z+1}}{(2+z)}\cdots\frac{k^{z+1}}{(k-1)^z(k+z)} = \frac{1}{z+1}\prod_{n=2}^{k}\frac{n^{z+1}}{(n-1)^z(n+z)} \tag{10.6}$$

The interesting observation from $\lim_{k\to\infty}\frac{\left(1+\frac{1}{k}\right)^{z+1}}{1+\frac{z+1}{k}} = 1$ in (10.5) is that the limit $\lim_{k\to\infty}\Pi(k,z)$ exists for all z, which Gauss demonstrates[1] after taking the logarithm of both sides, while the first functional equation (10.4) indicates that $\lim_{k\to\infty}\Pi(k,z)$ satisfies the functional equation $\Pi(z+1) = (z+1)\Pi(z)$. Combining both, Gauss is led to

$$\Pi(z) = \lim_{k\to\infty}\Pi(k,z) = \lim_{k\to\infty}\frac{\Pi(k)\Pi(z)}{\Pi(k+z)}k^z \tag{10.7}$$

which is equivalent with $\Gamma(z+1) = \Pi(z)$ to

$$\Gamma(z) = \lim_{k\to\infty}\frac{k!\Gamma(z)}{\Gamma(k+1+z)}k^z = \lim_{k\to\infty}\frac{k^z k!}{z(z+1)\ldots(z+k)} \tag{10.8}$$

Rewriting Gauss's definition in (10.7) as $\Gamma(z) = \lim_{k\to\infty}\Pi(k,z-1)$ and introducing (10.6) yields

$$\Gamma(z) = \lim_{k\to\infty}\frac{1}{z}\prod_{n=2}^{k}\frac{n^z}{(n-1)^{z-1}(n+z-1)} = \lim_{k\to\infty}\frac{1}{z}\prod_{n=1}^{k-1}\frac{(n+1)^z}{n^{z-1}(n+z)}$$

Finally[2], we arrive at the *Euler-Gauss infinite product*[3] for the Gamma

---

[1] As an application of the general theory for the convergence of products, Jordan (1909, art. 325) proves that the product $\Pi(k,z)$ for $k\to\infty$ converges absolutely. A similar proof is given by Jensen in **art**. 144.

[2] Gauss (1813, p. 148) proceeds further and derives the reflection formula from his classical result (Abramowitz and Stegun, 1968, 15.1.20) for the hypergeometric series at $z=1$, for $c\neq -k$ (k integer) and $\mathrm{Re}(c-a-b)>0$,

$$F(a,b;c;1) = \frac{\Gamma(c)\Gamma(c-a-b)}{\Gamma(c-a)\Gamma(c-b)} = \frac{\Gamma(c)}{\Gamma(a)\Gamma(b)}\sum_{n=0}^{\infty}\frac{\Gamma(a+n)\Gamma(b+n)}{\Gamma(c+n)n!} \tag{10.9}$$

He also deduces his multiplication formula, compares his theory with Stirling and Euler's logarithmic expansion in terms of Bernoulli numbers, studies the digamma function, derives his exact formula (12.23) for fractional arguments of the digamma function, deduces an integral for the digamma function and complements Euler's computations. In short, an amazing sequence of beautifully derived deep results that constitute our current basis of the Gamma function. In line with his genius, Gauss even laid the basis of prime factors of an entire function of which Weierstass has given the functional theory (Titchmarsh, 1964, Chapter VIII).

[3] Both Klein (1933, p. 74) and Whittaker and Watson (1996, p. 237) mention that Euler has given (10.10) in a letter to Goldbach in 1729 (see also Section 9.4), but that Gauss has provided the first rigorous analysis in Gauss (1813).

function

$$\Gamma(z) = \frac{1}{z}\prod_{n=1}^{\infty}\left(1+\frac{1}{n}\right)^{z}\left(1+\frac{z}{n}\right)^{-1} \tag{10.10}$$

which converges for all complex z, except for the integers at $z = 0, -1, -2, \ldots$, at which $\Gamma(z)$ has simple poles. The inverse of the product (10.10) shows that $\frac{1}{\Gamma(z)}$ is an entire function (i.e. without poles) and, consequently, that $\Gamma(z)$ has no zeros in the finite complex plane.

142. *Discussion of Gauss's approach.* Of course, the proposal of (10.3) by Gauss, which is essentially equivalent via (10.6) to Euler's proposal (9.9), was the crucial idea leading to the elegant product (10.10). Gauss posited (10.3) without providing intuition, which we give below, following Rademacher (1973, p. 31). Perhaps the most convincing argument for Gauss's starting point (10.3) is given by Klein (1933, p. 71), who gives three definitions of the Gamma function, of which the third is also discussed by Gauss himself (Gauss, 1813, p. 151). Klein (1933, p. 74) starts from the Beta-integral, studied by Euler and Binet and valid for $\operatorname{Re}(p) > 0$ and $\operatorname{Re}(q) > 0$,

$$B(p,q) = \frac{\Gamma(p)\Gamma(q)}{\Gamma(p+q)} = \int_0^1 u^{p-1}(1-u)^{q-1}\,du \tag{10.11}$$

Klein considers integer values for $q = k+1$ and makes the substitution $u = \frac{v}{k}$,

$$B(p,k+1) = \int_0^1 u^{p-1}(1-u)^{k}\,du = k^{-p}\int_0^k v^{p-1}\left(1-\frac{v}{k}\right)^{k}dv$$

Using $\lim_{k\to\infty}\left(1-\frac{v}{k}\right)^{k} = e^{-v}$ and Euler's integral (10.2), Klein (1933, p. 74) arrives for $\operatorname{Re}(z) > 0$ at

$$\Gamma(z) = \int_0^\infty v^{z-1}e^{-v}dv = \lim_{k\to\infty}\left(k^{z}B(z,k+1)\right) = \lim_{k\to\infty}\frac{\Gamma(z)\Gamma(k+1)}{\Gamma(z+k+1)}k^{z}$$

which is Gauss's definition (10.7). Alternatively, $\lim_{k\to\infty}\frac{\Gamma(k+1)}{\Gamma(z+k+1)}k^{z} = 1$ shows that $\frac{\Gamma(x)}{\Gamma(z+x)} \sim x^{-z}$ for large x, an asymptotic result that is made more precise in **art.** 198. In contrast to Euler's integral (10.2) for $\operatorname{Re}(z) > 0$, the functional equation (10.5) is valid for all z and so is the Euler-Gauss product (10.10). The demonstration of the Beta function $B(p,q) = \frac{\Gamma(p)\Gamma(q)}{\Gamma(p+q)}$ as a product of Gamma functions is given in Whittaker and Watson (1996, p. 254 (art. 12.41)), Binet (1839, p. 202, art. 15.) and here in **art.** 238.

10.3 Various other approaches

143. *Rademacher's approach.* As mentioned earlier, the definition $m! = 1.2\ldots m$, where m is a positive integer, is difficult to generalize to real or complex numbers. We present a formal approach towards the Euler-Gauss product (10.10) by Rademacher (1973, p. 31). Iterating the functional equation (10.1) N times for the integer $z = m$ gives

$$m! = \frac{(m+N)!}{(m+1)(m+2)\cdots(m+N)}$$
$$= \frac{1}{m+1}\frac{2}{m+2}\cdots\frac{N}{m+N}(N+1)(N+2)\cdots(N+m)$$

which we rewrite as

$$m! = \prod_{n=1}^{N} \frac{n}{m+n}(N+1)^m \frac{(N+1)(N+2)\cdots(N+m)}{(N+1)^m}$$

We could have replaced $(N+1)^m$ by $(N+a)^m$, where a is any positive real number, because the main reason to rewrite the iterated functional equation is that the last factor tends to 1 if $N \to \infty$,

$$m! = \lim_{N\to\infty}(N+1)^m \prod_{n=1}^{N}\frac{n}{m+n}$$

However, the next step invokes the identity $N+1 = \prod_{n=1}^{N}\frac{n+1}{n}$ and motivates the choice $a = 1$. Thus, we obtain

$$m! = \prod_{n=1}^{\infty}\frac{n}{m+n}\left(\frac{n+1}{n}\right)^m$$

which leads to the Euler-Gauss product (10.10) after replacing the integer m by an arbitrary complex number z and defining $z! = \Gamma(z+1)$. The remaining step, for which we refer to Rademacher (1973, pp. 31-32), is the demonstration that the Euler-Gauss product (10.10) is convergent for any z.

144. *Jensen's approach.* In the article of Jensen and Gronwall (1916), translated and commented by Gronwall in Princeton, Jensen proves Gauss's limit (10.8) and the Euler-Gauss infinite product (10.10) similarly as Gauss and Rademacher. Jensen starts, by considering $z = m$ is an integer, from the functional equation

$$\Gamma(m+n) = (n+m-1)(n+m-2)\cdots n(n-1)!$$
$$= n^m\left(1+\frac{m-1}{n}\right)\left(1+\frac{m-2}{n}\right)\cdots 1.(n-1)!$$

from which

$$1 = \lim_{n\to\infty}\frac{\Gamma(m+n)}{n^m\left(1+\frac{m-1}{n}\right)\left(1+\frac{m-2}{n}\right)\cdots 1.(n-1)!} = \lim_{n\to\infty}\frac{\Gamma(m+n)}{n^m(n-1)!} \tag{10.12}$$

Then, Jensen writes

$$\Gamma(z) = \frac{\Gamma(z+n)}{z(z+1)\dots(z+n-1)} = \frac{\Gamma(z+n)}{n^z(n-1)!}\frac{n^z(n-1)!}{z(z+1)\dots(z+n-1)}$$

and concludes that Gauss's limit (10.8)

$$\begin{aligned}\Gamma(z) &= \lim_{n\to\infty}\frac{\Gamma(z+n)}{n^z(n-1)!}\frac{n^z(n-1)!}{z(z+1)\dots(z+n-1)}\\ &= \lim_{n\to\infty}\frac{\Gamma(z+n)}{n^z(n-1)!}\lim_{n\to\infty}\frac{n^z(n-1)!}{z(z+1)\dots(z+n-1)}\end{aligned}$$

after invoking the limit (10.12) is

$$\Gamma(z) = \lim_{n\to\infty}\frac{n^z(n-1)!}{z(z+1)\dots(z+n-1)}$$

which is correct, at least for integer $z = m$. Since $\frac{n^z(n-1)!}{z(z+1)\dots(z+n-1)} = \frac{n^z}{z}\prod_{k=1}^{n-1}\frac{k}{z+k}$ and $n^z = \prod_{k=1}^{n-1}\left(\frac{k+1}{k}\right)^z$, because $\prod_{k=1}^{n-1}\left(\frac{k+1}{k}\right)^z = \frac{\prod_{k=1}^{n-1}(k+1)^z}{\prod_{k=1}^{n-1}k^z} = \frac{\prod_{k=2}^{n}k^z}{\prod_{k=1}^{n-1}k^z} = n^z$, it holds that

$$\frac{n^z(n-1)!}{z(z+1)\dots(z+n-1)} = \frac{1}{z}\prod_{k=1}^{n-1}\frac{\left(1+\frac{1}{k}\right)^z}{\left(1+\frac{z}{k}\right)}$$

which is the finite form (10.6) of the Euler-Gauss product (10.10). Jensen then shows that the limit $n\to\infty$ converges uniformly for any bounded z, that is different from $z = -k \in \mathbb{N}$. Indeed, the product $\prod_{k=1}^{\infty}(1+a_k)$ converges if series $\sum_{k=1}^{\infty}a_k$ converges absolutely (Titchmarsh, 1964, pp. 13-19). Here,

$$a_k = \frac{\left(1+\frac{1}{k}\right)^z}{\left(1+\frac{z}{k}\right)} - 1 = \frac{\sum_{j=0}^{\infty}\binom{z}{j}\frac{1}{k^j} - 1 - \frac{z}{k}}{1+\frac{z}{k}} = \frac{1}{k^2}\frac{\binom{z}{2}+\sum_{j=1}^{\infty}\binom{z}{j+2}\frac{1}{k^j}}{1+\frac{z}{k}}$$

with $|z| < k$ tends with $k\to\infty$ to $a_k = O\left(\frac{1}{k^2}\right)$, indicating for any finite z that $\sum_{k=1}^{\infty}a_k$ converges absolutely and uniformly. Indeed, Euler was the first to sum the series $\sum_{k=1}^{\infty}\frac{1}{k^2} = \frac{\pi^2}{6} = \zeta(2)$, where the Riemann Zeta-function $\zeta(z)$ is defined in (9.5). From the Euler-Gauss product (10.10), Jensen proceeds to derive the Weierstrass product (11.1) similarly as in **art.** 148 below. Jensen remarks that the Weierstrass product (11.1) was discovered by Schlomilch in 1844 and Newman in 1848 and further adds interesting historical references and notations of the Gamma function.

145. *The approach of Laugwitz and Rodewald.* Laugwitz and Rodewald (1987) start by their idea[4] that the function $L(n+x)$ is infinitesimally close to a linear function in the real $x \geq 0$, where $L(n) = \log n!$ for integers $n \geq 1$ and $L(0) = 1$. If $x = m$ is a positive integer, then

$$\begin{aligned}L(n+m) &= \log(n+m)! = \log n! + \sum_{k=1}^{m}\log(n+k)\\ &= L(n) + m\log(n+1) + \sum_{k=1}^{m}\log\left(1+\frac{k-1}{n+1}\right)\end{aligned}$$

[4] Laugwitz and Rodewald (1987) mention that the idea is due to Euler (1755), but I could not find it directly in Euler's book.

which suggests that

$$L(n+x) = L(n) + x\log(n+1) + r_n(x) \tag{10.13}$$

where the remainder $r_n(x)$ obeys $\lim_{n\to\infty} r_n(x) = 0$. Moreover, the function $L(x)$ obeys the functional equation

$$L(x+1) = L(x) + \log(x+1) \tag{10.14}$$

which leads, after n iterations with $L(0) = 1$, to

$$L(x+n) = L(x) + \sum_{k=1}^{n} \log(x+k) \tag{10.15}$$

Laugwitz and Rodewald (1987) combine (10.13) and (10.15) to

$$\begin{aligned} L(x) &= L(n) - \sum_{k=1}^{n} \log(x+k) + x\log(n+1) + r_n(x) \\ &= \sum_{k=1}^{n} (\log k - \log(x+k)) + x\log(n+1) + r_n(x) \end{aligned} \tag{10.16}$$

Generalizing the definition (9.3) of the Euler constant γ to the constants γ_n with $\lim_{n\to\infty} \gamma_n = \gamma$, Laugwitz and Rodewald (1987) introduce $\log(n+1) = -\gamma_n + \sum_{k=1}^{n} \frac{1}{k}$ in (10.16) and find

$$L(x) = -\gamma_n x + \sum_{k=1}^{n} \left(\frac{x}{k} - \log\left(1+\frac{x}{k}\right)\right) + r_n(x)$$

Since for $k > x \geq 0$,

$$0 \leq \frac{x}{k} - \log\left(1+\frac{x}{k}\right) = \frac{x}{k} - \sum_{j=1}^{\infty} \frac{(-1)^{j+1}}{j}\frac{x^j}{k^j} \leq \frac{x^2}{2k^2}$$

because $\sum_{j=2}^{\infty} \frac{(-1)^j}{j}\frac{x^j}{k^j}$ is an alternating series with decreasing (in absolute value) terms, the series $\sum_{k=1}^{n} \left(\frac{x}{k} - \log\left(1+\frac{x}{k}\right)\right)$ converges for $n \to \infty$ and

$$L(x) = -\gamma x + \sum_{k=1}^{\infty} \left(\frac{x}{k} - \log\left(1+\frac{x}{k}\right)\right) \tag{10.17}$$

is a uniquely determined real-valued function for $x \geq 0$.

In summary and formulated as a characterization of the Gamma function alternative to those in Section 10.4, Laugwitz and Rodewald (1987) have shown that there exists a unique function $\Gamma(x) > 0$ for $x \geq 1$, such that $L(x) = \log\Gamma(x+1)$ possesses the following properties (a) $L(0) = 1$, (b) the functional equation (10.17) and (c) the asymptotically "linear in x" relation (10.13) with $\lim_{n\to\infty} r_n(x) = 0$.

Exponentiation of (10.17), $e^{L(x)} = \Gamma(x+1) = e^{-\gamma x}\prod_{k=1}^{\infty} \frac{e^{\frac{x}{k}}}{1+\frac{x}{k}}$, leads to Weierstrass's product (11.1), which is equivalent (**art.** 148) to the Euler-Gauss product (10.10) and holds for any complex z.

Next to the infinite products of Euler & Gauss in (10.10) and Weierstrass in (11.1), we will deduce other infinite product for the Gamma function: Gilbert's infinite product (13.43) in **art.** 184, Artin's infinite product (14.16) in **art.** 199 and ours (12.65) in **art.** 172.

10.4 Artin's approach and Wielandt's characterization

Artin (1964) has simplified the earlier proof of Bohr and Mollerup. Webster (1997) has generalized the Bohr-Mollerup-Artin framework, which is based on convexity (**art.** 102) and log-convexity (**art.** 103).

146. *Generalized functional equation.* The functional equation (10.1) of the Gamma function is the special case for $g(x) = x$ of the functional equation

$$f(x+1) = g(x) f(x) \tag{10.18}$$

with initial condition $f(1) = 1$. The functional equation $f(x+1) = sf(x) - r(x)$, where $r(x)$ is a polynomial, is solved in Jensen and Gronwall (1916, art. 11). After n times iterating the functional equation (10.18), we have

$$f(x+n+1) = g(x+n) \cdots g(x) f(x) \tag{10.19}$$

Theorem 1 *Let the positive function $g(x)$ on the positive real axis satisfy, for any positive real $w > 0$,*

$$\frac{g(x+w)}{g(x)} \to 1 \qquad \text{if } x \to \infty \tag{10.20}$$

If the positive function $f(x)$ is log-convex on the positive real axis and satisfies the functional equation (10.18) with $f(1) = 1$, then $f(x)$ is determined by

$$f(x) = \lim_{n\to\infty} \frac{g(n)\cdots g(1)}{g(n+x)\cdots g(x)} g^x(n) \qquad \text{for } x > 0 \tag{10.21}$$

Proof: We first rewrite the definition (6.2) of log-convexity for $x < y < z$ on an interval I of the positive real axis, after choosing $\lambda = \frac{z-y}{z-x}$, as

$$f^{z-x}(y) \leq f^{z-y}(x) f^{y-x}(z) \tag{10.22}$$

The log-convexity inequality (10.22) for f on the interval $[n+m, \infty)$, where $m = \lfloor x \rfloor$ and n is an integer, shows that

$$f(n+x+1) \leq f^{m+1-x}(n+m+1) f^{x-m}(n+m+2) \tag{10.23}$$

and

$$f(n+m+1) \leq f^{x-m}(n+x) f^{m+1-x}(n+x+1) \tag{10.24}$$

The functional equation (10.18) also shows that

$$f(n+m+2) = g(n+m+1) f(n+m+1)$$

so that the right-hand side of the first inequality (10.23) becomes

$$\begin{aligned} r_1 &= f^{m+1-x}(n+m+1)\, f^{x-m}(n+m+2) \\ &= f^{m+1-x}(n+m+1)\,(g(n+m+1)\, f(n+m+1))^{x-m} \\ &= f(n+m+1)\, g^{x-m}(n+m+1) \end{aligned}$$

and inequality (10.23) becomes

$$f(n+x+1) \le f(n+m+1)\, g^{x-m}(n+m+1) \tag{10.25}$$

Similarly for the right-hand side of the second inequality (10.24),

$$\begin{aligned} r_2 &= f^{x-m}(n+x)\, f^{m+1-x}(n+x+1) \\ &= f^{x-m}(n+x)\,(g(n+x)\, f(n+x))^{m-x}\, f(n+x+1) \\ &= f(n+x+1)\, g^{m-x}(n+x) \end{aligned}$$

and the second log-convex inequality (10.24) becomes

$$f(n+m+1) \le f(n+x+1)\, g^{m-x}(n+x)$$

which we rewrite as

$$f(n+m+1)\, g^{x-m}(n+x) \le f(n+x+1)$$

Adding (10.25) to the above inequality bounds $f(n+x+1)$ from above and below,

$$\frac{f(n+m+1)}{g^m(n+x)} g^x(n+x) \le f(n+x+1) \le \frac{f(n+m+1)}{g^m(n+m+1)} g^x(n+m+1)$$

We substitute $f(n+x+1)$ by the iterated solution (10.19), divide all sides by $g(n)\cdots g(1)\, g^x(n)$ and obtain

$$\begin{aligned} \mathbb{G} &= \frac{g(n+m)\cdots g(n+1)}{g(n+x)\cdots g(n+x)} \left(\frac{g(n+x)}{g(n)}\right)^x \\ &\le \frac{g(n+x)\cdots g(x)}{g(n)\cdots g(1)\, g^x(n)} f(x) \\ &\le \frac{g(n+m)\cdots g(n+1)}{g(n+m+1)\cdots g(n+m+1)} \left(\frac{g(n+m+1)}{g(n)}\right)^x \end{aligned}$$

Using the limit property (10.20) of $g(x)$ illustrates that both lower and upper bound of the middle inequality containing $f(x)$ tend to 1 if $n \to \infty$, demonstrating the solution (10.21) for $f(x)$. □

The Gauss limit $\Gamma(x) = \lim_{n\to\infty} \frac{n^x n!}{x(x+1)\dots(x+n)}$ follows from the solution (10.21) in Theorem 1 for $g(x) = x$. The solution (10.21) is, however, not unique, but by sharpening the condition (10.20) on $g(x)$ to $\lim_{x\to\infty} g(x) = 1$, Webster (1997) shows that $f(x) = \lim_{n\to\infty} \frac{g(n)\cdots g(1)}{g(n+x)\cdots g(x)}$ is unique. Thus, the Gamma function $\Gamma(x)$ is log-convex[5] for real $x > 0$.

[5] Since $\frac{d^2}{dx^2} \log \Gamma(x) = \psi^{(1)}(z)$, the series (12.11) illustrates that $\psi^{(1)}(z) > 0$ for real, positive x. Hence, $\Gamma(x)$ is log-convex on the positive real x-axis (**art.** 103).

The log-convexity property is thus sufficient to define the Gamma function and a class of Gamma-like functions that satisfy Theorem 1. That insight was first established by the Bohr-Mollerup-Artin theorem, the special case of Theorem 1 with $g(x) = x$, which provides an additional justification for the Euler-Gauss product.

147. *Wielandt's characterization.* Next to the Bohr-Mollerup-Artin characterization in **art**. 146, Remmert (1996) presents another, more function theoretic characterization of the Gamma function, based on Wielandt's Theorem:

Theorem 2 *Let $f(z)$ be an analytic function for $\operatorname{Re}(z) > 0$ satisfying (a) $f(z+1) = zf(z)$ for all $\operatorname{Re}(z) > 0$ and (b) $f(z)$ is bounded in the strip $1 \leq \operatorname{Re}(z) \leq 2$. Then, $f(z) = a\Gamma(z)$ with $a = f(1)$.*

Remmert (1996) nicely proves Wielandt's Theorem 2, from which he verifies the Weierstrass product (11.1), Gauss's multiplication formula (11.8), the Beta function property $B(p,q) = \frac{\Gamma(p)\Gamma(q)}{\Gamma(p+q)}$ and the definition (13.1) of the Binet function $\mu(z)$. Srinivasan (2007) gives other applications of Wielandt's Theorem 2.

We end this Chapter 10 by pointing to the recent work of Pérez-Marco (2022), who discusses the many viewpoints and definitions of the Gamma function in an educational and historical way.

11
The Euler-Gauss product for $\Gamma(z)$

Many direct consequences of the Euler-Gauss product (10.10) for the Gamma function $\Gamma(z)$ are derived in this Chapter 11: Weierstrass's product (11.1) from Gauss's limit (10.7) in **art**. 148, the reflection formula (11.4) and its consequences in **art**. 149 and **art**. 150, Gauss's multiplication formula (11.8) in **art**. 151 and products of ratio's of Gamma functions in **art**. 153.

148. *Weierstrass's product.* Weierstrass's product (11.1) can be obtained from Gauss's definition $\Gamma(z+1) = \lim_{k\to\infty} \Pi(k,z)$. Following Erdélyi *et al.* (1953b, p. 2), Gauss's definition of the Gamma function can be written, with (10.3), as

$$\begin{aligned}\frac{1}{\Gamma(z+1)} &= \lim_{k\to\infty} (1+z)\left(1+\frac{z}{2}\right)\cdots\left(1+\frac{z}{k}\right)e^{-z\log k}\\ &= \lim_{k\to\infty} (1+z)\,e^{-z}\left(1+\frac{z}{2}\right)e^{-\frac{z}{2}}\cdots\left(1+\frac{z}{k}\right)e^{-\frac{z}{k}}e^{z\left(\sum_{n=1}^{k}\frac{1}{n}-\log k\right)}\\ &= \lim_{k\to\infty}\prod_{n=1}^{k}\left(1+\frac{z}{n}\right)e^{-\frac{z}{n}}\lim_{k\to\infty} e^{z\left(\sum_{n=1}^{k}\frac{1}{n}-\log k\right)}\end{aligned}$$

Introducing Euler's constant $\gamma = \lim_{k\to\infty}\left(\sum_{n=1}^{k}\frac{1}{n} - \log k\right)$ in (9.3) leads to Weierstrass's product,

$$\frac{1}{\Gamma(z+1)} = e^{\gamma z}\prod_{n=1}^{\infty}\left(1+\frac{z}{n}\right)e^{-\frac{z}{n}} \tag{11.1}$$

which illustrates that Euler's constant γ plays a fundamental role in the theory of the Gamma function.

149. *Reflection formula.* Gauss (1813, p. 148) derives the reflection formula for the Gamma function via his contiguous relations of the hypergeometric function, that, for specific parameters, reduce to the sinus function. Gauss then finds the infinite product of the sinus function (Abramowitz and Stegun, 1968, 4.3.89)

$$\sin(\pi z) = \pi z\prod_{k=1}^{\infty}\left(1-\frac{z^2}{k^2}\right) \tag{11.2}$$

There are several different[1] proofs of the infinite product (11.2). Jordan (1909, art. 372) derives (11.2) as the limit $m \to \infty$ of the identity, valid for any odd integer m,

$$\sin(\pi z) = m \sin\left(\frac{\pi z}{m}\right) \prod_{n=1}^{\frac{m-1}{2}} \left(1 - \frac{\sin^2\left(\frac{\pi z}{m}\right)}{\sin^2\left(\frac{\pi n}{m}\right)}\right)$$

which follows either elementary as in Bourbaki (2004, p. 286) or from Liouville's theorem (i.e. an analytic function that is bounded everywhere in the complex plane is a constant). Jordan (1909, art. 373) also demonstrates that

$$\sin(\pi z) = \pi z \prod_{k=-\infty}^{\infty} \left(1 + \frac{z}{k}\right) e^{-\frac{z}{k}} \tag{11.3}$$

which is a consequence of Hadamard's factorization theorem (1.18). With (11.2), the Euler-Gauss infinite product (10.10) for $\Gamma(z)$ shows that

$$\frac{1}{\Gamma(z)\Gamma(-z)} = -z^2 \prod_{n=1}^{\infty} \left(1 - \frac{z^2}{n^2}\right) = -z \frac{\sin(\pi z)}{\pi}$$

which establishes the reflection formula of the Gamma function, valid for all z,

$$\Gamma(z)\Gamma(1-z) = \frac{\pi}{\sin \pi z} \tag{11.4}$$

For example, if $z = \frac{1}{2}$, then the reflection formula (11.4) shows that $\Gamma\left(\frac{1}{2}\right) = \sqrt{\pi}$.

150. *Absolute value of the Gamma function.* Slightly more general than the previous **art.** 149, the Euler-Gauss infinite product (10.10) indicates that

$$\begin{aligned}\Gamma(\sigma + it)\Gamma(\sigma - it) &= \frac{1}{\sigma^2 + t^2} \prod_{n=1}^{\infty} \frac{\left(1 + \frac{1}{n}\right)^{2\sigma}}{\left(1 + \frac{\sigma}{n} + \frac{it}{n}\right)\left(1 + \frac{\sigma}{n} - \frac{it}{n}\right)} \\ &= \frac{1}{\sigma^2 + t^2} \prod_{n=1}^{\infty} \frac{\left(1 + \frac{1}{n}\right)^{2\sigma}}{\left(\left(1 + \frac{\sigma}{n}\right)^2 + \left(\frac{t}{n}\right)^2\right)}\end{aligned}$$

---

[1] After integration of the Taylor series of $\pi \cot(\pi z) = \frac{1}{z} - 2\sum_{n=1}^{\infty} \zeta(2n)\, z^{2n-1}$, valid for $|z| < 1$ and where $\zeta(2n) = \sum_{k=1}^{\infty} \frac{1}{k^{2n}}$ in (9.5) is the Riemann Zeta-function, leading to $\log \frac{\pi x}{\sin(\pi z)} = 2\sum_{n=1}^{\infty} \frac{\zeta(2n)}{2n} z^{2n}$, one deduces that

$$\log \frac{\pi z}{\sin(\pi z)} = \sum_{k=1}^{\infty} \sum_{n=1}^{\infty} \frac{1}{n} \left(\frac{z}{k}\right)^{2n} = -\sum_{k=1}^{\infty} \log\left(1 - \frac{z^2}{k^2}\right)$$

from which (11.2) follows. Although derived under the restriction $|z| < 1$, (11.2) can be shown to hold for any complex z. Euler (1748, art. 156) derives the product (11.2) differently, ingeniously and surely worth reading!

Since $\Gamma(z)$ is real on the real axis, the principle of reflection (Titchmarsh, 1964, p. 155) shows that $\Gamma(\sigma+it)\Gamma(\sigma-it)=\Gamma(\sigma+it)\overline{\Gamma(\sigma+it)}$ and

$$|\Gamma(\sigma+it)|^2=\frac{1}{\sigma^2+t^2}\frac{\prod_{n=1}^{\infty}\left(1+\frac{1}{n}\right)^{2\sigma}}{\prod_{n=1}^{\infty}\left(1+\frac{\sigma}{n}\right)^2+\left(\frac{t}{n}\right)^2}$$

Hence, with $|\Gamma(\sigma)|^2=\frac{1}{\sigma^2}\frac{\prod_{n=1}^{\infty}\left(1+\frac{1}{n}\right)^{2\sigma}}{\prod_{n=1}^{\infty}\left(1+\frac{\sigma}{n}\right)^2}$, we obtain

$$|\Gamma(\sigma+it)|^2=|\Gamma(\sigma)|^2\frac{\sigma^2}{\sigma^2+t^2}\prod_{n=1}^{\infty}\frac{\left(1+\frac{\sigma}{n}\right)^2}{\left(1+\frac{\sigma}{n}\right)^2+\left(\frac{t}{n}\right)^2}$$

and, finally,

$$|\Gamma(\sigma+it)|^2=\frac{\Gamma^2(\sigma)}{\prod_{n=0}^{\infty}\left(1+\left(\frac{t}{n+\sigma}\right)^2\right)} \tag{11.5}$$

which again illustrates for any real $t\neq0$ that $|\Gamma(\sigma+it)|<|\Gamma(\sigma)|$ for any real σ.

Moreover, for an integer value of $\sigma=k>0$, the product in (11.5) can be evaluated with (11.2):

$$\prod_{n=0}^{\infty}\left(1+\left(\frac{t}{n+k}\right)^2\right)=\prod_{n=k}^{\infty}\left(1+\left(\frac{t}{n}\right)^2\right)=\frac{\prod_{n=1}^{\infty}\left(1+\left(\frac{t}{n}\right)^2\right)}{\prod_{n=1}^{k-1}\left(1+\left(\frac{t}{n}\right)^2\right)}$$
$$=\frac{\sinh(\pi t)}{\pi t\prod_{n=1}^{k-1}\left(1+\left(\frac{t}{n}\right)^2\right)}$$

resulting in $|\Gamma(k+it)|^2=\frac{\pi t(k-1)!^2}{\sinh(\pi t)}\prod_{n=1}^{k-1}\left(1+\left(\frac{t}{n}\right)^2\right)$ and

$$|\Gamma(k+it)|^2=\frac{\pi t}{\sinh(\pi t)}\prod_{n=1}^{k-1}\left(n^2+t^2\right) \tag{11.6}$$

For example, for $k=1$ in (11.6), we find $|\Gamma(1+it)|^2=\frac{\pi t}{\sinh(\pi t)}$. The reflection formula (11.4) and (11.6) immediately yield

$$|\Gamma(1-k+it)|^2=\frac{\pi}{t\sinh(\pi t)\prod_{n=1}^{k-1}\left(n^2+t^2\right)} \tag{11.7}$$

For $k=1$, (11.7) simplifies to

$$|\Gamma(it)|^2=\frac{\pi}{t\sinh(\pi t)}$$

If $\sigma=\frac{1}{2}$ in (11.5), then the infinite product $\cos(\pi z)=\prod_{n=1}^{\infty}\left(1-\left(\frac{2z}{2n-1}\right)^2\right)$, derived by Jordan (1909, art. 375) from (11.3) and $\cos(\pi z) = \frac{\sin(2\pi z)}{2\sin(\pi z)}$, indicates that

$$\left|\Gamma\left(\frac{1}{2}+it\right)\right|^2 = \frac{\left|\Gamma\left(\frac{1}{2}\right)\right|^2}{\prod_{n=0}^{\infty}\left(1+\left(\frac{2t}{2n+1}\right)^2\right)} = \frac{\pi}{\cosh(\pi t)}$$

151. *Multiplication formula.* Gauss (1813, p. 149) derives his elegant[2] multiplication formula

$$\Gamma(nz) = (2\pi)^{\frac{1}{2}(1-n)} n^{nz-\frac{1}{2}} \prod_{k=0}^{n-1} \Gamma\left(z+\frac{k}{n}\right) \tag{11.8}$$

of which Legendre's duplication formula for $n=2$ is perhaps the most used application,

$$\Gamma(2z) = \frac{2^{2z-1}}{\sqrt{\pi}}\Gamma(z)\Gamma\left(z+\frac{1}{2}\right) \tag{11.9}$$

Gauss observes that

$$\frac{n^{nz}\prod_{j=0}^{n-1}\Pi\left(k, z-\frac{j}{n}\right)}{\Pi(nk, nz)} = \frac{(\Gamma(k+1))^n}{\Gamma(nk+1)\,k^{\frac{(n-1)}{2}}} \tag{11.10}$$

does not depend on z. However, Gauss does not give the tedious derivation of (11.10), but we do. Multiplying the Gauss factors (10.3),

$$\begin{aligned}\prod_{j=0}^{n-1}\Pi\left(k, z-\frac{j}{n}\right) &= \prod_{j=0}^{n-1}\frac{\Pi(k)\,\Pi\left(z-\frac{j}{n}\right)}{\Pi\left(k+z-\frac{j}{n}\right)}k^{z-\frac{j}{n}} \\ &= (\Pi(k))^n\,k^{nz}k^{-\frac{1}{n}\sum_{j=1}^{n-1}j}\prod_{j=0}^{n-1}\frac{\Pi\left(z-\frac{j}{n}\right)}{\Pi\left(z+k-\frac{j}{n}\right)}\end{aligned}$$

and changing the notation $\Pi(z) = \Gamma(z+1)$ and with $\sum_{j=1}^{n-1} j = \frac{n(n-1)}{2}$, we have

$$\prod_{j=0}^{n-1}\Pi\left(k, z-\frac{j}{n}\right) = (\Gamma(k+1))^n\frac{k^{nz}}{k^{\frac{(n-1)}{2}}}\prod_{j=0}^{n-1}\frac{\Gamma\left(z+1-\frac{j}{n}\right)}{\Gamma\left(z+k+1-\frac{j}{n}\right)}$$

Introducing Gauss's infinite product (10.10) yields

$$\begin{aligned}\prod_{j=0}^{n-1}\frac{\Gamma\left(z+1-\frac{j}{n}\right)}{\Gamma\left(z+k+1-\frac{j}{n}\right)} &= \prod_{j=0}^{n-1}\prod_{m=1}^{\infty}\frac{m^k}{(m+1)^k}\prod_{m=0}^{\infty}\frac{(nm+nz+nk+n-j)}{(nm+nz+n-j)} \\ &= \prod_{m=1}^{\infty}\frac{m^{nk}}{(m+1)^{nk}}\prod_{m=0}^{\infty}\prod_{l=1}^{n}\frac{(nm+nz+nk+l)}{(nm+nz+l)} \\ &= \prod_{m=1}^{\infty}\frac{m^{nk}}{(m+1)^{nk}}\prod_{m=0}^{\infty}\prod_{l=1}^{n}\left(1+\frac{nk}{nm+nz+l}\right)\end{aligned}$$

[2] Gauss writes "Unde habemus theorema elegans".

Because $r = nm + l$ runs over all integers (see the progression property in **art.** 12), we observe that

$$\prod_{m=0}^{\infty}\prod_{l=1}^{n}\left(1+\frac{nk}{nm+l+nz}\right)=\prod_{r=1}^{\infty}\left(1+\frac{nk}{r+nz}\right)$$

and we arrive at

$$\prod_{j=0}^{n-1}\Pi\left(k,z-\frac{j}{n}\right)=\left(\Gamma\left(k+1\right)\right)^{n}\frac{k^{nz}}{k^{\frac{(n-1)}{2}}}\prod_{m=1}^{\infty}\frac{m^{nk}}{(m+1)^{nk}}\prod_{r=1}^{\infty}\left(1+\frac{nk}{r+nz}\right)$$

Gauss divides $\prod_{j=0}^{n-1}\Pi\left(k,z-\frac{j}{n}\right)$ by $\Pi\left(nk,nz\right)$, which equals

$$\begin{aligned}\Pi\left(nk,nz\right)&=\frac{\Pi\left(nk\right)\Pi\left(nz\right)}{\Pi\left(nk+nz\right)}n^{nz}k^{nz}=\frac{\Gamma\left(nk+1\right)\Gamma\left(nz+1\right)}{\Gamma\left(nk+nz+1\right)}n^{nz}k^{nz}\\&=n^{nz}k^{nz}\Gamma\left(nk+1\right)\prod_{m=1}^{\infty}\frac{m^{nk}}{(m+1)^{nk}}\prod_{m=1}^{\infty}\frac{m+nz+nk}{m+nz}\\&=n^{nz}k^{nz}\Gamma\left(nk+1\right)\prod_{m=1}^{\infty}\frac{m^{nk}}{(m+1)^{nk}}\prod_{r=1}^{\infty}\left(1+\frac{nk}{r+nz}\right)\end{aligned}$$

resulting in (11.10) and demonstrating that the left-hand side of (11.10) is independent of z. As follows from the Gauss factors (10.3), $\Pi\left(k,0\right)=\Pi\left(nk,0\right)=1$ and the choice of $z=0$ in the left-hand side of (11.10) is

$$\frac{n^{nz}\prod_{j=0}^{n-1}\Pi\left(k,z-\frac{j}{n}\right)}{\Pi\left(nk,nz\right)}=\prod_{j=0}^{n-1}\Pi\left(k,-\frac{j}{n}\right)$$

After taking the limit of $k\to\infty$ of both sides, the definition $\Pi\left(z\right)=\Gamma\left(z+1\right)=\lim_{k\to\infty}\Pi\left(k,z\right)$ leads to

$$\frac{n^{nz}\prod_{j=0}^{n-1}\Gamma\left(z+1-\frac{j}{n}\right)}{\Gamma\left(nz+1\right)}=\prod_{j=1}^{n-1}\Gamma\left(1-\frac{j}{n}\right)$$

Let $l=n-j$, then $\prod_{j=1}^{n-1}\Gamma\left(1-\frac{j}{n}\right)=\prod_{j=1}^{n-1}\Gamma\left(\frac{n-j}{n}\right)=\prod_{l=1}^{n-1}\Gamma\left(\frac{l}{n}\right)$ so that

$$\left(\prod_{j=1}^{n-1}\Gamma\left(1-\frac{j}{n}\right)\right)^{2}=\prod_{j=1}^{n-1}\Gamma\left(1-\frac{j}{n}\right)\Gamma\left(\frac{j}{n}\right)=\prod_{j=1}^{n-1}\frac{\pi}{\sin\pi\frac{j}{n}}=\frac{\pi^{n-1}}{\prod_{j=1}^{n-1}\sin\frac{\pi j}{n}}$$

where the reflection formula (11.4) has been invoked. Using Euler's formula[3]

$$\sin x=2^{n-1}\prod_{k=0}^{n-1}\sin\frac{(\pi k+x)}{n}\tag{11.11}$$

[3] The polynomial z^n-1 has as zeros the n-th roots of unity, $z^n-1=(z-1)\prod_{k=1}^{n-1}\left(1-z\,e^{-\frac{2\pi ik}{n}}\right)$. Choosing $z=e^{\frac{2ix}{n}}$ leads, after some manipulations, to (11.11).

for $x \to 0$ gives $\prod_{k=1}^{n-1} \sin \frac{\pi k}{n} = \frac{n}{2^{n-1}}$ and finally leads to

$$\Gamma(nz+1) = (2\pi)^{-\frac{n-1}{2}} n^{nz+\frac{1}{2}} \prod_{k=1}^{n} \Gamma\left(z+\frac{k}{n}\right) \tag{11.12}$$

that equals (11.8), because

$$\prod_{k=1}^{n} \Gamma\left(z+\frac{k}{n}\right) = \Gamma(z+1) \prod_{k=1}^{n-1} \Gamma\left(z+\frac{k}{n}\right) = z \prod_{k=0}^{n-1} \Gamma\left(z+\frac{k}{n}\right)$$

Example Evaluation of (11.12) for $z=0$ yields

$$\prod_{k=1}^{n-1} \Gamma\left(\frac{k}{n}\right) = \frac{\left(\sqrt{2\pi}\right)^{n-1}}{\sqrt{n}}$$

that is related to Gauss's formula (12.25) and its inverse (12.23) for the digamma function.

152. *More about Gauss' multiplication formula.* Binet (1839, pp. 213-222) discusses Gauss' multiplication formula (11.8), applies (11.8) to the Beta functions and presents the corresponding multiplication formula for the Beta function. Gilbert (1876, art. 29) gives another proof of Gauss's multiplication formula (11.8) based on his integral

$$\mu(z) = \frac{1}{2}\sum_{k=1}^{\infty} \frac{\cos(2\pi k z)}{k} + \frac{1}{2}\int_0^{\infty} \left(\frac{1}{x} - \cot x\right) \frac{\sin(2zx)}{x} dx$$

of the Binet function $\mu(z) = \log\Gamma(z) - \left(z-\frac{1}{2}\right)\log z + z - \frac{1}{2}\log(2\pi)$ in Chapter 13 and Euler's formula (11.11). Nielsen (1956, pp. 196-198) derives the generalization due to Schobloch, for integers n and p,

$$\frac{\prod_{k=0}^{n-1} \Gamma\left(pz+\frac{pk}{n}\right)}{\prod_{j=0}^{p-1} \Gamma\left(nz+\frac{nj}{p}\right)} = (2\pi)^{\frac{n-p}{2}} \left(\frac{p}{n}\right)^{npz+\frac{np-n-p}{2}}$$

which reduces for $p=1$ to Gauss' multiplication formula (11.8). Aycock (2021) argues that Euler already knew the multiplication formula (11.8).

Art. 159 shows that Gauss's multiplication formula (11.8) is a consequence of the progression concept (**art.** 12).

153. *Products of Gamma functions.* Using the Gauss product (10.10), Jensen and Gronwall (1916, p. 131) mention

$$\frac{\Gamma(a)\Gamma(b)}{\Gamma(a+s)\Gamma(b-s)} = \frac{(a+s)}{a}\frac{(b+s)}{b} \prod_{n=1}^{\infty} \frac{1+\frac{a+s}{n}}{1+\frac{a}{n}} \frac{1+\frac{b-s}{n}}{1+\frac{b}{n}}$$

Further, with $\frac{1+\frac{a+s}{n}}{1+\frac{a}{n}} \frac{1+\frac{b-s}{n}}{1+\frac{b}{n}} = \frac{n+a+s}{n+a}\frac{n+b-s}{n+b} = \left(1+\frac{s}{n+a}\right)\left(1-\frac{s}{n+b}\right)$, we arrive at

$$\frac{\Gamma(a)\Gamma(b)}{\Gamma(a+s)\Gamma(b-s)} = \prod_{n=0}^{\infty} \left(1+\frac{s}{n+a}\right)\left(1-\frac{s}{n+b}\right) \tag{11.13}$$

For $b = a$, the simplified version

$$\frac{\Gamma^2(a)}{\Gamma(a+s)\Gamma(a-s)} = \prod_{n=0}^{\infty}\left(1-\left(\frac{s}{n+a}\right)^2\right)$$

immediately leads to the absolute value formula (11.5) by letting $s = it$. If $a = b = 1$, then (11.13) reduces with (11.2) to the reflection formula (11.4). As a curiosity, emphasizing the difference with the Euler-Gauss product (10.10), Weierstrass's product $\frac{1}{\Gamma(z+1)} = e^{\gamma z}\prod_{n=1}^{\infty}\left(1+\frac{z}{n}\right)e^{-z/n}$ in (11.1) shows that

$$\begin{aligned}\frac{\Gamma(a+b+1)}{\Gamma(a+s+1)\Gamma(b-s+1)} &= \frac{e^{\gamma(a+b)}\prod_{n=1}^{\infty}\left(1+\frac{a+s}{n}\right)\left(1+\frac{b-s}{n}\right)e^{-\frac{a+b}{n}}}{e^{\gamma(a+b)}\prod_{n=1}^{\infty}\left(1+\frac{a+b}{n}\right)e^{-\frac{a+b}{n}}} \\ &= \prod_{n=1}^{\infty}\frac{\left(1+\frac{a+s}{n}\right)\left(1+\frac{b-s}{n}\right)}{\left(1+\frac{a+b}{n}\right)}\end{aligned}$$

Since

$$\frac{\left(1+\frac{a+s}{n}\right)\left(1+\frac{b-s}{n}\right)}{\left(1+\frac{a+b}{n}\right)} = \frac{n+a+s}{n}\,\frac{n+b-s}{n+a+b} = \frac{\left(1+\frac{s}{n+a}\right)\left(1-\frac{s}{n+b}\right)}{\left(1+\frac{b}{n+a}\right)\left(1-\frac{b}{n+b}\right)}$$

we find

$$\frac{\Gamma(a+b+1)}{\Gamma(a+s+1)\Gamma(b-s+1)} = \frac{\prod_{n=1}^{\infty}\left(1+\frac{s}{n+a}\right)\left(1-\frac{s}{n+b}\right)}{\prod_{n=1}^{\infty}\left(1+\frac{b}{n+a}\right)\left(1-\frac{b}{n+b}\right)}$$

which also can be deduced from (11.13).

Furthermore, it follows from Weierstrass's product (11.1) that

$$\begin{aligned}\prod_{k=1}^{m}\frac{\Gamma(c_k z+d_k)}{\Gamma(a_k z+b_k)} &= \prod_{k=1}^{m}\frac{a_k z+b_k}{c_k z+d_k}\,e^{\gamma\sum_{k=1}^{m}(a_k-c_k)z+(b_k-d_k)} \\ &\times\prod_{n=1}^{\infty}\prod_{k=1}^{m}\frac{\left(1+\frac{(a_k z+b_k)}{n}\right)}{\left(1+\frac{(c_k z+d_k)}{n}\right)}\,e^{\frac{\sum_{k=1}^{m}(c_k-a_k)z+(d_k-b_k)}{n}}\end{aligned}$$

If $b_k = d_k = \alpha$ for all indices k, then

$$\prod_{k=1}^{m}\frac{\Gamma(c_k z+\alpha)}{\Gamma(a_k z+\alpha)} = \prod_{k=1}^{m}\frac{a_k z+\alpha}{c_k z+\alpha}\,e^{\gamma z\sum_{k=1}^{m}(a_k-c_k)}\prod_{n=1}^{\infty}\prod_{k=1}^{m}\frac{\frac{n+a_k z+\alpha}{n+c_k z+\alpha}}{e^{\frac{z}{n}\sum_{k=1}^{m}(a_k-c_k)}}$$

If also $c_k = 0$ for all indices k, then

$$\frac{\Gamma^m(\alpha)}{\prod_{k=1}^{m}\Gamma(a_k z+\alpha)} = e^{\gamma z\sum_{k=1}^{m}a_k}\prod_{k=1}^{m}\left(1+a_k\frac{z}{\alpha}\right)\prod_{n=1}^{\infty}\prod_{k=1}^{m}\frac{\left(1+a_k\frac{z}{n+\alpha}\right)}{e^{\frac{z}{n}\sum_{k=1}^{m}a_k}}$$

We denote the polynomial $p_m(x) = \prod_{k=1}^{m}(1+a_k x) = \sum_{j=0}^{m}A_j x^j$, which possesses zeros at $x = -\frac{1}{a_k}$ for each $k \in [1, m]$. Then, $p_m(0) = 1 = A_0$. Further, we deduce (Van Mieghem, 2023a) that

$$q_m(x) = x^m p_m\left(\frac{1}{x}\right) = \prod_{k=1}^{m}(x + a_k) = \sum_{l=0}^{m} A_{m-l}x^l = x^m + \sum_{l=0}^{m-1} A_{m-l}x^l$$

is a monic polynomial with zeros at $x = -a_k$, with coefficients $A_1 = \sum_{k=1}^{m} a_k$ and $A_m = \prod_{k=1}^{m} a_k$. The product of Gamma functions, written in terms of the polynomial $q_m(x)$, becomes

$$\frac{\Gamma^n(\alpha)}{\prod_{k=1}^{m}\Gamma(a_k z + \alpha)} = e^{-A_1\gamma z} q_m\left(\frac{z}{\alpha}\right) \prod_{n=1}^{\infty} q_m\left(\frac{z}{\alpha + n}\right) e^{\frac{z}{n}A_1} \tag{11.14}$$

For the polynomial $q_m(x) = 1 - x^m$, with zeros on the unit circle at $a_k = -e^{i\frac{2\pi(k-1)}{m}}$ for $1 \le k \le m$, (11.14) simplifies to

$$\frac{\Gamma^n(\alpha)}{\prod_{k=1}^{m}\Gamma\left(\alpha - e^{i\frac{2\pi(k-1)}{m}} z\right)} = \prod_{n=0}^{\infty}\left(1 - \left(\frac{z}{\alpha + n}\right)^m\right)$$

and, further with $\alpha = 1$, to Liouville's formula (Jensen and Gronwall, 1916, p. 131),

$$\frac{1}{\prod_{k=0}^{m-1}\Gamma\left(1 - e^{i\frac{2\pi k}{m}} z\right)} = \prod_{n=1}^{\infty}\left(1 - \left(\frac{z}{n}\right)^m\right) \tag{11.15}$$

Liouville's formula (11.15) for $m = 2$ and the infinite product (11.2),

$$\frac{1}{\Gamma(1-z)\Gamma(1+z)} = \prod_{n=1}^{\infty}\left(1 - \left(\frac{z}{n}\right)^2\right) = \frac{\sin(\pi z)}{\pi z}$$

lead again to the reflection formula (11.4).

12
The digamma function $\psi(z)$

Both the Euler-Gauss (10.10) and Weierstass's (11.1) infinite product for the Gamma function directly suggest to consider the logarithm of the Gamma function. Chapter 12 studies the digamma function in depth. Section 12.1 deduces the important properties and relates the digamma function and its derivatives, called polygamma functions, to the Hurwitz Zeta-function $\zeta(s,z)$ and the Lerch transcendent in **art**. 156. We prove Lerch's formula $\log\Gamma(z) = \zeta'(0,z) - \zeta'(0)$ in (12.16) in **art**. 158, because Lerch's formula (12.16) connects Gauss's multiplication formula (11.8) to the progression property in **art**. 12, which is important in the theory of the Mittag-Leffler function. Section 12.2 derives Gauss's remarkable, closed form of the digamma function (12.23) for fractions $z = \frac{j}{m} \in \mathbb{Q}$. We generalize Gauss's approach with an exact closed form (12.32) of the generating function $\sum_{j=1}^{m} \psi\left(\frac{j}{m}\right) z^{j-1}$ over the fractions in **art**. 162. Section 12.3 of Chapter 12 derives several integrals for $\log\Gamma(z)$, that open the way towards Binet's function $\mu(z) = \log\Gamma(z) - \left(z - \frac{1}{2}\right)\log z + z - \frac{1}{2}\log(2\pi)$ in (13.1), investigated in Chapter 13. The final Section 12.4 on interpolation and differences presents Newton series for the digamma function $\psi(z)$ in **art**. 170 and for $\log\Gamma(z)$, Hermite's elegant approach in **art**. 171 and ours in **art**. 172.

12.1 Properties of the digamma function

154. *The digamma function.* The digamma function $\psi(z) = \frac{d}{dz}\log\Gamma(z) = \frac{\Gamma'(z)}{\Gamma(z)}$ is the logarithmic derivative of the Gamma function $\Gamma(z)$. The functional equation (10.1) of the Gamma function and the definition $\psi(z) = \frac{d}{dz}\log\Gamma(z)$ directly lead to the functional equation of the digamma function

$$\psi(z+1) = \psi(z) + \frac{1}{z} \tag{12.1}$$

The logarithm of Weierstrass's infinite product (11.1) is

$$\log \Gamma(z+1) = -\gamma z - \sum_{n=1}^{\infty} \left(\log\left(1+\frac{z}{n}\right) - \frac{z}{n} \right) \tag{12.2}$$

and differentiation yields

$$\psi(z+1) = \frac{d}{dz} \log \Gamma(z+1) = -\gamma + \sum_{n=1}^{\infty} \left(\frac{1}{n} - \frac{1}{z+n} \right) \tag{12.3}$$

Similarly, after differentiating the logarithm of the Euler-Gauss product (10.10),

$$\log \Gamma(z+1) = \sum_{n=1}^{\infty} \left(z \log\left(1+\frac{1}{n}\right) - \log\left(1+\frac{z}{n}\right) \right)$$

we find

$$\psi(z+1) = \sum_{n=1}^{\infty} \left(\log\left(1+\frac{1}{n}\right) - \frac{1}{z+n} \right) \tag{12.4}$$

Both (12.2) for $z=1$ and (12.4) for $z=0$ show that

$$\psi(1) = -\gamma = -\sum_{n=1}^{\infty} \left(\frac{1}{n} - \log\left(1+\frac{1}{n}\right) \right) \tag{12.5}$$

Since Weierstrass's form (12.3) of the digamma function is easier than Gauss's form (12.4), due to the logarithm, the theory of the digamma function usually starts from Weierstrass's form (12.3).

The functional equation (12.1) of digamma function $\psi(z)$ shows that Weierstrass's form (12.3) becomes

$$\psi(z) = -\gamma - \frac{1}{z} + z \sum_{n=1}^{\infty} \frac{1}{(z+n)\,n} \tag{12.6}$$

The series (12.6) converges for all z, except at non-positive real numbers $z=-k$ for integer $k \geq 0$. The digamma function $\psi(z)$ possesses simple poles at the integers $z=-k$.

After n iterations of functional equation (12.1),

$$\psi(z) = \psi(z+n) - \sum_{k=0}^{n-1} \frac{1}{z+k} \tag{12.7}$$

and taking the limit $n \to \infty$, we formally obtain $\psi(z) = \lim_{n\to\infty} \psi(z+n) - \lim_{n\to\infty} \sum_{k=0}^{n} \frac{1}{z+k}$. However, the latter series does not converge[1], implying that $\lim_{n\to\infty} \psi(z+n) = \infty$.

[1] Indeed, $\sum_{k=0}^{n} \frac{1}{z+k} = \frac{1}{z} + \sum_{k=1}^{n_0} \frac{1}{k\left(1+\frac{z}{k}\right)} + \sum_{k=1+n_0}^{n} \frac{1}{k\left(1+\frac{z}{k}\right)}$. We can choose $n_0 > |z|$, so that $0 < \left|1+\frac{z}{k}\right| < 2$ and $\left|\sum_{k=1+n_0}^{n} \frac{1}{k\left(1+\frac{z}{k}\right)}\right| > \frac{1}{2} \sum_{k=1+n_0}^{n} \frac{1}{k} \to \infty$, because the harmonic series $\sum_{k=1}^{n} \frac{1}{k}$ diverges (**art.** 136) for $n \to \infty$.

The iterated functional equation in (12.7), with $z = 1$ and $\psi(1) = -\gamma$, indicates that

$$\psi(n) = \sum_{k=1}^{n-1} \frac{1}{k} - \gamma \tag{12.8}$$

Hence, the harmonic series in (9.1) equals

$$H_n = \sum_{k=1}^{n} \frac{1}{k} = \gamma + \psi(n+1) \tag{12.9}$$

Since $\psi(1) = -\gamma < 0$, $\psi(2) = 1 - \gamma > 0$ and $\psi(x)$ is strictly increasing on the real axis (because $\frac{d\psi(x)}{dx} = \psi^{(1)}(x) > 0$ in (12.11)), there is a zero $\xi \approx 1.46163...$ of $\psi(x)$ in the interval $(1, 2)$, which corresponds to the minimum of the Gamma function $\Gamma(x)$ on the real, positive axis.

The reflection formula (11.4) of the Gamma function similarly translates, for all z, to the reflection formula for the digamma function

$$\psi(1-z) = \psi(z) + \pi \cot \pi z \tag{12.10}$$

As Gauss (1813, p. 155) pointed out, after introducing the series (12.3) or (12.4), we find a partial fractional expansion for the cotangens,

$$\pi \cot \pi z = \psi(1-z) - \psi(z) = \sum_{n=0}^{\infty} \left(\frac{1}{z+n} - \frac{1}{1-z+n} \right)$$

which is rewritten in the more symmetrical form (Olver *et al.*, 2010, 4.22.3) as

$$\pi \cot \pi z = \frac{1}{z} + 2z \sum_{n=1}^{\infty} \frac{1}{z^2 - n^2}$$

The reflection formula (12.10) is a particular case of the difference quotient

$$\frac{\psi(z) - \psi(w)}{z - w} = \sum_{n=0}^{\infty} \frac{1}{(w+n)(z+n)}$$

which follows from Weierstrass's form (12.3) and the functional equation (12.1).

155. *Polygamma functions.* The polygamma functions, defined in Olver *et al.* (2010, 5.15), as $\psi^{(n)}(z) = \frac{d^{n+1} \log \Gamma(z)}{dz^{n+1}}$ with $\psi^{(0)}(z) = \psi(z)$, follow immediately from (12.6) and possess the convergent series, for any integer $n \geq 1$,

$$\psi^{(n)}(z) = (-1)^{n+1} n! \sum_{k=0}^{\infty} \frac{1}{(z+k)^{n+1}} \tag{12.11}$$

from which, for $n \geq 1$, we find that

$$\psi^{(n)}(1) = (-1)^{n+1} n! \sum_{k=1}^{\infty} \frac{1}{k^{n+1}} = (-1)^{n+1} n! \zeta(n+1)$$

while $\psi(1) = -\gamma$. In contrast to the digamma function $\psi(z)$, the series (12.11) indicates that $\lim_{z\to\infty}\psi^{(n)}(z) = 0$ for $n > 0$.

156. *The Hurwitz Zeta-function.* The "fractional extension to complex s" of the polygamma functions is the Hurwitz Zeta-function (Whittaker and Watson, 1996, Chap. XIII),

$$\zeta(s,z) = \sum_{k=0}^{\infty} \frac{1}{(z+k)^s} \qquad \text{for } \operatorname{Re}(s) > 1 \tag{12.12}$$

which reduces for $z = 1$ to the Riemann-Zeta function $\zeta(s,1) = \sum_{k=1}^{\infty} \frac{1}{k^s}$ in **art.** 138. The Hurwitz Zeta-function obeys the functional equation in the parameter z,

$$\zeta(s,z+1) = \zeta(s,z) - \frac{1}{z^s} \tag{12.13}$$

The functional equation (12.13) is a generalization of the functional equation $\psi(z+1) = \psi(z) + \frac{1}{z}$ of the digamma function in (12.1).

A further generalization of the Hurwitz Zeta-function is the Lerch transcendent

$$\Phi(w,s,z) = \sum_{k=0}^{\infty} \frac{w^k}{(z+k)^s} \tag{12.14}$$

which, as shown by Lerch (1887), is an entire function in s. Many impressive formulae and series of the Lerch transcendent $\Phi(w,s,z)$ are derived in Lerch (1894).

157. *Taylor series of the digamma function around z_0.* Since the digamma function is analytic, except at the poles $z = -k$ for integer $k \geq 0$, the digamma function possesses a Taylor series around a point z_0, that is different from a pole,

$$\psi(z) = \psi(z_0) + \sum_{k=1}^{\infty} \psi_k(z_0)(z-z_0)^k$$

The Taylor coefficient $\psi_k(z_0)$ around z_0 equals $\psi_k(z_0) = \frac{1}{k!} \left.\frac{d^k\psi(z)}{dz^k}\right|_{z=z_0} = \frac{1}{k!}\psi^{(k)}(z_0)$, specified in (12.11) or in terms of the Hurwitz Zeta-function (12.12), $\psi^{(n)}(z) = (-1)^{n+1} n! \zeta(n+1,z)$. The Taylor series then becomes

$$\psi(z) = \psi(z_0) + \sum_{k=1}^{\infty} (-1)^{k+1} \zeta(k+1,z_0)(z-z_0)^k \tag{12.15}$$

In particular, for the expansion point $z_0 = 1$, we find[2] with $\psi(1) = -\gamma$ and

[2] Via our characteristic coefficients in Appendix A, the zero $\xi \approx 1.46163...$ of $\psi(x)$ in the interval $(1,2)$ can be expressed as a Lagrange series, whose coefficients are functions of $\psi_k(1)$. Thus, a relatively rapidly converging series for the zero ξ exists in terms of the Euler constant γ and Riemann Zeta-function evaluations at positive integers $k > 1$.

$|z| < 1$,

$$\psi(z) = -\gamma - \sum_{k=1}^{\infty} \zeta(k+1)(1-z)^k$$

which also follows, after differentiation, from the Taylor series (12.44) of $\log\Gamma(z+1)$ around $z_0 = 0$ below.

158. *Lerch's formula.* We present a nice relation between the Hurwitz Zeta-function and the Gamma function, due to Matyáš Lerch (1894, eq. (22a)) and valid for complex z,

$$\log\Gamma(z) = \zeta'(0,z) - \zeta'(0) \tag{12.16}$$

where $\zeta'(0,z) = \left.\frac{\partial\zeta(s,z)}{\partial s}\right|_{s=0}$. We follow the proof in Berndt (1985).

Applying the basic summation formula (3.39) to the Hurwitz Zeta-function (12.12) yields, for $\text{Re}(s) > 1$, with $f(x) = \frac{1}{(z+x)^s}$, $a = 0$ and $b \to \infty$,

$$\sum_{k=1}^{\infty} \frac{1}{(z+k)^s} = \frac{z^{1-s}}{s-1} - \frac{1}{2z^s} - s\int_0^{\infty} \left(x - \lfloor x \rfloor - \frac{1}{2}\right) \frac{dx}{(z+x)^{s+1}}$$

Since $\sum_{k=1}^{\infty} \frac{1}{(z+k)^s} = \sum_{k=0}^{\infty} \frac{1}{(z+1+k)^s} = \zeta(s, z+1)$, the functional equation (12.13) of Hurwitz Zeta-function leads to

$$\zeta(s,z) = \frac{z^{1-s}}{s-1} + \frac{z^{-s}}{2} - s\int_0^{\infty} \frac{\left(x - \lfloor x \rfloor - \frac{1}{2}\right)}{(z+x)^{s+1}} dx \tag{12.17}$$

By analytic continuation, (12.17) holds for $\text{Re}(s) > -1$, because the integral exist for $\text{Re}(s) > -1$.

After differentiating (12.17) with respect to s

$$\frac{\partial\zeta(s,z)}{\partial s} = -\frac{z\log z}{(s-1)z^s} - \frac{z^{1-s}}{(s-1)^2} - \frac{\log z}{2z^s} - \int_0^{\infty} \frac{\left(x - \lfloor x \rfloor - \frac{1}{2}\right) dx}{(z+x)^{s+1}}$$
$$+ s\int_0^{\infty} \frac{\left(x - \lfloor x \rfloor - \frac{1}{2}\right) dx}{(z+x)} \log(z+x)\, e^{-s\log(z+x)}$$

and evaluating the result at $s = 0$, we find that

$$\left.\frac{\partial\zeta(s,z)}{\partial s}\right|_{s=0} = z\log z - z - \frac{\log z}{2} - \int_0^{\infty} \frac{\left(x - \lfloor x \rfloor - \frac{1}{2}\right) dx}{z+x}$$

Berndt considers the integral

$$\int_0^{n} \frac{\left(x - \lfloor x \rfloor - \frac{1}{2}\right)}{z+x} dx = \int_0^{n} \frac{\left(x + z - z - \lfloor x \rfloor - \frac{1}{2}\right)}{z+x} dx$$
$$= \int_0^{n} \frac{z+x}{z+x} dx - \int_0^{n} \frac{\lfloor x \rfloor}{z+x} dx - \left(z + \frac{1}{2}\right) \int_0^{n} \frac{dx}{z+x}$$

which is

$$\int_0^n \frac{\left(x-\lfloor x\rfloor-\frac{1}{2}\right)}{z+x}dx = n-\sum_{k=0}^{n-1}k\int_k^{k+1}\frac{dx}{z+x}-\left(z+\frac{1}{2}\right)\log\left(\frac{z+n}{z}\right)$$
$$= n-\sum_{k=1}^{n-1}k\log(z+k+1)+\sum_{k=1}^{n-1}k\log(z+k)$$
$$-\left(z+\frac{1}{2}\right)(\log(z+n)-\log(z))$$

Berndt finds, after elementary manipulations,

$$\int_0^n \frac{\left(x-\lfloor x\rfloor-\frac{1}{2}\right)}{z+x}dx = n+\sum_{k=0}^{n}\log(z+k)-\left(z+n+\frac{1}{2}\right)\log(z+n)$$
$$+\left(z-\frac{1}{2}\right)\log(z)$$

so that

$$\left.\frac{\partial\zeta(s,z)}{\partial s}\right|_{s=0} = z\log z-z-\frac{\log z}{2}-\left(z-\frac{1}{2}\right)\log(z)$$
$$-\lim_{n\to\infty}\left(n+\sum_{k=0}^{n}\log(z+k)-\left(z+n+\frac{1}{2}\right)\log(z+n)\right)$$

After simplification, Berndt arrives at

$$\zeta'(0,z) = \lim_{n\to\infty}\left(-(z+n)-\sum_{k=0}^{n}\log(z+k)+\left(z+n+\frac{1}{2}\right)\log(z+n)\right) \tag{12.18}$$

Evaluating (12.18) at $z=1$ and recalling that $\zeta(s,1)=\zeta(s)$ indicates that

$$\zeta'(0) = \lim_{n\to\infty}\left(-(1+n)-\sum_{k=0}^{n}\log(1+k)+\left(1+n+\frac{1}{2}\right)\log(1+n)\right)$$

Subtraction from (12.18) gives

$$\zeta'(0,z)-\zeta'(0) = \lim_{n\to\infty}\left(\begin{array}{c} -z-n+1+n+\sum_{k=1}^{n+1}\log(k)+ \\ -\sum_{k=0}^{n}\log(z+k)+\left(z+n+\frac{1}{2}\right)\log(z+n) \\ -\left(1+n+\frac{1}{2}\right)\log(1+n)\end{array}\right)$$
$$= \lim_{n\to\infty}\left(\begin{array}{c} z\log(1+n)+\sum_{k=1}^{n}\log(k)+1-z \\ -\sum_{k=0}^{n}\log(z+k)+\left(z+n+\frac{1}{2}\right)\log\frac{z+n}{1+n}\end{array}\right)$$

With $\log\frac{z+n}{1+n}=\log\frac{1+\frac{z}{n}}{1+\frac{1}{n}}$, Taylor expansion to first order shows, for large n, that

$$\left(z+n+\frac{1}{2}\right)\log\frac{z+n}{1+n} = n\left(1+\frac{z}{n}+\frac{1}{2n}\right)\left(\frac{z-1}{n}+O\left(\frac{1}{n^2}\right)\right)$$

Hence, for $\operatorname{Re}(z) > 0$, we arrive at

$$\zeta'(0,z) - \zeta'(0) = \lim_{n\to\infty}\left(z\log(1+n) + \sum_{k=1}^{n}\log(k) - \sum_{k=0}^{n}\log(z+k)\right)$$

Berndt observes from Gauss's limit (10.8) that

$$\begin{aligned}\log\Gamma(z) &= \lim_{n\to\infty}\left(\log\frac{n^z n!}{z(z+1)\dots(z+n)}\right)\\ &= \lim_{n\to\infty}\left(z\log(n+1) + \sum_{k=1}^{n}\log(k) - \sum_{k=0}^{n}\log(z+k)\right)\end{aligned}$$

which finally proves Lerch's formula (12.16) for $\operatorname{Re}(z) > 0$ and, by analytic continuation, for all complex z.

159. *Multiplication formula for polygamma functions.* We apply the progression concept (**art.** 13) to (12.12) and sum over $k = qm + j$ with integer period m, where the integer $0 \le j < m$ and $0 \le q$,

$$\sum_{k=0}^{\infty}\frac{1}{(z+k)^s} = \sum_{q=0}^{\infty}\sum_{j=0}^{m-1}\frac{1}{(z+qm+j)^s} = \frac{1}{m^s}\sum_{j=0}^{m-1}\sum_{q=0}^{\infty}\frac{1}{(q+\frac{j+z}{m})^s}$$

we obtain, after replacing z by mz, for $\operatorname{Re}(s) > 1$,

$$\zeta(s,mz) = m^{-s}\sum_{j=0}^{m-1}\zeta\left(s, z+\frac{j}{m}\right) \tag{12.19}$$

By analytic continuation, the Hurwitz-Zeta progression (12.19) holds for all complex s.

Since $(-1)^{n+1}\frac{\psi^{(n)}(z)}{n!} = \zeta(n+1,z)$ for $n > 0$, the Hurwitz-Zeta progression formula (12.19) directly yields the multiplication formula for polygamma functions (Abramowitz and Stegun, 1968, 6.4.8) for $n > 0$,

$$\psi^{(n)}(mz) = m^{-n-1}\sum_{j=0}^{m-1}\psi^{(n)}\left(z+\frac{j}{m}\right) \tag{12.20}$$

which can also be derived from Gauss's multiplication formula (11.8) for the Gamma function as well as the multiplication formula for the digamma function

$$\psi(mz) = \log m + \frac{1}{m}\sum_{j=0}^{m-1}\psi\left(z+\frac{j}{m}\right) \tag{12.21}$$

but the reverse, i.e. deducing Gauss's multiplication formula (11.8) from (12.20), is not obvious.

However, Lerch's formula (12.16) in **art.** 158 indicates that, after derivation of the Hurwitz-Zeta progression formula (12.19) with respect to s,

$$m^s\log m\ \zeta(s,mz) + m^s\zeta'(s,mz) = \sum_{j=0}^{m-1}\zeta'\left(s, z+\frac{j}{m}\right)$$

evaluation at $s=0$,

$$\log m\ \zeta(0,mz)+\zeta'(0,mz)=\sum_{j=0}^{m-1}\zeta'\left(0,z+\frac{j}{m}\right)$$

using $\zeta(0,z)=\frac{1}{2}-z$, that follows from (12.17), and replacing $\zeta'(0,z)=\log\Gamma(z)+\zeta'(0)$ in (12.16) yields

$$\log\Gamma(mz)=\sum_{j=0}^{m-1}\log\Gamma\left(z+\frac{j}{m}\right)+(m-1)\zeta'(0)+\left(mz-\frac{1}{2}\right)\log m \tag{12.22}$$

Finally, with $\zeta'(0)=-\frac{1}{2}\log(2\pi)$ (see e.g. Titchmarsh and Heath-Brown (1986)) and after exponentiation, we find Gauss's multiplication formula (11.8). The derivation, in fact, shows that Gauss's multiplication formula (11.8) is a consequence of the progression concept (**art.** 12).

12.2 Gauss's formula for the digamma function at fractions

160. *Digamma function for rational numbers.* For a rational number $z=\frac{j}{m}$, with integers $0<j<m$, Gauss (1813, formula 75, p. 157) has shown[3] that

$$\psi\left(\frac{j}{m}\right)=-\gamma-\log m-\frac{\pi}{2}\cot\left(\frac{\pi j}{m}\right)+\sum_{k=1}^{m-1}\cos\left(\frac{2\pi jk}{m}\right)\frac{\log\left(2-2\cos\left(\frac{2\pi k}{m}\right)\right)}{2} \tag{12.23}$$

or, equivalently,

$$\psi\left(\frac{j}{m}\right)=-\gamma-\log(2m)-\frac{\pi}{2}\cot\left(\frac{\pi j}{m}\right)+\sum_{k=1}^{m-1}\cos\left(\frac{2\pi jk}{m}\right)\log\left(\sin\frac{\pi k}{m}\right) \tag{12.24}$$

The functional equation $\psi(z+1)=\psi(z)+\frac{1}{z}$ in (12.1) demonstrates that all rational numbers $r\in\mathbb{Q}$ can be expressed in a finite, closed expression via Gauss's formula (12.23). Blagouchine (2015, p. 584) gives another rewriting of (12.24) due to Malmsten

$$\psi\left(\frac{j}{m}\right)=-\gamma-\log(2\pi m)-\frac{\pi}{2}\cot\left(\frac{\pi j}{m}\right)-2\sum_{k=1}^{m-1}\cos\left(\frac{2\pi jk}{m}\right)\log\left(\Gamma\left(\frac{k}{m}\right)\right)$$

obtained from (12.24) by using the reflection formula $\Gamma\left(\frac{k}{m}\right)\Gamma\left(1-\frac{k}{m}\right)=\frac{\pi}{\sin\pi\frac{k}{m}}$ in (11.4).

The multiplication formula (12.21) and the functional equation (12.1) lead to

$$\psi(mz+1)=\log m+\frac{\psi(z+1)}{m}+\frac{1}{m}\sum_{j=1}^{m-1}\psi\left(z+\frac{j}{m}\right)$$

[3] Gauss wrote that (12.23) "theorema sane maxime est memorabile" is a memorable theorem without any doubt!

which simplifies, for $z = 0$ and $\psi(1) = -\gamma$ in (12.5), to

$$\sum_{j=1}^{m-1} \psi\left(\frac{j}{m}\right) = -\gamma(m-1) - m\log m \tag{12.25}$$

For example, Gauss (1813, art. 32, p. 155) evaluates[4] (12.25) $\psi\left(\frac{1}{2}\right) = -\gamma - 2\log 2$ for $m = 2$. For $m = 3$, (12.25) is $\psi\left(\frac{1}{3}\right) + \psi\left(\frac{2}{3}\right) = -2\gamma - 3\log 3$ and with the reflection formula (12.10), $\psi\left(\frac{2}{3}\right) = \psi\left(\frac{1}{3}\right) + \pi\cot\frac{\pi}{3}$, we find

$$\psi\left(\frac{1}{3}\right) = -\gamma - \frac{3}{2}\log 3 - \frac{\pi}{2}\cot\frac{\pi}{3} = -\gamma - \frac{3}{2}\log 3 - \frac{\pi}{6}\sqrt{3}$$

Determining $\psi\left(\frac{j}{m}\right)$ at all fractions $\frac{j}{m}$, with $1 < j < m-1$, by solving (12.25) is awkward and ends with $m = 5$. Gauss's formula (12.23) is thus a kind of inverse of the summation (12.25), similarly to Möbius inversion (7.9). We first give the proof of (12.23) by Jensen (1891, pp. 52-54), also in Nielsen (1956, pp. 20-22), and later in **art.** 162 discuss Gauss' proof of (12.23).

161. *Jensen's proof of Gauss's formula (12.23).* The series (12.3), rewritten with the functional equation (12.1), as,

$$\psi(z) = -\gamma + \sum_{n=0}^{\infty}\left(\frac{1}{1+n} - \frac{1}{z+n}\right) \tag{12.26}$$

becomes, for $z = \frac{j}{m}$,

$$\psi\left(\frac{j}{m}\right) = -\gamma + \sum_{n=0}^{\infty}\left(\frac{1}{1+n} - \frac{m}{j+mn}\right)$$

Jensen (1891, p. 52) defines the generalization

$$\varphi_{j,m}(t) = -\gamma + \sum_{n=0}^{\infty}\left(\frac{1}{1+n} - \frac{m}{j+mn}\right)t^{j+nm}$$

which can be summed with $\log(1-t) = -\sum_{n=1}^{\infty}\frac{t^n}{n}$, provided that $|t| < 1$,

$$\begin{aligned}\varphi_{j,m}(t) &= -\gamma + \sum_{n=0}^{\infty}\frac{t^{j+nm}}{1+n} - m\sum_{n=0}^{\infty}\frac{t^{j+nm}}{j+mn} \\ &= -\gamma - t^{j-m}\log(1-t^m) - m\sum_{n=0}^{\infty}\frac{t^{j+nm}}{j+mn}\end{aligned} \tag{12.27}$$

Clearly, if $t = 1$, then $\varphi_{j,m}(1) = \psi\left(\frac{j}{m}\right)$. Let us denote the last series by $R_{j,m}(t) = \sum_{n=0}^{\infty}\frac{t^{j+nm}}{j+mn}$, then $R_{j,m}(0) = 0$ and

$$R'_{j,m}(t) = \sum_{n=0}^{\infty} t^{j+nm-1} = t^{j-1}\sum_{n=0}^{\infty}(t^m)^n = \frac{t^{j-1}}{1-t^m}$$

4 Gauss's definition for the digamma function $\Psi(z)$ stems from his definition $\Pi(z) = \Gamma(z+1)$ for the Gamma function, so that $\Psi(z) = \psi(z+1)$.

Integration shows that

$$R_{j,m}(t) = \int_0^t \frac{u^{j-1}}{1-u^m} du = \frac{1}{m} \int_0^{t^m} x^{\frac{j}{m}-1} (1-x)^{-1} dx \tag{12.28}$$

On the other hand, **art.** 12 indicates that $R_{j,m}(t)$ is the j-th m section (2.29) of $-\log(1-t) = \sum_{k=1}^{\infty} \frac{t^k}{k}$ and the inverse formula (2.30) gives

$$R_{j,m}(t) = -\frac{1}{m} \sum_{k=1}^{m} e^{-\frac{2\pi i k j}{m}} \log\left(1 - e^{\frac{2\pi i k}{m}} t\right) \tag{12.29}$$

Equating (12.28) and (12.29) results, for $|t| < 1$, in another generalization than (3.27) of the Beta-integral

$$\int_0^{t^m} x^{\frac{j}{m}-1} (1-x)^{-1} dx = -\sum_{k=1}^{m} e^{-\frac{2\pi i k j}{m}} \log\left(1 - e^{\frac{2\pi i k}{m}} t\right)$$

Substituting (12.29) in (12.27) yields Jensen's form, for $|t| < 1$,

$$\varphi_{j,m}(t) = -\gamma - t^{j-m} \log(1-t^m) + \sum_{k=1}^{m} e^{-\frac{2\pi i k j}{m}} \log\left(1 - e^{\frac{2\pi i k}{m}} t\right) \tag{12.30}$$

It remains to compute the limit $t \to 1$, which exists, because $\varphi_{j,m}(1) = \psi\left(\frac{j}{m}\right)$. Jensen rewrites

$$\begin{aligned}
\varphi_{j,m}(t) &= -\gamma - t^{j-m} \log(1-t^m) + \log(1-t) + \sum_{k=1}^{m-1} \frac{\log\left(1 - e^{\frac{2\pi i k}{m}} t\right)}{e^{\frac{2\pi i k j}{m}}} \\
&= -\gamma - t^{j-m} \log \frac{1-t^m}{1-t} + \left(1 - t^{j-m}\right) \log(1-t) \\
&\quad + \sum_{k=1}^{m-1} e^{-\frac{2\pi i k j}{m}} \log\left(1 - e^{\frac{2\pi i k}{m}} t\right)
\end{aligned}$$

and taking the limit $t \to 1$ of both sides results in Jensen's formula

$$\psi\left(\frac{j}{m}\right) = -\gamma - \log m + \sum_{k=1}^{m-1} e^{-\frac{2\pi i k j}{m}} \log\left(1 - e^{\frac{2\pi i k}{m}}\right) \tag{12.31}$$

Although the finite sum in (12.31) contains complex numbers, it must be real, because all remaining terms in (12.31) are real.

Jensen then proceeds with (12.31) as Gauss, by replacing j by $m - j$,

$$\psi\left(1 - \frac{j}{m}\right) = -\gamma - \log m + \sum_{k=1}^{m-1} e^{\frac{2\pi i k j}{m}} \log\left(1 - e^{\frac{2\pi i k}{m}}\right)$$

adding the latter to (12.31),

$$\psi\left(\frac{j}{m}\right) + \psi\left(1 - \frac{j}{m}\right) = -2\gamma - 2\log m + 2\sum_{k=1}^{m-1} \cos\left(\frac{2\pi k j}{m}\right) \log\left(1 - e^{\frac{2\pi i k}{m}}\right)$$

invoking the reflection formula (12.10) and dividing the result by 2,

$$\psi\left(\frac{j}{m}\right) = -\gamma - \log m - \frac{\pi}{2}\cot\frac{\pi j}{m} + \sum_{k=1}^{m-1}\cos\left(\frac{2\pi kj}{m}\right)\log\left(1-e^{\frac{2\pi ik}{m}}\right)$$

which finally leads to Gauss's formula (12.23), after taking the real part of

$$\operatorname{Re}\left(\log\left(1-e^{\frac{2\pi ik}{m}}\right)\right) = \frac{1}{2}\log\left(2-2\cos\frac{2\pi k}{m}\right)$$

The variant (12.24) follows from (12.23) by incorporating the formulae $\frac{1}{2}\log\left(2-2\cos\frac{2\pi k}{m}\right) = \log 2 + \log\left(\sin\frac{\pi k}{m}\right)$ and $\sum_{k=1}^{m-1}\cos\left(\frac{2\pi jk}{m}\right) = -1$.

162. *Generating function of the digamma function for rational numbers.* We generalize Jensen's method in **art.** 160 and compute the generating function $\sum_{j=1}^{m}\psi\left(\frac{j}{m}\right)z^{j-1}$.

After multiplying both sides in Jensen's formula (12.30) by z^j and summing over $1 \le j \le m$, we obtain

$$\sum_{j=1}^{m}\varphi_{j,m}(t)\,z^j = -\gamma z\frac{1-z^m}{1-z} - \frac{\log(1-t^m)}{t^m}zt\frac{1-(zt)^m}{1-zt} - m\sum_{n=0}^{\infty}\sum_{j=1}^{m}\frac{z^j t^{j+nm}}{j+mn}$$

The last series $L(t,z) = \sum_{n=0}^{\infty}\sum_{j=1}^{m}\frac{z^j t^{j+nm}}{j+mn}$ satisfies $L(0,z) = 0$. Its derivative equals

$$L'(t) = \sum_{n=0}^{\infty}\sum_{j=1}^{m} z^j t^{j+nm-1} = \sum_{n=0}^{\infty} t^{nm-1}\sum_{j=1}^{m}(zt)^j = \frac{z}{1-t^m}\frac{1-(zt)^m}{1-zt}$$

Integration $\int_0^t L'(u,z)\,du = L(t,z)$ indicates that

$$L(t,z) = \int_0^t \frac{z}{1-zu}\frac{1-(zu)^m}{1-u^m}du$$

On the other hand, multisectioning shows that

$$L(t,z) = \sum_{j=1}^{m} z^j \sum_{n=0}^{\infty}\frac{t^{j+nm}}{j+mn} = \sum_{j=1}^{m} z^j R_{j,m}(t)$$

and with (12.29)

$$L(t,z) = -\frac{1}{m}\sum_{k=1}^{m} ze^{-\frac{2\pi ik}{m}}\frac{1-z^m}{1-ze^{-\frac{2\pi ik}{m}}}\log\left(1-e^{\frac{2\pi ik}{m}}t\right)$$

Hence, for $|t| < 1$, we find the integral

$$\int_0^t \frac{z}{1-zu}\frac{1-(zu)^m}{1-u^m}du = -\frac{1-z^m}{m}\sum_{k=1}^{m}\frac{ze^{-\frac{2\pi ik}{m}}}{1-ze^{-\frac{2\pi ik}{m}}}\log\left(1-e^{\frac{2\pi ik}{m}}t\right)$$

as well as

$$\sum_{j=1}^{m}\varphi_{j,m}(t)\,z^{j-1}=-\gamma\frac{1-z^m}{1-z}-\frac{\log(1-t^m)}{t^{m-1}}\frac{1-(zt)^m}{1-zt}$$
$$+(1-z^m)\sum_{k=1}^{m}\frac{e^{-\frac{2\pi ik}{m}}}{1-ze^{-\frac{2\pi ik}{m}}}\log\left(1-e^{\frac{2\pi ik}{m}}t\right)$$

After taking the limit $t\to\infty$ as in **art.** 160 and recalling that $\varphi_{j,m}(1)=\psi\left(\frac{j}{m}\right)$, we arrive at

$$\sum_{j=1}^{m}\psi\left(\frac{j}{m}\right)z^{j-1}=\frac{z^m-1}{1-z}(\log m+\gamma)+\sum_{k=1}^{m-1}\frac{(1-z^m)\,e^{-\frac{2\pi ik}{m}}}{1-ze^{-\frac{2\pi ik}{m}}}\log\left(1-e^{\frac{2\pi ik}{m}}\right) \tag{12.32}$$

Equating corresponding powers z in (12.32) determines $\psi\left(\frac{j}{m}\right)$, which again leads to Gauss's formula (12.23). For $z=1$, (12.32) simplifies to (12.25). On the other hand, if $z=e^{\frac{2\pi il}{m}}$ for integer $1\le l\le m-1$, then $z^m=1$ and (12.32) becomes

$$\lim_{z\to e^{\frac{2\pi il}{m}}}\sum_{j=1}^{m}\psi\left(\frac{j}{m}\right)z^{j-1}=\sum_{k=1}^{m-1}\lim_{z\to e^{\frac{2\pi il}{m}}}\frac{(1-z^m)\,e^{-\frac{2\pi ik}{m}}}{1-ze^{-\frac{2\pi ik}{m}}}\log\left(1-e^{\frac{2\pi ik}{m}}\right)$$

and

$$\sum_{j=1}^{m}\psi\left(\frac{j}{m}\right)e^{\frac{2\pi ilj}{m}}=m\log\left(1-e^{\frac{2\pi il}{m}}\right) \tag{12.33}$$

Taking the complex conjugate of (12.33) and multiplying by $1=e^{\frac{2\pi ilj}{m}m}$, shows that

$$\sum_{j=1}^{m}\psi\left(\frac{j}{m}\right)e^{-\frac{2\pi il}{m}j}=\sum_{j=1}^{m}\psi\left(\frac{j}{m}\right)e^{\frac{2\pi il}{m}(m-j)}=m\log\left(1-e^{-\frac{2\pi il}{m}}\right)$$

Splitting-off the $j=m$ term in the sum, invoking $\psi(1)=-\gamma$ and taking the real part of the above leads to Gauss's formula (Gauss, 1813, formula 72, p. 156) with $\phi=\frac{2\pi l}{m}$,

$$\sum_{j=1}^{m-1}\psi\left(\frac{j}{m}\right)\cos j\phi=\sum_{j=1}^{m-1}\psi\left(\frac{j}{m}\right)\cos(m-j)\,\phi=\gamma+\frac{m}{2}\log(2-2\cos\phi) \tag{12.34}$$

Formula (12.34) complements (12.25). Formula (12.34) was the starting point of Gauss's proof of (12.23), which he has derived, perhaps, by genial insight from (12.4) after extensive numerical computations? Gauss then uses orthogonality formulae of $\cos j\phi$ to deduce (12.23) from (12.34), which is, in essence the same as our computations below.

Multiplying both sides in (12.33) by $e^{-\frac{2\pi irl}{m}}$ with $1\le r\le m-1$, after the split-off of the $j = m$ term,

$$\sum_{j=1}^{m-1} \psi\left(\frac{j}{m}\right) e^{\frac{2\pi i l(j-r)}{m}} = -\gamma e^{-\frac{2\pi i r l}{m}} + m e^{-\frac{2\pi i r l}{m}} \log\left(1 - e^{\frac{2\pi i l}{m}}\right)$$

summing over the validity range $1 \leq l \leq m-1$ yields

$$\sum_{j=1}^{m-1} \psi\left(\frac{j}{m}\right) \sum_{l=1}^{m-1} e^{\frac{2\pi i l(j-r)}{m}} = -\gamma \sum_{l=1}^{m-1} e^{-\frac{2\pi i r l}{m}} + m \sum_{l=1}^{m-1} e^{-\frac{2\pi i r l}{m}} \log\left(1 - e^{\frac{2\pi i l}{m}}\right)$$

Orthogonality via $\sum_{l=1}^{m} e^{\frac{2\pi i l(j-r)}{m}} = m\delta_{m|j-r} = m\delta_{j,r}$ requires us to consider

$$\sum_{j=1}^{m} \psi\left(\frac{j}{m}\right) \sum_{l=1}^{m} e^{\frac{2\pi i l(j-r)}{m}} = \sum_{j=1}^{m-1} \psi\left(\frac{j}{m}\right) + \gamma + m \sum_{l=1}^{m-1} e^{-\frac{2\pi i r l}{m}} \log\left(1 - e^{\frac{2\pi i l}{m}}\right)$$

Invoking (12.25) and applying $\sum_{l=1}^{m} e^{\frac{2\pi i l(j-r)}{m}} = m\delta_{j,r}$ leads to Jensen's formula (12.31), from which Gauss's formula (12.23) follows as in **art.** 161.

163. *A Fourier series for* $\psi\left(\frac{j}{m}\right)$. Blagouchine (2015, Appendix B) substitutes[5], for $1 \leq j \leq m-1$,

$$\sum_{k=1}^{m-1} k \sin\left(\frac{2\pi k j}{m}\right) = -\frac{m}{2} \cot\left(\frac{\pi j}{m}\right) \tag{12.35}$$

into (12.24) and finds the finite Fourier series

$$\psi\left(\frac{j}{m}\right) = -\gamma - \log(2m) + \frac{\pi}{m} \sum_{k=1}^{m-1} k \sin\left(\frac{2\pi k j}{m}\right) + \sum_{k=1}^{m-1} \log\left(\sin\frac{\pi k}{m}\right) \cos\left(\frac{2\pi j k}{m}\right) \tag{12.36}$$

Blagouchine (2015, Appendix B) studies the finite $m-1$ points Fourier series, for an integer $r = 1, 2, \ldots m-1$,

$$f_m(r) = a_m(0) + \sum_{l=1}^{m-1} \left(a_m(l) \cos\frac{2\pi r l}{m} + b_m(l) \sin\frac{2\pi r l}{m}\right)$$

computes a variant of the coefficients

$$\begin{cases} \sum_{r=1}^{m-1} f_m(r) \cos\frac{2\pi r k}{m} = -\sum_{l=0}^{m-1} a_m(l) + \frac{m}{2}\left(a_m(k) + a_m(m-k)\right) \\ \sum_{r=1}^{m-1} f_m(r) \sin\frac{2\pi r k}{m} = \frac{m}{2}\left(b_m(k) - b_m(m-k)\right) \end{cases}$$

---

5 Since $\sum_{k=1}^{m-1} k z^k = z\frac{d}{dz}\left(\sum_{k=0}^{m-1} z^k\right) = z\frac{d}{dz}\left(\frac{z^m-1}{z-1}\right) = \frac{(m-1)z^{m+1} - m z^m + z}{(z-1)^2}$, we find, for $z = e^{\frac{2\pi i j}{m}}$, that $\sum_{k=1}^{m-1} k e^{\frac{2\pi i j k}{m}} = -\frac{m}{2} \frac{\sin\left(\frac{\pi j}{m}\right) + i\cos\left(\frac{\pi j}{m}\right)}{\sin\left(\frac{\pi j}{m}\right)}$ after some manipulations. Taking the imaginary part yields (12.35).

as well as Parceval's equality

$$\sum_{r=1}^{m-1} f_m^2(r) = (m-1)\,a_m^2(0) - 2a_m(0)\sum_{l=1}^{m-1} a_m(l) - \left(\sum_{l=1}^{m-1} a_m(l)\right)^2$$
$$+\frac{m}{2}\sum_{l=1}^{m-1}\left(a_m^2(l) + a_m(l)\,a_m(m-l) + b_m^2(l) - b_m(l)\,b_m(m-l)\right)$$

Then, Blagouchine (2015, Appendix B.2) applies these general Fourier relations to (12.36) and finds, for integers $0 < k < m$,

$$\sum_{r=1}^{m-1}\psi\left(\frac{r}{m}\right)\cos\frac{2\pi rk}{m} = \gamma + m\log\left(2\sin\frac{k\pi}{m}\right)$$

which is Gauss's formula (12.34), but also

$$\sum_{r=1}^{m-1}\psi\left(\frac{r}{m}\right)\sin\frac{2\pi rk}{m} = \frac{\pi}{2}(2k-m)$$

and

$$\sum_{r=1}^{m-1}\psi^2\left(\frac{r}{m}\right) = (m-1)\,\gamma^2 + m\,(2\gamma + \log(4m))\log m - m\,(m-1)\log^2 2$$
$$+\frac{\pi^2\left(m^2-3m+2\right)}{12} + m\sum_{l=1}^{m-1}\log^2\left(\sin\frac{l\pi}{m}\right)$$

Blagouchine (2015, pp. 587-588) also adds other variants of the above formulae.

12.3 Integrals for the digamma function and $\log\Gamma(z)$

164. *Gauss's integral for the digamma function.* We will derive Gauss's integral (Gauss, 1813, p. 160, formula 78)

$$\psi(z) = \frac{d}{dz}\log\Gamma(z) = \int_0^\infty\left(\frac{e^{-t}}{t} - \frac{e^{-zt}}{1-e^{-t}}\right)dt \tag{12.37}$$

valid for $\mathrm{Re}(z) > 0$. Instead of following Gauss's deduction, a more elegant derivation (Whittaker and Watson, 1996, p. 247) is obtained from Weierstrass's infinite product (11.1).

Substituting $\int_0^\infty e^{-t(z+n)}dt = \frac{1}{z+n}$, valid for $\mathrm{Re}(z) > 0$ and integers $n \geq 0$, into the digamma series (12.6), written as

$$\psi(z) = -\gamma - \frac{1}{z} - \lim_{k\to\infty}\sum_{n=1}^{k}\left(\frac{1}{z+n} - \frac{1}{n}\right)$$

yields, for $\mathrm{Re}(z) > 0$,

$$\psi(z) = -\gamma - \int_0^\infty e^{-tz}dt - \lim_{k\to\infty}\sum_{n=1}^{k}\left(\int_0^\infty e^{-t(z+n)}dt - \int_0^\infty e^{-tn}dt\right)$$

and

$$\psi(z) = -\gamma - \int_0^\infty e^{-tz}dt - \lim_{k\to\infty}\int_0^\infty \left(e^{-tz}-1\right)\sum_{n=1}^{k} e^{-tn}dt$$

Substitution of the geometric series $\sum_{n=1}^{k} e^{-tn} = e^{-t}\frac{1-e^{-kt}}{1-e^{-t}}$ in (1.16) indicates that

$$\begin{aligned}\psi(z) &= -\gamma - \int_0^\infty e^{-tz}dt + \lim_{k\to\infty}\int_0^\infty \frac{\left(e^{-t}-e^{-t(z+1)}\right)\left(1-e^{-kt}\right)}{1-e^{-t}}dt \\ &= -\gamma - \int_0^\infty \frac{dt}{e^{tz}} + \int_0^\infty \frac{e^{-t}-e^{-t(z+1)}}{1-e^{-t}}dt - \lim_{k\to\infty}\int_0^\infty \frac{1-e^{-tz}}{1-e^{-t}}\frac{dt}{e^{(k+1)t}}\end{aligned}$$

and

$$\psi(z) = -\gamma + \int_0^\infty \frac{e^{-t}-e^{-tz}}{1-e^{-t}}dt + \lim_{k\to\infty}\int_0^\infty \frac{1-e^{-tz}}{1-e^{-t}}e^{-(k+1)t}dt$$

For large k, the last integral vanishes,

$$\begin{aligned}\left|\int_0^\infty \frac{1-e^{-tz}}{1-e^{-t}}e^{-(k+1)t}dt\right| &\le \int_0^\infty \left|\frac{1-e^{-tz}}{1-e^{-t}}\right| e^{-(k+1)t}dt \\ &\le \max_{t\ge 0}\left|\frac{1-e^{-tz}}{1-e^{-t}}\right|\frac{1}{k+1} \to 0\end{aligned}$$

because the function $\frac{1-e^{-tz}}{1-e^{-t}}$ is bounded for $t \ge 0$, and we arrive at

$$\psi(z) = -\gamma + \int_0^\infty \frac{e^{-t}-e^{-tz}}{1-e^{-t}}dt \tag{12.38}$$

Similarly as above, we rewrite Euler's constant γ in (9.3) by invoking again $\frac{1}{s} = \int_0^\infty e^{-st}dt$ for $\mathrm{Re}(t) > 0$ and by using $\log x = \int_1^x \frac{1}{s}ds = \int_1^x\int_0^\infty e^{-st}dtds = \int_0^\infty \frac{e^{-t}-e^{-xt}}{t}dt$,

$$\begin{aligned}\sum_{n=1}^{k}\frac{1}{n} - \log k &= \int_0^\infty e^{-t}\frac{1-e^{-kt}}{1-e^{-t}}dt - \int_0^\infty \frac{e^{-t}-e^{-kt}}{t}dt \\ &= \int_0^\infty \left(\frac{1}{1-e^{-t}}-\frac{1}{t}\right)e^{-t}dt - \int_0^\infty \left(\frac{e^{-t}}{1-e^{-t}}-\frac{1}{t}\right)e^{-kt}dt\end{aligned}$$

The generating function $\frac{1}{e^t-1} = \frac{1}{t} - \frac{1}{2} + \sum_{n=1}^\infty B_{2n}\frac{t^{2n-1}}{(2n)!}$ in (A.15) of the Bernoulli numbers B_n shows that $\frac{e^{-t}}{1-e^{-t}} - \frac{1}{t}$ is continuous at $t = 0$ and bounded $-\frac{1}{2} \le \frac{e^{-t}}{1-e^{-t}} - \frac{1}{t} \le 0$ for $t \ge 0$, leading for $k \to \infty$ to an integral for Euler's constant

$$\gamma = \int_0^\infty \left(\frac{1}{1-e^{-t}}-\frac{1}{t}\right)e^{-t}dt \tag{12.39}$$

and to Gauss's integral (12.37) via (12.38). The analysis also illustrates that the harmonic series $\sum_{n=1}^{k}\frac{1}{n} > \log k + \gamma$ for any $k \geq 1$. A slightly sharper upper bound

$$\sum_{k=1}^{n}\frac{1}{k} > \log n + \gamma + \frac{1}{2n}$$

follows from the exact convergent fractional series (13.61) below.

165. *Jordan's integral for the digamma function.* Jordan (1913, art. 210) substitutes the integral $\log t = \int_0^\infty \frac{e^{-u}-e^{-tu}}{u}du$ in **art**. 164 into the derivative of Euler's integral (10.2)

$$\Gamma'(z) = \int_0^\infty e^{-t}t^{z-1}\log t\, dt = \int_0^\infty dt \int_0^\infty du\, e^{-t}t^{z-1}\frac{e^{-u}-e^{-tu}}{u}$$

proves (Jordan, 1913, art. 211-212) that reversal of integration is allowed

$$\Gamma'(z) = \int_0^\infty \frac{du}{u}\int_0^\infty e^{-t}t^{z-1}\left(e^{-u}-e^{-tu}\right)dt$$

invokes Euler's integral (10.2), so that $\Gamma'(z) = \int_0^\infty \frac{du}{u}\left(e^{-u}\Gamma(z) - \frac{\Gamma(z)}{(1+u)^z}\right)$, and arrives (Jordan, 1913, art. 213) at an integral for the digamma function,

$$\psi(z) = \frac{\Gamma'(z)}{\Gamma(z)} = \int_0^\infty \frac{du}{u}\left(e^{-u} - \frac{1}{(1+u)^z}\right) \tag{12.40}$$

Integration of (12.40) from 1 to z yields

$$\log\Gamma(z) = \int_0^\infty \frac{du}{u}\left((z-1)e^{-u} - \frac{(1+u)^{-1}-(1+u)^{-z}}{\log(1+u)}\right)$$

Jordan (1913, art. 213) notices for $z=2$ that $\int_0^\infty \frac{du}{u}\left(e^{-u} - \frac{u(1+u)^{-2}}{\log(1+u)}\right) = \log\Gamma(2) = 0$ and subtracts $(z-1)$ times this null integral, resulting in

$$\log\Gamma(z) = \int_0^\infty \frac{du}{\log(1+u)}\left(\frac{z-1}{(1+u)^2} - \frac{(1+u)^{-1}-(1+u)^{-z}}{u}\right)$$

Finally, after making Binet's substitution $t = \log(1+u)$ on p. 234, we arrive at the integral (12.54) in **art**. 171, which follows after direct integration of Gauss's integral (12.37).

166. *Towards Binet's function* $\mu(z) = \log\Gamma(z) - \left(z-\frac{1}{2}\right)\log z + z - \frac{1}{2}\log(2\pi)$ *in (13.1).* Inspired by the Bernoulli generating function (A.15), we rewrite Gauss's integral (12.37) for $z \to z+1$,

$$\begin{aligned}\psi(z+1) &= \int_0^\infty \left(\frac{e^{-t}}{t} - \frac{e^{-zt}}{e^t-1}\right)dt \\ &= \int_0^\infty \left(\frac{e^{-t}-e^{-zt}}{t} + \frac{e^{-zt}}{2} - e^{-zt}\left(\frac{1}{e^t-1} - \frac{1}{t} + \frac{1}{2}\right)\right)dt\end{aligned}$$

and obtain, invoking $\log z = \int_0^\infty \frac{e^{-t}-e^{-zt}}{t}dt$,

$$\psi(z+1) = \log z + \frac{1}{2z} - \int_0^\infty e^{-zt}\left(\frac{1}{e^t-1} - \frac{1}{t} + \frac{1}{2}\right)dt \tag{12.41}$$

Since $0 \le \frac{1}{e^t-1} - \frac{1}{t} + \frac{1}{2} \le \frac{1}{2}$ is bounded for $t \ge 0$ and continuous at $t=0$, the last integral is uniformly convergent for $\mathrm{Re}(z) > 0$. Hence, both sides can be integrated from 1 to z, with $\psi(z+1) = \frac{d}{dz}\log\Gamma(z+1)$,

$$\log\Gamma(z+1) = z\log z - z + 1 + \frac{1}{2}\log z + \int_0^\infty \frac{e^{-zt}-e^{-t}}{t}\left(\frac{1}{e^t-1} - \frac{1}{t} + \frac{1}{2}\right)dt$$

With the functional equation (10.1), we find

$$\log\Gamma(z) = \left(z - \frac{1}{2}\right)\log z - z + 1 + \int_0^\infty \frac{e^{-zt}}{t}\left(\frac{1}{e^t-1} - \frac{1}{t} + \frac{1}{2}\right)dt$$
$$- \int_0^\infty \frac{e^{-t}}{t}\left(\frac{1}{e^t-1} - \frac{1}{t} + \frac{1}{2}\right)dt$$

The last integral can be evaluated (Whittaker and Watson, 1996, p. 249) as $1 - \frac{1}{2}\log(2\pi)$ and we arrive at Binet's first integral for $\mathrm{Re}(z) > 0$,

$$\log\Gamma(z) = \left(z - \frac{1}{2}\right)\log z - z + \frac{1}{2}\log(2\pi) + \int_0^\infty \frac{e^{-zt}}{t}\left(\frac{1}{e^t-1} - \frac{1}{t} + \frac{1}{2}\right)dt \tag{12.42}$$

The partial fraction expansion $\frac{1}{t}\left(\frac{1}{e^t-1} - \frac{1}{t} + \frac{1}{2}\right) = \sum_{n=1}^\infty \frac{2}{t^2+4\pi^2n^2}$ substituted in (12.42) yields

$$\log\Gamma(z) = \left(z - \frac{1}{2}\right)\log z - z + \frac{1}{2}\log(2\pi) + 2\int_0^\infty e^{-zt}\sum_{n-1}^\infty \frac{1}{t^2+4\pi^2n^2}dt$$

Since $\frac{1}{t^2+4\pi^2n^2} = \frac{1}{\left(1+\left(\frac{t}{2\pi n}\right)^2\right)(2\pi n)^2}$ and by changing the integration variable $u = \frac{t}{2\pi n}$ in the last term,

$$\int_0^\infty \sum_{n=1}^\infty \frac{e^{-zt}}{t^2+4\pi^2n^2}dt = \frac{1}{2\pi}\int_0^\infty \frac{du}{1+u^2}\sum_{n=1}^\infty \frac{e^{-2\pi nzu}}{n}$$
$$= -\frac{1}{2\pi}\int_0^\infty \frac{du}{1+u^2}\log\left(1 - e^{-2\pi zu}\right)$$

we find, after partial integration, Binet's second form (Whittaker and Watson (1996, p. 251), Sansone and Gerretsen (1960, p. 217)),

$$\log\Gamma(z) = \left(z - \frac{1}{2}\right)\log z - z + \frac{1}{2}\log(2\pi) + 2z\int_0^\infty \frac{\arctan t}{e^{2\pi tz}-1}dt \tag{12.43}$$

167. *Other integrals for* $\log\Gamma(z)$. The Taylor series of $\log\Gamma(z+1)$ around $z=0$, convergent for $|z|<1$,

$$\log\Gamma(z+1) = -\gamma z + \sum_{k=2}^\infty \frac{(-1)^k\zeta(k)}{k}z^k \tag{12.44}$$

follows directly from Weierstrass' product $\frac{1}{\Gamma(z+1)} = e^{\gamma z} \prod_{n=1}^{\infty} \left(1 + \frac{z}{n}\right) e^{-z/n}$ in (11.1). For $z = 1$, (12.44) converges and reduces to an alternating series with decreasing terms for the Euler constant

$$\gamma = \sum_{k=2}^{\infty} \frac{(-1)^k \zeta(k)}{k} \tag{12.45}$$

The lower and upper bounds $\sum_{k=2}^{2K+1} \frac{(-1)^k \zeta(k)}{k} < \gamma < \sum_{k=2}^{2K} \frac{(-1)^k \zeta(k)}{k}$ hold for integer $K \geq 1$.

Substituting in (12.44) the Riemann-Zeta function integral $\zeta(s)\Gamma(s) = \int_0^\infty \frac{t^{s-1}}{e^t-1} dt$, valid for $\text{Re}(s) > 1$,

$$\log \Gamma(z+1) = -\gamma z + \sum_{k=2}^{\infty} \frac{(-z)^k}{k!} \int_0^\infty \frac{t^{k-1}}{e^t - 1} dt$$

reversing summation and integration, allowed by absolute convergence, yields

$$\log \Gamma(z+1) = -\gamma z + \int_0^\infty \frac{\sum_{k=2}^{\infty} \frac{(-zt)^k}{k!}}{t\,(e^t - 1)} dt = -\gamma z + \int_0^\infty \frac{e^{-zt} - 1 + zt}{t\,(e^t - 1)} dt$$

With $\gamma = \int_0^\infty \left(\frac{e^t}{e^t-1} - \frac{1}{t}\right) e^{-t} dt = \int_0^\infty \frac{t-1+e^{-t}}{t(e^t-1)} dt$ in (12.39), the form (12.54) below appears.

Titchmarsh (1964, p. 151) gives the form

$$\log \Gamma(z) = \left(z - \frac{1}{2}\right) \log z - z + \frac{1}{2} \log(2\pi) + \int_0^\infty \frac{\lfloor t \rfloor - t + \frac{1}{2}}{t + z} dt$$

where $\lfloor t \rfloor$ is the nearest integer smaller than or equal to t and which also follows from Lerch's formula (12.16) and the integral (12.17). Rademacher (1973, pp. 34-36) and, similarly, Edwards (1974, pp. 106-114) derive

$$\log \Gamma(z) = \left(z - \frac{1}{2}\right) \log z - z + \frac{1}{2} \log(2\pi) + \sum_{j=1}^{m} \frac{B_{2j}}{(2j-1)(2j)} \frac{1}{z^{2j-1}} - \frac{1}{2m} \int_0^\infty \frac{B_{2m}(t - \lfloor t \rfloor)}{(t+z)^{2m}} dt \tag{12.46}$$

from the Euler-Maclaurin summation (3.40). Blagouchine (2016) lists seven series for $\log \Gamma(z)$ of which most are derived here in Part II.

168. *Limit case of multiplication formulae.* When the integer m grows without bound, the right-hand side sum over values of a continuous function $f(z)$,

$$\lim_{m \to \infty} \frac{1}{m} \sum_{j=0}^{m-1} f\left(z + \frac{j}{m}\right) = \int_0^1 f(z + x)\, dx \tag{12.47}$$

tends to an integral. Many multiplication formulae in **art.** 159 contain a sum of the type of the left-hand side in (12.47).

Example Application of (12.47) to the multiplication formula (12.21) of the digamma function

$$\int_0^1 \psi(z+x)\,dx = \lim_{m\to\infty}(\psi(mz) - \log m)$$

and substituting the variant $\psi(z) = \log z + \frac{1}{z} - \int_0^\infty e^{-zt}\left(\frac{1}{e^t-1} - \frac{1}{t} + \frac{1}{2}\right)dt$ of Gauss's integral (12.41)

$$\int_0^1 \psi(z+x)\,dx = \lim_{m\to\infty}\left(\log z + \frac{1}{mz} - \int_0^\infty e^{-mzt}\left(\frac{1}{e^t-1} - \frac{1}{t} + \frac{1}{2}\right)dt\right)$$

results in

$$\int_0^1 \psi(z+x)\,dx = \log z \tag{12.48}$$

Example The logarithm (12.22) of Gauss's multiplication formula (11.8) becomes with (12.47)

$$\int_0^1 \log\Gamma(z+x)\,dx = \lim_{m\to\infty}\left(\begin{array}{c}\frac{1}{m}\log\Gamma(mz) + \frac{1}{2}\log(2\pi)\left(1-\frac{1}{m}\right)\\ -\left(z-\frac{1}{2m}\right)\log m\end{array}\right)$$

Introducing the Binet's integral (12.43) for $\log\Gamma(z)$,

$$\begin{aligned}\int_0^1 \log\Gamma(z+x)\,dx &= \lim_{m\to\infty}\left(\begin{array}{c}\left(z-\frac{1}{2m}\right)\log z - z + 2z\int_0^\infty \frac{\arctan t}{e^{2\pi t m z}-1}dt\\ +\frac{\log(2\pi)}{2}\end{array}\right)\\ &= z\log z - z + \frac{\log(2\pi)}{2}\end{aligned}$$

results in the integral of Raabe (1840)

$$\int_z^{z+1} \log\Gamma(x)\,dx = z\log z - z + \frac{\log(2\pi)}{2} \tag{12.49}$$

12.4 Interpolation

169. *Interpolation theory.* Interpolation consists of constructing a polynomial that passes through a set of n distinct points, defined by their finite coordinates (x_j, y_j) for $1 \le j \le n$. In many cases, each ordinate $y_j = f(x_j)$ is a known value at x_j of a function $f(x)$, that we want to approximate by a polynomial. Since the Gamma function reduces to the factorial at integer x_j points, we apply equidistant interpolation and refer for its theory to Van Mieghem (2023a, Sec. 11.3). If the Taylor coefficients $\{f_k(z_0)\}_{k\ge 0}$ around z_0 are known, then Newton's series is

$$f(z) = \sum_{k=0}^{\infty} f_k(z_0)(z-z_0)^k = \sum_{k=0}^{\infty}\binom{z-z_0}{k}\Delta^k f_0(z_0) \tag{12.50}$$

where the k-th difference obeys $\Delta^k f_m = \Delta^{k-1} f_{m+1} - \Delta^{k-1} f_m$ for all $k \in \mathbb{N}_0$, thus $\Delta f_m = f_{m+1} - f_m$. By iteration, we have that

$$\Delta^k f_0(z_0) = (-1)^k \sum_{j=0}^{k} \binom{k}{j} (-1)^j f_j(z_0) \tag{12.51}$$

A major difficulty is to determine for which values of z Newton's series (12.50) converges.

170. *A Newton series for the digamma function.* We start with Gauss's integral (12.38), valid for $\mathrm{Re}(z) > 0$,

$$\psi(z) = -\gamma + \int_0^\infty \frac{e^{-t} - e^{-tz}}{1 - e^{-t}} dt$$

and invoke Binet's substitution $e^{-t} = 1 - u$ or $t = -\log(1-u)$ on p. 234,

$$\psi(z) = -\gamma + \int_0^1 \frac{1 - (1-u)^{z-1}}{u} du \tag{12.52}$$

After introducing Newton's binomial series $(1-u)^{z-1} = \sum_{k=0}^\infty \binom{z-1}{k} (-u)^k$ in (17.1), valid for $|u| < 1$, we obtain, after reversal summation and integration, the binomial or Newton series

$$\psi(z) = -\gamma + \sum_{k=1}^\infty \binom{z-1}{k} \frac{(-1)^{k+1}}{k} \tag{12.53}$$

converging for $\mathrm{Re}(z) > 0$. Davis (1959) mentions that the binomial series (12.53) was already proved in 1847 convergent for $\mathrm{Re}(z) > 0$ by M. A. Stern, a teacher of Riemann. A considerably longer proof of Stern's Newton series (12.53) is found in Srinivasan (2007).

The binomial series (12.53) can also be deduced from the Newton series (12.50). The k-th difference (12.51) becomes for the digamma function at $z_0 = 1$,

$$\Delta^k \psi(1) = (-1)^k \sum_{j=0}^{k} \binom{k}{j} (-1)^j \psi(1+j)$$

With (12.8) and the binomium $(1-x)^k = \sum_{j=0}^k \binom{k}{j} (-1)^j x^j$, we find

$$(-1)^k \Delta^k \psi(1) = \sum_{l=1}^{k} \binom{k}{l} (-1)^l \sum_{m=1}^{l} \frac{1}{m}$$

Introducing the binomial recursion $\binom{k}{l} = \binom{k-1}{l} + \binom{k-1}{l-1}$,

$$\begin{aligned}
(-1)^k \Delta^k \psi(1) &= \sum_{l=1}^{k} \binom{k-1}{l} (-1)^l \sum_{m=1}^{l} \frac{1}{m} + \sum_{l=1}^{k} \binom{k-1}{l-1} (-1)^l \sum_{m=1}^{l} \frac{1}{m} \\
&= \sum_{l=1}^{k-1} \binom{k-1}{l} (-1)^l \sum_{m=1}^{l} \frac{1}{m} - \sum_{l=0}^{k-1} \binom{k-1}{l} (-1)^l \sum_{m=1}^{l+1} \frac{1}{m}
\end{aligned}$$

leads to

$$(-1)^k \Delta^k \psi(1) = -\sum_{l=0}^{k-1} \binom{k-1}{l} \frac{(-1)^l}{l+1} = -\frac{1}{k}$$

where the last step is recognized as an instance the Beta-function expansion (17.2) for $p = 1$ and $q = k$. Combined with $\psi(1) = -\gamma$, the Newton series (12.50) then produces (12.53).

171. *Hermite's approach towards the Newton series of* $\log\Gamma(z+1)$. In a letter exchange with Pincherle, Hermite (1901) presents an elegant method. Hermite starts[6] from a direct integration from 1 to z of Gauss's integral (12.37) of the digamma function,

$$\log\Gamma(z) = \int_0^\infty \left(\frac{e^{-zt} - e^{-t}}{1 - e^{-t}} + e^{-t}(z-1) \right) \frac{dt}{t} \tag{12.54}$$

from which he deduces

$$\log\Gamma(a+z) - \log\Gamma(a) = \int_0^\infty \left(-\frac{e^{-at}\left(1 - e^{-zt}\right)}{1 - e^{-t}} + e^{-t}x \right) \frac{dt}{t} \tag{12.55}$$

Thus, the difference follows for $z = 1$ as

$$\Delta \log\Gamma(a) = \int_0^\infty \left(-e^{-at} + e^{-t} \right) \frac{dt}{t}$$

"After an easy[7] computation", Hermite (1901) gives, for $n > 1$,

$$\Delta^n \log\Gamma(a) = -\int_0^\infty e^{-at} \left(e^{-t} - 1 \right)^{n-1} \frac{dt}{t} \tag{12.56}$$

Next, Hermite (1901) invokes the binomial series

$$e^{-zt} = \left(1 + \left(e^{-t} - 1\right)\right)^z = \sum_{n=0}^{\infty} \binom{z}{n} \left(e^{-t} - 1\right)^n = 1 + \sum_{n=1}^{\infty} \binom{z}{n} \left(e^{-t} - 1\right)^n$$

which converges for any complex z and real $t \geq 0$, because $|e^{-t} - 1| < 1$ for $t \geq 0$, rewritten as

$$\begin{aligned} \frac{1 - e^{-zt}}{1 - e^{-t}} &= \sum_{n=1}^{\infty} \binom{z}{n} \left(e^{-t} - 1\right)^{n-1} = \sum_{n=0}^{\infty} \binom{z}{n+1} \left(e^{-t} - 1\right)^n \\ &= z + \frac{z(z-1)}{2} \left(e^{-t} - 1\right) + \frac{z(z-1)(z-2)}{2.3} \left(e^{-t} - 1\right)^2 + \cdots \end{aligned}$$

After introduction into (12.55),

$$\begin{aligned} \log\Gamma(a+z) = \log\Gamma(a) &+ z \int_0^\infty \left(-e^{-at} + e^{-t} \right) \frac{dt}{t} \\ &- \sum_{n=2}^{\infty} \binom{z}{n} \int_0^\infty e^{-at} \left(e^{-t} - 1 \right)^{n-1} \frac{dt}{t} \end{aligned}$$

[6] Actually, Hermite (1901) writes $\log\Gamma(x) = \int_{-\infty}^0 \left(\frac{e^{xy} - e^y}{e^y - 1} - e^y (x-1) \right) \frac{dy}{y}$.

[7] Indeed, $\Delta^2 \log\Gamma(a) = \Delta \log\Gamma(a+1) - \Delta \log\Gamma(a) = -\int_0^\infty e^{-at} \left(e^{-t} - 1\right) \frac{dt}{t}$ and continuation leads to (12.56).

and letting $a = 1$, Hermite (1901) finds the Newton series

$$\log\Gamma(z+1) = -\sum_{n=2}^{\infty}\binom{z}{n}\int_0^{\infty} e^{-t}\left(e^{-t}-1\right)^{n-1}\frac{dt}{t} \tag{12.57}$$

which is, after Binet's substitution $u = 1 - e^{-t}$ in the integral, precisely the same as (12.62) and (12.64) in **art.** 172.

172. *The Newton series for* $\log\Gamma(z+1)$. Apart from Hermite's elegant and short derivation in **art.** 171, we present another method based on Taylor series expansions, inspired by Binet's approach in **art.** 182. Since $\psi(z) = \frac{d\log\Gamma(z)}{dz}$, the integral (12.52) is written as

$$\frac{d\log\Gamma(z)}{dz} = -\gamma + \int_0^1 \frac{1-e^{(z-1)\log(1-u)}}{u}du$$

Integrating both sides with respect to z gives

$$\log\frac{\Gamma(z)}{\Gamma(z_0)} = -\gamma(z-z_0) + \int_0^1\frac{du}{u}\int_{z_0}^{z}\left(1-e^{(w-1)\log(1-u)}\right)dwdu$$

and

$$\int_{z_0}^{z}\left(1-e^{(w-1)\log(1-u)}\right)dw = (z-z_0) - \frac{e^{(z-1)\log(1-u)}-e^{(z_0-1)\log(1-u)}}{\log(1-u)}$$

It is convenient to choose $z_0 = 1$ and then replace z by $z+1$,

$$\log\Gamma(z+1) = -\gamma z + \int_0^1\frac{du}{u}\left(\frac{z\log(1-u)-(1-u)^z+1}{\log(1-u)}\right) \tag{12.58}$$

We expand the integrand in a Taylor expansion around $u = 0$. First,

$$\begin{aligned} z\log(1-u)-(1-u)^z+1 &= -z\sum_{k=1}^{\infty}\frac{u^k}{k} - \sum_{k=1}^{\infty}\binom{z}{k}(-1)^k u^k \\ &= -u^2\sum_{k=2}^{\infty}\left(\frac{z}{k+2}+\binom{z}{k+2}(-1)^k\right)u^k \end{aligned}$$

The Taylor series (13.33) for $n = 1$ reduces to

$$\frac{-u}{\log(1-u)} = 1 + \sum_{m=1}^{\infty}\left(\sum_{k=1}^{m}\frac{S_m^{(k)}}{k+1}\right)\frac{(-u)^m}{m!} = \sum_{m=0}^{\infty}G_m(-u)^m \tag{12.59}$$

where we defined for convenience $G_0 = 1$, while

$$G_m = \frac{1}{m!}\sum_{k=1}^{m}\frac{S_m^{(k)}}{k+1} \qquad \text{for } m \geq 1 \tag{12.60}$$

are known as Gregory's coefficients (Blagouchine, 2018). The first few are $G_1 = \frac{1}{2}$, $G_2 = -\frac{1}{12}$, $G_3 = \frac{1}{24}$, $G_4 = -\frac{19}{720}$ and $G_5 = \frac{3}{160}$. Davis (1957) calls G_m the logarithmic numbers and shows that $G_m = \int_0^1\binom{s}{m}ds$, which is a direct consequence of (13.36) below.

The Cauchy product of both Taylor series is

$$B = \frac{z\log(1-u)-(1-u)^z+1}{\log(1-u)} = \frac{-u^2\sum_{k=2}^{\infty}\left(\frac{z}{k+2}+\binom{z}{k+2}(-1)^k\right)u^k}{\log(1-u)}$$

$$= u\sum_{m=0}^{\infty} G_m(-1)^m u^m \sum_{k=0}^{\infty}\left(\frac{z}{k+2}+\binom{z}{k+2}(-1)^k\right)u^m$$

$$= u\sum_{m=0}^{\infty}\left[\sum_{k=0}^{m} G_{m-k}(-1)^{m-k}\left(\frac{z}{k+2}+\binom{z}{k+2}(-1)^k\right)\right]u^m$$

Substituting the Taylor series into the integral (12.58) and reversing sum and integral gives

$$\log\Gamma(z+1) = -\gamma z + \sum_{m=0}^{\infty}\sum_{k=0}^{m}\frac{G_{m-k}(-1)^{m-k}\left(\frac{z}{k+2}+\binom{z}{k+2}(-1)^k\right)}{m+1}$$

After interchanging the summations

$$\log\Gamma(z+1) = -\gamma z + \sum_{k=0}^{\infty}\sum_{m=k}^{\infty}\frac{G_{m-k}(-1)^{m-k}}{m+1}\left(\frac{z}{k+2}+\binom{z}{k+2}(-1)^k\right)$$

we obtain

$$\log\Gamma(z+1) = -\gamma z + z\sum_{k=0}^{\infty}\left(\frac{1}{k+2}\sum_{m=k}^{\infty}\frac{G_{m-k}(-1)^{m-k}}{m+1}\right) + \sum_{k=0}^{\infty}\binom{z}{k+2}\sum_{m=k}^{\infty}\frac{G_{m-k}(-1)^m}{m+1}$$

If $z=1$, then $\log\Gamma(z+1)=0$ and the latter series also vanishes, because $\binom{z}{k+2}=0$ for $z=1$ and $0\le k$. Hence, we find the Euler constant expressed as a series[8] in terms of Gregory coefficients,

$$\gamma = \sum_{k=0}^{\infty}\left(\frac{(-1)^k}{k+2}\sum_{m=k}^{\infty}\frac{G_{m-k}(-1)^m}{m+1}\right) \tag{12.61}$$

and

$$\log\Gamma(z+1) = \sum_{k=0}^{\infty}\binom{z}{k+2}\sum_{m=k}^{\infty}\frac{G_{m-k}(-1)^m}{m+1}$$

It remains to compute $\sum_{m=k}^{\infty}\frac{G_{m-k}(-1)^m}{m+1} = (-1)^k\sum_{m=0}^{\infty}\frac{G_m(-1)^m}{m+k+1}$. We find from the Taylor series (12.59) that $\int_0^1\frac{u^{k+1}}{\log(1-u)}du = -\sum_{m=0}^{\infty}\frac{G_m}{m+k+1}(-1)^m$, so that

$$\log\Gamma(z+1) = \sum_{k=0}^{\infty}\binom{z}{k+2}(-1)^{k+1}\int_0^1\frac{u^{k+1}}{\log(1-u)}du \tag{12.62}$$

---

[8] The series (12.61) converges extremely badly. Its simplification $\gamma = \frac{3}{2} - \frac{\pi^2}{12} + \sum_{m=2}^{\infty}\frac{G_m(-1)^m}{m-1}\sum_{k=2}^{m}\frac{1}{k}$, obtained after partial fraction decomposition in (12.61), converges slowly.

We end by evaluating the integral in (12.62). Using $\int_0^\infty e^{\log(1-u)x}dx = \frac{-1}{\log(1-u)}$, because $\log(1-u) < 0$ for $u \in (0,1)$, leads to

$$-\int_0^1 \frac{u^{k+1}}{\log(1-u)}du = \int_0^1 u^{k+1} \int_0^\infty (1-u)^x \, dxdu$$

Interchanging the integrals, allowed by absolute convergence, yields

$$-\int_0^1 \frac{u^{k+1}}{\log(1-u)}du = \int_0^\infty \left(\int_0^1 u^{k+1} (1-u)^x \, du \right) dx$$

Invoking the Beta-function integral (10.11) and the functional equation (10.1) of the Gamma function results in

$$\begin{aligned} -\int_0^1 \frac{u^{k+1}}{\log(1-u)}du &= \Gamma(k+2) \int_0^\infty \frac{\Gamma(x+1)}{\Gamma(k+x+3)}dx \\ &= \int_0^\infty \frac{(k+1)!dx}{(k+x+2)(k+x+1)\cdots(x+1)} \end{aligned}$$

Partial fraction expansion

$$\frac{(k+1)!}{(k+x+2)(k+x+1)\cdots(x+1)} = \frac{(k+1)!}{\prod_{j=0}^{k+1}(j+x+1)} = \sum_{j=0}^{k+1} \frac{a_j}{(j+x+1)}$$

with residue $a_j = \lim_{x\to-(j+1)} \frac{(k+1)!(j+x+1)}{\prod_{j=0}^{k+1}(m+x+1)}$, evaluated as

$$a_j = \frac{(k+1)!}{\prod_{m=0}^{j-1}(m-j)\prod_{m=j+1}^{k+1}(m-j)} = (-1)^j \binom{k+1}{j}$$

shows that

$$\int_0^\infty \frac{(k+1)!dx}{(k+x+2)(k+x+1)\cdots(x+1)} = \lim_{n\to\infty} \sum_{j=0}^{k+1} a_j \int_0^n \frac{dx}{j+x+1}$$

Rewriting the right-hand side with $\int_0^n \frac{dx}{j+x+1} = \log \frac{(n+j+1)}{j+1} = \log n + \log\left(\frac{1+\frac{j+1}{n}}{j+1}\right)$ and observing from the binomial series (17.1) that $\sum_{j=0}^{k+1} a_j = 0$ indicates that

$$\int_0^\infty \frac{(k+1)!dx}{(k+x+2)(k+x+1)\cdots(x+1)} = \lim_{n\to\infty} \sum_{j=0}^{k+1} a_j \log\left(\frac{1+\frac{j+1}{n}}{j+1}\right)$$

The limit $n \to \infty$ leads to

$$\int_0^\infty \frac{(k+1)!dx}{(k+x+2)(k+x+1)\cdots(x+1)} = -\sum_{j=0}^{k+1} (-1)^j \binom{k+1}{j} \log(j+1)$$

In summary, $(-1)^{k+1}$ times the k-th difference coefficient in Newton's

series (12.50) and (12.62) is

$$\int_0^1 \frac{u^{k+1}}{\log(1-u)}du = \sum_{j=1}^{k+1} (-1)^j \binom{k+1}{j} \log(j+1) = \log \prod_{j=1}^{k+1} (j+1)^{(-1)^j \binom{k+1}{j}}$$
$$= \log \left(\frac{\prod_{j=1}^{\left[\frac{k+1}{2}\right]} (2j+1)^{\binom{k+1}{2j}}}{\prod_{j=0}^{\left[\frac{k}{2}\right]} (2(j+1))^{\binom{k+1}{2j+1}}} \right) \tag{12.63}$$

Finally, we arrive at the Newton series

$$\log\Gamma(z+1) = \sum_{k=0}^{\infty} \binom{z}{k+2} \sum_{j=1}^{k+1} (-1)^{k+1+j} \binom{k+1}{j} \log(j+1) \tag{12.64}$$
$$= \binom{z}{2} \log(2) + \binom{z}{3} (\log 3 - 2\log 2)$$
$$+ \binom{z}{4} (3\log(2) - 3\log(3) - \log 4) + \cdots$$

The first few terms in (12.64) are readily verified from the Newton series (12.50). Also, formula (12.63) allows us to rewrite the Newton series (12.64), resulting in another product form

$$\Gamma(z+1) = \lim_{K\to\infty} \prod_{k=1}^{K} \left(\prod_{j=1}^{k} (j+1)^{(-1)^{k+j}\binom{k}{j}\binom{z}{k+1}} \right)$$

for the Gamma function for $\mathrm{Re}(z) > 0$ or

$$\Gamma(z+1) = \lim_{K\to\infty} \prod_{k=0}^{K} \left(\frac{\prod_{j=1}^{\left[\frac{k+1}{2}\right]} (2j+1)^{\binom{k+1}{2j}}}{\prod_{j=0}^{\left[\frac{k}{2}\right]} (2(j+1))^{\binom{k+1}{2j+1}}} \right)^{\binom{z}{k+2}(-1)^{k+1}} \tag{12.65}$$

Numerical computations indicate that evaluations of (12.65) for $K = 10$ for different real $z < K$ return about 4 correct digits.

As a by-product, we also find from (12.61) and (12.63) a slowly converging series for the Euler constant

$$\gamma = -\sum_{k=0}^{\infty} \frac{1}{k+2} \int_0^1 \frac{u^{k+1}}{\log(1-u)}du \tag{12.66}$$

and[9]

$$\begin{aligned}\gamma &= \sum_{k=0}^{\infty} \frac{1}{k+2} \sum_{j=1}^{k+1} (-1)^{j+1} \binom{k+1}{j} \log(j+1) \\ &= \frac{1}{2}\log(2) + \frac{1}{3}\log\left(\frac{4}{3}\right) + \frac{1}{4}\log\left(\frac{32}{27}\right) + \frac{1}{5}\log\left(\frac{4096}{3645}\right) + \cdots\end{aligned}$$

as well as

$$e^{\gamma} = \lim_{K\to\infty} \prod_{k=0}^{K} \left(\frac{\prod_{j=0}^{\left[\frac{k}{2}\right]} (2(j+1))^{\binom{k+1}{2j+1}}}{\prod_{j=1}^{\left[\frac{k+1}{2}\right]} (2j+1)^{\binom{k+1}{2j}}} \right)^{\frac{1}{k+2}}$$

We rewrite (12.66) as $\gamma = -\int_0^1 \frac{du}{u\log(1-u)} \sum_{k=0}^{\infty} \frac{u^{k+2}}{k+2}$. With $\sum_{k=0}^{\infty} \frac{u^{k+2}}{k+2} = \sum_{k=1}^{\infty} \frac{u^k}{k} - u = -\log(1-u) - u$, we have

$$\gamma = \int_0^1 \left(\frac{1}{u} + \frac{1}{\log(1-u)} \right) du \tag{12.67}$$

which equals (12.39) after Binet's substitution $e^{-t} = 1-u$. By introducing the Taylor series (12.59) in (12.67), we find for the Euler constant a faster converging series

$$\gamma = \lim_{K\to\infty} \sum_{m=1}^{K} (-1)^{m-1} \frac{G_m}{m} \tag{12.68}$$

than (12.61) with positive terms. Formula (12.68) was derived by execution of the integral (12.67) in Mascheronius (1790, pp. 21-23), in collaboration with Fontana (via letter exchange). The series (12.68) for the Euler constant γ was rediscovered several times as mentioned by Blagouchine (2016).

[9] Reversal of k- and j-sum is not allowed.

13
Binet's function $\mu(z)$

Chapter 13 investigates the Binet function $\mu(z)$ in (13.1), which can be regarded as the remainder in the asymptotic expansion of $\log\Gamma(z) = \left(z-\frac{1}{2}\right)\log z - z + \frac{1}{2}\log(2\pi) + \mu(z)$ for large z, investigated in the next Chapter 14 and known as Stirling's asymptotic expansion. Section 13.1 highlights Binet's work, on which Chapter 13 further builds; in particular, his two factorial series in (13.4) and (13.30) for $\mu(z)$, converging for $\text{Re}(z) > 0$ in contrast to Stirling's diverging expansion (14.1). Section 13.2 studies the properties of the complex function $\mu(z)$, while Section 13.3 presents and investigates a complex integral (13.23), whose integrand contains the Riemann Zeta-function $\zeta(z)$. Section 13.4 concentrates on the derivation of factorial series for $\mu(z)$: Binet's first factorial series in Section 13.4.1, Binet's second factorial series in Section 13.4.2, Gilbert's expansions in Section 13.4.3 that led to Gilbert's generalized factorial series in Section 13.4.4. Gilbert's factorial series in Theorem 5 in Section 13.4.5, containing a complex parameter α, illustrates, perhaps surprisingly, that there exist infinitely many different factorial series for the Binet function $\mu(z)$. Section 13.5 concludes Chapter 13 by deriving factorial series for the digamma and polygamma functions.

13.1 Binet's factorial series

In his truly comprehensive[1] "memoir", Jacques Binet (1839, Section 3, p. 223) defines his function $\mu(z)$ in relation to the Gamma function $\Gamma(z)$ as

$$\mu(z) = \log\Gamma(z) - \left(z-\frac{1}{2}\right)\log z + z - \frac{1}{2}\log(2\pi) \tag{13.1}$$

[1] The reference in Whittaker and Watson (1996, footnote, p. 248) with page numbers from 123 to 143 suggests an ordinary article, but the correct page numbers pp. 123-343 point to a book-size treatise. Apart from adding his own beautiful contributions, Binet has reviewed the knowledge about the Beta-function before 1839. The focal point at that time were the many properties of Euler's Beta integral (10.11) from which properties of the Gamma function were derived. Today, on the other hand and perhaps after Weierstrass's product and Hankel's contour integral, the theory concentrates on the Gamma function, with applications to the Beta function.

Binet has derived two integral representations of his function $\mu(z)$ for $\text{Re}(z) > 0$, which are deduced in **art.** 166 from the digamma function as (12.42)

$$\mu(z) = \frac{1}{2}\int_0^\infty \frac{e^{-zt}}{t}\left(\frac{1+e^{-t}}{1-e^{-t}} - \frac{2}{t}\right) dt \tag{13.2}$$

and as (12.43)

$$\mu(z) = 2\int_0^\infty \frac{\arctan\left(\frac{t}{z}\right)}{e^{2\pi t}-1} dt \tag{13.3}$$

Here, we merely concentrate on elegant, converging factorial series for $\text{Re}(z) > 0$, due to Binet (1839, p. 234), such as

$$\mu(z) = \sum_{m=1}^{\infty} \frac{\beta_m}{(z+1)(z+2)\cdots(z+m)} \tag{13.4}$$

where the coefficients[2] are

$$\beta_m = \frac{1}{m}\int_0^1 \left(u - \frac{1}{2}\right) u(u+1)\cdots(u+m-1)\, du \tag{13.7}$$

Explicitly, $\beta_1 = \beta_2 = \frac{1}{12}$, $\beta_3 = \frac{59}{360}$, $\beta_4 = \frac{29}{60}$, $\beta_5 = \frac{533}{280}$, $\beta_6 = \frac{1577}{168}$, $\beta_7 = \frac{280361}{5040}$, $\beta_8 = \frac{69311}{180}$, and β_m is positive and rapidly increasing in m. Nearly at the end of his memoir and somewhat hidden, Binet (1839, p. 342) has given a second factorial series (13.30), that is rederived slightly differently in Theorem 3 in Section 13.4.2 and generalized to Laplace transforms in Van Mieghem (2021).

Nemes (2013, Theorem 2.1) has generalized Binet's expansion (13.4), for $0 \leq a \leq 1$ and $\text{Re}(z) > 0$,

$$\log\Gamma(z+a) = \left(z + a - \frac{1}{2}\right)\log z - z + \frac{1}{2}\log(2\pi) + \sum_{m=1}^{\infty} \frac{c_m(a)}{\prod_{j=1}^{m}(z+j)} \tag{13.8}$$

where

$$c_m(a) = \int_0^1 \left(u + a - \frac{1}{2}\right)\frac{u\cdots(u+m-1)}{m} du - \int_0^a \frac{\prod_{j=1}^{m}(u-a+j)}{m} du \tag{13.9}$$

Clearly, if $a = 0$, we retrieve Binet's first factorial expansion (13.4) and

---

[2] In Binet's notation (Binet, 1839, p. 234),

$$2\mu(z) = \frac{I(1)}{z+1} + \frac{I(2)}{2(z+1)(z+2)} + \frac{I(3)}{3(z+1)(z+2)(z+3)} + \cdots \tag{13.5}$$

where the integral form of the coefficients (derived at Binet (1839, p. 238)) is

$$I(m) = \int_0^1 x(x+1)(x+2)\cdots(x+m-1)(2x-1)\, dx \tag{13.6}$$

Binet lists the first few coefficients, $I(1) = \frac{1}{6}$, $I(2) = \frac{1}{3}$, $I(3) = \frac{59}{60}$ and $I(4) = \frac{227}{60}$.

coefficients β_m in (13.7). Nemes' expansion (13.8), expressed in terms of Binet's function $\mu(z)$ with the definition (13.1),

$$\mu(z) = \left(z - \frac{1}{2}\right)\log\frac{z-a}{z} + a + \sum_{m=1}^{\infty}\frac{c_m(a)}{\prod_{k=1}^{m}(z-a+k)}$$

bears resemblance to

$$\mu(z) = \sum_{m=1}^{\infty}\frac{b_m(\alpha)}{\prod_{k=0}^{m-1}(z+\alpha+k)} \tag{13.10}$$

in our main Theorem 5, where the coefficients $b_m(\alpha)$ are called Binet polynomials, to honour Jacques Binet. Before Nemes, Hermite (1895) has deduced the corresponding generalized Stirling asymptotic expansion of (13.8)

$$\log\Gamma(z+a) \simeq \left(z+a-\frac{1}{2}\right)\log z - z + \frac{1}{2}\log(2\pi) + \sum_{k=1}^{K}\frac{(-1)^{k-1}B_{k+1}(a)}{k(k+1)z^k} \tag{13.11}$$

in terms of the Bernoulli polynomials $B_n(a)$. The series (13.11) reduces for $a = 0$ to Stirling's original asymptotic series (14.3).

13.2 Properties of Binet's function $\mu(z)$

The definition (13.1) of Binet's function $\mu(z)$ directly shows, for a positive integer $z = n$, that

$$\mu(n) = n + \log(n-1)! - \left(n - \frac{1}{2}\right)\log(n) - \frac{1}{2}\log(2\pi)$$

The sequence $\mu(1) \simeq 0.0811$, $\mu(2) \simeq 0.0413$, $\mu(3) \simeq 0.0277$, $\mu(4) \simeq 0.0208$, $\mu(5) \simeq 0.0166, \ldots$, $\mu(10) \simeq 0.0083$ demonstrates the slow decay roughly as $\mu(n) \approx \frac{1}{12n}$ in (13.13).

Binet's definition (13.1) illustrates that the complex conjugate $\mu^*(z) = \mu(z^*)$, by the reflection principle (Titchmarsh, 1964, p. 155).

173. *The Binet function $\mu(z)$ as a Laplace transform.* Binet's integral (13.2) can be rewritten as a Laplace transform

$$\mu(z) = \int_0^{\infty}\frac{e^{-zt}}{t}\left(\frac{1}{e^t-1} - \frac{1}{t} + \frac{1}{2}\right)dt \tag{13.12}$$

where $0 \leq \frac{1}{e^t-1} - \frac{1}{t} + \frac{1}{2} \leq \frac{1}{2}$ for real, non-negative t. Since the integrand is positive for real $z = x > 0$, we observe that $\mu(x) > 0$ for real, positive x. In addition, for a complex number $z = x + iy$, the above integral shows that $\mu(z)$ is analytic for $\text{Re}(z) > 0$ and that

$$|\mu(z)| \leq \int_0^{\infty}\frac{e^{-xt}}{t}\left(\frac{1}{e^t-1} - \frac{1}{t} + \frac{1}{2}\right)dt = \mu(x)$$

In other words, the maximum absolute value of Binet's function $\mu(z)$ for

$\mathrm{Re}(z) > 0$ is attained at the positive real axis. Moreover, $\mu(x)$ strictly decreases with $x = \mathrm{Re}(z)$.

174. *Bound derived from the Integral Mean Value Theorem.* The Integral Mean Value Theorem (**art.** 114) indicates that there exist a real number $\varphi \geq 0$, depending on real $x > 0$, such that

$$\mu(x) = \frac{1}{\varphi}\left(\frac{1}{e^{\varphi}-1} - \frac{1}{\varphi} + \frac{1}{2}\right)\int_0^\infty e^{-xt}dt = \frac{1}{x\varphi}\left(\frac{1}{e^{\varphi}-1} - \frac{1}{\varphi} + \frac{1}{2}\right)$$

Since the right-hand side is maximal for $\varphi = 0$ and decreases in φ, we find the bound

$$|\mu(z)| \leq \mu(x) \leq \frac{1}{x}\lim_{\varphi\to 0}\frac{1}{\varphi}\left(\frac{1}{e^{\varphi}-1} - \frac{1}{\varphi} + \frac{1}{2}\right)$$

The generating function (A.15) of the Bernoulli numbers B_n

$$\frac{1}{e^t-1} = \frac{1}{t} - \frac{1}{2} + \sum_{n=1}^{\infty} B_{2n}\frac{t^{2n-1}}{(2n)!} \qquad \text{for } |t| \leq 2\pi$$

illustrates that $\lim_{\varphi\to 0}\frac{1}{\varphi}\left(\frac{1}{e^{\varphi}-1} - \frac{1}{\varphi} + \frac{1}{2}\right) = \frac{B_2}{2} = \frac{1}{12}$. Hence, for $x = \mathrm{Re}(z) > 0$, we find (Whittaker and Watson, 1996, p. 249)

$$|\mu(z)| \leq \mu(x) \leq \frac{1}{12x} \tag{13.13}$$

Consequently, Binet's definition (13.1) leads to a sharp upper bound for the logarithm of the Gamma function for any $\mathrm{Re}(z) > 0$. In particular, the integral in (12.42), equal to (13.12), is positive for positive real $x > 0$, leading to the bounds

$$\left(x - \frac{1}{2}\right)\log x - x + \frac{\log(2\pi)}{2} \leq \log\Gamma(x) \leq \left(x - \frac{1}{2}\right)\log x - x + \frac{\log(2\pi)}{2} + \frac{1}{12x} \tag{13.14}$$

175. *The difference $\mu(z+1) - \mu(z)$.* The functional equation of the Gamma function $\Gamma(z+1) = z\Gamma(z)$ leads to

$$\mu(z+1) = \log\Gamma(z) + \log z - \left(z + \frac{1}{2}\right)\log(z+1) + z + 1 - \frac{1}{2}\log(2\pi)$$

The forward difference $\Delta\mu(z) = \mu(z+1) - \mu(z)$ equals

$$\mu(z+1) - \mu(z) = \left(z + \frac{1}{2}\right)\log\frac{z}{z+1} + 1 \tag{13.15}$$

which is valid for any complex z, except at the negative real axis that is a branch cut for $\mu(z)$. In particular, around $z = 0$, the forward difference (13.15) shows, with $\mu(1) = 1 - \frac{1}{2}\log(2\pi)$, that

$$\mu(z) \sim -\frac{1}{2}\log z - \frac{1}{2}\log(2\pi) \tag{13.16}$$

illustrating that Binet's function $\mu(z)$ possesses a logarithmic singularity at $z = 0$.

176. *An integral for the Binet function.* The introduction of the definition (13.1) of the Binet function into Raabe's integral (12.49) leads to the integral

$$\int_z^{z+1} \mu(t)\, dt = \frac{1}{2}\left(z(z+1)\log\left(\frac{z}{z+1}\right) + z + \frac{1}{2}\right) \tag{13.17}$$

Differentiation of (13.17) with respect to z again leads to the forward difference (13.15). For $z = 0$ in (13.17), we find $\int_0^1 \mu(t)\, dt = \frac{1}{4}$.

177. *Gudermann's series.* If we replace z by $z + k$ in the forward difference (13.15) and change the sign, then (13.15) becomes $\mu(z+k) - \mu(z+k+1) = \left(z+k+\frac{1}{2}\right)\log\left(1+\frac{1}{z+k}\right) - 1$. Summing over integer k results in a telescoping series $\sum_{k=0}^{K}\{\mu(z+k) - \mu(z+k+1)\} = \mu(z) - \mu(z+K+1)$ leading to

$$\mu(z) - \mu(z+K+1) = \sum_{k=0}^{K}\left\{\left(z+k+\frac{1}{2}\right)\log\left(1+\frac{1}{z+k}\right) - 1\right\}$$

The bound (13.13) shows that $\lim_{K\to\infty}\mu(z+K+1) = 0$ and we obtain the series in Gudermann (1845)

$$\mu(z) = \sum_{k=0}^{\infty}\left\{\left(z+k+\frac{1}{2}\right)\log\left(1+\frac{1}{z+k}\right) - 1\right\} \tag{13.18}$$

178. *Reflection formula of Binet's function $\mu(z)$.* We replace $z \to 1 - z$ in Binet's definition (13.1)

$$\mu(1-z) = \log\Gamma(1-z) - \left(\frac{1}{2} - z\right)\log(1-z) - z + 1 - \frac{1}{2}\log(2\pi)$$

which added to $\mu(z)$ in (13.1), yields

$$\mu(z) + \mu(1-z) = \log(\Gamma(z)\Gamma(1-z)) - \left(z - \frac{1}{2}\right)\log\frac{z}{1-z} + 1 - \log(2\pi)$$

After invoking the reflection formula $\Gamma(z)\Gamma(1-z) = \frac{\pi}{\sin\pi z}$ of the Gamma function, we find the corresponding "reflection" formula for Binet's function

$$\mu(1-z) = 1 - \mu(z) - \log(2\sin\pi z) - \left(z - \frac{1}{2}\right)\log\frac{z}{1-z} \tag{13.19}$$

which is valid for any complex number $z = x + iy$, with the exception of the negative real axis and odd integers $x = 2n + 1$ with $n \in \mathbb{Z}$ at any y. Since $\mu(z)$ is analytic for $\text{Re}(z) > 0$, the reflection formula (13.19) illustrates that Binet's function $\mu(z)$ has only logarithmic singularities at negative integers $z = -k$ (with $k \in \mathbb{N}$) including $z = 0$, as shown in (13.16).

179. *Duplication and multiplication formula for $\mu(z)$.* Combining the duplication formula $\Gamma(2z) = \frac{2^{2z-1}}{\sqrt{\pi}}\Gamma(z)\Gamma\left(z+\frac{1}{2}\right)$ in (11.9 for the Gamma function $\Gamma(z)$ and the definition (13.1) of Binet's function $\mu(z)$ leads to

$$\mu(2z) = \mu\left(z+\frac{1}{2}\right) + \mu(z) + z\log\left(1+\frac{1}{2z}\right) - \frac{1}{2} \tag{13.20}$$

which is generalized by Gauss's multiplication formula (11.8) as

$$\mu(nz) = \mu(z) + \sum_{k=1}^{n-1}\mu\left(z+\frac{k}{n}\right) + \sum_{k=1}^{n-1}\left(z+\frac{k}{n}-\frac{1}{2}\right)\log\left(z+\frac{k}{n}\right) \tag{13.21}$$
$$-(n-1)\left(\frac{1}{2}+z\log z\right)$$

After choosing $z=\frac{1}{n}$ in (13.21), we find for $n \geq 2$

$$\mu\left(\frac{1}{n}\right) = (n-1)\left(\frac{1}{2}-\frac{\log n}{n}\right) - \sum_{k=2}^{n-1}\left(\mu\left(\frac{k}{n}\right) + \left(\frac{k}{n}-\frac{1}{2}\right)\log\left(\frac{k}{n}\right)\right)$$

For example, for $n=2$, we obtain $\mu\left(\frac{1}{2}\right) = \frac{1}{2}(1-\log 2) \simeq 0.1534$ but, for $n=3$, $\mu\left(\frac{1}{3}\right) = 1-\mu\left(\frac{2}{3}\right) - \frac{\log(54)}{6}$, which is also immediate from Binet's reflection formula (13.19). The duplication formula (13.20) yields a closed expression for $z=n+\frac{1}{2}$ with integer $n \geq 1$,

$$\mu\left(n+\frac{1}{2}\right) = \sum_{k=n}^{2n-1}\log k - n\left(\log(4n+2)-1\right) + \frac{1}{2}\left(\log(2)+1\right)$$

13.3 Complex integral for Binet's function $\mu(z)$

180. *A complex integral for* $\mu(z)$. From Weierstrass's product (11.1) of the Gamma function, Whittaker and Watson (1996, p. 277) deduce

$$\log\frac{\Gamma(a)}{\Gamma(z+a)} = -z\frac{\Gamma'(a)}{\Gamma(a)} + \frac{1}{2\pi i}\int_{q-i\infty}^{q+i\infty}\frac{\pi}{\sin\pi s}\frac{\zeta(s,a)}{s}z^s ds \quad \text{with } 1<q<2 \tag{13.22}$$

valid for all a and z, where $\zeta(s,a) = \sum_{n=0}^{\infty}\frac{1}{(n+a)^s}$ is the Hurwitz Zeta-function (**art.** 156), which reduces for $a=1$ to the Riemann Zeta-function $\zeta(s)$. Thus, for $a=1$ and $\frac{\Gamma'(1)}{\Gamma(1)} = -\gamma$ in (12.5), where γ is the Euler constant, the integral (13.22) reduces to

$$\log\Gamma(z+1) = -\gamma z - \frac{1}{2\pi i}\int_{q-i\infty}^{q+i\infty}\frac{\pi}{\sin\pi s}\frac{\zeta(s)}{s}z^s ds \quad \text{with } 1<q<2$$

When moving the line of integration from $1<q<2$ to $-1<c<0$, we (Van Mieghem, 2021) find the complex integral

$$\mu(z) = -\frac{1}{2\pi i}\int_{c-i\infty}^{c+i\infty}\frac{\pi}{\sin\pi s}\frac{\zeta(s)}{s}z^s ds \quad \text{with } -1<c<0 \tag{13.23}$$

valid for any complex number $z=|z|e^{i\arg z}$ with $|\arg z|<\pi$.

We substitute the functional equation of the Riemann Zeta-function $\zeta(s) = 2(2\pi)^{s-1}\sin\frac{\pi s}{2}\Gamma(1-s)\zeta(1-s)$ in the integral (13.23), invoke the reflection formula $\Gamma(s)\Gamma(1-s) = \frac{\pi}{\sin\pi s}$ and obtain the variant of (13.23)

$$\mu(z) = -\frac{1}{4i}\int_{c-i\infty}^{c+i\infty}\frac{(2\pi z)^s}{\cos\frac{\pi s}{2}}\frac{\zeta(1-s)}{\sin\pi s\Gamma(1+s)}ds \qquad \text{with } -1<c<0 \tag{13.24}$$

The variant (13.24) allows the introduction of the Dirichlet series $\zeta(s) = \sum_{k=1}^{\infty}\frac{1}{k^s}$ for $\text{Re}(s) > 1$ and leads, after evaluating the resulting contour integrals, to Malmsten-Kummer's series (see e.g. Blagouchine (2016), Erdélyi *et al.* (1953b, p. 23)) for real $0 < x < 1$

$$\log\Gamma(x) = (\gamma + \log(2\pi))\left(\frac{1}{2} - x\right) - \frac{1}{2}\log\frac{\sin\pi x}{\pi} + \sum_{k=2}^{\infty}\frac{\log(k)\sin(2\pi xk)}{\pi k} \tag{13.25}$$

We evaluate the integral (13.23) along the line $s = c + it$, where $-1 < c < 0$,

$$\mu(z) = -\frac{1}{2}\int_{-\infty}^{\infty}\frac{z^{c+it}}{\sin\pi(c+it)}\frac{\zeta(c+it)}{c+it}dt$$

If we choose $c = -\frac{1}{2}$, then the integral simplifies to

$$\mu(z) = z^{-\frac{1}{2}}\int_{-\infty}^{\infty}\frac{e^{it\log z}}{\cosh\pi t}\frac{\zeta\left(-\frac{1}{2}+it\right)}{-1+2it}dt \tag{13.26}$$

Since $\zeta(x+iT) = O\left(T^{\frac{1}{2}-x}\right)$ for $x \leq 0$ and large T (see e.g. Titchmarsh and Heath-Brown (1986, Chapter V)), it holds that $\frac{\zeta\left(-\frac{1}{2}+it\right)}{-1+2it} = O(1)$ for large t. Hence, the integral (13.26) can be bounded as

$$|\mu(z)| \leq \left|z^{-\frac{1}{2}}\right|\int_{-\infty}^{\infty}\frac{\left|e^{it\log z}\right|}{\cosh\pi t}\left|\frac{\zeta\left(-\frac{1}{2}+it\right)}{-1+2it}\right|dt \leq C\left|z^{-\frac{1}{2}}\right|\int_{-\infty}^{\infty}\frac{e^{-t\arg(z)}}{\cosh\pi t}dt$$

where C is a positive real number, demonstrating existence for all complex $z = |z|e^{i\arg z}$ provided $|\arg z| < \pi$. In other words, the integral (13.23) defines Binet's function $\mu(z)$ everywhere in the complex plane, except at the negative real axis, where $\mu(z)$ possesses a branch cut. The integral

$$\int_{-\infty}^{\infty}\frac{e^{izu}}{\cosh\alpha u}du = \frac{\pi}{\alpha}\frac{1}{\cosh\frac{\pi z}{2\alpha}} \qquad \text{valid for } |\text{Im}\, z| < \alpha$$

shows that $\int_{-\infty}^{\infty}\frac{e^{it\log z}}{\cosh\pi t}dt = \frac{1}{\cosh\frac{\log z}{2}} = \frac{2z^{\frac{1}{2}}}{z+1}$ and, roughly, that the integral (13.26) can be estimated as $\mu(z) = O\left(\frac{1}{z}\right)$, which complements the bound (13.13) to complex z, except at the negative real axis.

181. *Branch cut along the negative real axis.* The complex integral (13.23) indicates with $z = re^{i\theta}$ that

$$\mu\left(re^{i\theta}\right) - \mu\left(re^{-i\theta}\right) = -\int_{c-i\infty}^{c+i\infty}\frac{\sin\theta s}{\sin\pi s}\frac{\zeta(s)}{s}r^s ds$$

is purely imaginary, because $\mu\left(re^{i\theta}\right) - \mu\left(re^{-i\theta}\right) = \mu\left(re^{i\theta}\right) - \mu^*\left(re^{i\theta}\right) = 2i\,\mathrm{Im}\,\mu\left(re^{i\theta}\right)$, and that

$$\lim_{\theta\to\pi}\left(\mu\left(re^{i\theta}\right) - \mu\left(re^{-i\theta}\right)\right) = -\int_{c-i\infty}^{c+i\infty}\frac{\zeta(s)}{s}r^s ds$$

By moving the line of integration from $c < 0$ to $c' > 1$, two poles at $s = 0$ and $s = 1$ are enclosed and Cauchy's residue theorem leads to

$$\begin{aligned}\Delta\mu &= \mu\left(re^{i\pi}\right) - \mu\left(re^{-i\pi}\right)\\ &= 2\pi i\left(\lim_{s\to 0}\zeta(s)\,r^s + \lim_{s\to 1}\frac{\zeta(s)(s-1)}{s}r^s\right) - \int_{c'-i\infty}^{c'+i\infty}\frac{\zeta(s)}{s}r^s ds\\ &\quad + \lim_{T\to\infty}\left(\int_c^{c'}\frac{\zeta(x+iT)}{x+iT}\frac{r^x dx}{r^{-iT}} - \int_c^{c'}\frac{\zeta(x-iT)}{x-iT}\frac{r^x dx}{r^{iT}}\right)\end{aligned}$$

Due to $\zeta(x+iT) = O\left(T^{\frac{1}{2}-x}\right)$ for $x \le 0$ and large T and since $c < 0$ can be chosen small enough, the limits $T \to \infty$ vanish. After using Perron's formula (Titchmarsh, 1964, p. 301), $\frac{1}{2\pi i}\int_{c'-i\infty}^{c'+i\infty}\frac{\zeta(s)}{s}r^s ds = \sum_{n=1}^{r} 1 = \lfloor r \rfloor$, we find that the difference at both sides of the branch cut is periodic in $r \ge 0$,

$$\mu\left(re^{i\pi}\right) - \mu\left(re^{-i\pi}\right) = 2\pi i\left(-\frac{1}{2} + r - \lfloor r \rfloor\right)$$

and vanishes at $r = \frac{1}{2} + n$ with integer $n \in \mathbb{N}$.

13.4 Factorial series for $\mu(z)$

13.4.1 Binet's first factorial series for $\mu(z)$

We review Binet's first expansion for $\mu(z)$ in Binet (1839, Section 3, pp. 223-229). After writing $\log\frac{z}{z+1} = \log\left(1 - \frac{1}{z+1}\right)$, Binet expands the right-hand side of the forward difference formula (13.15)

$$\mu(z+1) - \mu(z) = 1 + \left(z + \frac{1}{2}\right)\log\left(1 - \frac{1}{z+1}\right)$$

by introducing the Taylor series $\log(1-z) = -\sum_{n=1}^{\infty}\frac{z^n}{n}$ around $z_0 = 0$, convergent for $|z| < 1$, and obtains for $|z+1| > 1$,

$$\mu(z) - \mu(z+1) = \frac{1}{2}\sum_{n=1}^{\infty}\frac{n}{(n+2)(n+1)}\frac{1}{(z+1)^{n+1}} \tag{13.27}$$

Binet replaces $z \to z + k$ in (13.27), sums over all integer $k \ge 0$,

$$\sum_{k=0}^{N-1}\mu(z+k) - \mu(z+k+1) = \frac{1}{2}\sum_{n=1}^{\infty}\frac{n}{(n+2)(n+1)}\sum_{k=0}^{N-1}\frac{1}{(z+k+1)^{n+1}}$$

and rewrites the telescoping series at the left-hand side

$$\mu(z) - \mu(z+N) = \frac{1}{2}\sum_{n=1}^{\infty} \frac{n}{(n+2)(n+1)} \sum_{k=0}^{N-1} \frac{1}{(z+k+1)^{n+1}}$$

By observing that $\lim_{N\to\infty} \mu(z+N) = 0$, which follows e.g. from the bound (13.13), Binet (1839, eq. (58), p. 229) arrives at his first convergent expansion

$$\mu(z) = \frac{1}{2}\sum_{n=1}^{\infty} \frac{n}{(n+2)(n+1)} \sum_{k=0}^{\infty} \frac{1}{(z+k+1)^{n+1}} \tag{13.28}$$

Analogously, substitution of the Taylor series $\log(1+z) = \sum_{n=1}^{\infty} \frac{(-1)^{n-1} z^n}{n}$ for $|z| < 1$ in Gudermann's series (13.18) yields, after reworking,

$$\mu(z) = \frac{1}{2}\sum_{n=1}^{\infty} \frac{n}{(n+1)(n+2)} \sum_{k=0}^{\infty} \frac{(-1)^{n+1}}{(z+k)^{n+1}} \tag{13.29}$$

In terms of the polygamma functions (12.11), we rewrite the two variants (13.28) and (13.29) as

$$\mu(z) = \frac{1}{2}\sum_{n=1}^{\infty} \frac{n(-1)^{n+1}}{(n+2)!}\psi^{(n)}(z+1) \quad \text{and} \quad \mu(z) = \frac{1}{2}\sum_{n=1}^{\infty} \frac{n}{(n+2)!}\psi^{(n)}(z)$$

Binet (1839, art 20, pp. 232-234) then concentrates on the evaluation of the k-sum in (13.28), thus on the higher-order derivatives $\psi^{(n)}(z)$ of the digamma function $\psi(z)$ and presents (Binet, 1839, p. 234) his first factorial expansion (13.4). Binet (1839, art 21, p. 242) proceeds by constructing integrals for $\mu(z)$. Introducing Euler's Gamma integral $\frac{\Gamma(s)}{x^s} = \int_0^\infty t^{s-1} e^{-xt} dt$ in (15.9), valid for $\operatorname{Re}(s) > 0$ and $\operatorname{Re}(x) > 0$, into (13.28) yields, for $\operatorname{Re}(z) > 0$,

$$\mu(z) = \frac{1}{2}\sum_{n=1}^{\infty} \frac{n}{(n+2)(n+1)} \frac{1}{n!} \sum_{k=0}^{\infty} \int_0^\infty t^n e^{-(z+k+1)t} dt$$

After reversal of integral and k-summation, evaluating the k-series,

$$\mu(z) = \frac{1}{2}\sum_{n=1}^{\infty} \frac{n}{(n+2)(n+1)} \frac{1}{n!} \int_0^\infty e^{-zt} \frac{t^n}{e^t - 1} dt$$

summing the resulting n-series $\frac{1}{2}\sum_{n=1}^{\infty} \frac{n}{(n+2)(n+1)} \frac{t^n}{n!} = \frac{e^t t - 2e^t + 2 + t}{t^2}$ by using the partial fraction $\frac{n}{(n+2)(n+1)} = \frac{2}{n+2} - \frac{1}{n+1}$, then leads to Binet's integral (13.2). Via an integral due to Poisson, $\frac{e^t+1}{e^t-1} - \frac{1}{2t} = \frac{1}{4}\int_0^\infty \frac{\sin tx \, dx}{e^{2\pi x} - 1}$, Binet also derives (13.3). A different derivation by partial fraction expansion was given in (12.43) in **art.** 166.

13.4.2 Binet's second factorial series

We first give the theory in **art.** 182 and draw conclusions in **art.** 183.

182. *Derivation of Binet's second factorial series for the Binet function* $\mu(z)$.

Theorem 3 *A second convergent factorial series of Binet's function* $\mu(z)$ *is*

$$\mu(z) = \frac{1}{z}\sum_{m=1}^{\infty} \frac{c_m}{\prod_{k=1}^{m-1}(z+k)} \qquad \text{for } \operatorname{Re}(z) > 0 \tag{13.30}$$

where the rational coefficients are

$$c_m = \frac{(-1)^{m-1}}{2m}\sum_{k=1}^{m} \frac{kS_m^{(k)}}{(k+2)(k+1)} \tag{13.31}$$

and $S_m^{(k)}$ *is the Stirling number of the first kind.*

We essentially follow the steps in Binet's original proof in Binet (1839, p. 339). We formalize Binet's proof as a recipe in five steps.

Proof (Binet): Binet (1839, p. 339) substitutes $e^{-t} = 1-u$ or $t = -\log(1-u)$ in the integral (13.2), resulting in

$$2\mu(z) = -\int_0^1 \frac{(1-u)^{z-1}}{u}\left(\frac{2-u}{\log(1-u)} + \frac{2u}{\log^2(1-u)}\right)du \tag{13.32}$$

and proceeds to expand

$$\frac{2-u}{\log(1-u)} + \frac{2u}{\log^2(1-u)} = 2\left\{\frac{1}{\log(1-u)} + \frac{u}{\log^2(1-u)}\right\} - \frac{u}{\log(1-u)}$$

in a Taylor series around $u = 0$. Instead of following Binet, who has used integrals rather than Stirling numbers $S_m^{(k)}$, we invoke the Taylor expansion (13.33), derived in Van Mieghem (2021, Appendix B) and convergent for $|x| < 1$ and integer $n \geq 1$,

$$\frac{x}{\log^n(1+x)} = \frac{1}{n!} + \sum_{k=1}^{n-1}\frac{1}{k!\log^{n-k}(1+x)} + \sum_{m=1}^{\infty}\left(\sum_{k=1}^{m}\frac{k!S_m^{(k)}}{(k+n)!}\right)\frac{x^m}{m!} \tag{13.33}$$

to obtain the Taylor series, valid for $|u| < 1$,

$$\frac{2-u}{\log(1-u)} + \frac{2u}{\log^2(1-u)} = \sum_{m=1}^{\infty}\left(\sum_{k=1}^{m}\frac{kS_m^{(k)}}{(k+2)(k+1)}\right)\frac{(-u)^m}{m!} \tag{13.34}$$

Introducing (13.34) in Binet's function (13.32) and reversing summation and integral, justified because a Taylor series can be term-wise integrated within its radius of convergence yields

$$2\mu(z) = \sum_{m=1}^{\infty}\left(\sum_{k=1}^{m}\frac{kS_m^{(k)}}{(k+2)(k+1)}\right)\frac{(-1)^{m-1}}{m!}\int_0^1 (1-u)^{z-1}u^{m-1}du$$

We invoke the Beta-integral $\int_0^1 u^{p-1}(1-u)^{q-1}\,du = \frac{\Gamma(p)\Gamma(q)}{\Gamma(p+q)}$ in (10.11), valid for $\mathrm{Re}(p) > 0$ and $\mathrm{Re}(q) > 0$, and obtain a converging series, for $\mathrm{Re}(z) > 0$,

$$\mu(z) = \frac{1}{2}\sum_{m=1}^{\infty}\left(\sum_{k=1}^{m}\frac{kS_m^{(k)}}{(k+2)(k+1)}\right)\frac{(-1)^{m-1}}{m}\frac{\Gamma(z)}{\Gamma(z+m)}$$

With $\frac{\Gamma(z+m)}{\Gamma(z)} = \prod_{k=0}^{m-1}(z+k)$, we arrive at Binet's second factorial series (13.30). □

The first few coefficients c_m in (13.31) are $c_1 = \frac{1}{12}$, $c_2 = 0$, $c_3 = -\frac{1}{360}$, $c_4 = -\frac{1}{120}$, $c_5 = -\frac{5}{168}$, $c_6 = -\frac{11}{84}$, $c_7 = -\frac{3499}{5040}$, which are smaller in absolute value than 1, but $c_8 = -\frac{1039}{240}$, $c_9 = -\frac{369689}{11880}$ exceed 1 in absolute value. It holds that $|c_m| > 1$ for $m > 8$ as shown below after Theorem 4.

The generating function $\frac{\Gamma(x+1)}{\Gamma(x+1-m)} = \prod_{k=0}^{m-1}(x-k) = \sum_{k=0}^{m} S_m^{(k)} x^k$ in (A.5) of the Stirling numbers $S_m^{(k)}$ of the first kind indicates that $\sum_{k=1}^{m} S_m^{(k)} = \delta_{1m}$. Thus, we can add $a\sum_{k=1}^{m} S_m^{(k)}$ to $\sum_{k=1}^{m}\frac{kS_m^{(k)}}{(k+2)(k+1)}$ in (13.31) and find that the coefficient equals

$$c_m = \frac{(-1)^{m-1}}{2m}\sum_{k=1}^{m}\frac{ak^2+(3a-1)k+2a}{(k+1)(k+2)}S_m^{(k)} \qquad \text{for } m > 1 \text{ and any } a \in \mathbb{C}$$

For example, for $a = \frac{1}{6}$ and $m > 1$, we find

$$c_m = \frac{(-1)^{m-1}}{12m}\sum_{k=1}^{m}\frac{(k-1)(k-2)}{(k+1)(k+2)}S_m^{(k)}$$

illustrating that $c_2 = 0$.

Theorem 4 *The rational coefficients c_m in the second factorial series (13.30) of Binet's function $\mu(z)$ can be represented by an integral*

$$c_m = \frac{1}{m}\int_0^1\left(u-\frac{1}{2}\right)u\prod_{k=1}^{m-1}(k-u)du \tag{13.35}$$

Moreover, all coefficients c_m, except for the first two, are negative, i.e. $c_m < 0$ for all $m > 2$.

Proof: Multiplying both sides of the generating function (A.5) of the Stirling numbers $S_m^{(k)}$ by x^{q-1} and integrating yields

$$\int_a^b u^{q-1}\prod_{k=0}^{m-1}(u-k)du = \sum_{l=0}^{m} S_m^{(l)}\frac{b^{l+q}-a^{l+q}}{l+q} \tag{13.36}$$

Using $\frac{j}{(j+2)(j+1)} = \frac{2}{j+2} - \frac{1}{j+1}$, the coefficient c_m in (13.31) is written as

$$c_m = \frac{(-1)^m}{2m}\sum_{j=1}^{m}\frac{S_m^{(j)}}{j+1} - \frac{(-1)^m}{2m}\sum_{j=1}^{m}\frac{2S_m^{(j)}}{j+2}$$

After substituting (13.36) with $a = 0$ and $b = 1$, for $q = 1$ and $q = 2$, we obtain

$$c_m = \frac{(-1)^m}{2m} \int_0^1 (1 - 2u) \prod_{k=0}^{m-1} (u - k) du$$

from which (13.35) follows. If $m = 1$ in (13.35), then $c_1 = \int_0^1 \left(u - \frac{1}{2}\right) u du = \frac{1}{12}$.

In the second part, we will demonstrate that $c_m < 0$ for $m > 2$. Since $u \prod_{k=1}^{m-1} (k-u) \geq 0$ for $u \in [0,1]$, we split the integration interval in (13.35),

$$mc_m = \int_0^{\frac{1}{2}} \left(u - \frac{1}{2}\right) u \prod_{k=1}^{m-1} (k - u) du + \int_{\frac{1}{2}}^1 \left(u - \frac{1}{2}\right) u \prod_{k=1}^{m-1} (k - u) du$$

After making the substitution $u = 1 - w$ in the last integral, we arrive at

$$mc_m = \int_0^{\frac{1}{2}} \left(\frac{1}{2} - u\right) (1 - u) u \left\{ \prod_{k=2}^{m-1} (k - (1 - u)) - \prod_{k=2}^{m-1} (k - u) \right\} du$$

For $m = 2$, the right-hand side is zero, because the two products are equal to 1. Since $1 - u > u$ for $0 \leq u < \frac{1}{2}$, the product $\prod_{k=2}^{m-1} (k - (1 - u)) < \prod_{k=2}^{m-1} (k - u)$ for $u \in [0, \frac{1}{2})$ and for $m > 2$. Hence, we conclude that $c_m < 0$ for $m > 2$. □

183. *Behavior of Binet's function* $\mu(z)$. The important fact in Theorem 4, that all coefficients $c_m < 0$ for $m > 2$, implies that any truncation at $m = K > 2$ terms in (13.30) upper bounds Binet's function $\mu(z)$,

$$\mu(z) \lessapprox \frac{1}{12z} - \frac{1}{360z(z+1)(z+2)} + \sum_{m=4}^{K} \frac{c_m}{\prod_{k=0}^{m-1} (z + k)} \tag{13.37}$$

The truncated Binet factorial series (13.37) sharpens the bound $|\mu(z)| \leq \frac{1}{12 \operatorname{Re} z}$ in (13.13). Moreover, Binet's factorial series (13.37) can be written as

$$\mu(z) = \frac{\theta(z)}{12z} \tag{13.38}$$

where

$$\theta(z) = 1 - \frac{1}{30(z+1)(z+2)} - \frac{1}{10(z+1)(z+2)(z+3)} - \sum_{m=5}^{\infty} \frac{12|c_m|}{\prod_{k=1}^{m-1} (z + k)}$$

obeys, for real positive $z = x$, the bounds $0 < \theta(x) < 1$. We observe that the first two coefficients in Stirling's asymptotic expansion (14.1) in **art.** 194 for the Binet function

$$\mu(z) \simeq \frac{1}{12z} - \frac{1}{360z^3} + \frac{1}{1260z^5} + \sum_{m=4}^{K} \frac{B_{2m}}{(2m-1)(2m) z^{2m-1}}$$

and Binet's convergent (13.37) factorial series are the same.

The growth of the coefficients c_m with m and the asymptotic behavior of c_m for large m is computed in Van Mieghem (2021).

13.4.3 Gilbert's expansion (13.42) of $\mu(z)$

By substitution in Binet's integral $\mu(z) = \int_0^\infty \frac{e^{-zt}}{t}\left(\frac{1}{e^t-1} - \frac{1}{t} + \frac{1}{2}\right) dt$ in (13.12) of the partial fraction expansion[3]

$$\frac{1}{e^t-1} = \frac{1}{t} - \frac{1}{2} + 2t\sum_{k=1}^{\infty} \frac{1}{4k^2\pi^2 + t^2} \tag{13.39}$$

Gilbert (1876, art. 6) obtains

$$\mu(z) = 2\sum_{k=1}^{\infty} \int_0^\infty \frac{e^{-zt}}{4k^2\pi^2 + t^2} dt \tag{13.40}$$

due to Cauchy. Using the Laplace transform $\int_0^\infty e^{-zu}\sin(au)\,du = \frac{a}{a^2+z^2}$ for $\operatorname{Re}(z) > 0$, Gilbert (1876, art. 6) reformulates the integral

$$\begin{aligned}
\int_0^\infty \frac{e^{-zt}}{a^2+t^2} dt &= \frac{1}{a}\int_0^\infty dt\, e^{-zt} \int_0^\infty e^{-tu}\sin(au)\,du \\
&= \frac{1}{a}\int_0^\infty \left(\int_0^\infty dt\, e^{-(z+u)t}\right)\sin(au)\,du \\
&= \frac{1}{a}\int_0^\infty \frac{\sin(au)}{z+u} du
\end{aligned}$$

and obtains from (13.40)

$$\mu(z) = \frac{1}{\pi}\sum_{k=1}^{\infty} \frac{1}{k}\int_0^\infty \frac{\sin(2\pi k u)}{z+u} du \tag{13.41}$$

Since (13.41) can be written as $\mu(z) = \frac{1}{\pi}\int_0^\infty \frac{du}{z+u}\sum_{k=1}^\infty \frac{\sin(2\pi k u)}{k}$, Gilbert (1876, art. 8) introduces the Fourier series $\sum_{k=1}^\infty \frac{\sin(ku)}{k} = \frac{\pi-u}{2}$ for $u \in [0, 2\pi]$ and finds, after some manipulations, that

$$\mu(z) = \sum_{k=0}^{\infty} \int_0^1 \frac{\left(\frac{1}{2} - x\right) dx}{z+k+x} \tag{13.42}$$

Next, Gilbert (1876, art. 9) shows that (13.42) can be obtained from Gudermann's series (13.18). Moreover, (13.42) leads to Binet's series (13.28) by observing that $\int_0^1 \frac{\left(\frac{1}{2}-x\right)dx}{z+k+x} = \int_0^1 \frac{\left(u-\frac{1}{2}\right)du}{z+k+1-u}$, expanding the denominator into a geometric series and using the integral $\int_0^1 \left(z-\frac{1}{2}\right) z^n dz = \frac{n}{2(n+1)(n+2)}$.

Gilbert (1876, art. 14-16) derives from (13.41) the Stirling series (14.3) and the Malmsten-Kummer series (13.25). Analogous to the theory of

[3] Cauchy's integral $\frac{1}{e^t-1} = \frac{1}{2\pi i}\int_{C(t)} \frac{dw}{(w-t)(e^w-1)}$, where the contour $C(t)$ encloses only the point $w = t$, leads, after deforming the contour to enclose the entire plane except for a small region around $w = t$, to (13.39).

the Riemann Zeta-function (Titchmarsh and Heath-Brown, 1986), Gilbert (1876, art. 17) integrates the argument $f(t) = \frac{e^{-zt}}{t}\left(\frac{1}{e^t-1} - \frac{1}{t} + \frac{1}{2}\right)$ in Binet's integral (13.12) along a contour that starts at the origin of the complex t-plane along the real t-axis, until $t = R$, travels along a circle $t = Re^{i\theta}$ to the imaginary t-axis, from which the contour returns to the origin by passing the poles of $\frac{1}{e^t-1}$ at $t = 2mi\pi$ along small semicircles in the positive $\operatorname{Re}(t)$-plane. After letting $R \to \infty$ and invoking Cauchy's integral theorem, Gilbert obtains, after some manipulations,

$$\mu(z) = \frac{1}{2}\sum_{k=1}^{\infty} \frac{\cos(2\pi k z)}{k} + \frac{1}{2}\int_0^{\infty} \left(\frac{1}{x} - \cot x\right) \frac{\sin(2zx)}{x} dx$$

as well as $\sum_{k=1}^{\infty} \frac{\sin(2\pi k z)}{k} = \int_0^{\infty} \left(\frac{1}{x} - \cot x\right) \frac{\cos(2zx)}{x} dx$, from which he again deduces the Malmsten-Kummer series (13.25) in Gilbert (1876, art. 30).

184. *Gilbert's infinite product for* $\Gamma(z)$. Gilbert (1876, art. 3) starts from Gudermann's series (13.18) in **art.** 177 for the Binet function $\mu(z)$. After remarking that

$$\begin{aligned}
\mathbb{G} &= \sum_{k=0}^{K} \left(z + k + \frac{1}{2}\right) \log\left(1 + \frac{1}{z+k}\right) \\
&= \sum_{k=0}^{K} \left(z + k + \frac{1}{2}\right) (\log(z+k+1) - \log(z+k)) \\
&= \sum_{k=1}^{K+1} \left(z + k - \frac{1}{2}\right) \log(z+k) - \sum_{k=0}^{K} \left(z + k + \frac{1}{2}\right) \log(z+k)
\end{aligned}$$

and reworking the telescoping series,

$$\mathbb{G} = \left(z + K + \frac{1}{2}\right) \log(z+K+1) - \left(z + \frac{1}{2}\right) \log(z) - \sum_{k=1}^{K} \log(z+k)$$

Gilbert (1876, art. 3) finds

$$\begin{aligned}
\mu(z) - \mu(z+K+1) &= \sum_{k=0}^{K} \left\{\left(z + k + \frac{1}{2}\right) \log\left(1 + \frac{1}{z+k}\right) - 1\right\} \\
&= -\left(z + \frac{1}{2}\right) \log(z) + \left(z + K + \frac{1}{2}\right) \log(z+K+1) \\
&\quad - (K+1) - \sum_{k=1}^{K} \log(z+k)
\end{aligned}$$

Proceeding to the limit $K \to \infty$, using $\lim_{z\to\infty} \mu(z) = 0$ which follows from (13.13), yields

$$\mu(z) = -\left(z + \frac{1}{2}\right) \log z + \lim_{K\to\infty} \left(\begin{array}{c} \left(z + K + \frac{1}{2}\right) \log(z+K+1) \\ -(K+1) - \sum_{k=1}^{K} \log(z+k) \end{array} \right)$$

Since

$$f = \left(z+K+\frac{1}{2}\right)\log(z+K+1) = \left(z+K+\frac{1}{2}\right)\log\left(K\left(1+\frac{z+1}{K}\right)\right)$$
$$= \left(z+K+\frac{1}{2}\right)\log K + \left(z+\frac{1}{2}\right)\log\left(1+\frac{z+1}{K}\right) + K\log\left(1+\frac{z+1}{K}\right)$$

it holds that

$$\mu(z) = -\left(z+\frac{1}{2}\right)\log z + z + \lim_{K\to\infty}\left(\begin{array}{c}\left(z+K+\frac{1}{2}\right)\log K - K \\ -\sum_{k=1}^{K}\log(z+k)\end{array}\right)$$

Gilbert's derivation resembles Berndt's deduction of Lerch's formula (12.16) in **art.** 158. The definition $\log\Gamma(z) = \mu(z) + \left(z-\frac{1}{2}\right)\log z - z + \frac{1}{2}\log(2\pi)$ of Binet's function $\mu(z)$ in (13.1) then shows that

$$\log\Gamma(z) = \log\left(\sqrt{2\pi}\right) + \lim_{K\to\infty}\left(\left(z+K+\frac{1}{2}\right)\log K - K - \log\prod_{k=0}^{K}(z+k)\right)$$

After exponentiation, Gilbert (1876, art. 3) arrives at

$$\Gamma(z) = \sqrt{2\pi}\lim_{n\to\infty}\frac{n^{z+n+\frac{1}{2}}e^{-n}}{\prod_{k=0}^{n}(z+k)} \tag{13.43}$$

which is another product form than the Euler-Gauss product (10.10) and the Weierstrass product in (11.1). Gilbert (1876, art. 4) adds that Stirling's formula $n! = \sqrt{2\pi}n^{n+\frac{1}{2}}e^{-n}\left(1+O\left(\frac{1}{n}\right)\right)$ in (14.17) transforms his product (13.43) with $n^z = \prod_{k=1}^{n-1}\left(\frac{k+1}{k}\right)^z$ to the Euler-Gauss product (10.10). A fourth product form (14.16) with Stirling prefactor $\sqrt{2\pi}z^{z-\frac{1}{2}}e^{-z}$, due to Artin, is derived in **art.** 199.

13.4.4 Gilbert's generalized factorial series for Binet's function

We slightly generalize the derivations of Gilbert (1886) by starting from our general factorial series of $\frac{1}{z+a}$ in (13.44).

185. *A general factorial series for* $\frac{1}{z+a}$. In the identity $\frac{1}{z+a} = \frac{1}{z+b_k} + \frac{b_k-a}{(z+b_k)}\frac{1}{(z+a)}$ for arbitrary numbers b_k, we recursively replace $\frac{1}{z+a}$ in each iteration k and obtain, after p iterations, the finite factorial series

$$\frac{1}{z+a} = \sum_{l=1}^{p}\frac{1}{z+b_l}\prod_{j=1}^{l-1}\frac{b_j-a}{b_j+z} + \frac{1}{z+a}\prod_{j=1}^{p}\frac{b_j-a}{b_j+z} \tag{13.44}$$

If a, z and the set $\{b_j\}_{1\leq j\leq p}$ are positive real numbers, then (13.44) converges for $p\to\infty$, because $\prod_{j=1}^{p}\frac{b_j-a}{b_j+z}\to 0$, since $\left|\frac{b_j-a}{b_j+z}\right| < 1$. The factorial series (13.44) for $(z+a)^{-1}$ demonstrates that for functions, possessing a series $\sum_{k=0}^{\infty}a_k z^{-k}$ such as Laplace transforms, infinitely many factorial series are possible.

Leaving convergence considerations aside for an arbitrary set $\{b_j\}_{1\leq j}$

of complex numbers when $p \to \infty$ in (13.44), Cauchy's integral $f(z) = \frac{1}{2\pi i}\int_{C(z)} \frac{f(w)}{w-z}dw$ becomes with (13.44)

$$f(z) = \frac{1}{2\pi i}\sum_{l=1}^{\infty}\prod_{j=1}^{l-1}(b_j+z)\int_{C(z)} \frac{f(w)}{\prod_{j=1}^{l} b_j + w}dw$$

If the contour can be closed over the entire w-plane and all b_j are different, then $\frac{1}{2\pi i}\int_{C(z)} \frac{f(w)}{\prod_{j=1}^{l} b_j+w}dw = \sum_{k=1}^{l} \frac{f(-b_k)}{\prod_{j=1;j\neq k}^{l}(b_j-b_k)}$ and we formally arrive at a generalization of Taylor's series

$$f(z) = \sum_{l=1}^{\infty}\sum_{k=1}^{l} \frac{f(-b_k)}{\prod_{j=1;j\neq k}^{l}(b_j-b_k)}\prod_{j=1}^{l-1}(b_j+z) \tag{13.45}$$

Explicitly,

$$f(z) = f(-b_1) + \frac{f(-b_1)-f(-b_2)}{b_2-b_1}(b_1+z) + \sum_{k=1}^{3}\frac{f(-b_k)(b_1+z)(b_2+z)}{\prod_{j=1;j\neq k}^{3}(b_j-b_k)}$$
$$+\left(\sum_{k=1}^{4}\frac{f(-b_k)}{\prod_{j=1;j\neq k}^{4}(b_j-b_k)}\right)(b_1+z)(b_2+z)(b_3+z)+\cdots$$

We omit here the further exploration of (13.45).

186. *A multiple parameters factorial series for* $\mu(z)$. Substitution of the factorial series (13.44) with $z \to z+k$ and $a = x$ in $\mu(z) = \sum_{k=0}^{\infty}\int_0^1 \frac{\left(\frac{1}{2}-x\right)dx}{z+k+x}$ in (13.42) yields, for positive real z and b_j,

$$\mu(z) = \sum_{k=0}^{\infty}\sum_{l=2}^{p}\frac{\int_0^1\left(\frac{1}{2}-x\right)\prod_{j=1}^{l-1}(b_j-x)\,dx}{\prod_{j=1}^{l}(b_j+z+k)} + \sum_{k=0}^{\infty}\frac{\int_0^1 dx\frac{\left(\frac{1}{2}-x\right)\prod_{j=1}^{p}(b_j-x)}{z+k+x}dx}{\prod_{j=1}^{p}(b_j+z+k)}$$

because $\int_0^1\left(\frac{1}{2}-x\right)dx = 0$. If the last sum[4] vanishes for $p\to\infty$, we obtain, for any set $\{b_j\}_{1\leq j}$ of positive real numbers, the multiple parameters factorial series

$$\mu(z) = \sum_{l=2}^{\infty}\left(\int_0^1\left(\frac{1}{2}-x\right)\prod_{j=1}^{l-1}(b_j-x)\,dx\right)\sum_{k=0}^{\infty}\frac{1}{\prod_{j=1}^{l}(b_j+z+k)} \tag{13.46}$$

Since $\frac{1}{\prod_{j=1}^{l}(b_j+z+k)} = \frac{1}{(b_1+z+k)(b_l+z+k)\prod_{j=2}^{l-1}(b_j+z+k)}$ and with the partial fraction $\frac{1}{(b_1+z+k)(b_l+z+k)} = \frac{1}{(b_l-b_1)}\left(\frac{1}{(b_1+z+k)} - \frac{1}{(b_l+z+k)}\right)$, we rewrite the

[4] As shown in **art.** 188 below, for $b_j = j-1$, the series reduces to Binet's second series (13.30) and the remainder $R_p = \sum_{k=0}^{\infty}\frac{1}{\prod_{j=0}^{p-1}(j+z+k)}\int_0^1 dx\frac{\left(\frac{1}{2}-x\right)\prod_{j=0}^{p-1}(j-x)}{z+k+x}dx$ in the integer p is nicely bounded by Gilbert (1886) as $|R_p| < \frac{1}{64z^2}\frac{1}{1+\gamma z+z\log(p-1)}$, where γ is Euler's constant. Gilbert also presents another bound $|R_p| < \frac{\Gamma(z)}{64z}\frac{e^{\frac{1}{2p}\left(z+\frac{1}{6}\right)}}{p^z}$.

infinite k-sum as

$$\sum_{k=0}^{\infty} \frac{1}{\prod_{j=1}^{l}(b_j+z+k)} = \sum_{k=0}^{\infty} \frac{\frac{\left(\frac{1}{(b_1+z+k)} - \frac{1}{(b_l+z+k)}\right)}{(b_l-b_1)}}{\prod_{j=2}^{l-1}(b_j+z+k)}$$
$$= \sum_{k=0}^{\infty} \frac{\frac{1}{(b_l-b_1)}}{\prod_{j=1}^{l-1}(b_j+z+k)} - \sum_{k=0}^{\infty} \frac{\frac{1}{(b_l-b_1)}}{\prod_{j=2}^{l}(b_j+z+k)}$$
$$= \sum_{k=0}^{\infty} \frac{\frac{1}{(b_l-b_1)}}{\prod_{j=1}^{l-1}(b_j+z+k)} - \sum_{k=0}^{\infty} \frac{\frac{1}{(b_l-b_1)}}{\prod_{j=1}^{l-1}(b_{j+1}+z+k)}$$

If $b_{j+1} = b_j + m$, where m is an integer, then the second sum is

$$\sum_{k=0}^{\infty} \frac{\frac{1}{(b_l-b_1)}}{\prod_{j=1}^{l-1}(b_{j+1}+z+k)} = \sum_{k=m}^{\infty} \frac{\frac{1}{(b_l-b_1)}}{\prod_{j=1}^{l-1}(b_j+z+k)}$$

and a finite series is found

$$\sum_{k=0}^{\infty} \frac{1}{\prod_{j=1}^{l}(b_j+z+k)} = \frac{1}{(b_l-b_1)} \sum_{k=0}^{m-1} \frac{1}{\prod_{j=1}^{l-1}(b_j+z+k)}$$

Since the solution of the difference equation $b_{j+1} = b_j + m$ is $b_j = jm + b_0$, we conclude that only if b_j is linear in j with highest coefficient an integer, the infinite k-sum in (13.46) can be written as finite sum.

187. *Gilbert's general double sum factorial series.* Gilbert (1876, art. 13) has started from (13.44) with the choice $b_j = b + (j-1)p$, in which case

$$\sum_{k=0}^{\infty} \frac{1}{\prod_{j=1}^{l}(b_j+z+k)} = \frac{1}{(l-1)p} \sum_{k=0}^{p-1} \frac{1}{\prod_{j=0}^{l-2}(b+jp+z+k)}$$

Substituted into the general series (13.46), Gilbert arrives at a general, double sum factorial series

$$\mu(z) = \frac{1}{p} \sum_{l=1}^{\infty} \sum_{k=0}^{p-1} \frac{\frac{1}{l}\int_0^1 \left(\frac{1}{2}-x\right) \prod_{j=0}^{l-1}(b+jp-x)\, dx}{\prod_{j=0}^{l-1}(b+jp+z+k)} \tag{13.47}$$

Example For $b=1$ and $p=1$, Gilbert's factorial series (13.47) reduces to

$$\mu(z) = \sum_{l=1}^{\infty} \frac{\frac{1}{l}\int_0^1 \left(\frac{1}{2}-x\right) \prod_{j=0}^{l-1}(1+j-x)\, dx}{\prod_{j=0}^{l-1}(1+j+z)}$$

which is Binet's first factorial series (13.4). For $b=0$ and $p=1$, Gilbert's factorial series (13.47) reduces to

$$\mu(z) = \sum_{l=1}^{\infty} \frac{\frac{1}{l}\int_0^1 \left(\frac{1}{2}-x\right) \prod_{j=0}^{l-1}(j-x)\, dx}{\prod_{j=0}^{l-1}(j+z)}$$

which is Binet's second factorial series (13.30).

13.4.5 Gilbert's factorial series for the Binet function $\mu(z)$

Several forms and properties of the coefficients in Gilbert's single sum factorial series are deduced.

188. *Gilbert's general single sum factorial series.* We investigate Gilbert's double sum factorial series (13.47) for $p = 1$, which simplifies to a single sum factorial series.

Theorem 5 *Binet's function $\mu(z)$ possesses infinitely many factorial expansions in the complex parameter α, for* $\operatorname{Re}(z) > 0$ *and* $\operatorname{Re}(\alpha) > -\operatorname{Re}(z)$,

$$\mu(z) = \sum_{m=1}^{\infty} \frac{b_m(\alpha)}{\prod_{k=0}^{m-1}(z+\alpha+k)}$$

where the Binet polynomials in α are

$$b_m(\alpha) = \frac{(-1)^m}{m}\sum_{k=1}^{m}\left\{\frac{\alpha^{k+2}-(\alpha-1)^{k+2}}{k+2} + \frac{\alpha^{k+1}-(\alpha-1)^{k+1}}{\left(\frac{1}{2}-\alpha\right)^{-1}(k+1)}\right\}(-1)^k S_m^{(k)} \tag{13.48}$$

and $S_m^{(k)}$ is the Stirling number of the first kind. Another expression in terms of the coefficients $c_k = b_k(0)$ in (13.31) is

$$b_m(\alpha) = c_m + \alpha\sum_{k=1}^{m-1}\binom{m-1}{k-1}\prod_{j=1}^{m-k-1}(\alpha+j)c_k \tag{13.49}$$

The corresponding integral representation is

$$b_m(\alpha) = \frac{1}{m}\int_{\alpha-1}^{\alpha}\left(x+\left(\frac{1}{2}-\alpha\right)\right)\prod_{k=0}^{m-1}(k+x)dx \tag{13.50}$$

In particular, $b_1(\alpha) = \frac{1}{12}$ *and* $b_2(\alpha) = \frac{\alpha}{12}$.

Proof: We write Binet's integral in (13.32), valid for $\operatorname{Re}(z) > 0$, as

$$\mu(z) = \int_0^1 \frac{(1-u)^{z+\alpha-1}}{u}(1-u)^{-\alpha}\left(\frac{\frac{u}{2}-1}{\log(1-u)} - \frac{u}{\log^2(1-u)}\right)du$$

After substituting the Taylor series

$$g_\alpha(u) = (1-u)^{-\alpha}\left(\frac{\frac{u}{2}-1}{\log(1-u)} - \frac{u}{\log^2(1-u)}\right) = \sum_{m=1}^{\infty}\frac{b_m(\alpha)}{(m-1)!}u^m \tag{13.51}$$

and following the same steps as in the proof of Theorem 3, we arrive at (13.10). Executing the Cauchy product of the two Taylor series of $(1-u)^{-\alpha}$ and $g_0(u) = \sum_{m=1}^{\infty}\frac{c_m}{(m-1)!}u^m$ and equating corresponding powers in u, leads to the factorial expansion (13.49) of the Binet polynomial $b_m(\alpha)$ in terms of the coefficients $c_k = b_k(0)$ in (13.31).

We proceed by deducing (13.48). Introducing the series (13.10) in the difference formula (13.15), provides a factorial expansion for the function

$$1+\left(z+\frac{1}{2}\right)\log\frac{z}{z+1}=-\sum_{m=1}^{\infty}\frac{mb_m(\alpha)}{\prod_{k=0}^{m}(z+\alpha+k)} \tag{13.52}$$

which we rewrite, after denoting $y=z+\alpha$, as

$$\mathbb{Y}=1-\left(y+\frac{1}{2}-\alpha\right)\log\frac{\left(1-\frac{\alpha-1}{y}\right)}{\left(1-\frac{\alpha}{y}\right)}=-\sum_{m=1}^{\infty}\frac{mb_m(\alpha)}{\prod_{k=0}^{m}(y+k)}$$

We expand now both sides of (13.52) into powers of $\frac{1}{y}$. The Taylor series around $z_0=0$ of $\log(1-x)=-\sum_{k=1}^{\infty}\frac{x^k}{k}$, convergent for $|z|<1$, in the left-hand side of (13.52), leads, for $\max(|\alpha-1|,|\alpha|)<|y|$, to

$$\mathbb{Y}=-\sum_{k=2}^{\infty}\left(\frac{\alpha^{k+1}-(\alpha-1)^{k+1}}{k+1}+\frac{\left(\frac{1}{2}-\alpha\right)\left(\alpha^k-(\alpha-1)^k\right)}{k}\right)\frac{1}{y^k}$$

Nielsen (1956, band I, p. 68) derives

$$\frac{1}{\prod_{k=0}^{m}(z+k)}=\sum_{k=m}^{\infty}(-1)^{m-k}\,\mathcal{S}_k^{(m)}\frac{1}{z^{k+1}} \tag{13.53}$$

where $\mathcal{S}_k^{(n)}$ denotes the Stirling number of the second kind, which we use in the right-hand side of (13.52)

$$\begin{aligned}\sum_{m=1}^{\infty}\frac{mb_m(\alpha)}{\prod_{k=0}^{m}(y+k)}&=\sum_{m=1}^{\infty}mb_m(\alpha)\sum_{k=m}^{\infty}(-1)^{m-k}\,\mathcal{S}_k^{(m)}\frac{1}{y^{k+1}}\\&=\sum_{k=1}^{\infty}\sum_{m=1}^{k}mb_m(\alpha)\,(-1)^{m-k}\,\mathcal{S}_k^{(m)}\frac{1}{y^{k+1}}\end{aligned}$$

Equating corresponding powers in $\frac{1}{y}$ of both sides in (13.52) yields, for $k\geq 1$,

$$\sum_{m=1}^{k}mb_m(\alpha)\frac{\mathcal{S}_k^{(m)}}{(-1)^{k-m}}=\frac{\alpha^{k+2}-(\alpha-1)^{k+2}}{k+2}+\frac{\left(\frac{1}{2}-\alpha\right)\left(\alpha^{k+1}-(\alpha-1)^{k+1}\right)}{k+1} \tag{13.54}$$

Finally, after multiplying both sides in (13.54) by $(-1)^k\,S_j^{(k)}$, summing over $k\in[1,j]$, we have

$$\begin{aligned}\mathbb{B}&=\sum_{k=1}^{j}\sum_{m=1}^{k}mb_m(\alpha)\,(-1)^m\,S_j^{(k)}\mathcal{S}_k^{(m)}\\&=\sum_{k=1}^{j}\left\{\frac{\alpha^{k+2}-(\alpha-1)^{k+2}}{k+2}+\frac{\alpha^{k+1}-(\alpha-1)^{k+1}}{\left(\frac{1}{2}-\alpha\right)^{-1}(k+1)}\right\}(-1)^kS_j^{(k)}\end{aligned}$$

We reverse the k- and m-summation in the double sum at the left-hand side

$$\mathbb{B} = \sum_{m=1}^{j} m b_m(\alpha) (-1)^m \left(\sum_{k=m}^{j} S_j^{(k)} \mathcal{S}_k^{(m)} \right)$$

invoke the second orthogonality relation $\sum_{k=m}^{j} S_j^{(k)} \mathcal{S}_k^{(m)} = \delta_{mj}$ for the Stirling numbers in (2.63) and obtain $\mathbb{B} = j b_j(a) (-1)^j$, which demonstrates (13.48).

The corresponding integral representation of the coefficient $b_m(\alpha)$ is translated, via (13.36), as

$$m b_m(\alpha) = \int_{-(\alpha-1)}^{-\alpha} \left(u - \left(\frac{1}{2} - \alpha \right) \right) \prod_{k=0}^{m-1} (k-u) du$$

After substitution of $x = -u$, we arrive at (13.50). □

189. *Implications of Gilbert's general single sum factorial series.* Theorem 5 contains two particularly interesting cases of the Binet polynomial $b_m(\alpha)$ in (13.48): for $\alpha = 0$,

$$b_m(0) = c_m = \frac{1}{2m} \sum_{k=1}^{m} \left\{ \frac{k}{(k+1)(k+2)} \right\} (-1)^m S_m^{(k)}$$

but $\alpha = 1$ leads to the original Binet coefficients (13.7),

$$b_m(1) = \beta_m = \frac{1}{2m} \sum_{k=1}^{m} \left\{ \frac{k}{(k+1)(k+2)} \right\} (-1)^{k-m} S_m^{(k)}$$

Since $(-1)^{k-j} S_j^{(k)}$ is a non-negative integer, it follows that the original Binet coefficients $\beta_m = b_m(1)$ are all positive, in contrast to $b_m(0) = c_m$ in Theorem 4, whose sum (13.31) is alternating and does not obviously to lead to conclusions about the sign.

As mentioned earlier, any truncation at $m = K > 2$ terms in (13.30) upper bounds Binet's function $\mu(z)$, whereas any truncation of $m = K$ terms in Binet's original expansion (13.4) lower bounds $\mu(x)$ for real $x > 0$. Hence, for any finite integer $\dot{K} > 2$, it holds that

$$\sum_{m=1}^{K} \frac{b_m(1)}{\prod_{k=1}^{m} (x+k)} < \mu(x) < \sum_{m=1}^{K} \frac{b_m(0)}{\prod_{k=0}^{m-1} (x+k)}$$

Since $\mu_K(x;\alpha) = \sum_{m=1}^{K} \frac{b_m(\alpha)}{\prod_{k=0}^{m-1}(x+\alpha+k)}$ is continuous in $\alpha \in [0,1]$ and $\mu_K(x;0) < \mu(x) < \mu_K(x;1)$, there must exist a value for $\alpha = \alpha_{x,K} \in (0,1)$ where $\mu_K(x;\alpha_{x,K}) = \mu(x)$, for any finite $K > 2$ and real x. In other words, for such a value $\alpha_{x,K}$, depending on real $x > 0$ and integer $K > 2$, the factorial series (13.10) contains a *finite number of terms*.

It follows from (13.48)

$$b_m(\alpha) = \frac{(-1)^m}{m} \sum_{k=1}^{m} \left\{ \frac{\alpha^{k+2} - (\alpha-1)^{k+2}}{k+2} + \frac{\alpha^{k+1} - (\alpha-1)^{k+1}}{\left(\frac{1}{2}-\alpha\right)^{-1}(k+1)} \right\} (-1)^k S_m^{(k)}$$

that

$$b_m(1-\alpha) = \frac{(-1)^m}{m} \sum_{k=1}^{m} \left\{ \frac{\alpha^{k+2} - (\alpha-1)^{k+2}}{k+2} + \frac{\alpha^{k+1} - (\alpha-1)^{k+1}}{\left(\frac{1}{2}-\alpha\right)^{-1}(k+1)} \right\} S_m^{(k)}$$

illustrating, with $(-1)^{m-k} S_m^{(k)} \geq 0$, the absence of symmetry around $\alpha = \frac{1}{2}$. A second observation of (13.48) for $b_m(\alpha)$ and the fact, that Stirling numbers are integers, is that, if $\alpha \in \mathbb{Q}$ is a rational number, i.e. $\alpha = \frac{l}{k}$ for integers l and k, then the Binet polynomial $b_m(\alpha)$ is also rational.

190. *The Binet polynomial $b_m(\alpha)$.* The Binet polynomial $b_m(\alpha)$ is further investigated in Van Mieghem (2021). Without proofs, we list the following properties:

Property 1 *The Binet polynomial $b_m(\alpha)$ in (13.48) is a polynomial of degree $m-1$ in α,*

$$b_m(\alpha) = \sum_{j=0}^{m-1} p_j(m)\, \alpha^j \tag{13.55}$$

where the coefficients

$$p_j(m) = \frac{1}{j!} \frac{d^j b_m(\alpha)}{d\alpha^j}\bigg|_{\alpha=0} = \frac{(-1)^{m-1-j}}{2m\, j!} \sum_{k=j+1}^{m} \frac{k!\,(k-j)}{(k+2-j)!} S_m^{(k)} \tag{13.56}$$

from which $p_{m-1}(m) = \frac{1}{12}$ and $p_{m-2}(m) = \frac{1}{12}\binom{m-1}{2}$. An integral form is

$$p_j(m) = \frac{1}{mj!} \int_{-1}^{0} \left(u + \frac{1}{2}\right) \frac{d^j}{du^j} \prod_{k=0}^{m-1} (k+u)du \tag{13.57}$$

Property 2 *The Binet polynomial*

$$b_m(\alpha) = \sum_{j=0}^{m-1} p_j(m)\, \alpha^j = \frac{1}{12} \prod_{j=1}^{m-1} (\alpha - \xi_j(m))$$

has $m-1$ real, distinct zeros $\xi_1(m) > \xi_2(m) > \cdots > \xi_{m-1}(m)$.

We end this section with Property 3 that relates the Binet polynomials $b_m(\alpha)$ to the Bernoulli polynomials $B_n(\alpha)$ in **art.** 242,

Property 3 *The Binet polynomial $b_m(\alpha)$ in (13.48) of the generalized factorial series (13.10) of Binet's function $\mu(z)$ for* $\text{Re}(z) > 0$ *can be*

expressed in terms of Bernoulli polynomials $B_n(\alpha)$ *as*

$$b_m(\alpha) = (-1)^{m-1} \sum_{k=0}^{m-1} \frac{S_{m-1}^{(k)} (-1)^k}{(k+1)(k+2)} \left(B_{k+2}(\alpha) - \alpha^{k+2} + \frac{1}{2}(k+2)\alpha^{k+1} \right) \tag{13.58}$$

13.5 Fractional series for digamma and polygamma functions

We present the corresponding factorial series for the digamma function $\psi(z) = \frac{\Gamma'(z)}{\Gamma(z)}$ and for the polygamma function $\psi^{(n)}(z) = \frac{d^{n+1} \log \Gamma(z)}{dz^{n+1}}$. We confine ourselves to the case $\alpha = 0$ in Theorem 5.

191. *Factorial series for the digamma function.* Differentiation of (13.1) with respect to z expresses the digamma function $\psi(z) = \frac{\Gamma'(z)}{\Gamma(z)}$ in terms of the Binet function $\mu(z)$ as

$$\psi(z) = \log z - \frac{1}{2z} + \mu'(z)$$

The derivative $\mu'(z)$ follows from fractional series (13.10) and (13.56) as

$$\psi(z) = \log z - \frac{1}{2z} - \sum_{m=2}^{\infty} \frac{p_1(m)}{\prod_{k=0}^{m-1}(z+k)} \tag{13.59}$$

where

$$p_1(m) = \frac{(-1)^m}{2m} \sum_{k=2}^{m} \frac{(k-1)}{(k+1)} S_m^{(k)}$$

Explicitly, we compute a few coefficients $p_1(m)$ in (13.59) for $m \geq 2$,

$$\begin{array}{llll} p_1(2) = p_1(3) = \frac{1}{12} & p_1(4) = \frac{19}{120} & p_1(5) = \frac{9}{20} & p_1(6) = \frac{863}{504} \\ p_1(7) = \frac{1375}{168} & p_1(8) = \frac{33953}{720} & p_1(9) = \frac{57281}{180} & p_1(10) = \frac{3250433}{1320} \end{array}$$

resulting in

$$\psi(z) = \log z - \frac{1}{2z} - \frac{1}{12z(z+1)} - \frac{1}{12z(z+1)(z+2)} - \sum_{m=4}^{\infty} \frac{p_1(m)}{\prod_{k=0}^{m-1}(z+k)}$$

The fractional series (13.59) is the convergent companion for $\text{Re}(z) > 0$ of the asymptotic series (Abramowitz and Stegun, 1968, 6.3.18), obtained after differentiation from (14.1),

$$\psi(z) \sim \log z - \frac{1}{2z} + \sum_{m=1}^{\infty} \frac{B_{2m}}{2mz^{2m}} = \log z - \frac{1}{2z} - \frac{1}{12z^2} + \frac{1}{120z^4} - \frac{1}{252z^6} + \cdots \tag{13.60}$$

Since $p_1(m) > 0$, truncation of $\sum_{m=2}^{K} \frac{p_1(m)}{\prod_{k=0}^{m-1}(z+k)}$ in (13.59) after any $K \geq 2$ provides an upper bound for the digamma function $\psi(z)$ for real

$z > 0$. For integer $z = n$ in (13.59) and recalling $\psi(n) = \sum_{k=1}^{n-1} \frac{1}{k} - \gamma$ in (12.8), the harmonic series H_n is

$$\sum_{k=1}^{n} \frac{1}{k} = \log n + \gamma + \frac{1}{2n} - \lim_{K \to \infty} \sum_{m=2}^{K} \frac{(n-1)! p_1(m)}{(n-1+m)!} \tag{13.61}$$

and the m-sum converges rapidly, also for relatively small n, but increasingly fast for larger n. For example, with $K = 5$ terms evaluated in (13.61), the error is less that 10^{-6} for $n = 10$ and less than 10^{-11} for $n = 100$.

Differentiating Binet's second form (12.43) results in a companion of Gauss's integral (12.37) for the digamma function,

$$\psi(z) = \log z - \frac{1}{2z} - 2\int_0^\infty \frac{t}{(e^{2\pi t} - 1)} \frac{dt}{(z^2 + t^2)} \tag{13.62}$$

Euler's constant $\gamma = -\psi(1)$ can be evaluated from (13.62) as $\gamma = \frac{1}{2} + 2\int_0^\infty \frac{t}{(e^{2\pi t}-1)} \frac{dt}{(1+t^2)}$. The integrand is positive, which leads to $\gamma > \frac{1}{2}$. Since $\int_0^\infty \frac{t}{(e^{2\pi t}-1)} \frac{dt}{(1+t^2)} < \int_0^\infty \frac{tdt}{e^{2\pi t}-1}$ and $\Gamma(s)\zeta(s) = \int_0^\infty \frac{t^{s-1}dt}{e^t-1}$, an upper bound follows as $\gamma < \frac{1}{2} + \frac{2}{(2\pi)^2}\zeta(2) = 0.5833$. For integer values of $z = n$, combining (12.8) and (13.62) results in the integral version of (13.61) for the harmonic series

$$\sum_{k=1}^{n} \frac{1}{k} = \log n + \gamma + \frac{1}{2n} - 2\int_0^\infty \frac{t}{(e^{2\pi t} - 1)} \frac{dt}{(n^2 + t^2)} \tag{13.63}$$

192. *Factorial series for polygamma functions.* We can construct fractional series for polygamma functions $\psi^{(n)}(z)$ with $n \geq 1$. Indeed, in terms of the Binet function and starting from $\psi^{(0)}(z) = \psi(z) = \log z - \frac{1}{2z} + \mu'(z)$, it holds for $n \geq 1$ that

$$\psi^{(n)}(z) = (-1)^{n-1}(n-1)! z^{-n} + \frac{1}{2}(-1)^{n-1} n! z^{-n-1} + \frac{d^{n+1}}{dz^{n+1}}\mu(z)$$

Computing the derivative from the fractional series (13.10) of the Binet function $\mu(z)$, valid for $n \geq 1$, yields

$$\psi^{(n)}(z) = \frac{(n-1)!}{(-1)^{n-1}} \left(\frac{1}{z^n} + \frac{1}{2}\frac{n}{z^{n+1}} + (n+1)n \sum_{m=n+2}^{\infty} \frac{p_{n+1}(m)}{\prod_{k=0}^{m-1}(z+k)} \right) \tag{13.64}$$

However, the factorial series for $\psi^{(n)}(z)$ converges slower for larger n, is more complicated and only valid for $\mathrm{Re}(z) > 0$ in contrast to $\psi^{(n)}(z) = n!(-1)^{n-1}\sum_{k=0}^{\infty} \frac{1}{(z+k)^{n+1}}$ in (12.11) that converges for all complex z, except for the poles at $z = -k$ for integer $k \geq 0$.

193. *Factorial series for* $\psi^{(1)}(z) = \frac{d^2 \log \Gamma(z)}{dz^2}$. The factorial series for $n = 1$ in (13.64) is

$$\psi^{(1)}(z) = \frac{1}{z} + \frac{1}{2z^2} + 2\sum_{m=3}^{\infty} \frac{p_2(m)}{\prod_{k=0}^{m-1}(z+k)} \tag{13.65}$$

and $p_2(m)$ follows from (13.56) as

$$p_2(m) = \frac{(-1)^{m-1}}{4m}\sum_{k=3}^{m}(k-2)\,S_m^{(k)} = \frac{(m-2)!}{4}\left(1-\frac{2}{m}\right)$$

The factorial series (13.64) is not "completely" factorial, due to two first terms. Ruben (1976) derives the "complete" factorial series from Gauss's hypergeometric function $F(a,b,c;z) = \frac{\Gamma(c)}{\Gamma(a)\Gamma(b)}\sum_{n=0}^{\infty}\frac{\Gamma(a+n)\Gamma(b+n)}{\Gamma(c+n)n!}z^n$ in (1.17). Gauss's classical product (10.9) for the hypergeometric series (Olver *et al.*, 2010, 15.4.20) at $z=1$, for $c \neq -k$ with $k \in \mathbb{N}$ and $\operatorname{Re}(c-a-b) > 0$, is

$$F(a,b;c;1) = \frac{\Gamma(c)\Gamma(c-a-b)}{\Gamma(c-a)\Gamma(c-b)} = \frac{\Gamma(c)}{\Gamma(a)\Gamma(b)}\sum_{n=0}^{\infty}\frac{\Gamma(a+n)\Gamma(b+n)}{\Gamma(c+n)n!}$$

Ruben (1976) writes $\psi^{(1)}(z) = \frac{d^2\log\Gamma(z)}{dz^2}$ as a limit

$$\psi^{(1)}(z) = \lim_{h\to 0}\frac{\log\Gamma(z+h) - 2\log\Gamma(z) + \log\Gamma(z-h)}{h^2}$$

After taking the logarithm of Gauss's product (10.9), choosing $a=b$,

$$\log F(a,a;c;1) = \log\Gamma(c) - 2\log\Gamma(c-a) + \log\Gamma(c-2a)$$

and further letting $c = z+h$, $c-a = z$ and $c-2a = z-h$, Ruben (1976) finds

$$\psi^{(1)}(z) = \lim_{h\to 0}\frac{\log F(h,h;z+h;1)}{h^2} = \lim_{h\to 0}\frac{\log\left(1+\sum_{n=1}^{\infty}\frac{\Gamma(z+h)\Gamma^2(h+n)}{\Gamma(z+h+n)\Gamma^2(h)n!}\right)}{h^2}$$

The functional equation $\Gamma(h) = \frac{\Gamma(1+h)}{h}$ in (10.1) together with the Taylor series $\log(1+x) = x + O(x^2)$ lead to

$$\psi^{(1)}(z) = \sum_{n=1}^{\infty}\frac{\Gamma(z)\Gamma^2(n)}{\Gamma(z+n)n!}$$

Hence, Ruben (1976) arrives at the beautiful factorial series, for $\operatorname{Re}(z) > 0$,

$$\psi^{(1)}(z) = \sum_{n=1}^{\infty}\frac{(n-1)!}{n}\frac{1}{\prod_{k=0}^{n-1}(z+k)} \tag{13.66}$$

Numerical computations reveal that the factorial series (13.65) converges faster than the more elegant (13.66) factorial series of Ruben.

14
Asymptotic behavior of $\log \Gamma(z)$ and $\Gamma(z)$

Section 14.1 discusses the asymptotic expansion for $\log \Gamma(z)$, attributed to Stirling, for large z. The consequences in Section 14.2 of Stirling's asymptotic formula (14.1) up to first-order $O\left(\frac{1}{z}\right)$ for large complex $z = re^{i\theta}$ emphasizes the three highly different θ-regimes in the complex z-plane, with a special attention to $|\Gamma(b \pm ir)| \sim \sqrt{2\pi} r^{b-\frac{1}{2}} e^{-\frac{\pi}{2}r}$ in (14.8) for a point r far on a line $z = b \pm ir$ parallel with the imaginary z-axis, because integrands containing $\Gamma(b \pm ir)$ often occur in inverse Laplace transforms and in the theory of the Mittag-Leffler function. We derive the asymptotic expansion of $\frac{\Gamma(a+z)}{\Gamma(b+z)}$ in (14.12) in **art**. 198.

Section 14.3 concentrates on the asymptotic behavior of $\Gamma(z)$ for large z. We compute the Stirling variant for $\Gamma(z+1) \simeq \sqrt{2\pi z}\left(\frac{z}{e}\right)^z \left(1 + \sum_{m=1}^{K} \frac{c_m}{z^m}\right)$ up to arbitrary K in (14.17) via the steepest descent method and list its coefficients c_m exactly only up to $m = 15$ terms in **art**. 200. **Art**. 201 gives the approach of Watson, Copson & Marsaglia, that leads to precisely the same asymptotic expansion (14.17). We end Chapter 14 with converging, factorial series for $\Gamma(z)$: **art**. 202 derives the factorial series (14.26) of Cornelius Lanczos, a beautiful, converging counter part of Stirling's expansion (14.17), and **art**. 203 presents the factorial series (14.28) of Rubinstein.

14.1 Stirling's asymptotic formula

One of the focal expansions in the theory of the Gamma function is Stirling's asymptotic formula (14.1), which we discuss in this Section 14.1. Although Stirling's expansion does not converge (**art**. 195), it is one of the most famous and useful asymptotic series. There are many derivations of Stirling's asymptotic series of which only a few are deduced here.

194. *Stirling's asymptotic series.* Invoking the generating function of the Bernoulli numbers (A.15) in (12.42) and ignoring the convergence restriction $|t| < 2\pi$ yield Stirling's approximation (Abramowitz and Stegun, 1968,

6.1.40),

$$\log\Gamma(z) \simeq \left(z-\frac{1}{2}\right)\log z - z + \frac{1}{2}\log(2\pi) + \lim_{K\to\infty}\sum_{n=1}^{K}\frac{B_{2n}}{2n(2n-1)z^{2n-1}} \tag{14.1}$$

$$\simeq \left(z-\frac{1}{2}\right)\log z - z + \frac{1}{2}\log(2\pi) + \frac{1}{12z} - \frac{1}{360z^3} + O\left(\frac{1}{z^5}\right)$$

In terms of Binet's function $\mu(z) = \log\Gamma(z) - \left(z-\frac{1}{2}\right)\log z + z - \frac{1}{2}\log(2\pi)$, Stirling's asymptotic series (14.1) is

$$\mu(z) \simeq \lim_{K\to\infty}\sum_{n=1}^{K}\frac{B_{2n}}{2n(2n-1)z^{2n-1}}$$

Substituting in Binet's second form (12.43), the Taylor series $\arctan t = \sum_{n=0}^{\infty}\frac{(-1)^n}{2n+1}t^{2n+1}$ around $t=0$, valid for $|t|<1$, reversing the integral and summation while ignoring the restriction $|t|<1$, leads, with $B_{2n} = (-1)^{n-1}4n\int_0^\infty \frac{t^{2n-1}}{e^{2\pi t}-1}dt$ for $n\geq 1$, again to Stirling's approximation (14.1).

Rademacher's form (12.46) of $\log\Gamma(z)$ is the convergent variant of Stirling's asymptotic series (14.1) with the Euler-Maclaurin integral as remainder term.

195. *Stirling's asymptotic series via Euler-Maclaurin summation.* Perhaps the simplest, but definitely the first derivation of Stirling's asymptotic series starts from the Euler-Maclaurin summation formula (3.40) in **art.** 53, with the choice $f(t) = \log t$, $\alpha = 1$ and $\beta = m \in \mathbb{N}$,

$$\sum_{n=2}^{m}\log(n) = \int_1^m \log(t)\,dt + \frac{1}{2}\log m + \sum_{n=2}^{N}\frac{(n-2)!B_n}{n!}\left[\frac{1}{m^{n-1}}-1\right] + R_N$$

Taking into account that the odd Bernoulli numbers $B_{2n+1}=0$ for $n\geq 1$, we find

$$\log(m!) = \left(m+\frac{1}{2}\right)\log m - m + C + \sum_{n=1}^{N}\frac{B_{2n}}{2n(2n-1)}\frac{1}{m^{2n-1}} + R_N \tag{14.2}$$

However, (14.2) suffers from the complication[1] to show that $C = 1 - \sum_{n=1}^{N}\frac{B_{2n}}{2n(2n-1)} \overset{?}{=} \frac{1}{2}\log(2\pi)$ to align with (14.1). The underlying reason

[1] Diaconis and Freedman (1986) mention that the first proofs of Stirling's approximation (14.1) by de Moivre (1730) and Stirling (1730) around 1730 were based on what is now called the Euler-Maclaurin summation. De Moivre could initially not determine the constant C, but acknowledged in an addendum in 1731 that Stirling was able to determine C using Wallis' product (9.11), that also helped Euler in 1729 (see Section 9.4).
Euler (1755, Pars posterior, Chapt. 6, art. 158, pp. 446-447) computes the constant $C = \frac{1}{2}\log(2\pi) = 0.9189385332046727417803297$ up to 25 digits correct from his variant (3.41) of the Euler-Maclaurin formula for $x=1$ and using Wallis' product (9.11), which he has derived in Euler (1748, art. 185, p. 146). Also Rademacher (1973, Section 15) gives the entire derivation of Stirling's asymptotic series (14.1) using the Euler-Maclaurin summation (3.40).

of the difficulty lies in the divergence of the sum $\sum_{n=1}^{N}\frac{B_{2n}x^{2n}}{2n(2n-1)}$ for any x when N increases unboundedly. Divergence is immediate from Euler's famous formula $B_{2n} = 2\frac{(2n)!(-1)^{n-1}}{(2\pi)^{2n}}\zeta(2n)$ in (A.18) for the even Bernoulli numbers, illustrating for large n that $|B_{2n}| \sim 2\frac{(2n)!}{(2\pi)^{2n}}$. The argument emphasizes that Stirling's asymptotic formula (14.1) diverges for any z if $N \to \infty$.

196. *Stirling's asymptotic series reviewed.* We cannot close the contour in

$$\mu(z) = -\frac{1}{2\pi i}\int_{c-i\infty}^{c+i\infty}\frac{\pi}{\sin \pi s}\frac{\zeta(s)}{s}z^s ds \qquad \text{with } -1 < c < 0$$

in (13.23) over the entire $\text{Re}(s) < 0$ -plane, because the functional equation of the Riemann Zeta-function $\zeta(s) = 2(2\pi)^{s-1}\sin\frac{\pi s}{2}\Gamma(1-s)\zeta(1-s)$ indicates that $\zeta(-s) = O(\Gamma(s))$. However, neglecting this restriction and using $\zeta(-k) = \frac{(-1)^k}{k+1}B_{k+1}$ at the poles $s = -k \le -1$ and the odd Bernoulli numbers $B_{2k+1} = 0$, for $k > 0$, leads to Stirling's asymptotic approximation[2] in the Poincaré sense (see e.g. Paris and Kaminski (2001))

$$\mu(z) \simeq \sum_{k=1}^{K}\frac{(-1)^k\zeta(-k)}{kz^k} = \sum_{k=1}^{K}\frac{B_{k+1}}{k(k+1)z^k} = \sum_{m=1}^{K}\frac{B_{2m}}{(2m-1)(2m)z^{2m-1}} \tag{14.3}$$

Although (14.3) diverges if $K \to \infty$, Stirling's asymptotic approximation (14.3) is surprisingly accurate up to some finite $K \le K^*(z)$, where $K^*(z)$ depends upon z and is roughly equal to the minimum k-value of $\frac{|B_{k+1}|}{k(k+1)|z|^k}$. We refer to Paris (2014) for a study of the Stokes phenomenon (**art.** 71) in Stirling's asymptotic series and the exponentially improved expansion for $\log\Gamma(z)$ valid in $|\arg z| < \pi$.

On the other hand for $|z| < 1$, the contour in (13.23) can be closed over the $\text{Re}(s) > 0$ -plane, where two double poles at $s = 0$ and $s = 1$ are encountered whose residues are computed in Van Mieghem (2021, Appendix A), resulting, with the Taylor series (12.44) of $\log\Gamma(z+1)$, in

$$\mu(z) = \sum_{k=2}^{\infty}\frac{(-1)^k\zeta(k)}{k}z^k - z(\log z - 1 + \gamma) - \frac{1}{2}\log z - \frac{1}{2}\log(2\pi) \tag{14.4}$$

In contrast to Stirling's series $\sum_{k=1}^{K}\frac{(-1)^k\zeta(-k)}{k}z^{-k}$ for some finite $K \le K^*(z)$ in (14.3), violation of the restriction $|z| < 1$ in the Taylor series in (14.4) leads to useless results.

A convergent companion of Stirling's asymptotic series (14.3) is derived

---

[2] By introducing the generating function (A.15) of the Bernoulli numbers in Binet's integral (13.2), ignoring the convergence restriction $|t| < 2\pi$,

$$\mu(z) = \int_0^\infty \frac{e^{-zt}}{t}\left(\frac{1}{e^t-1} - \frac{1}{t} + \frac{1}{2}\right)dt = \int_0^\infty e^{-zt}\left(\sum_{m=1}^{\infty}B_{2m}\frac{t^{2m-2}}{(2m)!}\right)dt$$

and reversing sum and integral, we obtain again Stirling's asymptotic series (14.3).

in Van Mieghem (2021)

$$\mu(z) = \sum_{m=1}^{\infty} g_m(1) \sum_{v=1}^{m-1} (v-1)!(-1)^{m-1-v} \mathcal{S}_m^{(v+1)} \left(\frac{1}{z}\right)^v \tag{14.5}$$

where $\mathcal{S}_m^{(k)}$ is the Stirling number of second kind and $g_m(1)$ are the Taylor coefficients of the Taylor series $(s-1)\zeta(s) = \sum_{m=0}^{\infty} g_m(1)(s-1)^m$ around $s_0 = 1$. The Taylor coefficients (Abramowitz and Stegun, 1968, (23.2.5)) for $k \geq 0$

$$g_{m+1}(1) = \frac{(-1)^m}{m!} \lim_{K\to\infty} \left(\sum_{n=1}^{K} \frac{\ln^m n}{n} - \frac{\ln^{m+1} K}{m+1} \right) \tag{14.6}$$

are attributed to Stieltjes, with $g_0(1) = 1$ and $g_1(1) = \gamma$ by (9.3). However, computationally, Stieltjes expression (14.6) is less suited; fast converging series for $g_m(1)$ are presented in Van Mieghem (2021, Appendix).

14.2 Asymptotic behavior of $\log\Gamma(z)$

197. *Asymptotic behavior of* $\log\Gamma(z)$. For large r and $\theta \neq \pi$, Stirling's formula (14.1) shows that

$$\log\Gamma\left(b+re^{i\theta}\right) = \left(b+re^{i\theta}-\frac{1}{2}\right)\log\left(b+re^{i\theta}\right) - b - re^{i\theta} + \frac{\log(2\pi)}{2} + O\left(\frac{1}{r}\right)$$

With

$$\begin{aligned}\log\left(b+re^{i\theta}\right) &= \log\left(re^{i\theta}\left(1+r^{-1}e^{-i\theta}b\right)\right) = \log re^{i\theta} + \log\left(1+r^{-1}e^{-i\theta}b\right)\\ &= \log r + i\theta + r^{-1}e^{-i\theta}b + O\left(\frac{1}{r^2}\right)\end{aligned}$$

we find

$$\begin{aligned}\log\Gamma\left(b+re^{i\theta}\right) &= \left(b-\frac{1}{2}+r\cos\theta\right)\log r - \theta\left(r\sin\theta\right) - r\cos\theta + \frac{1}{2}\log(2\pi)\\ &\quad + i\left\{r\ln r\sin\theta + \left(b-\frac{1}{2}+r\cos\theta\right)\theta - r\sin\theta\right\} + O\left(\frac{1}{r}\right)\end{aligned}$$

Thus, for any $z = re^{i\theta}$ with large modulus $r = |z|$ and fixed b, the asymptotic behavior of $\Gamma(b+z)$ is

$$\Gamma\left(b+re^{i\theta}\right) = \sqrt{2\pi}r^{b-\frac{1}{2}} \frac{e^{(r(\log r-1)\cos\theta - \theta r\sin\theta)}}{e^{-i\left(r(\log r-1)\sin\theta + r\theta\cos\theta + \left(b-\frac{1}{2}\right)\theta\right)}} \left(1+O\left(\frac{1}{r}\right)\right) \tag{14.7}$$

In particular, for $\theta = \pm\frac{\pi}{2}$ and large r, (14.7) simplifies to

$$\Gamma(b\pm ir) = \sqrt{2\pi}r^{b-\frac{1}{2}\pm ir}e^{-\frac{\pi}{2}r}e^{\pm i\left(\left(b-\frac{1}{2}\right)\frac{\pi}{2}-r\right)}\left(1+O\left(\frac{1}{r}\right)\right) \tag{14.8}$$

illustrating an exponential decrease for large r. More generally, it follows from (14.7) that

$$\frac{1}{|\Gamma(b+re^{i\theta})|}=\frac{r^{\frac{1}{2}-b}}{\sqrt{2\pi}}e^{-r(\ln(r)-1)\cos\theta+\theta r\sin\theta}\left(1+O\left(\frac{1}{r}\right)\right) \tag{14.9}$$

which indicates the different asymptotic regimes for large r, depending on (i) $\cos\theta>0$, (ii) $\cos\theta<0$ and (iii) $\cos\theta=0$, similar to (14.8). If $-\frac{\pi}{2}<\theta<\frac{\pi}{2}$ for regime (i), then (14.9) shows that $\frac{1}{|\Gamma(b+re^{i\theta})|}$ fastly *decreases* as $O\left(e^{-r\ln r}\right)$. If $\frac{\pi}{2}<\theta<\frac{3\pi}{2}$ for regime (ii), then (14.9) shows that $\frac{1}{|\Gamma(b+re^{i\theta})|}$ fastly *increases* as $O\left(e^{r\ln r}\right)$. In the intermediate regime (iii), $\frac{1}{|\Gamma(b\pm ir)|}=\frac{r^{\frac{1}{2}-b}}{\sqrt{2\pi}}e^{\frac{\pi}{2}r}\left(1+O\left(\frac{1}{r}\right)\right)$ increases purely exponentially.

198. *Asymptotic behavior of* $\frac{\Gamma(a+z)}{\Gamma(b+z)}$. Relation (14.9) shows, for any complex number $z=re^{i\theta}$ with large modulus $r=|z|$, that

$$\frac{|\Gamma(a+z)|}{|\Gamma(b+z)|}=|z|^{a-b}\left(1+O\left(\frac{1}{|z|}\right)\right)$$

A shorter, but less accurate derivation was presented in **art.** 142; yet another will appear in **art.** 239.

We will derive an asymptotic expansion of $\frac{\Gamma(a+z)}{\Gamma(b+z)}$ for large z. We start from the Beta-integral (10.11),

$$\frac{\Gamma(a+z)}{\Gamma(b+z)}=\frac{1}{\Gamma(b-a)}\int_0^1 u^{z+a-1}(1-u)^{b-a-1}\,du\quad\text{for }\operatorname{Re}(b)\geq\operatorname{Re}(a)>0 \tag{14.10}$$

$$=\frac{1}{\Gamma(b-a)}\int_0^\infty e^{-(z+a)t}\left(1-e^{-t}\right)^{b-a-1}dt$$

Repeating[3] the method in Titchmarsh (1964, p. 58),

$$\frac{\Gamma(a+z)\,\Gamma(b-a)}{\Gamma(b+z)}=\frac{\Gamma(b-a)}{z^{b-a}}-\int_0^\infty\frac{t^{b-a-1}}{e^{zt}}\left(1-e^{-at}\left(\frac{1-e^{-t}}{t}\right)^{b-a-1}\right)dt \tag{14.11}$$

requires us to deduce the Taylor series of the function $e^{-at}\left(\frac{1-e^{-t}}{t}\right)^{b-a-1}$, which is entire if $b-a-1>0$. The generating function $\left(\frac{t}{1-e^{-t}}\right)^{x+1}=$

[3] Titchmarsh (1964, p. 58) starts from the Beta-function integral (10.11), substitutes $u=e^{-t}$ and obtains

$$\frac{\Gamma(z)\,\Gamma(b)}{\Gamma(z+b)}=\int_0^\infty e^{-zt}\left(1-e^{-t}\right)^{b-1}dt$$

$$=\int_0^\infty t^{b-1}e^{-zt}dt-\int_0^\infty e^{-zt}\left(t^{b-1}-\left(1-e^{-t}\right)^{b-1}\right)dt$$

where the last step constitutes the essential trick. By bounding, Titchmarsh proves that the second integral is $O\left(z^{-b-1}\right)$ for large z and $\operatorname{Re}(z)>0$, while the first integral equals $\Gamma(b)\,z^{-b}$. Hence, $\frac{\Gamma(z)}{\Gamma(z+b)}=z^{-b}+O\left(z^{-b-1}\right)$.

$\sum_{m=0}^{\infty} \mathbb{S}_m(x)\frac{t^m}{m!}$ in (A.9) of the Stirling polynomials $\mathbb{S}_m(x)$, with $\mathbb{S}_0(z) = 1$, leads to the Cauchy product,

$$\begin{aligned} e^{-at}\left(\frac{1-e^{-t}}{t}\right)^{b-a-1} &= \sum_{k=0}^{\infty}\frac{(-a)^k}{k!}t^k \sum_{k=0}^{\infty}\frac{\mathbb{S}_k(-(b-a))}{k!}t^k \\ &= \sum_{k=0}^{\infty}\left(\sum_{m=0}^{k}\frac{\mathbb{S}_m(-(b-a))}{m!}\frac{(-a)^{k-m}}{(k-m)!}\right)t^k \\ &= 1+\sum_{k=1}^{\infty}\left(\sum_{m=0}^{k}\binom{k}{m}(-a)^{k-m}\,\mathbb{S}_m(-(b-a))\right)\frac{t^k}{k!} \end{aligned}$$

Substitution of the Taylor series, that converges for all t if $b-a>1$, into the integral (14.11),

$$\frac{\Gamma(a+z)}{\Gamma(b+z)} = \frac{1}{z^{b-a}} + \sum_{k=1}^{\infty}\frac{\sum_{m=0}^{k}\binom{k}{m}(-a)^{k-m}\,\mathbb{S}(-(b-a))\,\Gamma(b-a+k)}{k!z^{b-a+k}\Gamma(b-a)}$$

results, for $K\to\infty$, in

$$\frac{\Gamma(a+z)}{\Gamma(b+z)}z^{b-a} = 1+\sum_{k=1}^{K}\frac{\sum_{m=0}^{k}\binom{k}{m}(-a)^{k-m}\,\mathbb{S}_m(-(b-a))}{k!}\frac{\Gamma(b-a+k)}{\Gamma(b-a)\,z^k}$$

With the explicit form (A.12) of the Stirling polynomials $\mathbb{S}_m(-z)$, for $m>0$, while $\mathbb{S}_0(z)=1$, we arrive at the asymptotic, thus diverging for $K\to\infty$, expansion (Olver *et al.*, 2010, 5.11) for large z,

$$\frac{\Gamma(a+z)}{\Gamma(b+z)}z^{b-a} = 1+\lim_{K\to\infty}\sum_{k=1}^{K}\frac{G_k(a,b)}{z^k} \tag{14.12}$$

where

$$\begin{aligned} G_k(a,b) &= \frac{\Gamma(b-a+k)}{\Gamma(b-a)}\frac{(-a)^k}{k!} \\ &\quad+\sum_{m=1}^{k}\frac{(-a)^{k-m}}{(k-m)!}\frac{\Gamma(b-a+k)}{\Gamma(b-a-m-1)}\sum_{r=1}^{m}\frac{\frac{(-1)^{r+1}}{(m+r)!(m-r)!}\mathcal{S}_{m+r}^{(r)}}{1-(b-a)+r} \end{aligned} \tag{14.13}$$

For a few values of k, we list the explicit form (14.13) of the coefficient $G_k(a,b)$:

$$\begin{aligned} \frac{G_1(a,b)}{a-b} &= \frac{1}{2}(a+b-1) \\ \frac{G_2(a,b)}{a-b} &= \frac{1}{24}(a-b-1)\left(6ab+3a^2+3b^2-7a-5b+2\right) \\ \frac{G_3(a,b)}{a-b} &= \frac{1}{48}(a-b-2)(a-b-1)(a+b-1)\left(a^2+2ba-3a+(b-1)b\right) \end{aligned}$$

If $a = 0$, then (14.12) simplifies to

$$\frac{\Gamma(z)}{\Gamma(z+b)} z^b = 1 + \lim_{K\to\infty} \sum_{k=1}^{K} \frac{1}{z^k} \frac{\Gamma(b+k)}{\Gamma(b-k-1)} \sum_{r=1}^{k} \frac{\frac{(-1)^{r+1}}{(k+r)!(k-r)!} \mathcal{S}_{k+r}^{(r)}}{1-b+r} \tag{14.14}$$

The right-hand side converges fast for already small z; for $z = 3$ and $K = 4$ terms, (14.14) is accurate to 10^{-5} for $|b| \le 1$. After a sufficiently high K, divergence starts.

The binomial series (17.2) for the Beta function in (14.10) leads to

$$\begin{aligned}\frac{\Gamma(a+z)}{\Gamma(b+z)} &= \sum_{k=0}^{\infty} \frac{(-1)^k}{k!\Gamma(b-a-k)} \frac{1}{z+a+k} \\ &= \frac{1}{\Gamma(b-a)} \frac{1}{z+a} - \frac{(b-a-1)}{\Gamma(b-a)(z+a+1)} + \cdots\end{aligned}$$

which differs from the asymptotic expansion $\frac{|\Gamma(a+z)|}{|\Gamma(b+z)|} \sim |z|^{a-b}$ in (14.12).

14.3 Series expansions for $\Gamma(z)$

So far, most attention was devoted to the asymptotic behavior of $\log\Gamma(z)$ rather than of the Gamma function $\Gamma(z)$ itself.

199. *Artin's product form.* The exponentiation of Binet's function (13.1) is

$$\Gamma(z) = \sqrt{2\pi} z^{z-\frac{1}{2}} e^{-z+\mu(z)} \tag{14.15}$$

For real x, the Binet function $\mu(x) = \frac{\theta(x)}{12x}$ with $0 < \theta(x) < 1$ in (13.38) illustrates that

$$\Gamma(x) = \sqrt{2\pi} x^{x-\frac{1}{2}} e^{-x+\frac{\theta(x)}{12x}} \qquad \text{for } 0 < \theta(x) < 1$$

After transforming Gudermann's series (13.18) into a product,

$$\mu(z) = \log\left(\prod_{k=0}^{\infty} \frac{1}{e} \left(1 + \frac{1}{z+k}\right)^{\left(z+k+\frac{1}{2}\right)} \right)$$

and substitution into (14.15), we arrive at the product form for the Gamma function with Stirling prefactor, mentioned in Artin (1964, eq. (3.8) on p. 24),

$$\Gamma(z) = \sqrt{2\pi} z^{z-\frac{1}{2}} e^{-z} \prod_{k=0}^{\infty} \frac{1}{e} \left(1 + \frac{1}{z+k}\right)^{\left(z+k+\frac{1}{2}\right)} \tag{14.16}$$

200. *Steepest descent.* The principle of the steepest descent or saddle-point method seems to be invented by Laplace and later refined by several other scholars. Its essence stems from the analytic character of a complex function $f(z)$, explained e.g. in Morse and Feshbach (1978). Copson (1965) nicely explains, among others, the method of Laplace, of the stationary phase, of the steepest descents and the saddle-point method.

The traditional introduction of the steepest descent method starts from the Euler's integral (10.2), after making the substitution $t = \alpha u$,

$$\Gamma(z+1) = \alpha^{z+1}\int_0^\infty u^z e^{-\alpha u}du = \alpha^{z+1}\int_0^\infty e^{-\alpha u + z\log u}du$$

If we choose $\alpha = z$, then

$$\Gamma(z+1) = z^{z+1}\int_0^\infty e^{-z(u-\log u)}du$$

where the function $f(u) = u - \log u$ has a minimum at $u = 1$ equal to $f(1) = 1$. The Taylor series of $f(u) =$ around $u_0 = 1$ is

$$f(u) = 1 + (u-1) - \log(1 + (u-1)) = 1 + \sum_{k=2}^{\infty}(-1)^k\frac{(u-1)^k}{k}$$

and only converges for $|u-1| < 1$ or $|u| < 1$. Approximating $f(u) \approx 1 + \frac{(u-1)^2}{2}$ and ignoring the validity $|u| < 1$ requirement yields

$$\Gamma(z+1) \approx z^{z+1}e^{-z}\int_0^\infty e^{-\frac{z}{2}(u-1)^2}du$$

A second approximation, $\int_0^\infty e^{-\frac{z}{2}(u-1)^2}du \approx \int_{-\infty}^\infty e^{-\frac{z}{2}(u-1)^2}du = \sqrt{\frac{2\pi}{z}}$, finally leads to

$$\Gamma(z+1) \approx \sqrt{2\pi}z^{z+\frac{1}{2}}e^{-z} = \sqrt{2\pi z}\left(\frac{z}{e}\right)^z$$

which is also called Stirling's approximation and which follows from (14.1) by ignoring the series as

$$\Gamma(z+1) = ze^{\log\Gamma(z)} \approx ze^{\left(z-\frac{1}{2}\right)\log(z) - z + \frac{1}{2}\log(2\pi)} = \sqrt{2\pi}z^{\left(z+\frac{1}{2}\right)}e^{-z}$$

Instead of limiting $f(u) = u - \log u = \sum_{m=0}^{\infty} f_m(1)(u-1)^2$ with Taylor coefficients $f_0(1) = 1$, $f_1(1) = 0$ and $f_m(1) = \frac{(-1)^m}{m}$ for $m \geq 2$ up to $O\left((u-1)^2\right)$, the method of the steepest descent can be extended formally, to any desired order (Van Mieghem, 2026),

$$\int_{-\infty}^{\infty} e^{-zf(u)}du = \sqrt{\frac{2\pi}{z}}e^{-z}\left(1 + \sum_{m=1}^{\infty}\frac{\sum_{k=1}^{2m}\frac{(-1)^k\Gamma(k+m+\frac{1}{2})2^k\, s^{**}[k,2m](1)}{\sqrt{\pi}k!}}{2^{-m}z^m}\right)$$

where $s^{**}[k,n](z_0) = s[k,n](z_0)|_{\forall m\,:\, f_m(z_0)\to f_{m+2}(z_0)}$ is a characteristic coefficient (Appendix A). For our particular choice $f(u) = u - \log u$ with all known Taylor coefficients around $u_0 = 1$, we arrive at the asymptotic expansion

$$\Gamma(z+1) \simeq \sqrt{2\pi z}\left(\frac{z}{e}\right)^z\left(1 + \lim_{K\to\infty}\sum_{m=1}^{K}\frac{\widetilde{c}_m}{z^m}\right) \tag{14.17}$$

where

$$\widetilde{c}_m = \sum_{k=1}^{2m} (-1)^k \frac{\Gamma\left(k+m+\frac{1}{2}\right)}{k!\sqrt{\pi}} 2^{k+m}\, s^{**}[k,2m](1)$$

The characteristic coefficients are easily computed by the recursion (A.4). We list the first 16 values of $\widetilde{c}_m$, that are all rational numbers,

$$\begin{array}{ll}
\widetilde{c}_1 = \frac{1}{12} & \widetilde{c}_9 = \frac{432261921612371}{514904800886784000} \\
\widetilde{c}_2 = \frac{1}{288} & \widetilde{c}_{10} = \frac{6232523202521089}{86504006548979712000} \\
\widetilde{c}_3 = -\frac{139}{51840} & \widetilde{c}_{11} = -\frac{25834629665134204969}{13494625021640835072000} \\
\widetilde{c}_4 = -\frac{571}{2488320} & \widetilde{c}_{12} = -\frac{1579029138854919086429}{9716130015581401251840000} \\
\widetilde{c}_5 = \frac{163879}{209018880} & \widetilde{c}_{13} = \frac{7465908699626516022203151}{116593560186976815022080000} \\
\widetilde{c}_6 = \frac{5246819}{75246796800} & \widetilde{c}_{14} = \frac{1511513601028097903631961}{2798245444487443560529920000} \\
\widetilde{c}_7 = -\frac{534703531}{902961561600} & \widetilde{c}_{15} = -\frac{8849272268392873147705987190261}{299692087104605205332754432000000} \\
\widetilde{c}_8 = -\frac{4483131259}{86684309913600} & \widetilde{c}_{16} = -\frac{142801712490607530608130701097701}{57540880724084199423888850944000000}
\end{array}$$

Thus, up to $m = 5$, we obtain

$$\Gamma(z+1) \simeq \sqrt{2\pi z}\left(\frac{z}{e}\right)^z \left(1 + \frac{1}{12z} + \frac{1}{288z^2} - \frac{139}{51840z^3} - \frac{571}{2488320z^4} + \frac{163879}{209018880z^5} + O\left(\frac{1}{z^6}\right) \right)$$

The table illustrates the alternating sign in $\widetilde{c}_m = (-1)^{\left\lfloor \frac{m-1}{2} \right\rfloor} |\widetilde{c}_m|$. Moreover, the numerical values $\log|\widetilde{c}_m|$ show two "parabolic" curves, where the curve $\log|\widetilde{c}_{2m+1}|$ resembles the upwards shifted curve $\log|\widetilde{c}_{2m}|$. From $m = 8$ on, the coefficients $|\widetilde{c}_m|$ increase with m and point towards divergence. The asymptotic series (14.17) indeed diverges due to ignoring the validity range $|u| < 1$ of the Taylor series $u - \log u$ around $u_0 = 1$.

201. *Approach of Watson, Copson & Marsaglia.* Marsaglia and Marsaglia (1990)[4] manipulate Euler's integral (10.2) for real $x > 0$ as

$$\Gamma(x+1) = \int_0^\infty e^{-t} t^x dt = x^x e^{-x} \int_0^\infty \left(\frac{e^{1-\frac{t}{x}} t}{x} \right)^x dt$$

and make a change in the integration variable $y = \frac{t}{x}$ resulting in

$$\Gamma(x+1) = x^{x+1} e^{-x} \int_0^\infty \left(y e^{1-y} \right)^x dy$$

The function $g(y) = y e^{1-y}$, with derivative $g'(y) = e^{1-y}(1-y)$, monotonically increases towards the maximum $g(1) = 1$ at $y = 1$ and monotonically

[4] I have found the paper of Marsaglia and Marsaglia (1990) in the review of Borwein and Corless (2018), who also included the letter to the editor of Grossman (1991). Grossman remarked that Marsaglia and Marsaglia (1990) have nearly completely followed the derivation of Copson (1965, art. 25). Indeed, the resemblance is striking. Copson, in his turn, mentions his inspiration from Watson (1929). We have followed Marsaglia and Marsaglia (1990), but blended with the ideas of Copson (1965, art. 25) and Watson (1929).

decreases for $y > 1$, which suggests[5] to split the integration interval into two parts $[0,1]$ and $[1,\infty)$,

$$\Gamma(x+1) = x^{x+1}e^{-x}\left(\int_0^1 \left(ue^{1-u}\right)^x du + \int_1^\infty \left(ve^{1-v}\right)^x dv\right) \tag{14.18}$$

The main ingenuity[6] in Marsaglia and Marsaglia (1990) is the transformation $g(v) = ve^{1-v} = e^{-\frac{1}{2}r^2} = h(r)$ in (14.18), which maps real numbers $v \in [1,\infty)$ to real numbers $r \in [0,\infty)$. We can explicitly write $r = \pm\sqrt{-2\log\left(ve^{1-v}\right)}$, i.e. $r = h^{-1}(g(v))$, but the inverse function $v = g^{-1}(h(r))$ cannot be written in terms of elementary functions[7].

However, differentiating $g(v) = ve^{1-v}$ with respect to r gives

$$\frac{dg(v)}{dr} = e^{1-v}(1-v)\frac{dv}{dr} = \frac{g(v)}{v}(1-v)\frac{dv}{dr}$$

while $g(v) = e^{-\frac{1}{2}r^2}$ shows that $\frac{dg(v)}{dr} = -re^{-\frac{1}{2}r^2} = -rg(v)$. Combining both leads to the differential equation

$$(1-v)\frac{dv}{dr} + rv = 0 \tag{14.19}$$

which can be solved by series expansion assuming that $v(r) = \sum_{k=0}^\infty a_k r^k$, where $v(0) = a_0 = 1$ and

$$\lim_{r\to 0}\frac{dv}{dr} = \lim_{r\to 0}\frac{-rv(r)}{1-v(r)} = \lim_{r\to 0}\frac{v(r)+rv'(r)}{-v'(r)} = \frac{1}{-v'(0)}$$

leading to $(v'(0))^2 = 1$, so that $a_1 = \pm 1$. Introducing the series into the differential equation (14.19)

$$\sum_{k=0}^\infty a_{k+1}r^k \sum_{k=0}^\infty (k+1)\,a_{k+1}r^k = \sum_{k=0}^\infty a_k r^{k+1}$$

[5] Copson (1965, art. 25) basically starts by Euler's modified integral (15.9), valid for $\mathrm{Re}(z) > 0$ and $\mathrm{Re}(p) > 0$, with the choice $z+1$ instead of z, followed by replacing z by p, $\frac{\Gamma(p)}{p^p} = \int_0^\infty \left(xe^{-x}\right)^p dx$. Copson (1965, art. 25) as well as Watson (1929) give precisely the same arguments to split the integration interval as Marsaglia and Marsaglia (1990) that led to (14.18).

[6] Watson (1929) makes the transformation $g(v) = ve^{1-v} = e^{-t}$, while the transformation of Copson (1965, art. 25) is somewhat more involved.

[7] Indeed, the inverse function of $g(v) = ve^{1-v}$, written as $-e^{-1}g(v) = -ve^{-v} = z$, is the Lambert function $v = -W_k(z)$. The Lambert function $y = W_k(z)$ satisfies $ye^y = z$ and is a multi-valued complex function (Corless *et al.*, 1996) in the integer k, which is complicating, but for real v and z, only two branches matter, $k = 0$ and $k = -1$. Hence, in the first integral, the transform is $u = -W_0\left(-e^{-1}e^{-\frac{1}{2}w^2}\right)$ and in the second, $v = -W_{-1}\left(-e^{-1}e^{-\frac{1}{2}w^2}\right)$, the branch $k = -1$ must be chosen. From $\Gamma(x+1) = x^{x+1}\left(\int_0^1 \left(ye^{-y}\right)^x dy + \int_1^\infty \left(ye^{-y}\right)^x dy\right)$ substituting $w = ye^{-y}$ and invoking the Lambert function, we find

$$\frac{\Gamma(x)}{x^{x+1}} = -e^{-x}\sum_{k=1}^\infty \frac{k^{k-1}e^{-k}}{k!\,(k+x)} - \int_0^{\frac{1}{e}} w^{x-1}W_{-1}(-w)\,dw$$

The positive series converges slowly and is dominated by the negative integral.

executing the Cauchy product

$$\sum_{k=0}^{\infty}\left(\sum_{m=0}^{k}(m+1)\,a_{k-m+1}a_{m+1}\right)r^k=\sum_{k=0}^{\infty}a_k r^k$$

and equating corresponding powers in r results in $a_k=\sum_{m=1}^{k+1}ma_{k+2-m}a_m$ from which we obtain the recursion for $k\geq 0$,

$$a_{k+1}=\frac{1}{(k+2)\,a_1}\left(a_k-\sum_{m=2}^{k}ma_{k+2-m}a_m\right) \tag{14.20}$$

For example[8], let $a_1(\theta)=\theta$, then iterating the recursion (14.20) produces the table of coefficients $a_k(\theta)$ as a function of θ

$$\begin{array}{lll}
a_0(\theta)=1 & a_5(\theta)=\frac{1}{4320\theta^3} & a_{10}(\theta)=\frac{-281}{151559100\theta^8}\\
a_1(\theta)=\theta & a_6(\theta)=\frac{1}{17010\theta^4} & a_{11}(\theta)=\frac{163879}{2172751257600\theta^9}\\
a_2(\theta)=\frac{1}{3} & a_7(\theta)=\frac{-139}{5443200\theta^5} & a_{12}(\theta)=\frac{-5221}{354648294000\theta^{10}}\\
a_3(\theta)=\frac{1}{36\theta} & a_8(\theta)=\frac{1}{204120\theta^6} & a_{13}(\theta)=\frac{5246819}{101684758855680000\theta^{11}}\\
a_4(\theta)=\frac{-1}{270\theta^2} & a_9(\theta)=\frac{-571}{2351462400\theta^7} & a_{14}(\theta)=\frac{5459}{7447614174000\theta^{12}}
\end{array}$$

For $a_1=\theta=-1$, only the coefficients with odd indices change sign. There is no simple formula for the general coefficient $a_n(\theta)$, but Watson (1929, p. 297) has shown, for large n, that

$$a_n(1)\sim(-1)^n\frac{\cos\frac{\pi}{4}(n+1)}{2^{n-\frac{3}{2}}n^{\frac{3}{2}}\pi^{\frac{n}{2}}}$$

The case $a_1=\theta=-1$ applies to the first integral in (14.18) with $u\in[0,1]$, which transforms to

$$\begin{aligned}
\int_0^1\left(ue^{1-u}\right)^x du &= \int_{\infty}^{0}\left(e^{-\frac{1}{2}r^2}\right)^x d\left(\sum_{k=0}^{\infty}a_k(-1)\,r^k\right)\\
&= -\sum_{k=1}^{\infty}ka_k(-1)\int_0^{\infty}e^{-\frac{1}{2}xr^2}r^{k-1}dr
\end{aligned}$$

The case $a_1=\theta=1$ holds for the second integral on $v\in[1,\infty)$, which becomes

$$\begin{aligned}
\int_1^{\infty}\left(ve^{1-v}\right)^x dv &= \int_0^{\infty}\left(e^{-\frac{1}{2}r^2}\right)^x d\left(\sum_{k=0}^{\infty}a_k(1)\,r^k\right)\\
&= \sum_{k=1}^{\infty}ka_k(1)\int_0^{\infty}e^{-\frac{1}{2}xr^2}r^{k-1}dr
\end{aligned}$$

After adding both, taken into account the sign change for odd indices,

[8] Copson (1965, p. 56) derives the same recursion (14.20) and gives the same coefficients (with $\theta=1$) up to index 5, i.e. up to $a_5=\frac{1}{4320}$.

i.e. $a_k(-1) = (-1)^k a_k(1)$ and $\int_0^\infty e^{-\frac{1}{2}xr^2} r^{2k} dr = \frac{1}{2}\Gamma\left(k+\frac{1}{2}\right)\left(\frac{2}{x}\right)^{k+\frac{1}{2}}$, Marsaglia and Marsaglia (1990) find

$$\Gamma(x+1) = x^{x+1}e^{-x}\left(\sum_{k=1}^{\infty} k\left\{(-1)^{k+1}+1\right\} a_k(1)\int_0^\infty e^{-\frac{1}{2}xr^2} r^{k-1} dr\right)$$
$$= x^{x+1}e^{-x}\left(\sum_{k=1}^{\infty}(2k-1)\,a_{2k-1}(1)\,\Gamma\left(k-\frac{1}{2}\right)\left(\frac{2}{x}\right)^{k-\frac{1}{2}}\right)$$

Finally, $\Gamma\left(k+\frac{1}{2}\right) = 2^{-2k}\sqrt{\pi}\frac{(2k)!}{k!}$ obtained from the duplication formula (11.9) gives Stirling's asymptotic expansion

$$\Gamma(x+1) = \sqrt{2\pi x}\left(\frac{x}{e}\right)^x\left(1+\sum_{m=1}^{\infty} a_{2m+1}(1)\frac{(2m+2)!}{2^{m+1}(m+1)!}\frac{1}{x^m}\right) \tag{14.21}$$

whose coefficients are *precisely* the same as in (14.17). In particular, it holds that $\widetilde{c}_m = a_{2m+1}(1)\frac{(2m+2)!}{2^{m+1}(m+1)!}$ for $m \geq 1$.

Copson (1965, p. 57) states (and sketches the idea of the proof) that the asymptotic expansion (14.21) holds for complex x with $|\arg x| < \pi$.

202. *Lanczos' approach.* The reflection formula (11.4) of the Gamma function, written as $\Pi(z)\Pi(-z) = \Gamma(1+z)\Gamma(1-z) = \frac{\pi z}{\sin \pi z}$, indicates that, if $\Pi(z) = \Gamma(z+1)$ is known for $\text{Re}(z) \geq 0$, any other value of $\Gamma(z)$ can be computed. Lanczos (1964) starts from Euler's integral (10.2) and prefers Gauss's notation over Legendre's,

$$\Pi(z) = \Gamma(z+1) = \int_0^\infty t^z e^{-t} dt \qquad \text{for } \text{Re}(z) \geq 0$$

which he transforms as follows. First and similar to **art.** 200, let $t = \alpha u$ for $\alpha > 0$, then

$$\Pi(z) = \alpha^{z+1}\int_0^\infty u^z e^{-\alpha u} du$$

and choose $\alpha = 1 + \rho z$,

$$\Pi(z) = (1+\rho z)^{z+1}\int_0^\infty \left(ue^{-\rho u}\right)^z e^{-u} du$$

Subsequently, Lanczos (1964) multiplies $(e\rho)^{-z}$ in front and its reciprocal after the integral,

$$\Pi(z) = (1+\rho z)^{z+1}(e\rho)^{-z}\int_0^\infty \left(\rho u e^{1-\rho u}\right)^z e^{-u} du$$

Let $v = e^{1-\rho u}$, then $\Pi(z) = \left(z+\frac{1}{\rho}\right)^{z+1} e^{-\left(z+\frac{1}{\rho}\right)}\int_0^e (v(1-\log v))^z v^{\frac{1}{\rho}-1} dv$ and replace $\rho = \frac{1}{q+1}$ and $z \to z - \frac{1}{2}$ to finally arrive at

$$\Pi\left(z-\frac{1}{2}\right) = \Gamma\left(z+\frac{1}{2}\right) = \left(z+\frac{1}{2}+q\right)^{z+\frac{1}{2}} e^{-\left(z+\frac{1}{2}+q\right)} L_q(z) \tag{14.22}$$

where $q \geq 0$ is a tunable parameter and where Lanczos' integral is

$$L_q(z) = \int_0^e (v(1-\log v))^{z-\frac{1}{2}} v^q dv \tag{14.23}$$

Lanczos (1964) investigates the integral (14.23) further. The prefactor in the asymptotic series $\Gamma(z+1) \simeq \sqrt{2\pi} z^{z+\frac{1}{2}} e^{-z}$ in (14.17) shows that (i) $\Gamma\left(z+\frac{1}{2}\right) \simeq \sqrt{2\pi}\left(z-\frac{1}{2}\right)^z e^{-\left(z-\frac{1}{2}\right)}$, while (ii) $\Gamma\left(z+\frac{1}{2}\right) = \frac{\Gamma\left(z+\frac{3}{2}\right)}{\left(z+\frac{1}{2}\right)} \simeq \sqrt{2\pi}\left(z+\frac{1}{2}\right)^z e^{-\left(z+\frac{1}{2}\right)}$. The case (i) in (14.22) suggests that $q=-1$ and $L_{-1}(z) = O\left(\frac{1}{\sqrt{z}}\right)$, while case (ii) in (14.22) favors $q=0$ and $L_0(z) = O\left(\frac{1}{\sqrt{z}}\right)$.

The function $g(v) = v(1-\log v)$, with $g'(v) = -\log v$, vanishes at $v=0$ and $v=e$, is positive within the integration interval $[0,e]$ and attains a maximum at $v=1$. Moreover, $g(v)$ has a branch cut at the negative real v-axis. The sequel of the paper of Lanczos (1964) is ingenious. Lanczos makes the substitution $1-x^2 = g(v) = v(1-\log v)$ in (14.23), where $v=0$ corresponds to $x=-1$, $v=1$ to $x=0$ and $v=e$ to $x=1$,

$$L_q(z) = \int_{-1}^1 \left(1-x^2\right)^{z-\frac{1}{2}} v^q \frac{dv}{dx} dx \tag{14.24}$$

Further, taking the differential of both sides $1-x^2 = v(1-\log v)$ yields $2xdx = \log v dv$ and

$$\frac{dv}{dx} = \frac{2x}{\log v} = \frac{2xv}{v-(1-x^2)}$$

which is rewritten as a first-order non-linear differential equation for the function $v(x)$

$$\frac{1}{2}\frac{dv^2}{dx} - \left(1-x^2\right)\frac{dv}{dx} - 2xv = 0 \tag{14.25}$$

with initial condition $v(0)=1$. Further $v(-1)=0$ and $v(1)=e$; we find that $\left.\frac{dv}{dx}\right|_{x=-1} = \left.\frac{2x}{\log v(x)}\right|_{x=-1} = 0$, while

$$\left.\frac{d^2v}{dx^2}\right|_{x=-1} = \left.\frac{2}{\log v}\left(1-\left(\frac{2x}{\log v}\right)^2\frac{1}{2v}\right)\right|_{x=-1} = -\infty$$

and all $\left.\frac{d^m v}{dx^m}\right|_{x=-1}$ are infinite for $m \geq 2$, due to the logarithmic singularity at $x=-1$. Hence, $v(x)$ is analytic in a disk around $x=0$ with radius $|x|<1$. Consequently, $v(x)$ possesses a Taylor series $v(x) = \sum_{m=0}^\infty v_m x^m$, whose Taylor coefficients are obtained from the differential equation, apart from $v_0 = 1$. After substitution of the Taylor series $v(x) = \sum_{m=0}^\infty v_m x^m$ into the differential equation (14.25), we have

$$\sum_{m=1}^\infty \left(\frac{m+1}{2}\sum_{k=1}^m v_{m+1-k}v_k\right)x^m + \sum_{m=2}^\infty (m-3)\, v_{m-1}x^m - 2x = 0$$

Equating corresponding powers in x yields, $v_1 = \sqrt{2}$ and, for $m \geq 2$,

$$\frac{m+1}{2}\sum_{k=1}^{m} v_{m+1-k}v_k + (m-3)\,v_{m-1} = 0$$

from which the recursion relation follows

$$v_m = -\frac{1}{(m+1)\,v_1}\left((m-3)\,v_{m-1} + \frac{m+1}{2}\sum_{k=2}^{m-1} v_{m+1-k}v_k\right)$$

We list the first few coefficients

$$\begin{array}{ll} v_0 = 1 & v_1 = \sqrt{2} \\ v_2 = \frac{1}{3} & v_3 = -\frac{\sqrt{2}}{36} \\ v_4 = \frac{2}{135} & v_5 = -\frac{23\sqrt{2}}{432160} \\ v_6 = \frac{38}{8505} & v_7 = -\frac{11237\sqrt{2}}{5443200} \\ v_8 = \frac{52}{25515} & v_9 = -\frac{2482411\sqrt{2}}{2351462400} \end{array}$$

The coefficients $|v_m|$ decrease monotonously with m; the even $v_{2m} > 0$ are positive, while the odd coefficients $v_{2m+1} < 0$ are negative and $\frac{v_{2m+1}}{\sqrt{2}} \in \mathbb{Q}$. After introducing the Taylor series into Lanczos' integral (14.24) for $q = 0$, we obtain

$$L_0(z) = \int_{-1}^{1} \left(1-x^2\right)^{z-\frac{1}{2}} \frac{dv}{dx}dx = \sum_{m=1}^{\infty} m v_m \int_{-1}^{1} \left(1-x^2\right)^{z-\frac{1}{2}} x^{m-1}dx$$

Since

$$\int_{-1}^{1} \left(1-x^2\right)^{z-\frac{1}{2}} x^{m-1}dx = \left(1+(-1)^{m-1}\right)\int_{0}^{1} \left(1-x^2\right)^{z-\frac{1}{2}} x^{m-1}dx$$

and $\int_0^1 \left(1-x^2\right)^{z-\frac{1}{2}} x^{m-1}dx = \frac{1}{2}\int_0^1 (1-t)^{z-\frac{1}{2}}\, t^{\frac{m}{2}-1}dt$, invoking the Beta-function integral (10.11),

$$\int_{-1}^{1} \left(1-x^2\right)^{z-\frac{1}{2}} x^{m-1}dx = \frac{\left(1+(-1)^{m-1}\right)}{2}\frac{\Gamma\left(z+\frac{1}{2}\right)\Gamma\left(\frac{m}{2}\right)}{\Gamma\left(z+\frac{1+m}{2}\right)}$$

illustrating that the integral vanishes for even m, we obtain

$$L_0(z) = \sum_{m=0}^{\infty} (2m+1)\,v_{2m+1}\frac{\Gamma\left(z+\frac{1}{2}\right)\Gamma\left(m+\frac{1}{2}\right)}{\Gamma(z+m+1)}$$

Via $\Gamma\left(z+\frac{1}{2}\right) = \left(z+\frac{1}{2}\right)^{z+\frac{1}{2}} e^{-\left(z+\frac{1}{2}\right)} L_0(z)$ in (14.22), we return to the Gamma function

$$\Gamma\left(z+\frac{1}{2}\right) = \left(z+\frac{1}{2}\right)^{z+\frac{1}{2}} e^{-\left(z+\frac{1}{2}\right)} \sum_{m=0}^{\infty} (2m+1)\,v_{2m+1}\frac{\Gamma\left(z+\frac{1}{2}\right)\Gamma\left(m+\frac{1}{2}\right)}{\Gamma(z+m+1)}$$

Finally, after multiplying both sides by $\frac{\Gamma(z+1)}{\Gamma\left(z+\frac{1}{2}\right)}$, substituting $\frac{\Gamma\left(m+\frac{1}{2}\right)}{\sqrt{\pi}} =$

$\frac{(2m)!}{2^{2m}m!}$ from the duplication formula (11.9) and with $\frac{\Gamma(z+m)}{\Gamma(z)} = \prod_{k=0}^{m-1}(z+k)$, we arrive at Lanczos' converging factorial series

$$\Gamma(z+1) = \sqrt{2\pi}\frac{\left(z+\frac{1}{2}\right)^{z+\frac{1}{2}}}{e^{\left(z+\frac{1}{2}\right)}}\left(1+\sum_{m=1}^{\infty}\frac{(2m+1)!}{2^{2m}m!}\frac{v_{2m+1}}{\sqrt{2}}\frac{1}{\prod_{k=1}^{m}(z+k)}\right) \tag{14.26}$$

which is a converging counterpart of Stirling's asymptotic and divergent series (14.17). Explicitly, up to $m=5$, Lanczos' factorial series is

$$\Gamma(z+1) = \sqrt{2\pi}e^{-z-\frac{1}{2}}\left(z+\frac{1}{2}\right)^{z+\frac{1}{2}}\left(1-\frac{\frac{1}{24}}{(z+1)}-\frac{\frac{23}{1152}}{(z+1)(z+2)}\right.$$
$$-\frac{\frac{11237}{414720}}{(z+1)(z+2)(z+3)}-\frac{\frac{2482411}{39813120}}{(z+1)(z+2)(z+3)(z+4)}$$
$$\left.-\frac{\frac{272785979}{1337720832}}{(z+1)(z+2)(z+3)(z+4)(z+5)}+\cdots\right)$$

and illustrates that each term in the m-sum in (14.26) is negative. Although Lanczos mentions that his factorial series (14.26) converges "too slowly to be of practical value", our numerical computations contradict Lanczos' statement for large z. If z is small, say $z=1$, then $K=10$ terms give a 4 digits accuracy, but $K=30$ only 5 digits. Hence, for small z, Lanczos' converging factorial series (14.26) indeed converges rather slowly. We may consider other, integral values of q in $L_q(z)$, which will lead to different converging factorial series, similarly as our discovery of infinitely many factorial series for Binet's function $\mu(z)$ in Van Mieghem (2021). In fact, exponentiation of a factorial series of $\mu(z)$ is expected to lead to a Lanczos factorial series for $\Gamma(z)$.

In the remainder of his paper, Lanczos (1964, p. 92) concentrates on devising numerically better series for $L_q(z)$ by replacing the Taylor series, which is extrapolating, by a Fourier series in which Chebyshev's polynomials appear, that are interpolating.

203. *The factorial series of Rubinstein.* The two related papers, Rubinstein (2012) and Rubinstein (2013), illustrate Rubinstein's basic idea, consisting of the powerful substitution $e^{-x} = 1-u$ of Binet in Mellin-type integrals, followed by Taylor expansion. **Art.** 182 illustrates how Binet's recipe of five steps leads to nice, convergent factorial series for the Binet function $\mu(z)$. Instead of copying the derivation of Rubinstein, we apply our more general method for a Mellin transform in Appendix A.5.

Rubinstein (2013) performs Binet's substitution $x = -\log(1-u)$ in Euler's, slightly generalized integral $\Gamma(z)p^{-z} = \int_0^\infty e^{-px}x^{z-1}dx$ in (15.9), valid for $\text{Re}(p) > 0$ and $\text{Re}(z) > 0$, and obtains

$$\Gamma(z)p^{-z} = \int_0^1 (1-u)^{p-1}(-\log(1-u))^{z-1}du$$

which was, for $p = 1$, Euler's first integral for $\Gamma(z)$, as mentioned in Section 9.4.

Since $f(x) = e^{-px}$ in the Mellin transform (B.3), the general series (A.34) in Appendix A.5 reduces to

$$\Gamma(z) = p^z \sum_{m=0}^{\infty} \frac{\mathbb{S}_m(z+m-1)}{m!} \frac{\Gamma(p+1)\Gamma(z+m)}{\Gamma(p+z+m+1)} \tag{14.27}$$

Explicitly from (A.36) for integer $p = N$, (14.27) is recognized as a factorial series

$$\Gamma(z) = \frac{N^z \Gamma(N+1)}{\prod_{l=0}^{N}(z+l)} \left(1 + \sum_{m=1}^{\infty} \sum_{r=1}^{m} \frac{\frac{(-1)^{r+m} \mathcal{S}_{m+r}^{(r)}}{(m+r)!(m-r)!} \prod_{l=0}^{2m}(z+l)}{(z+m+r) \prod_{l=N+1}^{N+m}(z+l)}\right) \tag{14.28}$$

Rubinstein (2013) defines the generating function $\left(\frac{-\log(1-t)}{t}\right)^{x-1} = \sum_{m=0}^{\infty} \alpha_m(x) t^m$, where his coefficients $\alpha_m(z)$ are unfortunately different from the standard Stirling polynomial $\mathbb{S}_m(x)$ of degree m in x in Section A.2. Although Rubinstein (2013) concentrates most on the Riemann and Hurwitz Zeta-function (**art.** 156), he shows, for $\text{Re}(p) > 0$, that (14.27) converges uniformly for all z and that (14.27) is an analytic continuation of $\Gamma(z)$ to all complex z, except for its poles. Rubinstein (2012) also presents the factorial series for the Riemann Zeta-function integral $\Gamma(z)\zeta(z) = \int_0^\infty \frac{x^{z-1}}{e^x-1}dx$, valid for $\text{Re}(z) > 1$. Since $f(x) = \frac{e^{-x}}{1-e^{-x}}$ in the Mellin transform (B.3), we find from (A.33) that

$$\Gamma(z)\zeta(z+1) = \sum_{m=0}^{\infty} \frac{\mathbb{S}_m(z+m-1)}{m!(z+m)^2} \tag{14.29}$$

which, compared to (14.27) with $p = 1$,

$$\Gamma(z) = \sum_{m=0}^{\infty} \frac{\mathbb{S}_m(z+m-1)}{m!(1+z+m)(z+m)}$$

only differs by a factor $(z+m)$ versus $(z+1+m)$ in the factorial series of $\Gamma(z)$. These two series converge very slowly: $K = 2000$ terms in (14.29) barely give 2 correct digits for $z = 2$, while $K = 1500$ terms give 4 correct digits for $z = 1$!

The series (14.27) converges slowly for small z and p, but increasingly fast for large p. Rubinstein (2013, Section 3) shows that, for large N in (14.28), the m-sum is dominated by the first term and that

$$\Gamma(z) = \lim_{N\to\infty} \frac{N^z N!}{\prod_{l=0}^{N}(z+l)}$$

is precisely Gauss's limit (10.8). Hence, the converging factorial series (14.28), valid for any z except at the poles, is, for sufficiently large N, valuable to compute the Gamma function at large z.

15
The entire function $\frac{1}{\Gamma(z)}$

From a function theoretical point, the entire function $\frac{1}{\Gamma(z)}$ is more interesting than the Gamma function $\Gamma(z)$ itself: both the Hankel integral for $\frac{1}{\Gamma(z)}$, studied in Section 15.1, and the Taylor series around any point z_0 hold for any complex number z. Hankel's contour integral (15.1) generalizes Euler's integral $\Gamma(z) = \int_0^\infty e^{-t}t^{z-1}dt$ to all complex z. Section 15.2 studies different ways to deduce the Taylor coefficients c_k around $z_0 = 0$ of the entire function $\frac{1}{\Gamma(z)} = \sum_{k=0}^{\infty} c_k z^k$. The order ρ of the entire function $\frac{1}{\Gamma(z)}$ is one, computed from the definition in Section 1.4.3 as $\rho = \lim_{z\to\infty} \frac{\log(\log|\Gamma(z)|)}{\log z} = 1$. Indeed, the bounds (13.14) in **art.**174 show, for large z, that $\log(\log|\Gamma(z)|) = \log\left(z - \frac{1}{2}\right) + \log\log z + O\left(\frac{1}{\log z}\right)$.

15.1 The Hankel integral for $\frac{1}{\Gamma(z)}$

The complex integral due to Hankel is a cornerstone integral in complex function theory, that is further generalized for Mellin integrals and transforms (**art.** 245). The integration along Hankel's particular path is characteristic: Hankel's path follows the utmost boundaries – here around the branch cut of the function z^a, which is the negative z-axis – of a region, where the integrand is analytic. On the boundary of an analytic region the largest and smallest absolute values of the analytic function inside that region are achieved by the harmonic function principle or the Cauchy-Riemann equations (Titchmarsh, 1964, p. 67).

204. *Hankel's integral.* We derive Hankel's integral

$$\frac{1}{\Gamma(z)} = \frac{1}{2\pi i}\int_C w^{-z}e^w dw \tag{15.1}$$

where the contour C starts at $-\infty$ below the real axis, encircles the origin at $z = 0$ and returns above the negative real axis again to $-\infty$. Such a contour around the branch-cut (here the negative real axis) is "classical" in integrals containing w^α such as Mellin transforms (Titchmarsh, 1948). If ε is the radius of a circle at the origin, then evaluation of the integral

(15.1) along the contour C yields

$$\begin{aligned}\int_C w^{-z} e^w dw &= -\int_\infty^\varepsilon \left(xe^{-i\pi}\right)^{-z} e^{-x} dx + i\int_{-\pi}^{\pi} \left(\varepsilon e^{i\theta}\right)^{-z} e^{\varepsilon e^{i\theta}} \varepsilon e^{i\theta} d\theta \\ &\quad - \int_\varepsilon^\infty \left(xe^{i\pi}\right)^{-z} e^{-x} dx \\ &= \left(e^{i\pi z} - e^{-i\pi z}\right)\int_\varepsilon^\infty x^{-z} e^{-x} dx + i\varepsilon^{1-z}\int_{-\pi}^{\pi} e^{i\theta(1-z)} e^{\varepsilon e^{i\theta}} d\theta\end{aligned}$$

If $\operatorname{Re}(1-z) > 0$, then we take the limit $\varepsilon \to 0$ and find with Euler's integral (10.2)

$$\int_C w^{-z} e^w dw = 2i \sin \pi z \int_0^\infty x^{1-z-1} e^{-x} dx = 2i \sin \pi z \Gamma(1-z)$$

After replacing z by $1-z$, we obtain a contour integral for the Gamma function,

$$\Gamma(z) = \frac{\pi}{\sin \pi z} \frac{1}{2\pi i} \int_C w^{z-1} e^w dw \tag{15.2}$$

The reflection formula $\Gamma(z)\Gamma(1-z) = \frac{\pi}{\sin \pi z}$ in (11.4) of the Gamma function leads to Hankel's contour integral (15.1). Although the derivation was restricted to $\operatorname{Re}(z) > 0$ in (15.2), by analytic continuation (see e.g. Titchmarsh (1964, Chapter IV), Evgrafov (2019, Chapter III)), (15.1) as well as (15.2) hold for all $z \in \mathbb{C}$.

205. *Hankel's integral at integer values of* z. For negative integer values of $z = -n$ with $n \geq 0$, the branch cut of w^z disappears and the contour C in (15.2) encloses an analytic region with a pole of multiplicity $n+1$ at $w = 0$. Cauchy's integral (1.14) then states that

$$\frac{1}{2\pi i} \int_C \frac{e^w}{w^{n+1}} dw = \frac{1}{n!} \left.\frac{d^n e^w}{dw^n}\right|_{w=0} = \frac{1}{n!}$$

Hence, the contour integral (15.2) for the Gamma function demonstrates that $\Gamma(z)$ has simple poles at $z = -n$ for each integer $n \geq 0$ due to $\sin \pi z$, in agreement with Gauss's product (10.10).

If $z = n \geq 1$, then $\frac{1}{2\pi i}\int_C w^{n-1} e^w dw = 0$ as well as $\sin \pi n = 0$. Applying the rule of de l'Hospital,

$$\lim_{z \to n} \Gamma(z) = \lim_{z \to n} \frac{\pi \frac{1}{2\pi i}\int_C w^{z-1} e^w dw}{\sin \pi z} = \lim_{z \to n} \frac{\frac{1}{2\pi i}\int_C w^{z-1} (\log w) e^w dw}{\cos \pi z}$$

equivalent to

$$\Gamma(n) = \frac{(-1)^n}{2\pi i} \int_C w^{n-1} (\log w) e^w dw$$

leads, again following Hankel's contour due to the branch cut caused by

$\log w$, to

$$\begin{aligned}\mathbb{H} &= \int_C w^{n-1} (\log w) e^w dw = (-1)^n \left(\int_\infty^\varepsilon x^{n-1} \log\left(xe^{-i\pi}\right) \frac{dx}{e^x} \right. \\ &\quad \left. + \int_\varepsilon^\infty x^{n-1} \log\left(xe^{i\pi}\right) \frac{dx}{e^x} \right) + i \int_{-\pi}^{\pi} \left(\varepsilon e^{i\theta}\right)^{n-1} \log\left(\varepsilon e^{i\theta}\right) e^{\varepsilon e^{i\theta}} \varepsilon e^{i\theta} d\theta \\ &= 2i\pi (-1)^n \int_\varepsilon^\infty x^{n-1} e^{-x} dx + i\varepsilon^n \int_{-\pi}^{\pi} e^{i\theta n} (\log \varepsilon + i\theta) e^{\varepsilon e^{i\theta}} d\theta\end{aligned}$$

After the limit $\varepsilon \to 0$, we arrive at the familiar $\Gamma(n) = \int_0^\infty x^{n-1} e^{-x} dx$.

206. *Deforming the Hankel contour.* We can deform[1] the contour C into C_ϕ by tilting the straight line above the negative real axis over an angle ϕ and the straight line below the negative real axis over an angle $-\phi$. Those tilted straight half-lines intersect the real w-axis at a point $c \geq \rho > 0$, where ρ is the radius of the small circle around the origin $w = 0$.

Indeed, consider the contour L consisting of the contour C, the circle segment at infinity from the angle π to the angle ϕ, followed by the line $w = re^{i\phi}$, where r ranges from infinity towards ρ, the circle centered at $w = 0$ with radius ρ turning from angle ϕ towards $-\phi$ and complemented by the line $w = re^{-i\phi}$ and the infinite circle segment towards the begin of the contour C. The integral $\int_L w^{-z} e^w dw = 0$, because the contour L encloses an analytic region of the function $w^{-z} e^w$. If $\phi > \frac{\pi}{2}$, then the part of L along the circle segment at infinity, $\lim_{r\to\infty} ir^{1-z} \int_\pi^\phi e^{(1-z)i\theta} e^{re^{i\theta}} d\theta = 0$ for all z, because $\left|e^{re^{i\theta}}\right| = e^{r\cos\theta} \to 0$ for any $\pi \geq \theta \geq \phi > \frac{\pi}{2}$. Hence, combining all contributions results in

$$\frac{1}{\Gamma(z)} = \frac{1}{2\pi i} \int_{C_\phi} w^{-z} e^w dw \tag{15.3}$$

where the contour C_ϕ starts at infinity on the straight line at the angle $-\frac{\pi}{2} > \phi > -\pi$ below the real axis until the circle at the origin with radius ρ that turns over the angle $-\phi$ to ϕ until hitting the line $w = re^{i\phi}$ along which it passes towards infinity again. The Hankel contour in (15.1) is the particular case where $C = C_\pi$ and $\rho = \varepsilon$, where, in contrast to C_ϕ with $\phi < \pi$, the small circle with radius ε around the origin $w = 0$ still plays a role.

Hankel's integral in (15.3) can be written as

$$\frac{1}{\Gamma(z)} = \frac{1}{2\pi i} \int_{c-\infty e^{-i\phi}}^{c+\infty e^{i\phi}} w^{-z} e^w dw \qquad \text{with } c > 0 \text{ and } \frac{\pi}{2} < \phi < \pi \tag{15.4}$$

Since multiplying $c' = ac > 0$ provided $a > 0$, the map $w \to aw$ for any

[1] The same deformation holds for the contour integral (15.2) as well.

real, positive number a yields

$$\frac{a^{z-1}}{\Gamma(z)} = \frac{1}{2\pi i}\int_{c'-\infty e^{-i\phi}}^{c'+\infty e^{i\phi}} w^{-z} e^{aw}\,dw \qquad \text{with } c' > 0 \text{ and } \frac{\pi}{2} < \phi < \pi \tag{15.5}$$

207. *Laplace's integral for* $\frac{1}{\Gamma(z)}$. We evaluate the integral in (15.4) along the two straight half-lines with angle $\frac{\pi}{2} < \phi < \pi$ and $-\pi < -\phi < -\frac{\pi}{2}$, respectively,

$$\begin{aligned}\frac{1}{\Gamma(z)} &= \frac{e^{i\phi}}{2\pi i}\int_0^\infty \frac{e^{\left(c+re^{i\phi}\right)}}{\left(c+re^{i\phi}\right)^z}dr - \frac{e^{-i\phi}}{2\pi i}\int_0^\infty \frac{e^{\left(c+re^{-i\phi}\right)}}{\left(c+re^{-i\phi}\right)^z}dr \\ &= \frac{e^c}{2\pi i}\int_0^\infty e^{r\cos\phi}\left\{\frac{e^{i(\phi+r\sin\phi)}}{\left(c+re^{i\phi}\right)^z} - \frac{e^{-i(\phi+r\sin\phi)}}{\left(c+re^{-i\phi}\right)^z}\right\}dr\end{aligned}$$

Hence, for any $c > 0$, any z and $\frac{\pi}{2} < \phi < \pi$, we find that

$$\frac{1}{\Gamma(z)} = \frac{e^c}{\pi}\int_0^\infty e^{r\cos\phi}\frac{\sin\left(r\sin\phi + \phi - z\arctan\frac{r\sin\phi}{c+r\cos\phi}\right)}{\left(c^2+2rc\cos\phi+r^2\right)^{\frac{z}{2}}}dr \tag{15.6}$$

We emphasize that the angle $\phi < \pi$ in (15.6), (15.4) and (15.5), because, as mentioned in **art.** 206, the integral $i\varepsilon^{1-z}\int_{-\pi}^{\pi} e^{i\theta(1-z)}e^{\varepsilon e^{i\theta}}d\theta$ in Hankel's contour integral was omitted, which is only justified if $\varepsilon \to 0$ and $\text{Re}(z) < 1$. Indeed, if $\phi \to \pi$, then $c \to 0$ and (15.6) simplifies to

$$\frac{1}{\Gamma(z)} = \frac{\sin(z\pi)}{\pi}\int_0^\infty e^{-r}r^{-z}dr$$

and Euler's integral (10.2) additionally requires that $\text{Re}(z) < 1$. If $c \neq 0$ in (15.6) while $\phi = \pi$, then the right-hand side erroneously vanishes!

The integral (15.6) is difficult to simplify further, which may suggest that another contour[2] for C_ϕ might be more suitable. If $\phi = \frac{\pi}{2}$ in (15.6) and $c > 0$, then

$$\frac{1}{\Gamma(z)} = \frac{e^c}{\pi}\int_0^\infty \frac{\cos\left(r - z\arctan\frac{r}{c}\right)}{\left(c^2+r^2\right)^{\frac{z}{2}}}dr \tag{15.7}$$

but this integral only converges[3] for $\text{Re}(z) > 0$. Hence, only for $\text{Re}(z) > 0$,

[2] For example, a parabola $w = (c+it)^2$ with $c > 0$ and focus at the origin $w = 0$, as studied in Markushevich (1985, Vol. I, p. 134), maps (15.3) into

$$\frac{1}{\Gamma(z)} = \frac{1}{\pi}\int_{-\infty}^\infty (c+it)^{1-2z}e^{(c+it)^2}dt = \frac{e^{c^2}}{\pi}\int_{-\infty}^\infty (c+it)^{1-2z}e^{2ict}e^{-t^2}dt$$

[3] Partial integration of (15.8)

$$\frac{1}{\Gamma(z)} = \lim_{T\to\infty}\left\{(c+iT)^{-z}e^{c+iT} - (c-iT)^{-z}e^{c-iT}\right\} + \frac{z}{2\pi i}\int_{c-i\infty}^{c+i\infty} w^{-(z+1)}e^w dw$$

demonstrates that the requirement that $\text{Re}(z) > 0$ for the limits to vanish.

Hankel's integral (15.4) reduces to an integration along a line parallel to the imaginary axis, with $c > 0$,

$$\frac{1}{\Gamma(z)} = \frac{1}{2\pi i}\int_{c-i\infty}^{c+i\infty} w^{-z} e^{w} dw = \frac{e^{c}}{2\pi}\int_{-\infty}^{\infty} (c+it)^{-z}\, e^{it} dt \tag{15.8}$$

The integral (15.8) appeared[4] in Laplace (1814, p. 133). With the reflection formula (11.4) and replacing $z \to 1-z$, (15.8) becomes, with $c > 0$ and $\mathrm{Re}\,(z) < 1$,

$$\Gamma(z) = \frac{1}{2i\sin \pi z}\int_{c-i\infty}^{c+i\infty} w^{z-1} e^{w} dw = \frac{e^{c}}{2\sin \pi z}\int_{-\infty}^{\infty} (c+it)^{z-1}\, e^{it} dt$$

208. *Alternative derivations of the Hankel integral.* We present two alternative, short and elegant derivations of the Hankel-Laplace integral (15.8).

First, Euler's integral (10.2) becomes, after letting $t = px$ with $\mathrm{Re}\,(p) > 0$ and $\mathrm{Re}\,(z) > 0$,

$$\int_{0}^{\infty} e^{-px} x^{z-1} dx = \frac{\Gamma(z)}{p^{z}} \tag{15.9}$$

which is a Laplace transform (B.1) in the complex variable p of the function $f(x) = x^{z-1}$, converging for $\mathrm{Re}\,(p) = c > 0$. The inverse Laplace transform (B.2) is, for $c > 0$,

$$x^{z-1} = \frac{1}{2\pi i}\int_{c-i\infty}^{c+i\infty} \frac{\Gamma(z)}{p^{z}} e^{px} dp$$

The rewriting

$$\frac{x^{z-1}}{\Gamma(z)} = \frac{1}{2\pi i}\int_{c-i\infty}^{c+i\infty} w^{-z} e^{wx} dw$$

is an instance of (15.5) for the angle $\phi = \frac{\pi}{2}$ that limits the validity to $\mathrm{Re}\,(z) > 0$ and reduces, for $x = 1$, precisely to the Hankel-Laplace integral (15.8).

---

[4] Laplace (1814, Chapitre III) starts the investigation of Euler's integral (10.2) of the Gamma function $\Gamma(z)$, derives via a kind of the saddle-point method the Stirling expansion (14.17) with the coefficients $c_1 = \frac{1}{12}$ and $c_2 = \frac{1}{288}$ on p. 129. He then continues from (14.17) to derive the reflection formula (11.4) on p. 131 by ingenious manipulations and using the infinite product (11.2) of $\sin \pi z$. Next, he concentrates on the integral $\int \frac{e^{-x}}{x^{\mu}} dx$, makes an imaginary substitution $x = -\mu + \sqrt{-1}w$ (before Cauchy's integral theorem was known!), reasons that the limits must be from $-\infty$ to ∞ and arrives, in our modern notation to (15.8) on p. 133 (rather than p. 134 as mentioned in Whittaker and Watson (1996, p. 246)). Laplace (1814, p. 134) discusses the "Hankel-reflection type" relation

$$\int_{a}^{b} \frac{e^{-x}}{x^{\mu}} dx = \frac{2\pi\,(-1)^{\mu+\frac{1}{2}}}{\int_{0}^{\infty} x^{\mu-1} e^{-x} dx}$$

where a and b are imaginary numbers of x satisfying $\frac{e^{-x}}{x^{\mu}} = 0$, which (I assume) implies that $a = c' - i\infty$ and $b = c' + i\infty$.

Second, the Taylor coefficient (1.14) of the exponential function in integral form is, for integer $k > 0$,

$$\frac{1}{\Gamma(k)} = \frac{1}{2\pi i}\int_{C(0)} \frac{e^{\omega}}{\omega^k} d\omega$$

The contour $C(0)$ around the origin $\omega = 0$ can be deformed to enclose the entire $\text{Re}(\omega) < 0$-plane, because $\lim_{|\omega|\to\infty} e^{|\omega|e^{i\phi}} = 0$ for $\frac{\pi}{2} < |\phi| < \pi$, resulting, for $c > 0$, in

$$\frac{1}{\Gamma(k)} = \frac{1}{2\pi i}\int_{c-\infty e^{-i\phi}}^{c+\infty e^{i\phi}} \omega^{-k} e^{\omega} d\omega$$

Finally, as explained in Section 1.3, the "fractional extension" from integer k to complex number z requires us to exclude the branch cut on the negative real ω-axis, which results in Hankel's contour integral (15.1).

209. *Properties of the Hankel-Laplace integral (15.8) for* $\frac{1}{\Gamma(z)}$. Since $\frac{1}{2\pi i}\int_{c-i\infty}^{c+i\infty} w^{-z}e^{-w}dw = 0$ for $\text{Re}(z) > 0$ and $c > 0$, because the contour can be closed over the positive $\text{Re}(w)$-plane, the companion integral of (15.8)

$$\int_{-\infty}^{\infty} (c+it)^{-z}\, e^{-it} dt = 0 \qquad \text{with } c > 0 \text{ and } \text{Re}(z) > 0 \tag{15.10}$$

Only the signs of the argument in the exponential e^{it} in (15.8) and e^{-it} in (15.10) differ!

Laplace's integral (15.8) is written[5] as

$$\begin{aligned}\frac{1}{\Gamma(z)} &= \frac{e^c}{2\pi}\int_{-\infty}^{\infty} \left(c^2+t^2\right)^{-\frac{z}{2}} e^{i\left(t - z \arctan \frac{t}{c}\right)} dt \\ &= \frac{e^c}{2\pi}\int_{-\infty}^{\infty} \frac{\cos\left(t - z\arctan\frac{t}{c}\right)}{\left(c^2+t^2\right)^{\frac{z}{2}}} dt + \frac{ie^c}{2\pi}\int_{-\infty}^{\infty} \frac{\sin\left(t - z\arctan\frac{t}{c}\right)}{\left(c^2+t^2\right)^{\frac{z}{2}}} dt\end{aligned}$$

while its companion $0 = \int_{-\infty}^{\infty}\left(c^2+t^2\right)^{-\frac{z}{2}} e^{-i\left(t + z\arctan\frac{t}{c}\right)} dt$ in (15.10) shows that

$$\int_{-\infty}^{\infty} \frac{\cos(t)\cos\left(z\arctan\frac{t}{c}\right)}{\left(c^2+t^2\right)^{\frac{z}{2}}} dt = \int_{-\infty}^{\infty} \frac{\sin(t)\sin\left(z\arctan\frac{t}{c}\right)}{\left(c^2+t^2\right)^{\frac{z}{2}}} dt$$

Since $\frac{1}{\Gamma(z)}$ is real on the real z-axis, the imaginary term must be zero illustrating[6] that

$$\int_{-\infty}^{\infty} \frac{\sin\left(t \pm z\arctan\frac{t}{c}\right)}{\left(c^2+t^2\right)^{\frac{z}{2}}} dt = 0$$

implying that

$$\int_{-\infty}^{\infty} \frac{\sin(t)\cos\left(z\arctan\frac{t}{c}\right)}{\left(c^2+t^2\right)^{\frac{z}{2}}} dt = \int_{-\infty}^{\infty} \frac{\cos(t)\sin\left(z\arctan\frac{t}{c}\right)}{\left(c^2+t^2\right)^{\frac{z}{2}}} dt = 0$$

[5] Laplace (1814, p. 133) actually writes a variant.

[6] Also, an odd integrand integrated over an even interval results in a zero integral.

but $\int_0^\infty \frac{\sin(t)\cos\left(z\arctan\frac{t}{c}\right)}{(c^2+t^2)^{\frac{z}{2}}}dt \neq \int_0^\infty \frac{\cos(t)\sin\left(z\arctan\frac{t}{c}\right)}{(c^2+t^2)^{\frac{z}{2}}}dt$.

In summary, for $\text{Re}(z) > 0$ and $c > 0$, the Hankel-Laplace integral (15.8) and its companion (15.10) satisfy[7]

$$\frac{\pi e^{-c}c^{z-1}}{\Gamma(z)} = \int_0^\infty \frac{\cos(cx - z\arctan x)}{(1+x^2)^{\frac{z}{2}}}dx = \int_0^\infty \frac{\cos(cx)\cos(z\arctan x)}{(1+x^2)^{\frac{z}{2}}}dx \tag{15.11}$$

$$0 = \int_0^\infty \frac{\cos(cx + z\arctan x)}{(1+x^2)^{\frac{z}{2}}}dx \tag{15.12}$$

15.2 Taylor coefficients around $z_0 = 0$ of the entire function $\frac{1}{\Gamma(z)}$

Although the Gamma function plays a prominent role in complex analysis, the Taylor series

$$\frac{1}{\Gamma(z)} = \sum_{k=0}^{\infty} c_k z^k \qquad |z| < \infty \tag{15.13}$$

is less studied and its Taylor coefficients c_k are only *numerically* listed in the literature (Abramowitz and Stegun, 1968, p. 256). Since $\lim_{z\to 0}\frac{1}{\Gamma(z)} = \frac{z}{\Gamma(z+1)} = 0$ and $\lim_{z\to 0}\frac{1}{\Gamma(z+1)} = 1$, it holds that $c_0 = 0$, $c_1 = 1$ and $c_2 = \gamma$. The first 40 Taylor coefficients $\{c_k\}_{k\leq 40}$ up to 20 digits accurate are computed from (15.35) below with Mathematica:

$c_3 = -6.5587807152025388108\ 10^{-1}$	$c_{22} = 5.1003702874544759790\ 10^{-13}$
$c_4 = -4.2002635034095235529\ 10^{-2}$	$c_{23} = -2.0583260535665067832\ 10^{-14}$
$c_5 = 1.6653861138229148950\ 10^{-1}$	$c_{24} = -5.3481225394230179824\ 10^{-15}$
$c_6 = -4.2197734555544336748\ 10^{-2}$	$c_{25} = 1.2267786282382607902\ 10^{-15}$
$c_7 = -9.6219715278769735621\ 10^{-3}$	$c_{26} = -1.1812593016974587695\ 10^{-16}$
$c_8 = 7.2189432466630995424\ 10^{-3}$	$c_{27} = 1.1866922547516003326\ 10^{-18}$
$c_9 = -1.1651675918590651121\ 10^{-3}$	$c_{28} = 1.4123806553180317816\ 10^{-18}$
$c_{10} = -2.1524167411495097282\ 10^{-4}$	$c_{29} = -2.2987456844353702066\ 10^{-19}$
$c_{11} = 1.2805028238811618615\ 10^{-4}$	$c_{30} = 1.7144063219273374334\ 10^{-20}$
$c_{12} = -2.0134854780788238656\ 10^{-5}$	$c_{31} = 1.3373517304936931148\ 10^{-22}$
$c_{13} = -1.2504934821426706573\ 10^{-6}$	$c_{32} = -2.0542335517666727893\ 10^{-22}$
$c_{14} = 1.1330272319816958824\ 10^{-6}$	$c_{33} = 2.7360300486079998448\ 10^{-23}$
$c_{15} = -2.0563384169776071035\ 10^{-7}$	$c_{34} = -1.7323564459105166391\ 10^{-24}$
$c_{16} = 6.1160951044814158179\ 10^{-9}$	$c_{35} = -2.3606190244992872873\ 10^{-26}$
$c_{17} = 5.0020076444692229301\ 10^{-9}$	$c_{36} = 1.8649829417172944307\ 10^{-26}$
$c_{18} = -1.1812745704870201446\ 10^{-9}$	$c_{37} = -2.2180956242071972044\ 10^{-27}$
$c_{19} = 1.0434267116911005105\ 10^{-10}$	$c_{38} = 1.2977819749479936688\ 10^{-28}$
$c_{20} = 7.7822634399050712540\ 10^{-12}$	$c_{39} = 1.1806974749665284062\ 10^{-30}$
$c_{21} = -3.6968056186422057082\ 10^{-12}$	$c_{40} = -1.1245843492770880903\ 10^{-30}$

The table of the Taylor coefficients $\{c_k\}_{k\leq 40}$ shows, similarly to the Taylor coefficients of the Riemann Zeta-function $\zeta(z)$ around $z_0 = 1$, an irregular structure. The absolute value $|c_k|$ decreases from $7 \leq k \leq 27$, but $c_{28} > c_{27}$ and $|c_{32}| > c_{31}$. A plot of $\frac{-\log|c_k|}{k\log k}$ versus k shows, from $k = 5$ up to $k = 40$, a slow, concave increase in k from 0.22 to 0.49. Since $\frac{1}{\Gamma(z)}$ is an entire function of order $\rho = 1$, the ratio $\frac{-\log|c_k|}{k\log k}$ must eventually for

[7] Similar integrals as (15.11) and (15.12) are given in Nielsen (1956, p. 156).

$k \to \infty$ tend to 1 (Section 1.4.3), but its convergence towards 1 is very slowly. Thus, the Taylor series (15.13) of $\frac{1}{\Gamma(z)}$ needs more terms than that of e^{-z} to achieve a same accuracy for a given z.

After recalling recursions for c_k in **art.** 210, this Section 15.2 reviews several approaches to compute the Taylor coefficients c_k of $\frac{1}{\Gamma(z)}$ in (15.13): Bourguet's approach in **art.** 211, our closed integral form (15.17) in **art.** 212, Fehik-Ahmed's asymptotic approximation in **art.** 215 and Ulgenes's approach based on differences in **art.** 216. Finally, we determine in **art.** 217-218 the Taylor coefficients c_k exactly as well as their asymptotic behavior in (15.36), from Weierstrass's product $\frac{1}{\Gamma(z+1)} = e^{\gamma z} \prod_{n=1}^{\infty} \left(1 + \frac{z}{n}\right) e^{-z/n}$ in (11.1) based on characteristic coefficients (Appendix A).

210. *Recursions for the Taylor coefficient* c_k. Several recursion relations can be derived.

The identity $\frac{d}{dz} \frac{1}{\Gamma(z+1)} = \frac{\Gamma'(z+1)}{\Gamma(z+1)} \frac{1}{\Gamma(z+1)} = \frac{1}{\Gamma(z+1)} \frac{d}{dz} \log \Gamma(z+1)$ leads, with $c_0 = 0$, $c_1 = 1$ and $c_2 = \gamma$, to a recursion formula for the coefficients c_k for $k \geq 0$,

$$c_{k+2} = \frac{1}{k+1} \left\{ \gamma c_{k+1} + \sum_{m=0}^{k-1} (-1)^{k+m} c_{m+1} \zeta(k-m+1) \right\}$$

Without solving for c_k but to illuminate the linear structure in the coefficients c_k, the first few recursive relations are

$$\begin{aligned} c_3 &= \frac{1}{2} \{\gamma c_2 - \zeta(2)\} \\ c_4 &= \frac{1}{3} \{\gamma c_3 - c_2 \zeta(2) + \zeta(3)\} \\ c_5 &= \frac{1}{4} \{\gamma c_4 - c_3 \zeta(2) + c_2 \zeta(3) - \zeta(4)\} \end{aligned}$$

We list other recursive relations for c_k that do not involve the Riemann Zeta-function $\zeta(n)$: (1) Substitution of (15.13) into the functional relation $\frac{1}{\Gamma(z+1)} = \frac{1}{z\,\Gamma(z)}$ leads to

$$c_m = -m c_{m+1} - \sum_{k=m+2}^{\infty} \binom{k}{m} c_k \tag{15.14}$$

while (2) a substitution of (15.13) into the reflection formula $\frac{1}{\Gamma(z)\,\Gamma(-z)} = -\frac{z}{\pi} \sin \pi z$ results, for $k > 1$, in

$$c_{2k-1} = \frac{(-1)^{k-1} \pi^{2k-2}}{2(2k-1)!} + \sum_{m=2}^{k-1} (-1)^m c_m c_{2k-m} + (-1)^k \frac{c_k^2}{2} \tag{15.15}$$

Both recursion formulae have deficiencies: (15.14) needs an infinite amount of c_k and, hence, is not a recursion formula, while (15.15) is incomplete because it jumps only over odd values and assumes the even coefficients c_{2k} as known. Neither of the other functional relations such as the duplication formula *seems* to produce a useful recursion formula to determine all c_k, without the involvement of the Riemann Zeta-function $\zeta(n)$ at integers n.

211. *Bourguet's approach.* Bourguet (1883) has computed the first 23 coefficients c_k in the Taylor series (15.13) of $\frac{1}{\Gamma(z)}$. He started from

$$\frac{1}{\Pi(k,z)} = (1+z)\left(1+\frac{z}{2}\right)\left(1+\frac{z}{3}\right)\dots\left(1+\frac{z}{k}\right)e^{-z\log k}$$

where the Gauss product $\Pi(k,z)$ in (10.3) defines the Gamma function $\Gamma(z+1) = \lim_{k\to\infty}\Pi(k,z)$. Bourguet expands both $e^{-z\log k}$ and the product $\prod_{j=1}^{k}\left(1+\frac{z}{j}\right)$ into Taylor series around $z = 0$. From the generating function (A.5) of the Stirling numbers of the first kind for $k > 0$,

$$\prod_{j=1}^{k}\left(1+\frac{z}{j}\right) = \frac{1}{k!}\prod_{j=1}^{k}(z+j) = \frac{1}{k!}\sum_{j=0}^{k} S_{k+1}^{(j+1)}(-1)^{k-j}z^j$$

we obtain, using $S_k^{(j)} = 0$ if $j > k$,

$$\begin{aligned}\frac{1}{\Pi(k,z)} &= \frac{(-1)^k}{k!}\sum_{j=0}^{\infty}(-1)^j S_{k+1}^{(j+1)}z^j\sum_{n=0}^{\infty}\frac{(-1)^n\log^n k}{n!}z^n \\ &= \frac{(-1)^k}{k!}\sum_{n=0}^{\infty}\left((-1)^n\sum_{j=0}^{n}S_{k+1}^{(n-j+1)}\frac{(\log k)^j}{j!}\right)z^n\end{aligned}$$

Since $\frac{1}{\Gamma(z)} = \lim_{k\to\infty}\frac{z}{\Pi(k,z)} = \sum_{n=1}^{\infty}c_n z^n$, we find, for $n > 0$,

$$c_n = \lim_{k\to\infty}\frac{(-1)^{k-n-1}}{k!}\sum_{j=0}^{n-1}S_{k+1}^{(n-j)}\frac{(\log k)^j}{j!} \tag{15.16}$$

For example, $c_1 = \frac{(-1)^k}{k!}S_{k+1}^{(1)} = 1$, where $S_m^{(1)} = (-1)^{m-1}(m-1)!$ has been used. With also $\frac{S_m^{(2)}}{(m-1)!} = (-1)^m\sum_{k=1}^{m-1}\frac{1}{k}$ and (9.3), we find

$$c_2 = \lim_{k\to\infty}\frac{(-1)^{k-1}\left(S_{k+1}^{(2)} + S_{k+1}^{(1)}\log k\right)}{k!} = \lim_{k\to\infty}\left(\sum_{j=1}^{k}\frac{1}{j} - \log k\right) = \gamma$$

These computations reveal that Euler's constant γ plays a crucial role and, in addition, that all Stirling numbers must be known exactly, which is possible with characteristic coefficients (Appendix A). Bourguet (1883, on pp. 271-272) has followed another, less transparent approach without Stirling numbers, but he introduces the Euler constant γ in (15.16), whereafter he remarks that the result is independent of $\log k$.

212. *Closed-form expression for c_n.* Using $S_m^{(k)} = \frac{1}{k!}\frac{d^m}{dx^m}\log^k(1+x)\Big|_{x=0}$

from (A.7), we can rewrite (15.16) as

$$c_n = \lim_{k\to\infty} \frac{(-1)^{k-n-1}}{k!n!} \frac{d^{k+1}}{dx^{k+1}} \sum_{j=0}^{n-1} \binom{n}{j} \log^{n-j}(1+x)(\log k)^j \Bigg|_{x=0}$$
$$= \lim_{k\to\infty} \frac{(-1)^{k-n-1}}{k!n!} \frac{d^{k+1}}{dx^{k+1}} (\log k + \log(1+x))^n \Big|_{x=0}$$

Instead of the differential operator, we use Cauchy's integral in (1.14)

$$c_n = \frac{(-1)^{n-1}}{n!} \lim_{k\to\infty} \frac{(k+1)(-1)^k}{2\pi i} \int_{C(0)} \frac{\log^n(k(1+\omega))\,d\omega}{\omega^{k+2}}$$

Deforming the contour to enclose the entire ω-plane, without the point at $\omega = 0$ and the branch cut, which is the real negative ω-axis from $\omega = -1$ on, yields

$$\mathbb{L}= \int_{C(0)} \frac{\log^n(k(1+\omega))\,d\omega}{\omega^{k+2}}$$
$$= -\int_{\infty}^{0} \frac{\log^n\left(k\left(re^{-i\pi}\right)\right) d\left(re^{-i\pi}\right)}{\left(-1+re^{-i\pi}\right)^{k+2}} - \int_0^{\infty} \frac{\log^n\left(k\left(re^{i\pi}\right)\right) d\left(re^{i\pi}\right)}{\left(-1+re^{i\pi}\right)^{k+2}}$$
$$= -2i\,\mathrm{Im}\left(e^{i\pi}\int_0^{\infty} \frac{(\log(kr)+i\pi)^n\,dr}{\left(-1+re^{i\pi}\right)^{k+2}}\right)$$

After substitution of $u = kr$, the integral transforms to

$$\int_{C(0)} \frac{\log^n(k(1+\omega))\,d\omega}{\omega^{k+2}} = -2i\,\mathrm{Im}\left(\frac{(-1)^k e^{i\pi}}{k}\int_0^{\infty} \frac{(\log(u)+i\pi)^n\,du}{\left(1+\frac{u}{k}\right)^{k+2}}\right)$$

Returning the Taylor coefficient (15.16),

$$c_n = \frac{(-1)^n}{n!} \lim_{k\to\infty} \frac{(k+1)(-1)^k(-1)^k}{2\pi i k} 2i\,\mathrm{Im}\left(e^{i\pi}\int_0^{\infty} \frac{(\log(u)+i\pi)^n\,du}{\left(1+\frac{u}{k}\right)^{k+2}}\right)$$

taking the limit $k \to \infty$ and using $\lim_{k\to\infty}\left(1+\frac{u}{k}\right)^{k+2} = e^u$, results in an integral for the Taylor coefficient

$$c_n = \frac{(-1)^{n-1}}{n!}\frac{1}{\pi}\,\mathrm{Im}\left(\int_0^{\infty} e^{-u}(\log(u)+i\pi)^n\,du\right) \tag{15.17}$$
$$= \frac{(-1)^n}{n!}\frac{1}{\pi}\,\mathrm{Im}\left(\int_0^{\infty} e^{-u}(\log(u)-i\pi)^n\,du\right)$$

213. *Generalization of the integral for* c_n. For any complex number p, we expand the Hankel integrand in $\frac{e^{pz}}{\Gamma(z)} = \frac{1}{2\pi i}\int_C e^{z(p-\log w)}e^w dw$ in a Taylor series

$$\frac{e^{pz}}{\Gamma(z)} = \frac{1}{2\pi i}\int_C \sum_{k=0}^{\infty} \frac{z(p-\log w)^k}{k!} e^w dw$$

and reverse sum and integral, which is justified as a Taylor series can be term-wise integrated, provided the resulting integrals exist,

$$\frac{e^{pz}}{\Gamma(z)} = \sum_{k=0}^{\infty} \frac{z^k}{k!} \frac{1}{2\pi i} \int_C (p - \log w)^k e^w dw$$

After evaluating the Hankel contour C in (15.1) and observing that the integral around the circle at the origin $w = 0$ with radius ε,

$$\frac{1}{2\pi} \int_{-\pi}^{\pi} \left(p - \log \varepsilon e^{i\theta}\right)^k e^{\varepsilon e^{i\theta}} \varepsilon e^{i\theta} d\theta = \frac{\varepsilon}{2\pi} \int_{-\pi}^{\pi} (p - \log \varepsilon - i\theta)^k e^{\varepsilon e^{i\theta}} e^{i\theta} d\theta$$

vanishes for $\varepsilon \to 0$, because $\lim_{\varepsilon \to 0} \varepsilon \log^k \varepsilon = 0$ for any integer k, we obtain

$$\frac{1}{\Gamma(z)} = \frac{e^{-pz}}{2\pi i} \sum_{k=0}^{\infty} \frac{z^k}{k!} \int_0^{\infty} e^{-u} \left\{ (p - \log(u) + i\pi)^k - (p - \log(u) - i\pi)^k \right\} du \tag{15.18}$$

Hence, the Taylor coefficient $c_k(p) = \frac{1}{k!} \frac{d^k}{du^k} \frac{e^{pu}}{\Gamma(u)}\Big|_{u=0}$ of the Taylor series of the entire function $\frac{e^{pz}}{\Gamma(z)} = \sum_{k=0}^{\infty} c_k(p) z^k$ around $z_0 = 0$ is

$$c_k(p) = \frac{1}{2\pi i} \frac{1}{k!} \int_0^{\infty} e^{-u} \left\{ (p - \log(u) + i\pi)^k - (p - \log(u) - i\pi)^k \right\} du \tag{15.19}$$

which reduces, for $p = 0$, to (15.17).

A substitution of $u = qt$ with real $q > 0$ in (15.19) leads to a Laplace transform, converging for $\operatorname{Re}(q) > 0$,

$$\frac{c_k(p + \log q)}{q} = \frac{1}{2\pi i} \frac{1}{k!} \int_0^{\infty} e^{-qt} \left\{ (p - \log t + i\pi)^k - (p - \log t - i\pi)^k \right\} dt$$

so that

$$\frac{1}{\Gamma(z)} = \frac{e^{-pz} q^{1-z}}{2\pi i} \sum_{k=0}^{\infty} \frac{z^k}{k!} \int_0^{\infty} e^{-qt} \left\{ (p - \log t + i\pi)^k - (p - \log t - i\pi)^k \right\} dt$$

For complex p and q, with $\operatorname{Re}(q) > 0$, we write the modified Taylor series for $\frac{1}{\Gamma(z)}$ as

$$\frac{1}{\Gamma(z)} = e^{-pz} q^{1-z} \sum_{k=0}^{\infty} \widetilde{c}_k(p, q) z^k \tag{15.20}$$

where the Taylor coefficient is

$$\widetilde{c}_k(p, q) = \frac{1}{2\pi i} \frac{1}{k!} \int_0^{\infty} e^{-qt} \left\{ (p - \log t + i\pi)^k - (p - \log t - i\pi)^k \right\} dt \tag{15.21}$$

Clearly, it holds that $\widetilde{c}_k(p, 1) = c_k(p)$.

214. *The Taylor coefficient c_n and that of $\Gamma(z)$ around $z_0 = 1$.* Newton's binomium applied to (15.17) indicates that

$$\int_0^\infty e^{-u} \left(\log(u) + i\pi\right)^n du = \sum_{j=0}^{n} \binom{n}{j} (i\pi)^j \int_0^\infty e^{-u} \log^{n-j}(u)\, du$$
$$= \sum_{j=0}^{n} \binom{n}{j} (i\pi)^j \left. \frac{d^{n-j}\Gamma(x)}{dx^{n-j}} \right|_{x=1}$$

The integral $\left.\frac{d^m\Gamma(x)}{dx^m}\right|_{x=1} = \int_0^\infty e^{-u} \log^m(u)\, du$ converges fast. Splitting the j-sum in odd and even indices results in

$$\mathbb{G} = \sum_{j=0}^{n} \binom{n}{j} (i\pi)^j \left. \frac{d^{n-j}\Gamma(x)}{dx^{n-j}} \right|_{x=1}$$
$$= \sum_{j=0}^{\left[\frac{n}{2}\right]} \frac{\binom{n}{2j}}{(i\pi)^{-2j}} \left. \frac{d^{n-2j}\Gamma(x)}{dx^{n-2j}} \right|_{x=1} + \sum_{j=1}^{\left[\frac{n+1}{2}\right]} \frac{\binom{n}{2j-1}}{(i\pi)^{1-2j}} \left. \frac{d^{n-2j+1}\Gamma(x)}{dx^{n-2j+1}} \right|_{x=1}$$

Since only the imaginary part in (15.17) counts, we arrive at the expression in terms of derivatives

$$c_n = \frac{1}{n!} \frac{d^n}{dx^n} \left(\frac{1}{\Gamma(x)} \right)\Bigg|_{x=0} = \sum_{j=1}^{\left[\frac{n+1}{2}\right]} \frac{(-1)^{n-j}\, \pi^{2j-2}}{(2j-1)!\,(n-2j+1)!} \left. \frac{d^{n-2j+1}\Gamma(x)}{dx^{n-2j+1}} \right|_{x=1} \tag{15.22}$$

Formula (15.22) directly follows from the reflection formula (11.4) and Leibniz' rule,

$$\frac{d^n}{dx^n} \left(\frac{1}{\Gamma(x)} \right) = \frac{1}{\pi} \sum_{j=0}^{n} \binom{n}{j} \frac{d^{n-j}\Gamma(1-x)}{dx^{n-j}} \frac{d^j \sin \pi x}{dx^j}$$

Since $\frac{d^{n-j}\Gamma(1-x)}{dx^{n-j}} = (-1)^{n-j} \frac{d^{n-j}\Gamma(1+x)}{dx^{n-j}}$, $\frac{d^{2j}\sin\pi x}{dx^{2j}} = (-1)^j \pi^{2j} \sin \pi x$ and $\frac{d^{2j-1}\sin\pi x}{dx^{2j-1}} = (-1)^{j-1}\pi^{2j-1}\cos\pi x$, we split in even and odd indexed terms,

$$\frac{d^n}{dx^n} \left(\frac{1}{\Gamma(x)} \right) = \sin \pi x \sum_{j=0}^{\left[\frac{n}{2}\right]} \binom{n}{2j} \frac{d^{n-2j}\Gamma(x+1)}{(-1)^{n-j} dx^{n-2j}} \pi^{2j-1} + \cos \pi x \sum_{j=1}^{\left[\frac{n+1}{2}\right]} \binom{n}{2j-1} \frac{d^{n-2j+1}\Gamma(x+1)}{(-1)^{n-j} dx^{n-2j+1}} \pi^{2j-2} \tag{15.23}$$

evaluate at $x = 0$ and find (15.22).

215. *Fekih-Ahmed's approximation of the integral* (15.17). In the finishing phase of the book, I found that Fekih-Ahmed (2017) has also derived (15.17). Fekih-Ahmed (2017) started from the reflection formula (11.4) as $\frac{1}{\Gamma(z)} = \frac{\sin \pi z}{\pi} \Gamma(1-z) = \frac{\sin \pi(z-1)}{\pi(z-1)} \Gamma(2-z)$ and Euler's integral (10.2),

$$\frac{1}{\Gamma(z)} = \frac{\sin \pi(z-1)}{\pi(z-1)} \int_0^\infty x^{1-z} e^{-x} dx \qquad \text{valid for } \operatorname{Re}(z) < 2$$

followed by Taylor expansion of $x^{1-z} = e^{-(z-1)\log x}$, which is a similar idea that led to (15.19).

Fekih-Ahmed (2017) further approximates the integral in (15.17) rewritten as

$$\int_0^\infty e^{-u} (\log(u) - i\pi)^n \, du = n \int_0^\infty e^{-nt} (\log(nt) - i\pi)^n \, du$$
$$= n \int_0^\infty e^{n(-t+n\log(\log(nt)-i\pi))} du$$

by a saddle-point expansion (**art.** 200) and found the approximation

$$c_n \approx \frac{(-1)^n}{n!} \sqrt{\frac{2n}{\pi}} \operatorname{Im}\left(e^{-nt_0} \frac{t_0^{\frac{1}{2}-n}}{\sqrt{1+t_0}} \right) \tag{15.24}$$

where t_0 is the minimum of the function $f(t) = -t + n\log(\log(nt) - i\pi)$, which can be written in terms of the Lambert function $W_k(z)$ (**art.** 201) as $t_0 = -\frac{1}{n} e^{W_{-1}(-n)}$. Fekih-Ahmed (2017) did not expand the function $f(t) = -t + n\log(\log(nt) - i\pi)$ in a Taylor series around its minimum t_0 to compute higher-order terms as in **art.** 200, but gave an extensive numerical evaluation of his asymptotic approximation (15.24).

216. *Ulgenes's approach.* Ulgenes (2025) observes that the k-th difference (12.51) of $\frac{1}{\Gamma(z+1)}$ for $z_0 = 0$ is

$$\Delta^k \left(\frac{1}{\Gamma(z+1)} \right) = (-1)^k \sum_{j=0}^{k} \binom{k}{j} \frac{(-1)^j}{j!} \tag{15.25}$$

which is written with the Laguerre polynomial $L_n(z) = \sum_{j=0}^{k} \binom{k}{j} \frac{(-1)^j}{j!} z^j$ as $L_n(1)$. The Newton series for $\frac{1}{\Gamma(z+1)}$ immediately follows from Newton's general form (12.50) as

$$\frac{1}{\Gamma(z+1)} = \sum_{k=0}^{\infty} \binom{z}{k} (-1)^k \sum_{j=0}^{k} \binom{k}{j} \frac{(-1)^j}{j!} \tag{15.26}$$

After invoking the generating function $\binom{z}{k} = \frac{1}{k!} \sum_{n=0}^{k} S_k^{(n)} z^n$ in (A.5) of the Stirling numbers of the first kind, the Newton series (15.26) becomes

$$\frac{1}{\Gamma(z+1)} = \sum_{k=0}^{\infty} \frac{(-1)^k}{k!} \sum_{n=0}^{k} S_k^{(n)} z^n \sum_{j=0}^{k} \binom{k}{j} \frac{(-1)^j}{j!}$$

Reversing the k- and n- sum produces the Taylor series of $\frac{1}{\Gamma(z)}$ around $z_0 = 0$,

$$\frac{1}{\Gamma(z)} = \sum_{n-0}^{\infty} \left(\sum_{k=n}^{\infty} \frac{(-1)^k}{k!} S_k^{(n)} \sum_{j=0}^{k} \binom{k}{j} \frac{(-1)^j}{j!} \right) z^{n+1}$$

Hence, the Taylor coefficients c_n for $n > 0$ in $\frac{1}{\Gamma(z)} = \sum_{n=0}^{\infty} c_n z^n$ in (15.13) are

$$c_n = \sum_{k=n-1}^{\infty} \frac{(-1)^k}{k!} S_k^{(n-1)} \sum_{j=0}^{k} \binom{k}{j} \frac{(-1)^j}{j!} \tag{15.27}$$

Ulgenes formula (15.27) directly follows from (A.20), applied to the Taylor series $\sum_{k=0}^{\infty} c_{k+1} z^k$ of the entire function $\frac{1}{\Gamma(z+1)}$. Although (15.27) is an exact result, the summation converges extremely badly. Reversing the k- and j- sum in (15.27) is not allowed, as follows from (A.7).

217. *Weierstrass's product and characteristic coefficients.* Weierstrass's infinite product $\frac{1}{\Gamma(z+1)} = e^{\gamma z} e^{\sum_{n=1}^{\infty} \{\log(1+\frac{z}{n}) - \frac{z}{n}\}} = \sum_{k=0}^{\infty} c_{k+1} z^k$ indicates that the coefficient c_{k+1} in the Taylor series (15.13) equals the Cauchy product $c_{k+1} = \sum_{m=0}^{k} \frac{\gamma^{k-m}}{(k-m)!} g_m$, where g_m are the Taylor coefficients of $g(z) = \exp\left(\sum_{n=1}^{\infty} \left\{\log\left(1 + \frac{z}{n}\right) - \frac{z}{n}\right\}\right)$ around $z_0 = 0$, which we will compute with characteristic coefficients (Appendix A).

The Taylor series of $\Gamma^x(z+1) = e^{x \log \Gamma(z+1)}$ around z_0,

$$e^{x \log \Gamma(z+1)} = e^{x \log \Gamma(z_0+1)} e^{x\psi(z_0+1)(z-z_0)} \exp\left(x \sum_{k=2}^{\infty} \frac{\zeta(k, z_0)}{k} (z_0 - z)^k\right)$$

reduces, for $z_0 = 0$ and with the general expansion (A.1), to

$$\Gamma^x(z+1) = 1 - \gamma x z + \sum_{m=2}^{\infty} \frac{1}{m!} \left.\frac{d^m \Gamma^x(z+1)}{dz^m}\right|_{z=0} z^m$$

where the Taylor coefficient (Van Mieghem, 2026) is

$$\frac{1}{m!} \left.\frac{d^m \Gamma^x(z+1)}{dz^m}\right|_{z=0} = \frac{(-\gamma x)^m}{m!} + (-1)^m \sum_{q=2}^{m} \frac{(x\gamma)^{m-q}}{(m-q)!} C_x(q) \tag{15.28}$$

with combinatorial sum

$$C_x(q) = \sum_{j=1}^{\left[\frac{q}{2}\right]} \frac{\sum_{\sum_{i=1}^{j} l_i = q; l_i > 1} \prod_{i=1}^{j} \frac{\zeta(l_i)}{l_i}}{j!} x^j = \sum_{j=1}^{\left[\frac{q}{2}\right]} \frac{s[j,q]|_{\log \Gamma(z+1)+\gamma z}}{j!} x^j \tag{15.29}$$

The appearing characteristic coefficient around $z_0 = 0$,

$$s[k,m]|_{\log \Gamma(z+1)+\gamma z} = (-1)^m \sum_{\sum_{i=1}^{k} l_i = m; l_i > 1} \prod_{i=1}^{k} \frac{\zeta(l_i)}{l_i} \tag{15.30}$$

consists of a sum of positive terms $\prod_{i=1}^{k} \frac{\zeta(l_i)}{l_i}$, where $0 < \frac{\zeta(n)}{n} < 1$ decreases with $n \geq 2$.

If we choose $x = -1$ in (15.28), then we obtain the Taylor coefficients of $\frac{1}{\Gamma(z)} = \frac{z}{\Gamma(z+1)}$ in (15.13) as $c_m = \frac{1}{(m-1)!} \left.\frac{d^{m-1}\Gamma^{-1}(z+1)}{dz^{m-1}}\right|_{z=0}$. Thus, $c_0 = 0$,

$c_1 = 1$ and $c_2 = \gamma$, while for $m > 2$ and writing $C_{-1}(q)$ in (15.29) explicitly,

$$c_{m+1} = \frac{\gamma^m}{m!} + (-1)^m \sum_{l=0}^{m-2} \frac{(-\gamma)^l}{l!} \sum_{j=1}^{\left[\frac{m-l}{2}\right]} (-1)^j \frac{\sum_{\sum_{i=1}^{j} l_i = m-l; l_i > 1} \prod_{i=1}^{j} \frac{\zeta(l_i)}{l_i}}{j!} \tag{15.31}$$

illustrates that each coefficient c_m is a polynomial in the Euler constant γ with highest order term $\frac{\gamma^{m-1}}{(m-1)!}$, followed[8] by the next highest order term $\frac{-\gamma^{m-3}}{(m-3)!}\frac{\zeta(2)}{2}$ and lower powers in γ. Since $C_{-1}(q)$ contains negative terms, its sign is irregular, which complicates computations. For sufficiently large m, the value of the Taylor coefficient $c_m = O\left(\frac{1}{m!}\right)$ decreases, whereas (15.31) exhibits an increasing number of products $\prod_{i=1}^{j} \frac{\zeta(l_i)}{l_i}$ in the combinatorial sum $C_{-1}(q)$, which points towards a huge cancellation of the terms.

In addition to $c_0 = 0$, $c_1 = 1$ and $c_2 = \gamma$, we list the coefficients c_m up to $m = 10$,

$$\begin{aligned}
c_3 &= \frac{\gamma^2}{2} - \frac{\zeta(2)}{2} \\
c_4 &= \frac{\gamma^3}{3!} - \frac{\gamma\,\zeta(2)}{2} + \frac{\zeta(3)}{3} \\
c_5 &= \frac{\gamma^4}{4!} - \frac{\gamma^2\,\zeta(2)}{4} + \frac{\gamma\,\zeta(3)}{3} + \frac{\zeta^2(2)}{8} - \frac{\zeta(4)}{4} \\
c_6 &= \frac{\gamma^5}{5!} - \frac{\gamma^3\,\zeta(2)}{12} + \frac{\gamma^2\,\zeta(3)}{6} + \gamma\left(\frac{\zeta^2(2)}{8} - \frac{\zeta(4)}{4}\right) - \frac{\zeta(2)\,\zeta(3)}{6} + \frac{\zeta(5)}{5} \\
c_7 &= \frac{\gamma^6}{6!} - \frac{\gamma^4\,\zeta(2)}{48} + \frac{\gamma^3\,\zeta(3)}{18} + \gamma^2\left(\frac{\zeta^2(2)}{16} - \frac{\zeta(4)}{8}\right) + \gamma\left(\frac{\zeta(5)}{5} - \frac{\zeta(2)\,\zeta(3)}{6}\right) \\
&\quad - \frac{\zeta^3(2)}{48} + \frac{\zeta(2)\,\zeta(4)}{8} + \frac{\zeta^2(3)}{18} - \frac{\zeta(6)}{6} \\
c_8 &= \frac{\gamma^7}{7!} - \frac{\gamma^5\,\zeta(2)}{240} + \frac{\gamma^4\,\zeta(3)}{72} + \gamma^3\left(\frac{\zeta^2(2)}{48} - \frac{\zeta(4)}{24}\right) + \gamma^2\left(\frac{\zeta(5)}{10} - \frac{\zeta(2)\,\zeta(3)}{12}\right) \\
&\quad + \gamma\left(\frac{\zeta^2(3)}{18} - \frac{\zeta^3(2)}{48} + \frac{\zeta(2)\,\zeta(4)}{8} - \frac{\zeta(6)}{6}\right) + \frac{\zeta^2(2)\zeta(3)}{24} - \frac{\zeta(3)\,\zeta(4)}{12} \\
&\quad - \frac{\zeta(2)\,\zeta(5)}{10} + \frac{\zeta(7)}{7} \\
c_9 &= \frac{\gamma^8}{8!} - \frac{\gamma^6\,\zeta(2)}{1440} + \frac{\gamma^5\,\zeta(3)}{360} + \gamma^4\left(\frac{\zeta^2(2)}{192} - \frac{\zeta(4)}{96}\right) + \gamma^3\left(\frac{\zeta(5)}{30} - \frac{\zeta(2)\,\zeta(3)}{36}\right) \\
&\quad + \gamma^2\left(\frac{\zeta(2)\,\zeta(4)}{16} - \frac{\zeta^3(2)}{96} + \frac{\zeta^2(3)}{36} - \frac{\zeta(6)}{12}\right) \\
&\quad + \gamma\left(\frac{\zeta^2(2)\,\zeta(3)}{24} - \frac{\zeta(3)\,\zeta(4)}{12} - \frac{\zeta(2)\,\zeta(5)}{10} + \frac{\zeta(7)}{7}\right) \\
&\quad + \frac{\zeta^4(2)}{384} - \frac{\zeta(2)\,\zeta^2(3)}{36} - \frac{\zeta^2(2)\,\zeta(4)}{32} + \frac{\zeta^2(4)}{32} + \frac{\zeta(3)\,\zeta(5)}{15} + \frac{\zeta(2)\,\zeta(6)}{12} - \frac{\zeta(8)}{8}
\end{aligned}$$

[8] The combinatorial definition (A.3) of the characteristic coefficient $s[k,m]$ shows that $s[k,m]$ does not contain powers of f_1^{m-1}.

$$c_{10}=\frac{\gamma^9}{9!}-\frac{\gamma^7\,\varsigma(2)}{10080}+\frac{\gamma^6\,\varsigma(3)}{2160}+\gamma^5\left(\frac{\varsigma^2(2)}{960}-\frac{\varsigma(4)}{480}\right)+\gamma^4\left(\frac{\varsigma(5)}{120}-\frac{\varsigma(2)\,\varsigma(3)}{144}\right)$$
$$+\gamma^3\left(\frac{\varsigma^2(3)}{108}+\frac{\varsigma(2)\,\varsigma(4)}{48}-\frac{\varsigma^3(2)}{288}-\frac{\varsigma(6)}{36}\right)$$
$$+\gamma^2\left(\frac{\varsigma^2(2)\,\varsigma(3)}{48}+\frac{\varsigma(7)}{14}-\frac{\varsigma(3)\,\varsigma(4)}{24}-\frac{\varsigma(2)\,\varsigma(5)}{20}\right)$$
$$+\gamma\left(\frac{\varsigma(3)\,\varsigma(5)}{15}-\frac{\varsigma(8)}{8}+\frac{\varsigma(2)\,\varsigma(6)}{12}+\frac{\varsigma^4(2)}{384}+\frac{\varsigma^2(4)}{32}-\frac{\varsigma(2)\,\varsigma^2(3)}{36}-\frac{\varsigma^2(2)\,\varsigma(4)}{32}\right)$$
$$+\frac{\varsigma(3)^3}{162}+\frac{\varsigma(2)\,\varsigma(3)\,\varsigma(4)}{24}+\frac{\varsigma^2(2)\,\varsigma(5)}{40}-\frac{\varsigma(4)\,\varsigma(5)}{20}-\frac{\varsigma^3(2)\,\varsigma(3)}{144}-\frac{\varsigma(3)\,\varsigma(6)}{18}$$
$$-\frac{\varsigma(2)\,\varsigma(7)}{14}+\frac{\varsigma(9)}{9}$$

218. *Characteristic coefficients and the reflection formula.* The instance of $x=1$ in (15.28)

$$\frac{1}{m!}\left.\frac{d^m\Gamma(z+1)}{dz^m}\right|_{z=0}=\frac{(-\gamma)^m}{m!}+(-1)^m\sum_{q=2}^{m}\frac{\gamma^{m-q}}{(m-q)!}C_1(q)$$

is rewritten as

$$\frac{1}{m!}\left.\frac{d^m\Gamma(z+1)}{dz^m}\right|_{z=0}=(-1)^m\sum_{q=0}^{m}\frac{\gamma^{m-q}}{(m-q)!}A_q \tag{15.32}$$

with $A_0=1$, $A_1=0$ and, for $q\geq 2$,

$$A_q=C_1(q)=\sum_{j=1}^{\left[\frac{q}{2}\right]}\frac{\sum_{\sum_{i=1}^{j}l_i=q;l_i>1}\prod_{i=1}^{j}\frac{\varsigma(l_i)}{l_i}}{j!} \tag{15.33}$$

which, in contrast to $C_{-1}(q)$ in (15.31), contains only positive terms.

We list the first few values of A_q, which illustrates the combinatorial nature of the characterisic coefficient (A.3) to find all possibilities in which an integer $q=\sum_{i=1}^{j}l_i$ can be written as a sum of j integers $l_i>1$, all larger than 1. Besides, $A_2=\frac{\varsigma(2)}{2}$ and $A_3=\frac{\varsigma(3)}{3}$, we have

$$A_4=\frac{\varsigma(4)}{4}+\frac{\varsigma^2(2)}{8}$$
$$A_5=\frac{\varsigma(5)}{5}+\frac{\varsigma(2)\,\varsigma(3)}{6}$$
$$A_6=\frac{\varsigma(6)}{6}+\frac{\varsigma(2)\,\varsigma(4)}{8}+\frac{\varsigma^2(3)}{18}+\frac{\varsigma^3(2)}{48}$$
$$A_7=\frac{\varsigma(7)}{7}+\frac{\varsigma(2)\,\varsigma(5)}{10}+\frac{\varsigma(3)\,\varsigma(4)}{12}+\frac{\varsigma^2(2)\,\varsigma(3)}{24}$$
$$A_8=\frac{\varsigma(8)}{8}+\frac{\varsigma(2)\,\varsigma(6)}{12}+\frac{\varsigma(3)\,\varsigma(5)}{15}+\frac{\varsigma^2(4)}{32}+\frac{\varsigma^2(2)\,\varsigma(4)}{32}+\frac{\varsigma(2)\,\varsigma^2(3)}{36}+\frac{\varsigma^4(2)}{384}$$
$$A_9=\frac{\varsigma(9)}{9}+\frac{\varsigma(2)\,\varsigma(7)}{14}+\frac{\varsigma(3)\,\varsigma(6)}{18}+\frac{\varsigma(4)\,\varsigma(5)}{20}+\frac{\varsigma^2(2)\,\varsigma(5)}{40}+\frac{\varsigma(2)\,\varsigma(3)\,\varsigma(4)}{24}$$
$$+\frac{\varsigma^3(3)}{162}+\frac{\varsigma^3(2)\,\varsigma(3)}{144}$$

Into $c_n = \frac{1}{\pi} \operatorname{Im}\left(\sum_{k=0}^{n} \frac{(-1)^{n-k}}{(n-k)!} \left. \frac{d^{n-k}\Gamma(z+1)}{dz^{n-k}} \right|_{z=0} \frac{i^k \pi^k}{k!} \right)$ in (15.22), we introduce the Taylor coefficient (15.32)

$$c_n = \frac{1}{\pi} \operatorname{Im}\left(\sum_{k=0}^{n} \frac{i^k \pi^k}{k!} \sum_{q=0}^{n-k} \frac{\gamma^{n-k-q}}{(n-k-q)!} A_q \right)$$

Reversing the k- and q-sum,

$$c_n = \frac{1}{\pi} \operatorname{Im}\left(\sum_{q=0}^{n} A_q \sum_{k=0}^{n-q} \frac{i^k \pi^k}{k!} \frac{\gamma^{n-k-q}}{(n-k-q)!} \right)$$

invoking Newton's binomium $\sum_{k=0}^{n-q} \frac{i^k \pi^k}{k!} \frac{\gamma^{n-k-q}}{(n-k-q)!} = \frac{(\gamma+i\pi)^{n-q}}{(n-q)!}$,

$$c_n = \frac{1}{\pi} \operatorname{Im}\left(\sum_{q=0}^{n} A_q \frac{(\gamma + i\pi)^{n-q}}{(n-q)!} \right) \tag{15.34}$$

and taking the imaginary part, results in

$$c_n = \frac{(\gamma^2+\pi^2)^{\frac{n}{2}} \sin\left(n \arctan \frac{\pi}{\gamma}\right)}{\pi n!} + \sum_{q=2}^{n} A_q \frac{\sin\left((n-q) \arctan \frac{\pi}{\gamma}\right)}{\pi (n-q)! (\gamma^2+\pi^2)^{\frac{q-n}{2}}} \tag{15.35}$$

Hence, we have derived two, different exact forms (15.31) and (15.35) for the Taylor coefficient c_n. Formula (15.35) is computationally superior to (15.31).

The positive numbers A_q in (15.33), computed below up to 20 digits accurate,

$A_2 = 0.82246703342411321824$	$A_{11} = 0.56137827869562279815$
$A_3 = 0.40068563438653142847$	$A_{12} = 0.56142190068506750773$
$A_4 = 0.60880681896251523273$	$A_{13} = 0.56144010921225537886$
$A_5 = 0.53693627607822333830$	$A_{14} = 0.56144989554746618960$
$A_6 = 0.56510204177883261170$	$A_{15} = 0.56145467123323785254$
$A_7 = 0.55855782217326626982$	$A_{16} = 0.56145707984787610214$
$A_8 = 0.56127498296681845242$	$A_{17} = 0.56145828116364508897$
$A_9 = 0.56105334187349761554$	$A_{18} = 0.56145888239067632342$
$A_{10} = 0.56132773858121889297$	$A_{19} = 0.56145918295712041799$

rapidly tend to the limit $A_\infty = \lim_{q\to\infty} A_q = 0.5614594836\ldots$. The difference $A_q - A_\infty < 0$ is negative for $q > 7$ and $|A_{10} - A_\infty| \simeq 1.3\ 10^{-4}$, $|A_{20} - A_\infty| \simeq 1.5\ 10^{-7}$ and $|A_{30} - A_\infty| \simeq 10^{-10}$. Since the Taylor series of $\Gamma(z+1)$ around $z_0 = 0$ has a radius of convergence $R = 1$, Section 1.4.2 shows that its positive Taylor coefficient (15.32) satisfies, for large n,

$$1 = \lim_{n\to\infty} \left| \sum_{q=0}^{n} \frac{\gamma^{n-q}}{(n-q)!} A_q \right| = \lim_{n\to\infty} \sum_{l=0}^{n} \frac{\gamma^l}{l!} A_{n-l} = \lim_{n\to\infty} A_\infty \sum_{l=0}^{n} \frac{\gamma^l}{l!} = A_\infty e^\gamma$$

implying[9] that $A_\infty = e^{-\gamma}$.

[9] We write $\sum_{l=0}^{n} A_{n-l} f_l = A_\infty \sum_{l=0}^{n} f_l + R_n$ with the remainder $R_n = \sum_{l=0}^{n} (A_{n-l} - A_\infty) f_l$, where the partial sum $F_n = \sum_{l=0}^{n} f_l$ converges absolutely,

Thus, $c_n = \frac{1}{\pi}\operatorname{Im}\left(\sum_{l=0}^{n} A_{n-l}\frac{(\gamma+i\pi)^l}{l!}\right)$ in (15.34) tends[9], for large n, to

$$c_n \to \frac{A_\infty}{\pi}\operatorname{Im}\left(\sum_{l=0}^{n}\frac{(\gamma+i\pi)^l}{l!}\right) = \frac{A_\infty}{\pi}\operatorname{Im}\left(e^{\gamma+i\pi} - \sum_{l=n+1}^{\infty}\frac{(\gamma+i\pi)^l}{l!}\right)$$
$$= -\frac{A_\infty}{\pi}\operatorname{Im}\left(\sum_{l=n+1}^{\infty}\frac{(\gamma+i\pi)^l}{l!}\right)$$

In summary, for large n, the Taylor coefficient c_n of $\frac{1}{\Gamma(z)} = \sum_{n=0}^{\infty} c_n z^n$ behaves as

$$c_{n-1} = -\frac{e^{-\gamma}}{\pi}\frac{\left(\gamma^2+\pi^2\right)^{\frac{n}{2}}\sin\left(n\arctan\frac{\pi}{\gamma}\right)}{n!}\left(1+O\left(\frac{\cot\left(n\arctan\frac{\pi}{\gamma}\right)}{n}\right)\right) \tag{15.36}$$

where $\cot\left(n\arctan\frac{\pi}{\gamma}\right)$ only plays a role, if the integer n is close to $k\frac{\pi}{\arctan\frac{\pi}{\gamma}}$ for any (positive) integer k, and reflects the oscillatory behavior around the dominant, first term in (15.36).

so that $G = \sum_{l=0}^{\infty}|f_l|$ is finite and $f_l \to 0$ for $l \to \infty$ (see e.g. Van Mieghem (2014, Lemma 6.1.1)). We show (see e.g. Hardy (1948, Chapter III)) that $\sum_{l=0}^{n} A_{n-l}f_l \to A_\infty\sum_{l=0}^{n} f_l$ for large n, by demonstrating that the remainder $R_n \to 0$. Indeed, $|R_n| \le \sum_{l=0}^{n}|A_{n-l}-A_\infty||f_l|$ and

$$|R_n| \le \sum_{l=0}^{n-q_0}|A_{n-l}-A_\infty||f_l| + \sum_{l=n-q_0+1}^{n}|A_{n-l}-A_\infty||f_l|$$
$$= \sum_{q=q_0}^{n}|A_q-A_\infty||f_{n-q}| + \sum_{l=1}^{q_0}|A_{q_0-l}-A_\infty|\,|f_{l+n-q_0}|$$

The convergence of the numbers A_q towards A_∞ and, similarly, of f_l towards 0 implies that, for every real $\varepsilon_A > 0$ and $\varepsilon_f > 0$, there exist integers q_0 and l_0, such that $|A_q - A_\infty| < \varepsilon_A$ for any integer $q \ge q_0$ and $|f_l| < \varepsilon_f$ for any integer $l \ge l_0$. Hence, for $q \ge q_0$ and $l+n-q_0 \ge 1+n-q_0 \ge l_0$, implying that $n \ge q_0 + l_0$, we have

$$|R_n| < \varepsilon_A\sum_{q=q_0}^{n}|f_{n-q}| + \varepsilon_f\sum_{l=1}^{q_0}|A_{q_0-l}-A_\infty| < \varepsilon_A G + \varepsilon_f\sum_{j=0}^{q_0-1}|A_j - A_\infty|$$

Since the last sum $\sum_{j=0}^{q_0-1}|A_j-A_\infty|$ is finite, we can choose $\varepsilon_A = \frac{\varepsilon}{2G}$ and $\varepsilon_f = \frac{\varepsilon}{2\sum_{j=0}^{q_0-1}|A_j-A_\infty|}$ with corresponding q_0 and l_0 to arrive at $|R_n| < \frac{\varepsilon}{2}+\frac{\varepsilon}{2} = \varepsilon$ for $n \ge q_0 + l_0$, which proves the claim.

16
The entire incomplete Gamma function $\gamma^* (a, z)$

Section 16.1 defines the incomplete Gamma integral $\Gamma(a, z)$ in (16.1) and the entire incomplete Gamma function $\gamma^*(a, z)$ in (16.3), that possesses a Taylor series (16.7) and a Taylor series for $e^z \gamma^*(a, z)$ in (16.18), converging for all complex z. **Art.** 220 demonstrates the fundamental connection $\Gamma(a) = \Gamma(a, z) + z^a \Gamma(a) \gamma^*(a, z)$ in (16.8), valid for all complex a and $z = |z| e^{i \arg z}$ with $-\pi < \arg z < \pi$. Section 16.2 derives recursions for $\Gamma(a, z)$ and $\gamma^*(a, z)$, which are crucial to unravel properties, in particular for large z, as shown in Section 16.3. Section 16.4 relates the incomplete Gamma function to the logarithmic integral and the Mittag-Leffler function. The Laplace and Mellin transforms of $\Gamma(a, z)$ and $\gamma^*(a, z)$ in Section 16.5 lead to two complex integrals (16.33) and (16.34) for the incomplete Gamma integral $\Gamma(a, z)$.

16.1 Definition and basic properties

219. *Definitions.* The natural extension of the Gamma function from Euler's integral (10.2) is the incomplete Gamma function

$$\Gamma(a, z) = \int_z^\infty x^{a-1} e^{-x} dx \tag{16.1}$$

After substitution $x = zt$ with $\mathrm{Re}(z) > 0$ justified in **art.** 220, (16.1) becomes

$$\Gamma(a, z) = z^a \int_1^\infty t^{a-1} e^{-zt} dt \qquad \text{for } \mathrm{Re}(z) > 0 \tag{16.2}$$

which is, however, *not* an entire function due to $z^a = e^{a \log z}$. The derivative $\frac{d\Gamma(a,z)}{dz} = -z^{a-1} e^{-z}$ tells, for real a and real $z = x$, that $\Gamma(a, x)$ is strictly decreasing on the positive real x-axis. The strength of the change $\left| \frac{d\Gamma(a,z)}{dz} \right| = |z|^{\mathrm{Re}(a)-1} e^{-\mathrm{Im}(a) \arg z} e^{-\mathrm{Re}(z)}$ exponentially increases for $\mathrm{Re}(z) < 0$, but hardly alters for large $\mathrm{Re}(z) > 0$. The influence of $\left| \frac{d\Gamma(a,z)}{dz} \right|$ on $a \in \mathbb{C}$ varies exponentially with $\arg z \in (-\pi, \pi]$ and $\mathrm{Im}(a)$.

Gautschi (1998) mentions that Tricomi has defined the *entire incomplete* Gamma function for all complex a and z as (Abramowitz and Stegun, 1968, 6.5.3-6.5.4),

$$\gamma^*(a,z) = z^{-a}\left(1 - \frac{\Gamma(a,z)}{\Gamma(a)}\right) \tag{16.3}$$

and in integral form

$$\gamma^*(a,z) = z^{-a}\left(1 - \frac{1}{\Gamma(a)}\int_z^\infty x^{a-1}e^{-x}dx\right) \tag{16.4}$$

If $\operatorname{Re}(a) > 0$, Euler's integral (10.2) is applicable and then (16.4) becomes

$$\gamma^*(a,z) = \frac{z^{-a}}{\Gamma(a)}\int_0^z x^{a-1}e^{-x}dx = \frac{1}{\Gamma(a)}\int_0^1 t^{a-1}e^{-zt}dt \tag{16.5}$$

For example, if $a = 1$, then (16.5) shows that

$$\gamma^*(1,z) = \frac{1-e^{-z}}{z} \tag{16.6}$$

After substituting the Taylor series of e^{-z} in the integral (16.5), we end up for all z with (Abramowitz and Stegun, 1968, 6.5.29)

$$\gamma^*(a,z) = \frac{1}{\Gamma(a)}\sum_{k=0}^\infty \frac{(-1)^k z^k}{k!\,(a+k)} \tag{16.7}$$

which holds for all complex a, although the derivation above has assumed $\operatorname{Re}(a) > 0$. **Art.** 223 shows that $\gamma^*(a,z)$ is an entire function in z and that (16.7) converges for all a.

220. *Fundamental integral representation.* We demonstrate that the relation

$$\Gamma(a) = \Gamma(a,z) + z^a\Gamma(a)\gamma^*(a,z) \tag{16.8}$$

holds for all complex a and $z = |z|e^{i\arg z}$ with $-\pi < \arg z < \pi$, thus avoiding the negative real z-axis.

The contour integral $\int_L w^{a-1}e^{-w}dw$, where the contour L starts from the real point $w = \tau > 0$ towards $w = r > \tau$ along the real w-axis, a circle with center at $w = 0$ and radius r over an angle φ, returning along the line $w = re^{i\varphi}$ at angle φ until $r = \tau$, and back along a circle with center at $w = 0$ over angle $-\varphi$ with radius τ to the point $w = \tau$, leads to

$$\begin{aligned}\int_L w^{a-1}e^{-w}dw &= \int_\tau^r x^{a-1}e^{-x}dx + ir^a\int_0^\varphi e^{i\theta a}e^{-re^{i\theta}}d\theta \\ &\quad + e^{i\varphi a}\int_r^\tau u^{a-1}e^{-ue^{i\varphi}}du + i\tau^a\int_\varphi^0 e^{i\theta a}e^{-\tau e^{i\theta}}d\theta\end{aligned}$$

Since the region enclosed by the contour L is analytic for complex a, provided the angle lies between $-\pi < \varphi < \pi$ and does not cross the branch cut (i.e. negative real w-axis), the above simplifies to

$$\int_{\tau}^{r} x^{a-1}e^{-x}dx = e^{i\varphi a}\int_{\tau}^{r} u^{a-1}e^{-ue^{i\varphi}}du - ir^{a}\int_{0}^{\varphi} e^{i\theta a}e^{-re^{i\theta}}d\theta + i\tau^{a}\int_{0}^{\varphi} e^{i\theta a}e^{-\tau e^{i\theta}}d\theta$$

If a is an integer, then the negative real axis is not a branch cut and the contour L is analytic in the annulus with radii τ and r. In other words, the constraint on the angle φ does not apply for integer values of a.

The bound $\left|\int_0^{\varphi} e^{i\theta a}e^{-re^{i\theta}}d\theta\right| \leq \int_0^{\varphi} e^{-(\operatorname{Im} a)\theta}e^{-r\cos\theta}d\theta$ for any finite complex a illustrates that, if r tends to infinity, $\int_0^{\varphi} e^{i\theta a}e^{-re^{i\theta}}d\theta \to 0$ only if the angle is further constrained to $-\frac{\pi}{2} \leq \varphi < \frac{\pi}{2}$, resulting in

$$\int_{\tau}^{\infty} x^{a-1}e^{-x}dx = e^{i\varphi a}\int_{\tau}^{\infty} u^{a-1}e^{-ue^{i\varphi}}du + i\tau^{a}\int_{0}^{\varphi} e^{i\theta a}e^{-\tau e^{i\theta}}d\theta$$

Expanding $\int_0^{\varphi} e^{i\theta a}e^{-\tau e^{i\theta}}d\theta = \sum_{k=0}^{\infty}\frac{(-\tau)^k}{k!}\frac{e^{i\varphi(a+k)}-1}{i(a+k)}$ then shows that

$$\int_{\tau}^{\infty} x^{a-1}e^{-x}dx + \sum_{k=0}^{\infty}\frac{(-1)^k\tau^{k+a}}{k!\,(a+k)} = e^{i\varphi a}\int_{\tau}^{\infty} u^{a-1}e^{-ue^{i\varphi}}du + \sum_{k=0}^{\infty}\frac{(-1)^k\left(e^{i\varphi}\tau\right)^{k+a}}{k!\,(a+k)}$$

We express this relation in terms of the definition $\Gamma(a,z) = \int_z^{\infty} x^{a-1}e^{-x}dx$ and the Taylor series (16.7), that defines $\gamma^*(a,z)$, as

$$\Gamma(a,\tau) + \tau^a\Gamma(a)\gamma^*(a,\tau) = e^{i\varphi a}\int_{\tau}^{\infty} u^{a-1}e^{-ue^{i\varphi}}du + \left(e^{i\varphi}\tau\right)^a\Gamma(a)\gamma^*\left(a,e^{i\varphi}\tau\right)$$

which holds for all a and real $\tau > 0$. Moreover, the right-hand side holds for any angle $|\varphi| < \frac{\pi}{2}$, i.e. for any $z = \tau e^{i\varphi}$ with $\operatorname{Re}(z) > 0$, and reduces to the left-hand side for $\varphi = 0$.

The "formal" integral $\Gamma\left(a,e^{i\varphi}\tau\right) = \int_{e^{i\varphi}\tau}^{\infty e^{i\phi}} e^{-t}t^{a-1}dt$, where the point at infinity must[1] have argument $|\phi| < \frac{\pi}{2}$, is chosen *here* as

$$\Gamma\left(a,e^{i\varphi}\tau\right) = e^{i\varphi a}\int_{\tau}^{\infty} u^{a-1}e^{-ue^{i\varphi}}du \qquad \text{for } \tau > 0$$

which exists for all a provided $|\varphi| < \frac{\pi}{2}$. Then, for any $z = \tau e^{i\varphi}$ with $\operatorname{Re}(z) > 0$, we arrive at

$$\Gamma(a,\tau) + \tau^a\Gamma(a)\gamma^*(a,\tau) = \Gamma\left(a,e^{i\varphi}\tau\right) + \left(e^{i\varphi}\tau\right)^a\Gamma(a)\gamma^*\left(a,e^{i\varphi}\tau\right)$$

Only if $\operatorname{Re}(a) > 0$, then $\lim_{\tau\to 0}\Gamma(a,\tau) = \Gamma(a)$ and we find that

$$\Gamma(a) = e^{i\varphi a}\int_{0}^{\infty} u^{a-1}e^{-ue^{i\varphi}}du \qquad \text{for } \operatorname{Re}(a) > 0 \text{ and } |\varphi| < \frac{\pi}{2}$$

If $\varphi = \frac{\pi}{2}$, then $\Gamma(a,i\tau) = e^{i\frac{\pi}{2}a}\int_{\tau}^{\infty} u^{a-1}e^{-iu}du$ converges absolutely for $\operatorname{Re}(a) < 0$, but exists for $\operatorname{Re}(a) < 1$ as follows from partial integration. Cauchy's integral theorem tells us that any path from the point $z = e^{i\varphi}\tau$

[1] Partial integration directly demonstrates this condition.

towards infinity (with argument $|\phi| < \frac{\pi}{2}$) has the same value $\Gamma\left(a, e^{i\varphi}\tau\right)$. However, if we choose the imaginary axis as path, then the integral is confined to $\text{Re}(a) < 1$. Due to this limitation, the path along the imaginary axis is the worst to choose among all paths P with $\text{Re}(P) \geq 0$. As a consequence, the integral $\int_\tau^\infty u^{a-1}e^{-iu}du = e^{-i\frac{\pi}{2}a}\Gamma(a, i\tau)$ is usually studied separately (see Fresnel integrals in Abramowitz and Stegun (1968, Chapter 7), Olver *et al.* (2010, Chapter 7)).

In summary, we have shown that the function $\Gamma(a,z) + z^a\Gamma(a)\gamma^*(a,z)$ exists for all a and z with $\text{Re}(z) > 0$. Further, differentiation of $\Gamma(a,z) + z^a\Gamma(a)\gamma^*(a,z)$ with respect to z indicates that

$$-z^{a-1}e^{-z} + \sum_{k=0}^{\infty} \frac{(-1)^k z^{k+a-1}}{k!} = 0$$

implying that $\Gamma(a,z) + z^a\Gamma(a)\gamma^*(a,z)$ does not dependent upon z and must be a constant. For $\text{Re}(a) > 0$, we know that $\lim_{z\to 0}\Gamma(a,z) = \Gamma(a)$ and $\lim_{z\to 0} z^a\Gamma(a)\gamma^*(a,z) = 0$ and that the constant must be $\Gamma(a)$. Thus, for $\text{Re}(a) > 0$ and $\text{Re}(z) > 0$, we have established $\Gamma(a) = \Gamma(a,z) + z^a\Gamma(a)\gamma^*(a,z)$ in (16.8). Since $\gamma^*(a,z)$ is an entire function for all a and z and both z^a and $\Gamma(a,z) = \int_z^{\infty e^{i\phi}} e^{-t}t^{a-1}dt$ are analytic functions for all z, except for the negative real axis, we finally conclude that (16.8) holds by analytic continuation for all a and z, excluding the negative real axis where $\Gamma(a,z)$ and z^a have a branch cut.

16.2 Recursions

221. *Recursions for* $\Gamma(a,z)$. Partial integration of $\Gamma(a,z) = \int_z^\infty e^{-x}x^{a-1}dx$ leads to the fundamental recursion[2]

$$\Gamma(a,z) = z^{a-1}e^{-z} + (a-1)\Gamma(a-1,z) \tag{16.9}$$

After m iterations of the recursion (16.9), we find, for $|\arg z| < \pi$,

$$\Gamma(a,z) = z^{a-1}e^{-z}\sum_{k=0}^{m-1}\frac{\prod_{j=1}^{k}(a-j)}{z^k} + \Gamma(a-m,z)\prod_{j=1}^{m}(a-j) \tag{16.10}$$

Using $\frac{\Gamma(a)}{\Gamma(a-m)} = \prod_{j=1}^m (a-j)$ in (A.5), the recursion (16.10) is rewritten as

---

[2] Nielsen (1956, Kapitel II, p. 25) defines the function $\gamma(a,z) = z^a\Gamma(a)\gamma^*(a,z) = \sum_{k=0}^\infty \frac{(-1)^k z^{k+a}}{k!(a+k)}$ and derives from its series

$$a\gamma(a,z) = \sum_{k=0}^{\infty}\frac{(-1)^k z^{k+a}}{k!}\frac{a+k-k}{a+k} = z^a e^{-z} - \sum_{k=1}^{\infty}\frac{(-1)^k z^{k+a}}{(k-1)!(a+k)}$$
$$= z^a e^{-z} + \sum_{k=0}^{\infty}\frac{(-1)^k z^{k+1+a}}{k!(a+1+k)}$$

the recursion $a\gamma(a,z) = z^a e^{-z} + \gamma(a+1,z)$ in Abramowitz and Stegun (1968, 6.5.22), that is equivalent to the recursion (16.9).

$$\frac{\Gamma(a,z)}{\Gamma(a)} = z^{a-1}e^{-z}\sum_{k=0}^{m-1}\frac{1}{\Gamma(a-k)}\frac{1}{z^k} + \frac{\Gamma(a-m,z)}{\Gamma(a-m)} \tag{16.11}$$

Example Since $\Gamma(1,z) = \int_z^\infty e^{-x}dx = e^{-z}$, the recursion (16.10) leads for an integer $a = n \geq 1$ to

$$\Gamma(n,z) = (n-1)!e^{-z}\sum_{j=0}^{n-1}\frac{z^j}{j!} \tag{16.12}$$

from which follows

$$\Gamma(n+1,1) = \frac{n!}{e}\sum_{j=0}^{n}\frac{1}{j!} = \frac{\lfloor n!e \rfloor}{e}$$

where the last equality, in which $\lfloor x \rfloor$ denotes the integer smaller than or equal to x, is derived in Van Mieghem (2014, p. 362, eq. 15.3).

Example Rewriting (16.10) as

$$\Gamma(a-m,z) = \frac{1}{\prod_{j=1}^{m}(a-j)}\left(\Gamma(a,z) - z^{a-1}e^{-z}\sum_{k=0}^{m-1}\frac{\prod_{j=1}^{k}(a-j)}{z^k}\right)$$

choosing $a = 0$ and replacing the integer m by n, gives us

$$\Gamma(-n,z) = \frac{(-1)^n}{n!}\left(\Gamma(0,z) - \frac{e^{-z}}{z}\sum_{k=0}^{n-1}\frac{(-1)^k k!}{z^k}\right) \tag{16.13}$$

where $\Gamma(0,z) = -\gamma - \log z - \sum_{k=1}^{\infty}\frac{(-1)^k z^k}{k!k}$ in (16.29) is deduced in **art.** 227.

222. *Recursions for* $\gamma^*(a,z)$. The corresponding recursion for $\gamma^*(a,z)$ follows from the Taylor series (16.7) as

$$\gamma^*(a-1,z) = \frac{e^{-z}}{\Gamma(a)} + z\gamma^*(a,z) \tag{16.14}$$

For $a = 1$, we find from the recursion (16.14) with $\gamma^*(1,z) = \frac{1-e^{-z}}{z}$ in (16.6) that

$$\gamma^*(0,z) = 1 \tag{16.15}$$

Iterating m times the recursion (16.14) of $\gamma^*(a,z)$ gives

$$\gamma^*(a,z) = \frac{1}{z^m}\gamma^*(a-m,z) - \frac{e^{-z}}{z}\sum_{k=0}^{m-1}\frac{1}{\Gamma(a-k)}\frac{1}{z^k} \tag{16.16}$$

Using $\gamma^*(0,z) = 1$ in (16.15), positive integer values of $a = m$ follow from (16.16) as

$$\gamma^*(m,z) = \frac{1}{z^m}\left(1 - e^{-z}\sum_{j=0}^{m-1}\frac{z^j}{j!}\right)$$

which is the same as (16.32) below.

Iterating (16.14) in the other direction yields

$$\gamma^*(a,z) = e^{-z}\sum_{k=0}^{m-1}\frac{z^k}{\Gamma(a+1+k)} + z^m\gamma^*(a+m,z) \tag{16.17}$$

223. *Second series for* $\gamma^*(a,z)$ *besides the Taylor series (16.7).* Bounding the Taylor series (16.7) similarly to Nielsen (1956, p. 26),

$$|\gamma^*(a,z)| \le \frac{1}{|\Gamma(a)|}\sum_{k=0}^{\infty}\frac{|z|^k}{k!\,|a+k|} < \frac{\max_k \frac{1}{|a+k|}}{|\Gamma(a)|}\sum_{k=0}^{\infty}\frac{|z|^k}{k!} = \frac{e^{|z|}\max_k\frac{1}{|a+k|}}{|\Gamma(a)|}$$

shows that $\lim_{m\to\infty}|\gamma^*(a+m,z)| = 0$ for any complex z and a. Consequently, the definition (16.3) of $\gamma^*(a,z)$ leads to $\lim_{m\to\infty}\frac{\Gamma(a+m,z)}{\Gamma(a+m)} = 1$ for any z and a. After letting $m\to\infty$ in the recursion (16.17), we arrive at the converging series (Abramowitz and Stegun, 1968, 6.5.29) for all z and all a,

$$\gamma^*(a,z) = e^{-z}\sum_{k=0}^{\infty}\frac{z^k}{\Gamma(a+1+k)} \tag{16.18}$$

demonstrating that $\gamma^*(a,z)$ is an entire function in both a and z.

By executing the Cauchy product of the right-hand side in (16.18),

$$\gamma^*(a,z) = \sum_{k=0}^{\infty}\frac{(-1)^k z^k}{k!}\sum_{k=0}^{\infty}\frac{z^k}{\Gamma(a+1+k)}$$
$$= \sum_{k=0}^{\infty}\left(\sum_{n=0}^{k}\frac{(-1)^n}{(k-n)!\Gamma(a+1+n)}\right)(-z)^k$$

and invoking $\sum_{n=0}^{k}\frac{\Gamma(a)(-1)^n}{(k-n)!\Gamma(a+1+n)} = \frac{1}{k!(k+a)}$ in (17.5), which is valid for any a, we retrieve the Taylor series (16.7) and demonstrate the validity of (16.7) for all a.

16.3 Asymptotic expansions

224. *Asymptotic expansion for* $\Gamma(a,z)$. Invoking the reflection formula $\Gamma(s)\Gamma(1-s) = \frac{\pi}{\sin\pi s}$ in (11.4) into (16.11)

$$\frac{\Gamma(a,z)}{\Gamma(a)} = z^{a-1}e^{-z}\frac{\sin\pi a}{\pi}\sum_{k=0}^{m-1}\Gamma(1-a+k)\frac{(-1)^k}{z^k} + \frac{\Gamma(a-m,z)}{\Gamma(a-m)} \tag{16.19}$$

shows that, for any fixed a and z, the series for $m\to\infty$ diverges. Since the ratio of two consecutive terms $\frac{\Gamma(a-k+1)}{\Gamma(a-k)}\frac{z^{k+1}}{z^k} = (a-k)z$ in (16.11), the last term $k = m-1$ in the series in (16.19) dominates for large m. The left-hand side in (16.19) clearly does not vary with m, which leads us to conclude, for fixed a and z, that the last term $k = m-1$ is about $-\frac{\Gamma(a-m,z)}{\Gamma(a-m)}$ for large m, i.e.

$$\frac{\Gamma(a-m,z)}{\Gamma(a-m)} \sim (-1)^m z^{a-m} \frac{\sin \pi a}{\pi} e^{-z} \Gamma(m-a) \qquad \text{for } m \to \infty$$

while $\lim_{m\to\infty} \frac{\Gamma(a+m,z)}{\Gamma(a+m)} = 1$ as proved in **art**. 223. The reflection formula (16.11) thus shows that

$$\Gamma(a-m,z) \sim \frac{z^{a-m}e^{-z}}{m-a} \qquad \text{for } m \to \infty \tag{16.20}$$

Consequently, for any fixed $z \neq 0$ (nor real negative) and large, positive a, the asymptotic behavior (16.20) indicates that the integral

$$\Gamma(-a,z) = \int_z^{\infty e^{i\phi}} t^{-a-1}e^{-t}dt \sim \frac{z^{-a}e^{-z}}{a} \qquad \text{for } a \to \infty$$

Hence, $\Gamma(-a,z)$ for $z \neq 0$ tends, for large a, to zero less fast than $\Gamma(-a,0) = \Gamma(-a) = -\frac{\pi}{\sin \pi a}\frac{1}{\Gamma(a+1)}$.

We have shown that the last term in the series (16.11) has the same magnitude as $\left|\frac{\Gamma(a-m,z)}{\Gamma(a-m)}\right|$, that we can consider as a remainder. In other words, when we omit the remainder $\frac{\Gamma(a-m,z)}{\Gamma(a-m)}$ for fixed m, but we allow z to grow large enough to make the remainder small enough, then the m-times iterated recursion (16.11) can serve as an asymptotic series (see also **art**. 231) for large z and fixed a,

$$\Gamma(a,z) \sim z^{a-1}e^{-z}\left(1 + \frac{a-1}{z} + \frac{(a-1)(a-2)}{z^2} + \cdots\right) \tag{16.21}$$

The restriction $|\arg z| < \frac{3\pi}{2}$ is mentioned in Abramowitz and Stegun (1968, 6.5.31) due to enhanced validity[3] when considering multiple Riemann sheets if the branch cut at the negative real axis is crossed. When limiting to $z = re^{i\theta}$ where $\theta \in [-\pi,\pi]$, then the restriction is replaced by the exclusion of the negative real axis, i.e. $|\arg z| < \pi$.

225. *Asymptotic expansion for the entire function* $\gamma^*(a,z)$. Introducing the recursion (16.11) of $\frac{\Gamma(a,z)}{\Gamma(a)}$ in the function $\gamma^*(a,z) = z^{-a}\left(1 - \frac{\Gamma(a,z)}{\Gamma(a)}\right)$ in (16.3) or, equivalently, using definition (16.3) in (16.16), yields, for any integer $m \geq 0$,

$$\gamma^*(a,z) = z^{-a}\left(1 - \frac{\Gamma(a-m,z)}{\Gamma(a-m)}\right) - \frac{e^{-z}}{z}\sum_{k=0}^{m-1} \frac{1}{\Gamma(a-k)}\frac{1}{z^k} \tag{16.22}$$

After introducing $\frac{\Gamma(a-m,z)}{\Gamma(a-m)} \sim \frac{z^{a-m}e^{-z}}{\Gamma(a-m+1)}$ in (16.20), we obtain, similarly as argued in **art**. 224, the asymptotic expansion, for large z and fixed a,

$$\gamma^*(a,z) \sim z^{-a} - \frac{e^{-z}}{z\Gamma(a)}\left(1 + \frac{a-1}{z} + \frac{(a-1)(a-2)}{z^2} + O\left(\frac{a^3}{z^3}\right)\right) \tag{16.23}$$

[3] Priviate communication with the late Richard B. Paris.

which will reappear in **art.** 229.

226. *Converging variant of the asymptotic expansion (16.21) for* $\Gamma(a,z)$. We present a converging companion for (16.21), from which the asymptotic expansion (16.21) can be deduced. After substitution of $x = u + z$ in $\Gamma(a,z) = \int_z^\infty x^{a-1}e^{-x}dx$, we obtain

$$\Gamma(a,z) = e^{-z}\int_0^\infty (u+z)^{a-1} e^{-u}du = z^{a-1}e^{-z}\int_0^\infty \left(1+\frac{u}{z}\right)^{a-1} e^{-u}du$$

The binomial series $\left(1+\frac{u}{z}\right)^{a-1} = \sum_{k=0}^\infty \binom{a-1}{k}\frac{u^k}{z^k}$ in (17.1) is only convergent for $u < |z|$ and suggests to split the integral into two parts,

$$\begin{aligned}\Gamma(a,z) &= z^{a-1}e^{-z}\sum_{k=0}^\infty \binom{a-1}{k}\frac{1}{z^k}\int_0^z u^k e^{-u}du + e^{-z}\int_z^\infty (z+u)^{a-1} e^{-u}du \\ &= z^{a-1}e^{-z}\sum_{k=0}^\infty \frac{\Gamma(a)\,(k! - \Gamma(k+1,z))}{\Gamma(a-k)\,k!}\frac{1}{z^k} + \Gamma(a,2z)\end{aligned}$$

Hence, with (16.12), we obtain a duplication formula for the argument z,

$$\Gamma(a,2z) = \Gamma(a,z) - z^{a-1}e^{-z}\sum_{k=0}^\infty \frac{\Gamma(a)}{\Gamma(a-k)}\frac{\left(1 - e^{-z}\sum_{j=0}^k \frac{z^j}{j!}\right)}{z^k} \tag{16.24}$$

For large z, we may argue that $\Gamma(a,2z)$ is negligible with respect to $\Gamma(a,z)$ and, similarly, that $e^{-z}\sum_{j=0}^k \frac{z^j}{j!}$ is negligible with respect to 1 for fixed k, so that (16.24) reduces, for the first finite k terms, to the asymptotic expansion (16.21). Fortunately, we can be more precise. After replacing $\frac{1}{z^k}\int_0^z u^k e^{-u}du = z\int_0^1 t^k e^{-zt}dt$, the alternative form of (16.24) is

$$\Gamma(a,z) = z^a \sum_{k=0}^\infty \binom{a-1}{k}\int_0^1 t^k e^{-z(t+1)}dt + \Gamma(a,2z)$$

After m iterations, we obtain[4]

$$\Gamma(a,z) = z^a \sum_{k=0}^\infty \binom{a-1}{k}\int_0^1 t^k \left\{\sum_{l=0}^m 2^{la} e^{-2^l z(1+t)}\right\} dt + \Gamma\left(a, 2^{m+1}z\right)$$

If $m \to \infty$, then the l-sum converges quickly, while (16.2) shows that

[4] After evaluating the binomial k-series with (17.1),

$$\Gamma(a,z) = z\sum_{l=0}^m 2^{la}\left\{\int_0^1 (z(1+t))^{a-1} e^{-2^l z(1+t)}dt\right\} + \Gamma\left(a,2^{m+1}z\right)$$

and executing the integral $\int_0^1 (z(1+t))^{a-1} e^{-2^l z(1+t)}dt = \frac{\int_{2^l z}^{2^{l+1}z} u^{a-1}e^{-u}dt}{2^{la}z} = \frac{1}{2^{la}z}\left(\Gamma\left(a,2^l z\right) - \Gamma\left(a,2^{l+1}z\right)\right)$, we arrive at the telescoping series

$$\Gamma(a,z) = \sum_{l=0}^m \left(\Gamma\left(a,2^l z\right) - \Gamma\left(a,2^{l+1}z\right)\right) + \Gamma\left(a,2^{m+1}z\right)$$

which is an identity, proving the correctness of (16.25).

$\lim_{m\to\infty}\Gamma\left(a,2^{m+1}z\right)=0$ for $\operatorname{Re}(z)>0$. With $\Gamma(k+1,z)=k!e^{-z}\sum_{j=0}^{k}\frac{z^j}{j!}$ in (16.12), the integral becomes

$$\begin{aligned}\int_0^1 t^k e^{-2^l z(1+t)}dt &= \frac{e^{-2^l z}}{\left(2^l z\right)^{k+1}}\int_0^{2^l z} u^k e^{-u}du \\ &= \frac{e^{-2^l z}}{\left(2^l z\right)^{k+1}}\left(\Gamma(k+1)-\Gamma\left(k+1,2^l z\right)\right) \\ &= \frac{k!e^{-2^l z}}{\left(2^l z\right)^{k+1}}\left(1-e^{-2^l z}\sum_{j=0}^{k}\frac{\left(2^l z\right)^j}{j!}\right)\end{aligned}$$

Finally, we find the converging series for $K\to\infty$, for any finite integer m, for any a and z, except for negative real z,

$$\Gamma(a,z)=\sum_{k=0}^{K}\frac{\Gamma(a)\,z^{a-k-1}}{\Gamma(a-k)}\sum_{l=0}^{m}\frac{e^{-2^l z}}{2^{l(k-a+1)}}\left(1-\frac{\sum_{j=0}^{k}\frac{\left(2^l z\right)^j}{j!}}{e^{2^l z}}\right)+\Gamma\left(a,2^{m+1}z\right) \tag{16.25}$$

The asymptotic expansion (16.21) only corresponds to the $l=0$ term in (16.25).

Numerically, however, (16.25) behaves similar as the asymptotic series (16.21) due to the fast convergence in m and the numerically instable difference. Therefore, we replace the difference $1-e^{-z}\sum_{j=0}^{k}\frac{z^j}{j!}=e^{-z}\sum_{j=k+1}^{\infty}\frac{z^j}{j!}$ in (16.25), reverse the k- and j-sum to obtain, for $J\to\infty$,

$$\Gamma(a,z)=\sum_{l=0}^{m}\frac{2^{la}}{e^{2^{l+1}z}}\sum_{j=0}^{J}\left(\sum_{k=0}^{j}\frac{\Gamma(a)}{\Gamma(a-k)}\frac{z^{a-k-1}}{2^{l(k+1)}}\right)\frac{\left(2^l z\right)^{j+1}}{(j+1)!}+\Gamma\left(a,2^{m+1}z\right) \tag{16.26}$$

The series (16.26) converges rapidly in m, but slower in J. For example, for $z=0.1$ and $a=0.5$, only $m=3$ is sufficient, but $J=30$ produces an error about 10^{-3} in (16.26), whereas $K=30$ in (16.25) is off by 10^{46} (already for $m=1$). For $z=1$ and $z=10$ (with $a=0.5$ and $J=K=30$), the error in (16.26) reduces to 2.10^{-4} and 4.10^{-12}, while the error in (16.25) is 10^{8} and 10^{-11}, respectively.

16.4 The logarithmic integral and the Mittag-Leffler function

227. *Logarithmic and exponential integral.* The theory of the logarithmic integral is beautifully treated in the book of Niels Nielsen (1956, band II), who proceeds much deeper than the brief derivation here. Nielsen masterfully demonstrates that the logarithmic integral is a special case of Prym's function $Q_x(a)=\Gamma(a,x)$, now called the incomplete Gamma function, $\Gamma(a,x)=\int_x^\infty t^{a-1}e^{-t}dt$, connected to the Gamma function by the fundamental relation $\Gamma(a)-\Gamma(a,z)+z^a\Gamma(a)\gamma^*(a,z)$ in (16.8). Nielsen defines

$$\mathrm{li}\left(e^{-x}\right) = -\int_x^\infty \frac{e^{-t}}{t} dt = -\Gamma(0,x) \tag{16.27}$$

which illustrates that $\lim_{x\to 0}\mathrm{li}(e^{-x}) = -\infty$. Today, Nielsen's definition is known (Abramowitz and Stegun, 1968, 5.1.1) as $\mathrm{li}(e^{-x}) = -E_1(x)$, where the exponential integral is $E_1(x) = \int_x^\infty \frac{e^{-t}}{t} dt$ for $|\arg x| < \pi$. After substitution $u = e^{-t}$ in (16.27), we find the logarithmic integral

$$\mathrm{li}(x) = \int_0^x \frac{dt}{\log t} \tag{16.28}$$

whose integrand contains a logarithmic singularity at $t = 1$ and $\log t < 0$ for $t < 1$ and $\log t > 0$ for $t > 1$. The number of primes less than x can be written as $\pi(x) = \mathrm{li}(x) + \varepsilon(x)$, where $\varepsilon(x) = o\left(\frac{x}{\log x}\right)$ due to the Prime Number theorem (Titchmarsh and Heath-Brown, 1986). Nielsen (1956, band II) starts by computing the limit $\Gamma(0,x) = \lim_{a\to 0}\Gamma(a) - x^a\Gamma(a)\gamma^*(a,x)$, after rewriting (16.8) with (16.7) as

$$\Gamma(a,x) = \frac{\Gamma(a+1) - x^a}{a} - \sum_{k=1}^\infty \frac{(-1)^k x^{k+a}}{k!\,(a+k)}$$

Using the Euler constant $\gamma = -\Gamma'(1)$, he finds the series

$$\mathrm{li}\left(e^{-x}\right) = -\Gamma(0,x) = \gamma + \log x + \sum_{k=1}^\infty \frac{(-1)^k x^k}{k!k} \tag{16.29}$$

which appeared in (16.13).

228. *Differentiation of $\gamma^*(a,z)$ with respect to a.* By derivation of (16.18) with respect to a, we obtain

$$\frac{\partial\gamma^*(a,z)}{\partial a} = -e^{-z}\sum_{k=0}^\infty \frac{z^k}{\Gamma(a+1+k)}\psi(a+1+k)$$

from which we deduce, with $\psi(1+k) = -\gamma + \sum_{j=1}^k \frac{1}{j}$, that

$$\left.\frac{\partial\gamma^*(a,z)}{\partial a}\right|_{a=0} = -e^{-z}\sum_{k=0}^\infty \frac{z^k}{k!}\left(-\gamma + \sum_{j=1}^k \frac{1}{j}\right) = \gamma - e^{-z}\sum_{k=1}^\infty \frac{z^k}{k!}\sum_{j=1}^k \frac{1}{j}$$

From the Taylor series (16.7) of $\gamma^*(a,z)$, we find that

$$\begin{aligned}\frac{\partial\gamma^*(a,z)}{\partial a} &= -\frac{\psi(a)}{\Gamma(a)}\sum_{k=0}^\infty \frac{(-1)^k z^k}{k!\,(a+k)} - \frac{1}{\Gamma(a)}\sum_{k=0}^\infty \frac{(-1)^k z^k}{k!\,(a+k)^2} \\ &= -\frac{\psi(a) + \frac{1}{a}}{\Gamma(a+1)} - \frac{a\psi(a)\sum_{k=1}^\infty \frac{(-1)^k z^k}{k!(a+k)}}{\Gamma(a+1)} - \frac{a\sum_{k=1}^\infty \frac{(-1)^k z^k}{k!(a+k)^2}}{\Gamma(a+1)}\end{aligned}$$

The functional equation $\psi(a+1) = \psi(a) + \frac{1}{a}$ in (12.1) indicates that $\lim_{a\to 0} a\psi(a) = -1$. Combining both expressions yields

$$\gamma - \left.\frac{\partial \gamma^*(a,z)}{\partial a}\right|_{a=0} = \sum_{k=1}^{\infty} \frac{(-1)^{k-1} z^k}{k!k} = e^{-z} \sum_{k=1}^{\infty} \frac{z^k}{k!} \sum_{j=1}^{k} \frac{1}{j}$$

illustrating that $\gamma - \left.\frac{\partial \gamma^*(a,x)}{\partial a}\right|_{a=0} \geq 0$ for real x. For real $x > 0$, Nielsen's series (16.29) in **art.** 227 indicates that

$$\left.\frac{\partial \gamma^*(a,x)}{\partial a}\right|_{a=0} = -\Gamma(0,x) - \log x$$

229. *The entire incomplete Gamma function and the Mittag-Leffler function.* We introduce Hankel's integral (15.4), along a path with two lines under an angle ϕ and $-\phi$ as specified in **art.** 204, into (16.18)

$$\gamma^*(a,z) = e^{-z} \sum_{k=0}^{\infty} z^k \frac{1}{2\pi i} \int_{c-\infty e^{-i\phi}}^{c+\infty e^{i\phi}} w^{-a-1-k} e^w dw$$

Since $|w| > c > 0$, we must require that $c > |z|$, then $\sum_{k=0}^{\infty} \left(\frac{z}{w}\right)^k = \frac{1}{1-\frac{z}{w}}$ and the reversal of integration and sum yields, for $\frac{\pi}{2} < \phi < \pi$,

$$\gamma^*(a,z) = e^{-z} \frac{1}{2\pi i} \int_{c-\infty e^{-i\phi}}^{c+\infty e^{i\phi}} \frac{w^{-a} e^w}{w - z} dw \tag{16.30}$$

which is a special case of the *basic complex integral* in (4.4), with $c > |z|$ and $\frac{\pi}{2} < \phi < \pi$,

$$E_{a,b}(z) = \sum_{k=0}^{\infty} \frac{z^k}{\Gamma(b+ak)} = \frac{1}{2\pi i} \int_{c-\infty e^{-i\phi}}^{c+\infty e^{i\phi}} \frac{w^{a-b} e^w}{w^a - z} dw$$

of the Mittag-Leffler function $E_{a,b}(z)$. Indeed, **art.** 5 shows that $e^z \gamma^*(a,z) = E_{1,a+1}(z)$.

When $\operatorname{Re}(z) > 0$, we can move the line of integration in (16.30) to the left of $|z|$. We thus enclose the simple pole at $w = z$ and (16.30) becomes, with $c' < |z|$ and $\frac{\pi}{2} < \phi' < \pi$,

$$\gamma^*(a,z) = z^{-a} - e^{-z} \frac{1}{2\pi i} \int_{c'-\infty e^{-i\phi'}}^{c'+\infty e^{i\phi'}} \frac{w^{-a} e^w}{z - w} dw$$

The asymptotic expansion (16.23) now follows for large z, after using the geometric series $\frac{1}{z-w} = \frac{1}{z} \sum_{k=0}^{\infty} \left(\frac{w}{z}\right)^k$, interchanging integration and summation *and* ignoring the violation $|w| > |z|$ along the path of integration towards $\infty e^{i\phi}$. The latter violation determines that (16.23) is an asymptotic expansion that diverges eventually when an unbounded number of terms are added. Another important observation from the above integral is that, for $a=-n$, the integral can be closed over the $\operatorname{Re}(w)<0$ plane, where an entire analytic region is enclosed, which implies that

$$\gamma^*(-n,z)=z^n \qquad \text{for } n=0,1,2,\ldots \tag{16.31}$$

The companion of (16.31) for positive integers follows from the definition (16.3) of $\gamma^*(a,z)$ and (16.12) as

$$\gamma^*(n,z)=z^{-n}\left(1-\frac{\Gamma(n,z)}{\Gamma(n)}\right)=z^{-n}\left(1-e^{-z}\sum_{j=0}^{n-1}\frac{z^j}{j!}\right) \qquad \text{for } n\geq 1 \tag{16.32}$$

16.5 Mellin and Laplace transforms of $\Gamma(a,z)$ and $\Gamma(z)$

230. *The Mellin transform of the incomplete Gamma function* $\Gamma(a,z)$. The Mellin transform of $\Gamma(a,z)=\int_z^\infty e^{-x}x^{a-1}dx$ is (**art.** 245)

$$\mathcal{M}[\Gamma(a,z),s]=\int_0^\infty t^{s-1}\Gamma(a,t)\,dt=\int_0^\infty t^{s-1}\left(\int_t^\infty e^{-x}x^{a-1}dx\right)dt$$

Partial integration yields, for $c=\operatorname{Re}(s)>0$ and $\operatorname{Re}(a)>0$, the closed form

$$\mathcal{M}[\Gamma(a,z),s]=\frac{1}{s}\int_0^\infty e^{-t}t^{a+s-1}dt=\frac{\Gamma(a+s)}{s}$$

The inverse Mellin transform $\Gamma(a,z)=\mathcal{M}^{-1}\left[\frac{\Gamma(a+s)}{s},z\right]$ is, with (B.4),

$$\Gamma(a,z)=\frac{1}{2\pi i}\int_{c-i\infty}^{c+i\infty}\frac{\Gamma(a+s)}{s}z^{-s}ds \qquad (c>0 \text{ and } \operatorname{Re}(a)>0) \tag{16.33}$$

By closing the contour over the $\operatorname{Re}(s)<0$, we encounter simples poles at $s=0$ and $s=-k-a$, assuming that $a\neq 0$, else there is a double pole at $s=0$ which leads to the exponential integral $E_1(z)=\Gamma(0,z)$ in **art.** 227. By Cauchy residue theorem, we obtain

$$\Gamma(a,z)=\Gamma(a)+\sum_{k=0}^{\infty}\frac{(-1)^k}{k!\,(-k-a)}z^{a+k}$$

which equals with the definition (16.3) of $\gamma^*(a,z)$ again the first Taylor series (16.7).

231. *The Laplace transform of the incomplete Gamma function* $\Gamma(a,z)$. The single-sided Laplace transform of $\Gamma(a,z)=\int_z^\infty e^{-x}x^{a-1}dx$ for real $z\geq 0$ is (**art.** 244)

$$\mathcal{L}[\Gamma(a,z),s]=\int_0^\infty e^{-st}\Gamma(a,t)\,dt=\int_0^\infty e^{-st}\left(\int_t^\infty e^{-x}x^{a-1}dx\right)dt$$

Partial integration yields, for $c = \text{Re}(s) > -1$ (due to (16.21)) and for $\text{Re}(a) > 0$,

$$\mathcal{L}\left[\Gamma(a,z), s\right] = \frac{\Gamma(a)}{s}\left(1 - \frac{1}{(s+1)^a}\right)$$

The inverse Laplace transform (B.2)

$$\Gamma(a,z) = \mathcal{L}^{-1}\left[\frac{\Gamma(a)}{s}\left(1 - \frac{1}{(s+1)^a}\right), z\right]$$
$$= \frac{\Gamma(a)}{2\pi i}\int_{c-i\infty}^{c+i\infty} \frac{e^{sz}}{s}\left(1 - \frac{1}{(s+1)^a}\right) ds \qquad (-1 < c < 0)$$

leads to an integral, valid for $\text{Re}(z) \geq 0$. The first integral $\int_{c-i\infty}^{c+i\infty} \frac{e^{sz}}{s} ds = 0$ for $\text{Re}(z) > 0$. Evaluating the second integral for $\text{Re}(z) \geq 0$,

$$\frac{\Gamma(a,z)}{\Gamma(a)} = -\frac{1}{2\pi i}\int_{c-i\infty}^{c+i\infty} \frac{e^{sz}}{s}\frac{1}{(s+1)^a} ds$$

along the line $s = c + it$, allows us to make the substitution $w = s + 1$,

$$\frac{\Gamma(a,z)}{\Gamma(a)} = -\frac{1}{2\pi}\int_{-\infty}^{\infty} \frac{e^{(c+it)z}}{c+it}\frac{1}{(1+c+it)^a} dt$$
$$= -\frac{1}{2\pi}\int_{-\infty}^{\infty} \frac{e^{(1+c+it-1)z}}{(1+c+it)-1}\frac{1}{(1+c+it)^a} dt$$
$$= -\frac{1}{2\pi i}\int_{1+c-i\infty}^{1+c+i\infty} \frac{e^{(w-1)z}}{w-1} w^{-a} dw$$

where now $1 + c = \text{Re}(w) > 0$. Next, we close the contour over the $\text{Re}(w) < 0$-half plane, but avoid the branch cut at the real negative w-axis. The resulting contour is precisely Hankel's contour C in **art.** 204,

$$\frac{\Gamma(a,z)}{\Gamma(a)} = -\frac{1}{2\pi i}\int_C \frac{e^{(w-1)z}}{w-1} w^{-a} dw \tag{16.34}$$

Similarly as Hankel's contour integral (15.1), the integral (16.34) holds for all complex a and $\text{Re}(z) \geq 0$. Evaluation along Hankel's contour C yields

$$2\pi i\frac{\Gamma(a,z)}{\Gamma(a)} = -\int_\infty^\varepsilon \frac{e^{-(x+1)z}dx}{(x+1)(xe^{-i\pi})^a} - \varepsilon^{1-a}\int_{-\frac{\pi}{2}}^{\frac{\pi}{2}} \frac{e^{\left(-1+\varepsilon e^{i\theta}\right)z}}{\left(-1+\varepsilon e^{i\theta}\right)} e^{i(1-a)\theta} d\theta$$
$$- \int_\varepsilon^\infty \frac{e^{-(x+1)z}dx}{(x+1)(xe^{i\pi})^a}$$

If $\varepsilon \to 0$, then the second integral vanishes for $1 > \text{Re}(a)$ and we find, for $\text{Re}(z) \geq 0$ and $\text{Re}(a) < 1$, that

$$\frac{\Gamma(a,z)}{\Gamma(a)} = \frac{\sin \pi a}{\pi} e^{-z}\int_0^\infty \frac{x^{-a}e^{-xz}dx}{x+1} \tag{16.35}$$

After substitution of $t = xz$ in (16.35) for $\text{Re}(z) > 0$,

$$\frac{\Gamma(a,z)}{\Gamma(a)} = \frac{\sin \pi a}{\pi} e^{-z} z^{a-1}\int_0^\infty \frac{t^{-a}e^{-t}dt}{1+\frac{t}{z}}$$

and introduction of the geometric series (while ignoring the convergence constraints that $\left|\frac{t}{z}\right| < 1$), we obtain the asymptotic, but diverging series (16.19) for $\mathrm{Re}(a) < 1$ and large z.

232. *Mellin transform of a product of Gamma functions.* Let $p_1, p_2, \ldots, p_n$ be different real, positive numbers, then the Mellin transform

$$\prod_{j=1}^{n} \Gamma(s+p_j) = \int_0^\infty u^{s-1} g\left(u; \{p_j\}_{1\le j\le n}\right) du \tag{16.36}$$

has an inverse

$$g\left(u; \{p_j\}_{1\le j\le n}\right) = \frac{1}{2\pi i}\int_{c-i\infty}^{c+i\infty} \prod_{j=1}^{n} \Gamma(s+p_j)\, u^{-s} ds \qquad \text{with } c>0 \tag{16.37}$$

which can be evaluated. The contour in (16.37) can be closed over the negative $\mathrm{Re}(s)$-plane and encloses the simple poles at $s = -q_j - p_j$ for integers $q_j \ge 0$ for $1 \le j \le n$ on the negative real axis. Cauchy's residue theorem (Titchmarsh, 1964) leads to

$$g\left(u; \{p_j\}_{1\le j\le n}\right) = \sum_{j=1}^{n}\sum_{q_j=0}^{\infty} res_j \prod_{m=1;m\neq j}^{n} \Gamma(s+p_m)\, u^{-s}$$

where the residue res_j is, after iterating the functional equation (10.1) of the Gamma function (q_j+1)-times,

$$\begin{aligned} res_j &= \lim_{s\to -q_j-p_j} (s+q_j+p_j)\,\Gamma(s+p_j) \\ &= \lim_{s\to -q_j-p_j} \frac{\Gamma(s+p_j+q_j+1)(s+p_j+q_j)}{(s+p_j)(s+p_j+1)\ldots(s+p_j+q_j-1)(s+p_j+q_j)} \\ &= \frac{\Gamma(1)}{(-q_j)(-q_j+1)\ldots(-1)} = \frac{(-1)^{q_j}}{(q_j)!} \end{aligned}$$

Hence, we obtain

$$g\left(u; \{p_j\}_{1\le j\le n}\right) = \sum_{j=1}^{n}\sum_{q_j=0}^{\infty} \frac{(-1)^{q_j} u^{q_j+p_j}}{(q_j)!} \prod_{m=1;m\neq j}^{n} \Gamma(-q_j+p_m-p_j)$$

Using the reflection formula (11.4)

$$\begin{aligned} \mathbb{B} &= \prod_{m=1;m\neq j}^{n} \Gamma(-q_j+p_m-p_j) = \prod_{m=1;m\neq j}^{n} \frac{\pi(-1)^{q_j}}{\sin\pi(p_m-p_j)\,\Gamma(1+q_j+p_j-p_m)} \\ &= \frac{\pi^{n-1}(-1)^{(n-1)q_j}}{\prod_{m=1}^{j-1}\sin\pi(p_m-p_j)\prod_{m=j+1}^{n}\sin\pi(p_m-p_j)} \frac{1}{\prod_{m=1;m\neq j}^{n}\Gamma(1+q_j+p_j-p_m)} \end{aligned}$$

we arrive at the series

$$g\left(u;\{p_j\}_{1\leq j\leq n}\right)=\sum_{j=1}^{n}\frac{(-1)^{j-1}\,\pi^{n-1}u^{p_j}}{\prod_{m=1}^{j-1}\sin\pi\left(p_j-p_m\right)\prod_{m=j+1}^{n}\sin\pi\left(p_m-p_j\right)} \tag{16.38}$$

$$\times\sum_{q_j=0}^{\infty}\frac{(-1)^{nq_j}\,u^{q_j}}{(q_j)!\prod_{m=1;m\neq j}^{n}\Gamma\left(1+q_j+p_j-p_m\right)}$$

with the convention that $\prod_{m=a}^{b}f_m=1$ if $b<a$.

Examples If $n=1$, then (16.38) reduces to the classical Mellin transform of the pair e^{-u} and $\Gamma(s)$,

$$g\left(u;p_1\right)=u^{p_1}\sum_{q_1=0}^{\infty}\frac{(-1)^{q_1}\,u^{q_1}}{(q_1)!}=u^{p_1}e^{-u}$$

If $p_j=\frac{j}{n}$ for $1\leq j\leq n-1$ as in Gauss's multiplication formula (11.8) without $j=0$ factor, then (16.38), denoted as $g\left(u;\left\{\frac{j}{n}\right\}_{1\leq j\leq n-1}\right)=h_n(u)$ reduces to

$$h_n(u)=\pi^{n-2}\sum_{j-1}^{n-1}\frac{(-1)^{j-1}\,u^{\frac{j}{n}}}{\prod_{k=1}^{j-1}\sin\left(\frac{\pi k}{n}\right)\prod_{k=1}^{n-j-1}\sin\left(\frac{\pi k}{n}\right)} \tag{16.39}$$

$$\times\sum_{q_j=0}^{\infty}\frac{\left((-1)^{n-1}\right)^{q_j}\,u^{q_j}}{(q_j)!\prod_{k=1}^{j-1}\Gamma\left(1+q_j+\frac{k}{n}\right)\prod_{m=1}^{n-j-1}\Gamma\left(1+q_j-\frac{m}{n}\right)}$$

In particular for $n=3$, then

$$h_3(u)=\frac{\pi u^{\frac{1}{3}}}{\sin\left(\frac{\pi}{3}\right)}\sum_{q_1=0}^{\infty}\frac{u^{q_1}}{(q_j)!\Gamma\left(1+q_1-\frac{1}{3}\right)}-\frac{\pi u^{\frac{2}{3}}}{\sin\left(\frac{\pi}{3}\right)}\sum_{q_2=0}^{\infty}\frac{u^{q_2}}{(q_j)!\Gamma\left(1+q_2+\frac{1}{3}\right)}$$

We briefly summarize the theory of the modified Bessel function $K_\nu(z)$, but strongly advise to consult the monumental treatise of Watson (1995). The modified Bessel functions $I_\nu(z)$ and $K_\nu(z)$ are defined (Watson, 1995, pp. 77-78) as the two independent solutions of the modified Bessel differential equation

$$z^2\frac{d^2y}{dz^2}+z\frac{dy}{dz}-\left(z^2+\nu^2\right)y=0$$

Both $I_\nu(z)$ and $K_\nu(z)$ are entire functions in v for $z\neq 0$ and analytic in z, except for a cut along the negative real axis. The function $I_\nu(z)$ in z

possesses the Taylor series

$$I_\nu(z) = \sum_{k=0}^{\infty} \frac{\left(\frac{1}{2}z\right)^{\nu+2k}}{k!\Gamma(\nu+k+1)} \tag{16.40}$$

The modified Bessel functions $K_\nu(z)$ defined as

$$K_\nu(z) = \frac{\pi}{2\sin\pi\nu}\left(I_{-\nu}(z) - I_\nu(z)\right) \tag{16.41}$$

is even in the order ν, i.e. $K_\nu(z) = K_{-\nu}(z)$ for all $z \neq 0$, with Taylor series

$$K_\nu(z) = \frac{\pi}{2\sin\pi\nu}\left(\left(\frac{z}{2}\right)^{-\nu}\sum_{k=0}^{\infty}\frac{\left(\frac{1}{4}z^2\right)^k}{k!\Gamma(1-\nu+k)} - \left(\frac{z}{2}\right)^{\nu}\sum_{k=0}^{\infty}\frac{\left(\frac{1}{4}z^2\right)^k}{k!\Gamma(1+\nu+k)}\right) \tag{16.42}$$

The Taylor series (16.42) shows that

$$h_3(u) = 2u^{\frac{1}{2}}K_{\frac{1}{3}}(2\sqrt{u}) \tag{16.43}$$

On the other hand, the integral in Rademacher (1973, eq. (28.73), p. 55), valid for $\operatorname{Re}(s) > 0$ and $\operatorname{Re}(p) > 0$,

$$\Gamma(s)\Gamma(s+p) = 2\int_0^\infty K_p\left(2\sqrt{x}\right)x^{\frac{p}{2}+s-1}dx$$

is a special case of (16.36) and corresponds to $h_3(u)$ in (16.43) with $p = \frac{1}{3}$ and $s \to s + \frac{1}{3}$.

17
The Beta function

The Beta function was already defined in (10.11) by both the fraction of Gamma functions and its integral, valid for $\mathrm{Re}\,(p) > 0$ and $\mathrm{Re}\,(q) > 0$,

$$B\,(p, q) = \frac{\Gamma\,(p)\,\Gamma\,(q)}{\Gamma\,(p+q)} = \int_0^1 u^{p-1}\,(1-u)^{q-1}\,du$$

The Beta-function integral (10.11) is also called the *Eulerean integral of the first kind* (Jordan, 1913, art. 222). Its intimate connection with Newton's binomial series is shown in **art**. 233. Some transformations of the Beta integral are derived in **art**. 237. Chapter 17 terminates with the classical proof in **art**. 238 of $B\,(p,q) = \frac{\Gamma(p)\Gamma(q)}{\Gamma(p+q)}$, derived from Euler's integral $\Gamma\,(z) = \int_0^\infty e^{-t}t^{z-1}dt$ and with Nanjundiah's proof (**art**. 239). The product $B\,(p,q) = \frac{\Gamma(p)\Gamma(q)}{\Gamma(p+q)}$ illustrates the symmetry in $B\,(p,q) = B\,(q,p)$.

Mainly due to historical reasons, explained in Section 9.1 and **art**. 233, we include a short chapter on the Beta function. Ogilvy (1951) graphically shows where the zeros and poles of the $B\,(p,q)$ lie in the real p- and real q-plane. Clearly, $B\,(p,q) = 0$ if $p + q = -k$, where k is an integer, but p and q are not negative integers.

233. *The binomial Beta series.* Newton's famous generalization of the binomium $\sum_{k=0}^{n}\binom{n}{k}z^k = (1+z)^n$ is the binomial series, convergent for all complex s provided $|z| < 1$,

$$(1+z)^s = \sum_{k=0}^{\infty}\binom{s}{k}z^k \tag{17.1}$$

Euler (1738) computed the series of the Beta function by insertion of the binomium for integer q in the Beta integral (10.11). Substituting (17.1) for $\mathrm{Re}\,(q) > 0$ in the Beta integral (10.11) yields

$$\begin{aligned} B\,(p,q) &= \sum_{k=0}^{\infty}\binom{q-1}{k}\frac{(-1)^k}{p+k} = \sum_{k=0}^{\infty}\frac{\prod_{m=1}^{k}\left(1-\frac{q}{m}\right)}{p+k} \qquad (17.2) \\ &= \frac{1}{p} + \frac{1-q}{p+1} + \frac{(1-q)\left(1-\frac{q}{2}\right)}{p+2} + \frac{(1-q)\left(1-\frac{q}{2}\right)\left(1-\frac{q}{3}\right)}{p+3} + \cdots \end{aligned}$$

The binomial Beta series (17.2) converges for all $\mathrm{Re}(p) > 0$ and $\mathrm{Re}(q) > 0$.

Example For integer[1] $q = n + 1$, (17.2) reduces to a finite sum

$$\frac{\Gamma(p)\, n!}{\Gamma(p+n)} = \sum_{k=0}^{n} \binom{n}{k} \frac{(-1)^k}{p+k} \tag{17.3}$$

Binomial inversion (Riordan, 1968, Chapter 2)

$$a_n = \sum_{k=0}^{n} \binom{n}{k} b_k \Leftrightarrow b_n = (-1)^n \sum_{k=0}^{n} (-1)^k \binom{n}{k} a_k \tag{17.4}$$

applied to (17.3) leads to

$$\frac{1}{p+n} = \sum_{k=0}^{n} (-1)^k \binom{n}{k} \frac{\Gamma(p)\, k!}{\Gamma(p+k)} \tag{17.5}$$

Formula (17.5) is a special case of the multiple parameter factorial series (13.44) with $b_j = j$.

234. *Infinite product for the Beta-function.* The Euler-Gauss product (10.10) transforms the definition $B(p,q) = \frac{\Gamma(p)\Gamma(q)}{\Gamma(p+q)}$ of the Beta function into

$$B(p,q) = \frac{p+q}{pq} \prod_{n=1}^{\infty} \frac{\left(1 + \frac{p+q}{n}\right)}{\left(1+\frac{p}{n}\right)\left(1+\frac{q}{n}\right)} = \frac{p+q}{pq} \prod_{n=1}^{\infty} \frac{n(n+p+q)}{(n+p)(n+q)}$$

which we can write as

$$B(p,q) = \frac{p+q}{pq} \prod_{n=1}^{\infty} \left(1 - \frac{pq}{(n+p)(n+q)}\right)$$

or

$$B(p,q) = \frac{\frac{p+q}{pq}}{\prod_{n=1}^{\infty} \left(1 + \frac{pq}{n(n+p+q)}\right)} \tag{17.6}$$

For positive real p and q, both products indicate that $B(p,q) < \frac{p+q}{pq} = \frac{1}{p} + \frac{1}{q}$.

The Integral Mean Value Theorem (**art.** 114) states that there exist real values $\xi_q, \xi_p \in [0,1]$ so that the Beta integral in (10.11) equals

$$B(p,q) = \int_0^1 u^{p-1}(1-u)^{q-1}\, du = \xi^{p-1} \int_0^1 (1-u)^{q-1}\, du = \frac{\xi_p^{p-1}}{q} = \frac{\xi_q^{q-1}}{p}$$

where the latter follows from the symmetry $B(p,q) = B(q,p)$.

235. *Recursion of the Beta-function.* From the definition $B(p,q) = \frac{\Gamma(p)\Gamma(q)}{\Gamma(p+q)}$ and the functional equation of the Gamma function (10.1), we obtain the recursion

$$B(p+1,q) = \frac{p}{p+q} B(p,q) \tag{17.7}$$

[1] Euler (1738, art. 8) deduces from (17.3) for integer q the property $B(p,q) = \frac{\Gamma(p)\Gamma(q)}{\Gamma(p+q)}$.

and $B(1,q)=\frac{1}{q}$. Alternatively[2], the Beta integral (10.11) indicates that

$$B(p,q+1)=\int_0^1 u^{p-1}(1-u)^{q-1}(1-u)\,du=B(p,q)-B(p+1,q)$$

or

$$B(p,q)=B(p,q+1)+B(p+1,q) \tag{17.8}$$

After partial integration of the Beta integral (10.11), we obtain

$$B(p+1,q)=\frac{p}{q}\int_0^1 u^{p-1}(1-u)^q\,du=\frac{p}{q}B(p,q+1)$$

whose substitution into (17.8) again demonstrates the recursion (17.7).

The recursion (17.7) shows that, for real positive $p>0$ and $q>0$, the Beta-function decreases when p increases (and similarly for q by symmetry). The same is directly concluded from the Beta integral $B(p,q)=\int_0^1 u^{p-1}(1-u)^{q-1}\,du$ in (10.11) for real $p,q>1$, because both $u\in[0,1]$ and $1-u\in[0,1]$. In addition, the maximum of the integrand $f(u)=u^p(1-u)^q$ in the Beta integral of $B(p+1,q+1)$ occurs at $u=\frac{p}{p+q}$ with value $\mu=\frac{p^pq^q}{(p+q)^{p+q}}$, and hence, for real $p,q>1$,

$$B(p+1,q+1)<\frac{p^pq^q}{(p+q)^{p+q}}=\frac{\left(\frac{p}{q}\right)^p}{\left(1+\frac{p}{q}\right)^{p+q}}$$

For example, $B(p+1,p+1)<\frac{1}{2^{2p}}$ and (17.14) indicates that $B\left(\frac{1}{2},p\right)<2$ for $p>1$.

After m iterations of the recursion (17.7), we find

$$B(p+m,q)=\frac{p(p+1)\cdots(p+m-1)}{(p+q)(p+q+1)\cdots(p+q+m-1)}B(p,q)$$

Binet (1839) presents further the comprehensive theory of the Beta-function by incorporating among many others, the reflection formula (11.4) of the Gamma function as $B(p,1-p)=\frac{\pi}{\sin\pi p}$, which is now treated in the realm of the Gamma function.

236. *The product* $\frac{\Gamma(b+z_1)\Gamma(b+z_2)}{\Gamma(b+z_1+z_2)}$. Invoking the Weierstrass's product $\Gamma(z)=\frac{1}{z}e^{-\gamma z}\prod_{n=1}^\infty\frac{e^{\frac{z}{n}}}{\left(1+\frac{z}{n}\right)}$ in (11.1) to $\mathbb{G}=\frac{\Gamma(b+z_1)\Gamma(b+z_2)}{\Gamma(b+z_1+z_2)}$ gives

$$\begin{aligned}\mathbb{G}&=e^{-\gamma b}\frac{(b+z_1+z_2)}{(b+z_2)(b+z_1)}\prod_{n=1}^\infty\frac{\left(1+\frac{b+z_1+z_2}{n}\right)}{\left(1+\frac{b+z_1}{n}\right)\left(1+\frac{b+z_2}{n}\right)}e^{\frac{b}{n}}\\&=\frac{(b+z_1+z_2)}{(b+z_2)(b+z_1)}\prod_{n=1}^\infty\frac{\left(1+\frac{b+z_1+z_2}{n}\right)\left(1+\frac{b}{n}\right)}{\left(1+\frac{b+z_1}{n}\right)\left(1+\frac{b+z_2}{n}\right)}\left\{e^{-\gamma b}\prod_{n=1}^\infty\frac{e^{\frac{b}{n}}}{\left(1+\frac{b}{n}\right)}\right\}\\&=\Gamma(b)\,b\frac{(b+z_1+z_2)}{(b+z_2)(b+z_1)}\prod_{n=1}^\infty\frac{\left(1+\frac{b+z_1+z_2}{n}\right)\left(1+\frac{b}{n}\right)}{\left(1+\frac{b+z_1}{n}\right)\left(1+\frac{b+z_2}{n}\right)}\end{aligned}$$

[2] Jordan (1913, art. 16) deduces properties of the indefinite integral $\int u^p(1-u)^q\,du$.

With

$$t_n = \frac{\left(1+\frac{b+z_1+z_2}{n}\right)\left(1+\frac{b}{n}\right)}{\left(1+\frac{b+z_1}{n}\right)\left(1+\frac{b+z_2}{n}\right)} = \frac{(n+b+z_1+z_2)(n+b)}{(n+b+z_1)(n+b+z_2)}$$

$$= \frac{1}{1+\frac{z_1 z_2}{(n+b)^2+(z_1+z_2)(n+b)}}$$

and $t_0 = b\frac{b+z_1+z_2}{(b+z_1)(b+z_2)}$ for $n=0$, we find, for all complex b, z_1 and z_2, that

$$\frac{\Gamma(b+z_1)\Gamma(b+z_2)}{\Gamma(b+z_1+z_2)} = \frac{\Gamma(b)}{\prod_{n=0}^{\infty}\left(1+\frac{z_1 z_2}{(n+b)^2+(z_1+z_2)(n+b)}\right)} \tag{17.9}$$

The product $\frac{\Gamma(b+z_1)\Gamma(b+z_2)}{\Gamma(b+z_1+z_2)}$ is a generalization of the Beta-function. If $b=1$, then (17.9) simplifies to $\frac{\Gamma(1+z_1)\Gamma(1+z_2)}{\Gamma(1+z_1+z_2)} = z_2 B(1+z_1, z_2)$. The recursion (17.7) and (17.9) then lead again to the infinite product (17.6) of the Beta-function.

237. *Transformations of the Beta integral.* Binet (1839) makes the substitution $u=\frac{t}{1+t}$ and $t=\frac{u}{u-1}$ in the Beta-integral in (10.11), for $\operatorname{Re}(p)>0$ and $\operatorname{Re}(q)>0$,

$$B(p,q) = \int_0^1 u^{p-1}(1-u)^{q-1}\,du = \int_0^\infty \frac{t^{p-1}}{(1+t)^{p+q}}dt \tag{17.10}$$

and uses the symmetry $B(p,q)=B(q,p)$ so that

$$B(p,q) = \frac{1}{2}\int_0^\infty \frac{t^{p-1}+t^{q-1}}{(1+t)^{p+q}}dt$$

After splitting the integration range as

$$2B(p,q) = \int_0^1 \frac{t^{p-1}+t^{q-1}}{(1+t)^{p+q}}dt + \int_1^\infty \frac{t^{p-1}+t^{q-1}}{(1+t)^{p+q}}dt$$

and letting $t=\frac{1}{y}$ in the last integral, we obtain

$$2B(p,q) = \int_0^1 \frac{t^{p-1}+t^{q-1}}{(1+t)^{p+q}}dt + \int_0^1 \frac{y^{q-1}+y^{p-1}}{(1+y)^{p+q}}dy$$

In summary, Binet (1839, p. 136) first demonstrates, for $\operatorname{Re}(p)>0$ and $\operatorname{Re}(q)>0$, that

$$B(p,q) = \int_0^1 \frac{t^{p-1}+t^{q-1}}{(1+t)^{p+q}}dt = \frac{1}{2}\int_0^\infty \frac{t^{p-1}+t^{q-1}}{(1+t)^{p+q}}dt \tag{17.11}$$

Next, Binet makes the substitution $t=e^{2x}$ in $B(p,q)=\int_0^\infty \frac{t^{p-1}}{(1+t)^{p+q}}dt$ and obtains

$$B(p,q) = 2\int_{-\infty}^\infty \frac{e^{2px}}{(1+e^{2x})^{p+q}}dx = 2\int_{-\infty}^\infty \frac{e^{(p-q)x}}{(e^x+e^{-x})^{p+q}}dx$$

Invoking again symmetry $B(p,q) = \int_{-\infty}^{\infty} \frac{e^{(p-q)x}+e^{-(p-q)x}}{(e^{x}+e^{-x})^{p+q}}dx$, Binet arrives at

$$B(p,q) = 2^{2-(p+q)} \int_0^{\infty} \frac{\cosh(p-q)x}{\cosh^{p+q} x} dx \tag{17.12}$$

which is a sort of inverse of an integral due Cauchy, valid for $p+q>1$,

$$\int_0^{\frac{\pi}{2}} \cos^{p+q-2} w \, \cos(p-q) w dw = \frac{\pi}{(p+q-1)\, 2^{p+q-1} B(p,q)} \tag{17.13}$$

Binet (1839, p. 167) derives Cauchy's formula (17.13) and acknowledges Cauchy as its discoverer; the derivation is found in Nielsen (1956, art. 65, p. 157).

For $u = \frac{1+y}{2}$ in the Beta-function integral in (10.11), Binet (1839, p. 136) obtains

$$\begin{aligned} B(p,q) &= \int_0^1 u^{p-1}(1-u)^{q-1} du = \frac{1}{2^{p+q-1}} \int_{-1}^1 (1+y)^{p-1} (1-y)^{q-1} dy \\ &= \frac{\int_0^1 (1-y)^{p-1} (1+y)^{q-1} dy + \int_0^1 (1+y)^{p-1} (1-y)^{q-1} dy}{2^{p+q-1}} \end{aligned}$$

which simplifies for $q=p$ to $B(p,p) = \frac{1}{2^{2p-2}} \int_0^1 \left(1-y^2\right)^{p-1} dy$ and becomes after $x=y^2$,

$$B(p,p) = \frac{1}{2^{2p-1}} \int_0^1 (1-x)^{p-1} x^{-\frac{1}{2}} dx = \frac{1}{2^{2p-1}} B\left(\frac{1}{2}, p\right) \tag{17.14}$$

With the definition $B(p,q) = \frac{\Gamma(p)\Gamma(q)}{\Gamma(p+q)}$, (17.14) translates to $\Gamma(2p) = 2^{2p-1} \frac{\Gamma(p)\Gamma\left(p+\frac{1}{2}\right)}{\Gamma\left(\frac{1}{2}\right)}$, which is the duplication formula (11.9) of the Gamma function.

Another representation for $\operatorname{Re}(p) > 0$ and $\operatorname{Re}(q) > 0$ is obtained, after the substitution $u = \sin^2\varphi$ in the Beta-function integral in (10.11), as

$$B(p,q) = \int_0^1 u^{p-1}(1-u)^{q-1} du = 2\int_0^{\frac{\pi}{2}} \sin^{2p-1}\varphi \, \cos^{2q-1}\varphi \, d\varphi \tag{17.15}$$

238. *Classical integral proof of* $B(p,q) = \frac{\Gamma(p)\Gamma(q)}{\Gamma(p+q)}$ *in (10.11).* We start from the product $\Gamma(p)\Gamma(q)$ as e.g. in Jordan (1913, art. 221). For $\operatorname{Re}(p) > 0$ and $\operatorname{Re}(q) > 0$, the Euler integral (10.2) is valid,

$$\Gamma(p)\Gamma(q) = \int_0^{\infty} e^{-t} t^{p-1} dt \int_0^{\infty} e^{-u} u^{q-1} du$$

After substituting $t = x^2$ and $u = y^2$, we obtain

$$\begin{aligned} \Gamma(p)\Gamma(q) &= 4\int_0^{\infty} e^{-x^2} x^{2p-1} dx \int_0^{\infty} e^{-y^2} y^{2q-1} dy \\ &= 4\int_0^{\infty}\int_0^{\infty} e^{-\left(x^2+y^2\right)} x^{2p-1} y^{2q-1} dx dy \end{aligned}$$

The major idea to evaluate the double integral over the first quadrant in the xy-plane is a polar transformation with $x = r\cos\theta$ and $y = r\sin\theta$, whose Jacobian is $\begin{vmatrix} \frac{\partial x}{\partial r} & \frac{\partial x}{\partial \theta} \\ \frac{\partial y}{\partial r} & \frac{\partial y}{\partial \theta} \end{vmatrix} = r$, resulting in

$$\begin{aligned}\Gamma(p)\Gamma(q) &= 4\int_0^\infty \int_0^{\frac{\pi}{2}} e^{-r^2} r^{2(p+q)-2} \cos^{2p-1}\theta \sin^{2q-1}\theta \, r dr d\theta \\ &= 2\int_0^\infty e^{-r^2} r^{2(p+q)-1} dr \times 2\int_0^{\frac{\pi}{2}} \cos^{2p-1}\theta \sin^{2q-1}\theta \, d\theta\end{aligned}$$

After substituting $r = \sqrt{u}$, the first integral gets $2\int_0^\infty e^{-r^2} r^{2(p+q)-1} dr = \int_0^\infty e^{-u} u^{(p+q)-1} du = \Gamma(p+q)$, while the second follows from (17.15), leading to

$$\Gamma(p)\Gamma(q) = \Gamma(p+q) B(q,p)$$

which demonstrates (10.11), by symmetry of $B(q,p) = B(p,q)$.

239. *Nanjundiah's proof of* $B(p,q) = \frac{\Gamma(p)\Gamma(q)}{\Gamma(p+q)}$ *and Gauss's limit.* For any integer $n \geq 1$, Nanjundiah (1969) proves the identity

$$B(p,q) = \frac{B(p,q+n) B(q,n)}{B(p+q,n)} \tag{17.16}$$

by induction. Since $B(q,1) = \frac{1}{q}$, the identity (17.16) reduces for $n = 1$ to the recursion (17.7). Assume that (17.16) holds for n. Using the recursion (17.7) and the symmetry $B(p,q) = B(q,p)$, the identity (17.16) for $n+1$ is correct and the induction argument then proves (17.16) for all integer $n \geq 1$. Next, Nanjundiah (1969) derives the upper and lower bound for $B(p,q)$,

$$\frac{\Gamma(p)}{(p+q)^p} < B(p,q) < \frac{\Gamma(p)}{(q-1)^p} \tag{17.17}$$

by invoking the inequality $e^x > 1 + x$ for any real $x \neq 0$, which is proved in e.g. Van Mieghem (2014, Sec. 5.3). The upper bound follows from Euler's integral (10.2), for $\text{Re}(p) > 0$ and $\text{Re}(q) > 1$, as

$$\begin{aligned}\frac{\Gamma(p)}{(q-1)^p} &= \int_0^\infty x^{p-1} e^{-(q-1)x} dx > \int_0^1 x^{p-1} \left(e^{-x}\right)^{(q-1)} dx \\ &> \int_0^1 x^{p-1} (1-x)^{q-1} dx = B(p,q)\end{aligned}$$

From $\frac{\Gamma(p)}{(p+q)^p} = \int_0^\infty x^{p-1} e^{-(p+q)x} dx$ and $(e^x)^{-(p+q)} < (1+x)^{-(p+q)}$, the lower bound in (17.17) follows as

$$\frac{\Gamma(p)}{(p+q)^p} = \int_0^\infty x^{p-1} e^{-(p+q)x} dx < \int_0^\infty \frac{x^{p-1}}{(1+x)^{p+q}} dx = B(p,q)$$

where the last step follows from (17.10). The bounds (17.17) lead to the asymptotic relation

$$B(p,q) \sim \frac{\Gamma(p)}{q^p} \qquad \text{for } q \to \infty \tag{17.18}$$

Let $n \to \infty$ in the identity (17.16) and invoke the asymptotic relation (17.18), then we arrive at fundamental product

$$B(p,q) = \lim_{n\to\infty} \frac{B(p,q+n)\,B(q,n)}{B(p+q,n)} = \lim_{n\to\infty} \frac{\frac{\Gamma(p)}{n^p}\frac{\Gamma(q)}{n^q}}{\frac{\Gamma(p+q)}{n^{p+q}}} = \frac{\Gamma(p)\,\Gamma(q)}{\Gamma(p+q)}$$

The asymptotic relation (17.18) is transformed with $B(p,q) = \frac{\Gamma(p)\Gamma(q)}{\Gamma(p+q)}$ into the asymptotics for the Gamma function (**art**. 198) as

$$\frac{\Gamma(q)}{\Gamma(p+q)} \sim \frac{1}{q^p} \qquad \text{for } q \to \infty \tag{17.19}$$

Finally, combining the asymptotic relation (17.19) for integer $q = n$, rewritten as $\Gamma(p+n) \sim n^p \Gamma(n) = n^{p-1} n!$, with n-iterations of the functional equation (10.1)

$$\Gamma(p) = \lim_{n\to\infty} \frac{\Gamma(p+n)}{p(p+1)\dots(p+n-1)} = \lim_{n\to\infty} \frac{n^{p-1} n!}{p(p+1)\dots(p+n-1)}$$

establishes again the Gauss limit (10.8) for the Gamma function.

Appendix A
Characteristic coefficient of a complex function

A.1 Concept of a characteristic coefficient

We briefly introduce the concept of a characteristic coefficient, which is studied in depth in the still unpublished book (Van Mieghem, 2026). Although power series or Taylor series belong to the core of complex function theory, the Taylor series of a function $G(z)$ of a function $f(z)$, thus $G(f(z))$ is hardly studied. There are some studies, e.g. by Hadamard (1893), but it seems that the nice combinatorial structure is fairly unknown. More than 30 years ago, I discovered that $G(f(z))$ can be formally expanded as

$$G(f(z)) = G(f(z_0)) + \sum_{m=1}^{\infty} \left(\sum_{k=1}^{m} \frac{1}{k!} \left. \frac{d^k G(f)}{df^k} \right|_{f=f(z_0)} s[k,m]_{f(z)}(z_0) \right) (z-z_0)^m \tag{A.1}$$

in always re-appearing quantities $s[k,m]_{f(z)}(z_0)$, that I therefore called, the "characteristic coefficients" (chc) of the function $f(z)$. The k-sum in (A.1) contains the typical Taylor coefficients (1.13) of the function $G(z)$, evaluated at $z = f_0 = f(0)$, and the characteristic coefficient $s[k,m]_{f(z)}$ of the function $f(z)$.

240. *Definition of the characteristic coefficient.* The characteristic coefficient of a complex function $f(z)$ with Taylor series $\sum_{k=0}^{\infty} f_k(z_0)(z-z_0)^k$, first published in Van Mieghem (1996, Appendix), is defined by

$$s[k,m]_f(z_0) = \frac{1}{2\pi i} \int_{C(z_0)} \frac{[f(z) - f(z_0)]^k}{(z-z_0)^{m+1}} dz = \frac{1}{m!} \left. \frac{d^m}{dz^m} [f(z) - f(z_0)]^k \right|_{z=z_0} \tag{A.2}$$

The characteristic coefficient possesses a general combinatorial form

$$s[k,m]|_f(z_0) = \sum_{\sum_{i=1}^{k} j_i = m; j_i > 0} \prod_{i=1}^{k} f_{j_i}(z_0) \tag{A.3}$$

and obeys $s[k,m]|_f(z_0) = 0$ if $k < 0$ and $k > m$. The characteristic coefficient for $k = 1$ reduces to the Taylor coefficient $s[1,m]|_f(z_0) = f_m(z_0)$, while (A.3) indicates that $s[m,m]|_f(z_0) = f_1^m(z_0)$. The characteristic coefficient (chc) (A.3) is the explicit solution of a determinant. For numerical computations, the recursive set[1] for $s[k,m]$,

$$\begin{aligned} s[1,m] &= f_m \\ s[k,m] &= \sum_{j=k}^{m} f_{m-j+1}\, s[k-1,j-1] \qquad (k>1) \\ &= \sum_{j=1}^{m-k+1} f_j\, s[k-1,m-j] \qquad (k>1) \end{aligned} \tag{A.4}$$

is very valuable, in particular, in combination with symbolic computing software as Mathematica.

A.2 Generating functions

The Taylor coefficient $\frac{1}{m!}\frac{d^m G(f(z))}{du^m}\Big|_{u=z_0}$ in (A.1) generalizes "generating function theory" (see e.g. Wilf (1994)), that connects combinatorics (see e.g. Comtet (1974)) and function theory and that has amply shown its power since the beginning of calculus. In particular, Euler championed generating function theory.

241. *Generating function of the Stirling numbers.* Due to their importance in both the theory of the Gamma and the Mittag-Leffler function, we list (see e.g. Abramowitz and Stegun (1968, Sec. 24.1.3 and 24.1.4) and Olver *et al.* (2010, Chapter 26)) the generating function of the Stirling numbers of the first kind $S_m^{(k)}$,

$$\frac{\Gamma(x+1)}{\Gamma(x+1-m)} = \prod_{k=0}^{m-1}(x-k) = \sum_{k=0}^{m} S_m^{(k)}\, x^k \tag{A.5}$$

and the generating function of the Stirling numbers of the second kind $\mathcal{S}_m^{(k)}$,

$$x^m = \sum_{k=0}^{m} k!\, \mathcal{S}_m^{(k)} \binom{x}{k} \tag{A.6}$$

The infinite sum generating function of the Stirling numbers of the first kind $S_m^{(k)}$ is

$$\frac{\log^k(1+x)}{k!} = \sum_{m=k}^{\infty} S_m^{(k)} \frac{x^m}{m!} \qquad |x|<1 \tag{A.7}$$

and for the the Stirling numbers of the second kind $\mathcal{S}_m^{(k)}$,

$$\frac{(e^x-1)^k}{k!} = \sum_{m=k}^{\infty} \mathcal{S}_m^{(k)} \frac{x^m}{m!} \tag{A.8}$$

[1] In view of (A.2), we may define $s[k,0] = \delta_{k0}$ and $s[0,m] = \delta_{m0}$, such that (A.4) is already valid for $k>0$.

The Stirling polynomial $\mathbb{S}_m(x)$ is a polynomial of degree m in x and defined[2] by the generating function

$$\left(\frac{t}{1-e^{-t}}\right)^{x+1} = \sum_{m=0}^{\infty} \mathbb{S}_m(x) \frac{t^m}{m!} \tag{A.9}$$

Nielsen (1956, band I, pp. 72-77) studies the Stirling polynomials and also presents a second generating function, convergent for $|u| < 1$,

$$\left(\frac{-\log(1-t)}{t}\right)^{x} = x \sum_{m=0}^{\infty} \frac{\mathbb{S}_m(x+m-1)}{(x+m)} \frac{t^m}{m!} \tag{A.10}$$

The Stirling polynomial $\mathbb{S}_m(x)$ generalizes both Stirling numbers $S_m^{(k)}$ and $\mathcal{S}_m^{(k)}$. There exist many properties the Stirling polynomial $\mathbb{S}_m(x)$. We just list that $\mathbb{S}_0(z) = 1$ and, for $m > 0$,

$$\mathbb{S}_m(z) = \sum_{r=1}^{m} \frac{(-1)^{r+m} m! \mathcal{S}_{m+r}^{(r)}}{(m+r)!(m-r)!} \prod_{k=0;k\neq r}^{m} (1+z+k) \tag{A.11}$$

which is rewritten as

$$\mathbb{S}_m(-z) = \frac{\Gamma(z)}{\Gamma(z-m-1)} \sum_{r=1}^{m} \frac{(-1)^{r+1} m!}{(m+r)!(m-r)!} \frac{\mathcal{S}_{m+r}^{(r)}}{1-z+r} \tag{A.12}$$

242. *Generating function of the Bernoulli numbers.* The generating function of the Bernoulli polynomials $B_n(x)$ is (see e.g. Abramowitz and Stegun (1968, Chapter 23) and Rademacher (1973, Chapters 1 and 2))

$$\frac{t\, e^{xt}}{e^t - 1} = \sum_{n=0}^{\infty} B_n(x) \frac{t^n}{n!} \tag{A.13}$$

The Taylor series in (A.13) converges for $|t| < 2\pi$. The limit $t \to 0$ in (A.13) leads to $B_0(x) = 1$. The Bernoulli numbers are defined as $B_n = B_n(0)$, the first few being $B_0 = 1$, $B_1 = -\frac{1}{2}$, $B_2 = \frac{1}{6}$, $B_4 = -\frac{1}{30}$, $B_6 = -\frac{1}{42}$, $B_8 = B_4 = -\frac{1}{30}$ and $B_{10} = \frac{5}{66}$. All odd Bernoulli numbers $B_{2n+1} = 0$ for $n \geq 1$ as follows from the *formula for the negative argument in* x,

$$B_n(-x) = (-1)^n \left\{B_n(x) + nx^{n-1}\right\} \tag{A.14}$$

The Bernoulli numbers B_n play an important role in analysis and number theory. The generating function of the Bernoulli numbers is often written, for $|t| < 2\pi$, as

$$\frac{1}{e^t - 1} = \frac{1}{t} - \frac{1}{2} + \sum_{n=1}^{\infty} B_{2n} \frac{t^{2n-1}}{(2n)!} \tag{A.15}$$

[2] Nielsen (1956, band I, p. 72) defines the Stirling polynomial $\psi_n(x)$, for $|t| < 2\pi$, by

$$\left(\frac{t}{1-e^{-t}}\right)^{x+1} = 1 + (x+1) \sum_{k=0}^{\infty} \psi_k(x)\, t^{k+1} \text{ so that } \mathbb{S}_k(x) = k!\,(x+1)\,\psi_{k-1}(x)$$

The generating function (A.13) also leads, for $n \geq 1$, to

$$\frac{dB_n(x)}{dx} = nB_{n-1}(x) \tag{A.16}$$

as well as to the *reflection formula* for the Bernoulli polynomials for all $n \geq 0$,

$$B_n(x) = (-1)^n B_n(1-x) \tag{A.17}$$

By partial fraction expansion of the Cauchy integral $\frac{1}{2\pi i}\int_{C(t)} \frac{dz}{(z-t)} \frac{z\,e^{xz}}{e^z-1} = \frac{t\,e^{xt}}{e^t-1}$, where the contour $C(t)$ encloses the point t, the famous formula (A.18) for $n \geq 1$, due to Euler[3], is deduced

$$B_{2n} = 2\frac{(2n)!(-1)^{n-1}}{(2\pi)^{2n}} \sum_{k=1}^{\infty} \frac{1}{k^{2n}} = 2\frac{(2n)!(-1)^{n-1}}{(2\pi)^{2n}} \zeta(2n) \tag{A.18}$$

that relates the even Bernoulli numbers B_{2n} and the positive even values $\zeta(2n)$ of the Riemann Zeta-function. Euler's formula (A.18) shows that the even Bernoulli numbers are alternatingly negative and positive and that the scaling law for Bernoulli numbers for large n is $|B_{2n}| \sim 2\frac{(2n)!}{(2\pi)^{2n}}$ because $\zeta(2n) \to 1$ for large n. Only up to $k = 6$, the non-zero Bernoulli numbers B_k decrease in absolute value and for $k > 6$, they start growing unboundedly.

There exist an astonishing large list of beautiful formulae involving Stirling numbers and Bernoulli numbers, which we cannot treat here, unfortunately.

243. *Inversion of a Taylor series.* Consider an entire function with Taylor series $f(z) = \sum_{k=0}^{\infty} f_k z^k$, that converges for all complex z. We will formally determine its Taylor coefficients, given the set of function evaluations at integer values of z, i.e.

$$f(n) = \sum_{k=0}^{\infty} f_k n^k \quad \text{for } n = 0, 1, 2, \ldots$$

Introducing the generating function (A.6) of the Stirling numbers of the second kind $\mathcal{S}_k^{(j)}$ and reversing the k- and j-sum yields

$$f(n) = \sum_{k=0}^{\infty} f_k \sum_{j=0}^{\infty} j!\,\mathcal{S}_k^{(j)} \binom{n}{j} = \sum_{j=0}^{\infty} j! \left(\sum_{k=j}^{\infty} f_k \mathcal{S}_k^{(j)}\right) \binom{n}{j}$$

Multiplying both sides by $(-1)^n \binom{m}{n}$ and summing n from 0 to m,

$$\sum_{n=0}^{m} (-1)^n \binom{m}{n} f(n) = \sum_{j=0}^{\infty} j! \left(\sum_{k=j}^{\infty} f_k \mathcal{S}_k^{(j)}\right) \sum_{n=0}^{m} (-1)^n \binom{m}{n} \binom{n}{j}$$

[3] Euler (1755, Pars posterior, Chapt. 5, art. 124-126, pp. 422-425 and later in art. 153) derives (A.18) elegantly, based on series expansions of $\cot(z)$.

invoking the well-known "orthogonality" relation for the binomial coefficients (Riordan, 1968, Chapter 2),

$$\sum_{n=j}^{m}(-1)^n\binom{n}{j}\binom{m}{n}=(-1)^m\delta_{jm} \tag{A.19}$$

leads to

$$\frac{1}{m!}\sum_{n=0}^{m}(-1)^{n+m}\binom{m}{n}f(n)=\sum_{k=m}^{\infty}f_k\mathcal{S}_k^{(m)}$$

The first orthogonality condition $\sum_{m=j}^{k}S_m^{(j)}\mathcal{S}_k^{(m)}=\delta_{jk}$ in (2.49) of the Stirling numbers similarly shows, taking into account that $S_m^{(j)}=0$ and $\mathcal{S}_m^{(j)}=0$ if $j>m$, that

$$\sum_{m=j}^{\infty}\frac{S_m^{(j)}}{m!}\sum_{n=0}^{m}(-1)^{n+m}\binom{m}{n}f(n)=\sum_{m=j}^{\infty}\sum_{k=0}^{\infty}f_kS_m^{(j)}\mathcal{S}_k^{(m)}=\sum_{k=0}^{\infty}f_k\sum_{m=j}^{k}S_m^{(j)}\mathcal{S}_k^{(m)}$$

and we arrive at the combinatorial relation for the Taylor coefficients f_k of an entire function $f(z)$ in terms of functional evaluations $f(n)$ at integer values

$$f_k=\sum_{m=j}^{\infty}\frac{S_m^{(j)}}{m!}\sum_{n=0}^{m}(-1)^{n+m}\binom{m}{n}f(n) \tag{A.20}$$

where $\sum_{n=0}^{m}(-1)^{n+m}\binom{m}{n}f(n)=\Delta^m f(0)$ is the m-th difference (12.51).

A.3 Examples of characteristic coefficients

We first give two important examples of characteristic coefficients. The characteristic coefficient of the exponential function is

$$s[k,m]|_{e^z}=\frac{k!}{m!}\mathcal{S}_m^{(k)} \tag{A.21}$$

and the combinatorial definition (A.3) of the characteristic coefficient gives

$$\mathcal{S}_m^{(k)}=\frac{m!}{k!}\sum_{\sum_{i=1}^{k}j_i=m}\frac{1}{\prod_{i=1}^{k}j_i!} \tag{A.22}$$

The companion characteristic coefficient is

$$s[k,m]|_{\log(1+z)}=\frac{k!}{m!}S_m^{(k)} \tag{A.23}$$

while the combinatorial definition (A.3) gives an explicit form

$$S_m^{(k)}=\frac{m!}{k!}(-1)^{k+m}\sum_{\sum_{i=1}^{k}j_i=m;j_i>0}\frac{1}{\prod_{i=1}^{k}j_i} \tag{A.24}$$

Formulae (A.22) and (A.24) indicate, for non-negative integer k and m, that $\mathcal{S}_m^{(k)}$ and $(-1)^{k+m}\,S_m^{(k)}$ are always non-negative and that $(-1)^{k+m}\,S_m^{(k)}\geq$

$\mathcal{S}_m^{(k)}$. The appearance of the Stirling numbers of the first kind $S_m^{(k)}$ in (A.23) and of the second kind $\mathcal{S}_m^{(k)}$ in (A.21) emphasizes, together with the general form (A.3), the combinatorial nature of characteristic coefficients. Hence, function theory and combinatorics meet. For example, from the definition (A.2) of a characteristic coefficient and the generating functions (A.5) of $S_m^{(k)}$ and (A.6) of $\mathcal{S}_m^{(k)}$, we find the characteristic coefficient of the function $(1+z)^\alpha$,

$$s[k,m]|_{(1+z)^\alpha} = \sum_{j=1}^{k} (-1)^{j+k} \binom{k}{j} \binom{\alpha j}{m} = \frac{k!}{m!} \sum_{j=k}^{m} S_m^{(j)} \mathcal{S}_j^{(k)} \alpha^j \tag{A.25}$$

A.4 Examples of Taylor series of a function $G(z)$ of a function $f(z)$

We provide some examples of the general Taylor series $G(f(z))$ in (A.1). Lagrange series for the inverse function $f^{-1}(z)$ in terms of characteristic coefficients are shown in Van Mieghem (1996, Appendix A). If $G(z) = \log z$, then (A.1) becomes

$$\log f(z) = \log f_0(z_0) + \sum_{m=1}^{\infty} \left(\sum_{k=1}^{m} \frac{(-1)^{k-1}}{k\, f_0^k(z_0)} s[k,m](z_0) \right) (z - z_0)^m \tag{A.26}$$

The Taylor series of $\frac{1}{f(z)}$ around z_0, that follows from (A.1) as

$$\frac{1}{f(z)} = \frac{1}{f_0(z_0)} + \sum_{m=1}^{\infty} \left[\sum_{k=1}^{m} \frac{(-1)^k}{f_0^{k+1}(z_0)} s[k,m](z_0) \right] (z - z_0)^m \tag{A.27}$$

illustrates for $m \geq 1$ that

$$\frac{1}{m!} \frac{d^m}{dz^m} \left(\frac{1}{f(z)} \right) \Bigg|_{z=z_0} = \sum_{k=1}^{m} \frac{(-1)^k}{f_0^{k+1}(z_0)} s[k,m](z_0)$$

Consider a complex function $f(z)$ and a transform $z = g(y)$ with corresponding inverse transform $y = g^{-1}(z)$. We can first compute the general Taylor series $f(g(y))$ with (A.1) and then use the inverse transform to obtain

$$f(z) = f(g_0) + \sum_{m=1}^{\infty} \left(\sum_{k=1}^{m} \frac{1}{k!} \frac{d^k f(p)}{dp^k} \bigg|_{p=g_0} s[k,m]|_{g(z)} \right) \left(g^{-1}(z)\right)^m \tag{A.28}$$

Formula (A.28) is a generalization of a Taylor series expansion. An application of (A.28) is the case where $g(y) = \frac{\log(1+y)}{\beta}$ and $g^{-1}(z) = e^{\beta z} - 1$. The characteristic coefficients follow from (A.23) as $s[k,m]|_{\frac{\log(1+z)}{\beta}} = \frac{k!}{m!} \frac{S_m^{(k)}}{\beta^k}$. Since $g_0 = g(0) = 0$, we find

$$f(z) = f_0 + \sum_{m=1}^{\infty} \left(\sum_{k=1}^{m} f_k\, S_m^{(k)} \frac{k!}{\beta^k} \right) \frac{\left(e^{\beta z} - 1\right)^m}{m!} \tag{A.29}$$

The case dual to (A.29), where $g(y) = \frac{e^y}{\alpha}$ and $g^{-1}(z) = \log \alpha z$, uses $s[k,m]|_{\frac{e^z}{\alpha}} = \frac{k!}{m!}\mathcal{S}_m^{(k)}\,\alpha^{-k}$ in (A.21) leading with (A.28) to

$$f(z) = f\left(\frac{1}{\alpha}\right) + \sum_{m=1}^{\infty}\left(\sum_{k=1}^{m} \left.\frac{d^k f(p)}{dp^k}\right|_{p=\frac{1}{\alpha}} \frac{\mathcal{S}_m^{(k)}}{\alpha^k}\right) \frac{\log^m(\alpha\, z)}{m!} \tag{A.30}$$

Finally, we add that the concept of Taylor characteristic coefficient can be transferred to Dirichlet series and the Dirichlet characteristic coefficient, for which we refer to Van Mieghem (2026).

A.5 A factorial series of a Mellin transform

We apply the powerful substitution $e^{-x} = 1-u$ of Binet to the Mellin transform of a function $f(x)$, defined as $M_f(s) = \int_0^\infty x^{s-1} f(x)\,dx$ in **art.** 245, which becomes

$$M_f(s) = \int_0^1 (-\log(1-u))^{s-1} \frac{f(-\log(1-u))}{1-u} du \tag{A.31}$$

We follow Binet's recipe of 5 steps in **art.** 182. The next step for (A.31) is the computation of the Taylor series of the function $(-\log(1-u))^{s-1}$ around u_0. The general Taylor series (A.1) of $G(h(u))$ with $h(u) = -\log(1-u)$ and $G(x) = x^{s-1}$, with additionally $\frac{d^k G(x)}{dx^k} = \frac{(s-1)!}{(s-1-k)!}x^{s-1-k}$ and the characteristic coefficient[4] is $s[k,m]|_{-\log(1-z)}(z_0) = \frac{(-1)^{k+m}}{(1-z_0)^m}\frac{k!}{m!}S_m^{(k)}$, where $S_m^{(k)}$ is the Stirling number of the first kind (A.24), results in the Taylor series around u_0,

$$\left(\frac{\log(1-u)}{\log(1-u_0)}\right)^{s-1} - 1 + \sum_{m=1}^{\infty}\left(\sum_{k=1}^{m} \frac{\binom{s-1}{k}}{\log^k(1-u_0)} \frac{(-1)^m S_m^{(k)}}{(1-u_0)^m} \frac{k!}{m!}\right)(u-u_0)^m$$

Reversing the m- and k-sum and invoking the generating function (A.7) leads to the binomial series

$$\left(\frac{\log(1-u)}{\log(1-u_0)}\right)^{s-1} = 1 + \sum_{k=1}^{\infty}\binom{s-1}{k}\left(\frac{\log\left(\frac{1-u}{1-u_0}\right)}{\log(1-u_0)}\right)^k$$

that only converges provided $\left|\frac{\log\left(\frac{1-u}{1-u_0}\right)}{\log(1-u_0)}\right| < 1$. In other words, there is no expansion point u_0 for which the Taylor series converges for all $u \in (0,1)$, an unfortunate fact that has been known for a long time.

However, the generating function (A.10) of the Stirling polynomial $\mathbb{S}_m(x)$ is a convergent Taylor series around $u_0 = 0$ for $|u| < 1$. We slightly modify the integral (A.31) to

$$M_f(s+1) = \int_0^1 \left(\frac{-\log(1-u)}{u}\right)^s \frac{u^s f(-\log(1-u))}{1-u} du \tag{A.32}$$

---

[4] The chc generalization of (A.23) is $s[k,m]|_{A\log(\alpha+\beta z)}(z_0) = \frac{\beta^m A^k}{(\alpha+\beta z_0)^m}\frac{k!}{m!}S_m^{(k)}$.

Substituting the Taylor series (A.10) of the Stirling polynomial $\mathbb{S}_m(x)$ yields

$$M_f(s+1) = s\sum_{m=0}^{\infty} \frac{\mathbb{S}_m(s+m-1)}{(s+m)\,m!} \int_0^1 \frac{u^{s+m} f(-\log(1-u))}{1-u} du \quad \text{(A.33)}$$

where the reversal of integral and sum is allowed, because any Taylor series can be integrated with its radius of convergence.

Proceeding in full generality is difficult. But, if the function is of the form $f(x) = \sum_{j=1}^{N} f_j e^{-\beta_j x}$ with $\mathrm{Re}(\beta_j) > 0$, then, with the Beta-integral (10.11), the integral is

$$\int_0^1 \frac{u^{s+m} f(-\log(1-u))}{1-u} du = \sum_{j=1}^{N} f_j \int_0^1 u^{s+m}(1-u)^{\beta_j-1} du$$
$$= \sum_{j=1}^{N} \Gamma(\beta_j) f_j \frac{\Gamma(s+m+1)}{\Gamma(\beta_j+s+m+1)}$$

and the Mellin transform becomes

$$M_f(s+1) = s\sum_{j=1}^{N} \Gamma(\beta_j) f_j \sum_{m=0}^{\infty} \frac{\mathbb{S}_m(s+m-1)}{m!} \frac{\Gamma(s+m)}{\Gamma(\beta_j+s+m+1)} \quad \text{(A.34)}$$

In summary, if the function $f(x) = \sum_{j=1}^{N} f_j \left(e^{-\beta_j x}\right)$ consists of a convergent (if $N \to \infty$) sum of exponential terms $e^{-\beta_j x}$ with positive integers β_j, then we arrive at a factorial expansion

$$M_f(s+1) = s\sum_{j=1}^{N} \Gamma(\beta_j) f_j \sum_{m=0}^{\infty} \frac{1}{m!} \frac{\mathbb{S}_m(s+m-1)}{\prod_{l=0}^{\beta_j}(s+m+l)} \quad \text{(A.35)}$$

which is, explicitly with (A.11), more clearly shown as,

$$M_f(s+1) = s\sum_{j=1}^{N} \Gamma(\beta_j) f_j \left(\frac{1}{\prod_{l=0}^{\beta_j}(s+l)} + \sum_{m=1}^{\infty} \frac{\sum_{r=1}^{m} \frac{(-1)^{r+m} \mathcal{S}_{m+r}^{(r)}}{(s+m+r)(m+r)!(m-r)!}}{\prod_{l=m+1}^{\beta_j}(s+m+l)} \right) \quad \text{(A.36)}$$

Appendix B
The Laplace transform

Together with the Fourier transform, the Laplace and Mellin transforms of a function are invaluable tools in the theory of functions. Here, we only touch on a few aspects of the Laplace transform, but refer to the many books (see e.g. Titchmarsh (1948), Widder (1946), Evgrafov (2019, Chapter VII)) for the entire wealth of properties of Laplace transforms.

244. *Definition of the Laplace transform.* The Laplace transform for complex z is defined as

$$\varphi(z) = \mathcal{L}\left[f\left(t\right), z\right] = \int_0^\infty e^{-zt} f(t) dt \tag{B.1}$$

with the inverse Laplace transform,

$$f(t) = \mathcal{L}^{-1}\left[\varphi(z), t\right] = \frac{1}{2\pi i} \int_{c-i\infty}^{c+i\infty} \varphi(z) e^{zt} dz \tag{B.2}$$

where c is the smallest real value of $\mathrm{Re}(z)$ for which the integral in (B.1) converges.

Actually, the definition (B.1) is called the single-sided Laplace transform. The double-sided Laplace transform is the integral $\int_{-\infty}^{\infty} e^{-zt} f(t) dt$.

245. *Definition of the Mellin transform.* The Mellin transform for complex z is defined as

$$\phi\left(z\right) = \mathcal{M}\left[f\left(x\right), z\right] = \int_0^\infty x^{z-1} f\left(x\right) dx \tag{B.3}$$

with the inverse Mellin transform,

$$f(x) = \mathcal{M}^{-1}\left[\phi\left(z\right), x\right] = \frac{1}{2\pi i} \int_{c-i\infty}^{c+i\infty} \phi\left(z\right) x^{-z} dz \tag{B.4}$$

where c is the smallest real value of $\mathrm{Re}(z)$ for which the integral in (B.3) converges. If we make the substitution $x = -\log t$ in (B.3), then

$$\int_0^\infty x^{z-1} f\left(x\right) dx = \int_{-\infty}^{\infty} e^{-zt} f(e^{-t}) dt$$

illustrates that the Mellin transform of the function $f(x)$ is the double-sided Laplace transform of the function $f(e^{-t})$.

246. *Integral (B.2) of the inverse Laplace transform.* We evaluate the integral in (B.2) along the line $z = c + iw$,

$$f(t) = \frac{e^{tc}}{2\pi} \int_{-\infty}^{\infty} \varphi(c+iw) e^{itw} dw$$

After writing the integrand in a real and an imaginary part,

$$\begin{aligned} f(t) &= \frac{e^{tc}}{2\pi} \int_{-\infty}^{\infty} \{\operatorname{Re}\varphi(c+iw)\cos tw - \operatorname{Im}(\varphi(c+iw))\sin tw\}\, dw \\ &\quad + i\frac{e^{tc}}{2\pi} \int_{-\infty}^{\infty} \{\operatorname{Re}(\varphi(c+iw))\sin tw + \operatorname{Im}(\varphi(c+iw))\cos tw\}\, dw \end{aligned}$$

and separating the real and imaginary part, we find that

$$\begin{cases} f(t) = \frac{e^{tc}}{2\pi} \int_{-\infty}^{\infty} \{\operatorname{Re}\varphi(c+iw)\cos tw - \operatorname{Im}\varphi(c+iw)\sin tw\}\, dw \\ \quad 0 = \int_{-\infty}^{\infty} \{\operatorname{Re}\varphi(c+iw)\sin tw + \operatorname{Im}\varphi(c+iw)\cos tw\}\, dw \end{cases}$$

On the other hand, it follows from (B.1) that

$$\begin{cases} \operatorname{Re}\varphi(c+iw) = \int_0^{\infty} e^{-ct} f(t) \cos wt dt \\ \operatorname{Im}\varphi(c+iw) = -\int_0^{\infty} e^{-ct} f(t) \sin wt dt \end{cases}$$

and that $\operatorname{Re}\varphi(c+iw)\cos tw$ and $\operatorname{Im}\varphi(c+iw)\sin tw$ are even in w. Likewise, $\operatorname{Re}\varphi(c+iw)\sin tw$ and $\operatorname{Im}\varphi(c+iw)\cos tw$ are odd in w. Hence, we arrive at

$$\begin{cases} f(t) = \frac{e^{tc}}{\pi} \int_0^{\infty} \{\operatorname{Re}\varphi(c+iw)\cos tw - \operatorname{Im}\varphi(c+iw)\sin tw\}\, dw \\ \int_{-\infty}^{\infty} \operatorname{Re}\varphi(c+iw)\sin tw dw = \int_{-\infty}^{\infty} \operatorname{Im}\varphi(c+iw)\cos tw dw = 0 \end{cases} \tag{B.5}$$

The derivation (B.5) is a corrected version of Berberan-Santos (2005a).

247. *The approach of Berberan-Santos.* Berberan-Santos[1] suggests to proceed a step further by defining $f(t) = 0$ for $t < 0$. In that case, it follows from (B.5) that

$$0 = \int_0^{\infty} \{\operatorname{Re}\varphi(c+iw)\cos tw - \operatorname{Im}\varphi(c+iw)\sin tw\}\, dw \qquad \text{for } t < 0$$

which leads, after replacing $t \to -t$, to

$$\int_0^{\infty} \operatorname{Re}\varphi(c+iw)\cos tw dw = -\int_0^{\infty} \operatorname{Im}\varphi(c+iw)\sin tw dw \qquad \text{for } t > 0$$

The final resulting set of integral equations, subject to "$f(t) = 0$ for $t < 0$", simplifies to

$$\begin{cases} f(t) = \frac{2e^{tc}}{\pi} \int_0^{\infty} \operatorname{Re}\varphi(c+iw)\cos tw dw & \text{for } t > 0 \\ f(t) = -\frac{2e^{tc}}{\pi} \int_0^{\infty} \operatorname{Im}\varphi(c+iw)\sin tw dw & \text{for } t > 0 \\ \int_{-\infty}^{\infty} \operatorname{Re}\varphi(c+iw)\sin tw dw = \int_{-\infty}^{\infty} \operatorname{Im}\varphi(c+iw)\cos tw dw = 0 \end{cases} \tag{B.6}$$

[1] Private communication.

The general form (B.2) does not impose the restriction "$f(t) = 0$ for $t < 0$" and is continuous[2] at $t = 0$. The restriction "$f(t) = 0$ for $t < 0$" is only continuous at $t = 0$ if $f(0) = 0$. In spite of this concern, Berberan-Santos (2005a) evaluates various Laplace inverses via $f(t) = \frac{2e^{tc}}{\pi} \int_0^\infty \operatorname{Re} \varphi(c + iw) \cos tw dw$.

248. *When has the inverse Laplace integral the form of a Laplace transform?* In Fourier transforms, the inverse has a similar form as the transform itself and tables of Fourier transforms can thus be used in two directions. Gross (1950) has written[3] a note on the question when the inverse Laplace transform (B.2) is of the same form as the Laplace transform (B.1) itself. In particular, if $\varphi(z) = \int_0^\infty e^{-zt} f(t) dt = \mathcal{L}[f(t), z]$, then Gross (1950) asks when the inverse $f(z) = \mathcal{L}^{-1}[\varphi(z), t]$ in (B.2) is of the form

$$f(z) = \int_0^\infty e^{-zt} g(t) dt = \mathcal{L}[g(t)]$$

for real functions f and g. Formal substitution of the latter into the former and reversing the integrals yields a Stieltjes transform (see e.g. Widder (1946, Chapter VIII), Titchmarsh (1948, 9.15, p. 269))

$$\varphi(z) = \int_0^\infty \left(\int_0^\infty e^{-(z+s)t} dt \right) g(s) ds = \int_0^\infty \frac{g(s)}{z+s} ds \tag{B.7}$$

which is inverted, for real s,

$$g(s) = \frac{i}{2\pi} \left(\varphi(se^{i\pi}) - \varphi(se^{-i\pi}) \right) = \frac{1}{\pi} \operatorname{Im} \left(\varphi(se^{-i\pi}) \right)$$

as shown by Titchmarsh (1948, 11.8, pp. 317-318). Hence, if the Laplace

---

[2] Indeed,

$$\begin{aligned} |f(\varepsilon) - f(-\varepsilon)| &\le \frac{1}{2\pi} \int_{-\infty}^\infty |\varphi(c+iw)| \left| e^{\varepsilon(c+iw)} - e^{-\varepsilon(c+iw)} \right| dw \\ &= \frac{1}{\pi} \lim_{T\to\infty} \int_{-T}^T |\varphi(c+iw)| \left| \sinh(\varepsilon(c+iw)) \right| dw \end{aligned}$$

Since $|\sinh(\varepsilon(c+iw))| = |\sinh \varepsilon c \cos \varepsilon w + \sin \varepsilon w \cosh \varepsilon c| < |\sinh \varepsilon c| + |\sin \varepsilon w| + O(\varepsilon^2)$ and choosing $\varepsilon = T^{-2-\delta}$,

$$\begin{aligned} \mathbb{I} &= \int_{-T}^T |\varphi(c+iw)| \left| \sinh(\varepsilon(c+iw)) \right| dw \\ &\le cT^{-2-\delta} \int_{-T}^T |\varphi(c+iw)| dw + T^{-2-\delta} \int_{-T}^T |w\varphi(c+iw)| dw \\ &\le 2cT^{-1-\delta} \max |\varphi(c+iw)| + T^{-\delta} \max |\varphi(c+iw)| \le AT^{-\delta} \end{aligned}$$

which can be made arbitrarily small for large T so that $|f(\varepsilon) - f(-\varepsilon)| \to 0$ when $\varepsilon \to 0$.

[3] Professor Apelblat has informed me about this note. Later Professor Mainardi has pointed me to a related work of Gross and Levi (1946). However, it appears that Titchmarsh (1948, Chapter IX) has earlier addressed and solved the problem. In fact, Titchmarsh (1948, Chapter IX) studies the integral equation $f(x) = g(x) + \lambda \int_a^b k(x,y) f(y) dy$ in depth, which can be solved by means of Fourier integrals in certain cases, e.g. if the kernel obeys $k(x,y) = k(x-y)$.

transform and its inverse are of the same form, then it holds that

$$\varphi(z) = \mathcal{L}\left[f(t), z\right] \iff f(z) = \mathcal{L}\left[\tfrac{1}{\pi}\,\mathrm{Im}\left(\varphi(te^{-i\pi})\right), z\right] \tag{B.8}$$

Gross (1950) briefly and Titchmarsh (1948, 11.8, pp. 317-319) in more depth state conditions on the validity of (B.8), which essentially relate to Cauchy's integral theorem for entire functions, leading e.g. to integrals as in (3.37). The Stieltjes transform (B.7) with solution (B.8) also appears in Caputo and Mainardi (1971a, eq. (2.16) and eq. (2.17)) and in Mainardi *et al.* (2023). Berberan-Santos' set (B.6) of equations is thus a further development of Gross's Laplace pair (B.8).

Bibliography

Abramowitz, M. and Stegun, I. A. (1968). *Handbook of Mathematical Functions.* (Dover Publications, Inc., New York).
Apelblat, A. (2020). Differentiation of the Mittag-Leffler functions with respect to parameters in the Laplace transform approach. *Mathematics 8,* 657 (doi:10.3390/math8050657).
Artin, E. (1964). *The Gamma Function.* Vol. translation of Einführung in die Theorie der Gammafunktion, B.G. Teubner, Leipzig, 1931. (Holt, Rinehart and Winston, New York).
Ashcroft, N. W. and Mermin, N. D. (1981). *Solid State Physics.* (Holt-Saunders International Editions, Tokyo).
Askey, R. (1980). Ramanujan's extensions of the Gamma and Beta functions. *The American Mathematical Monthly 87,* 5, 346–359.
Aycock, A. (2021). Euler and the multiplication formula for the Gamma function. *Euleriana 1,* 2, 197–204.
Barabási, A. L. (2016). *Network Science.* (Cambridge University Press, Cambridge, U.K.).
Barnes, E. W. (1906). The asymptotic expansion of integral functions defined by Taylor's series. *Philosophical Transactions of the Royal Society of London. Series A 206,* 402-411, 249–297.
Berberan-Santos, M. (2005a). Analytical inversion of the Laplace transform without contour integration: application to luminescence decay laws and other relaxation functions. *Journal of Mathematical Chemistry 38,* 2 (August), 165–173.
Berberan-Santos, M. (2005b). Properties of the Mittag-Leffler relaxation function. *Journal of Mathematical Chemistry 38,* 4 (November), 629–635.
Berger, M. A. (1993). *An Introduction to Probabiliy and Stochastic Processes.* (Springer-Verlag, New York).
Berndt, B. C. (1985). The Gamma function and the Hurwitz Zeta-function. *The American Mathematical Monthly 92,* 2 (February), 126–130.
Bernstein, S. (1929). Sur les fonctions absolument monotones. *Acta Mathematica 52,* 1–66.
Berry, M. V. (1989). Uniform asymptotic smoothing of Stokes's discontinuities. *Proceedings of the Royal Society London. Series A, Mathematical and Physical Sciences 422,* 1862, 7–21.
Bharatiya, P. L. (1967). Convolution transform with the Mittag-Leffler function as a kernel. *The American Mathematical Monthly 74,* 1 (January), 38.
Bieberbach, L. (1945). *Lehrbuch der Funktionentheorie.* Vol. II. (Teubner in Leipzig, 1934; reprinted by Chelsea Publishing Company, New York).
Binet, J. P. M. (1839). Mémoire sur les intégrales définites Eulériennes et sur leur application à la théorie des suites ainsi qu' à l'évaluation des functions des grands nombres. *Journal de l'École Polytechnique XVI,* 123–343.
Blagouchine, I. V. (2015). A theorem for the closed-form evaluation of the first generalized Stieltjes constant at rational arguments and some related summations. *Journal of Number Theory 148,* 537–592.
Blagouchine, I. V. (2016). Two series expansions for the logarithm of the Gamma function involving Stirling numbers and containing only rational coefficients for certain arguments related to π^{-1}. *arXiv1408.3902v9.*

Blagouchine, I. V. (2018). Three notes on Ser's and Hasse's representations for the Zeta-functions. *INTEGERS, Electronic Journal of Combinatorial Number Theory 18A,* A3, 1–45.

Blakemore, J. S. (1962). *Semiconductor Statistics.* (Pergamon, New York).

Borwein, J. M. and Corless, R. M. (2018). Gamma and Factorial in the Monthly. *The American Mathematical Monthly 125,* 5, 400–424.

Bourbaki, N. (2004). *Functions of a Real Variable.* Elements of Mathematics, (Springer-Verlag, Berlin).

Bourguet, L. (1883). Sur les intégrales Eulériennes et quelques autres fonctions uniformes. *Acta Mathematica 2,* 261–295.

Caputo, M. and Mainardi, F. (1971a). Linear models of dissipation in anelastic solids. *Rivista del Nuovo Cimento 1,* 2 (Aprile-Giugno), 161–198.

Caputo, M. and Mainardi, F. (1971b). A new dissipation model based on memory mechanism. *Pure and Applied Geophysics 91,* 134–147.

Chorfi, S. E., Maniar, L., and Yamamoto, M. (2022). The backward problem for time fractional evalution equations. *arXiv:2211.16493v1.*

Comtet, L. (1974). *Advanced Combinatorics*, revised and enlarged edn. (D. Riedel Publishing Company, Dordrecht, Holland).

Copson, E. T. (1965). *Aymptotic Expansions.* Cambridge Tracts in Mathematics and Mathematical Physics, (Cambridge University Press).

Corless, R. M., Gonnet, G. H., Hare, D. E. G., Jeffrey, D. J., and Knuth, D. E. (1996). On the Lambert w function. *Advances in Computational Mathematics 5,* 329–359.

Courant, R. and Hilbert, D. (1953). *Methods of Mathematical Physics*, first English edition, translated and revised from the German original of Springer in 1937 edn. Wiley Classic Library, vol. I. (Interscience, New York).

D'Alessandro, M. and Van Mieghem, P. (2025). Fractional derivative in continuous-time Markov processes and applications to epidemics in networks. *Physical Review Research 7,* 1, 013017.

Davis, H. T. (1957). Approximation of Logarithmic Numbers. *The American Mathematical Monthly 64,* 8, Part 2, 11–18.

Davis, P. J. (1959). Leonhard Euler's integral: A historical profile of the Gamma function. *The American Mathematical Monthly 66*, 849–869.

de Moivre, A. (1730). *Miscellanea Analytica de Seriebus et Quadraturis.* (excudebant J. Tonson & J. Watts, London).

Diaconis, P. and Freedman, D. (1986). An elementary proof Stirling's formula. *The American Mathematical Monthly 93,* 2, 123–125.

Dingle, R. B. (1956). The Fermi-Dirac integrals. *Applied Scientific Research B 6,* 225–239.

Edwards, H. M. (1974). *Riemann's Zeta Function.* (Academic Press, New York).

Erdélyi, A., Magnus, W., Oberhettinger, F., and Tricomi, F. G. (1953a). *Higher Transcendental Functions.* California Institute of Technology Bateman Manuscript Project, vol. 3. (McGraw-Hill Book Company, New York).

Erdélyi, A., Magnus, W., Oberhettinger, F., and Tricomi, F. G. (1953b). *Higher Transcendental Functions.* California Institute of Technology Bateman Manuscript Project, vol. 1. (McGraw-Hill Book Company, New York).

Euler, L. (1738). De progressionibus transcendentibus seu quarum termini generales algebraice dari nequeunt. *Leonhardi Euleri Opera Omnia, B. G. Teubner- Verlag, Leipzig - Berlin, 1924; Also, Euler Archive-All Works.19. 14.*

Euler, L. (1748). *Introductio in Analysin Infinitorum, Thomus Primus.* (Marcus-Michaeles Bousquet & Socios, Lausanne).

Euler, L. (1755). *Institutiones Calculi Differentialis cum eius usu in Analysi Finitorum ac Doctrina Serierum.* (Academiae Imperialis Scientiarum Petropolitanae, Saint Petersburg, Russia).

Evgrafov, M. A. (2019). *Analytic Functions*, Dover 2019 edn. (W. B. Saunders Company, 1966; Reprinted by Dover Publications, Inc., New York).

Fekih-Ahmed, L. (2017). On the power series expansion of the reciprocal Gamma function. *arXiv:1407.5983v4.*

Feller, W. (1949). Fluctuation theory of recurrent events. *Transactions of the American Mathematical Society 67,* 98–119.

Feller, W. (1971). *An Introduction to Probability Theory and Its Applications*, 2nd edn. Vol. 2. (John Wiley & Sons, New York).

Ferreira, R. A. C. (2025). Complete monotonicity of discrete equations and related results. *preprint.*

Garrappa, R. (2015). Numerical evaluation of two and three parameter Mittag-Leffler functions. *SIAM Journal on Numerical Analysis 53*, 3, 1350–1369.

Garrappa, R., Gerhold, S., Popolizio, M., and Simon, T. (2024). On some inequalities for the two-parameter Mittag-Leffler function in the complex plane. *arXiv:2410.11852v3*.

Gauss, C. F. (1813). Disquisitiones generales circa seriem infinitam $1 + \frac{\alpha\beta}{1.\gamma}x + \frac{\alpha(1+\alpha)\beta(1+\beta)}{1.2\gamma(\gamma+1)}xx + \frac{\alpha(1+\alpha)(2+\alpha)\beta(1+\beta)(2+\beta)}{1.2.3\gamma(\gamma+1)(2+\gamma)}x^3$ etc., pars prior. *Commentationes societatis regiae scientiarum Gottingensis recentiores vol. II* in Gauss Werke, Band III, 125–162.

Gautschi, W. (1998). The incomplete Gamma functions since Tricomi. *Atti Convegni Lincei, 147, Accad. Naz. Lincei, Rome*, 203–237.

Gilbert, P. (1876). Recherches sur de développement de la function γ et sur certaines intégrals définies qui en dépendent. *Mémoires de l'Académie royale des Sciences, des Lettres et des Beaux-Arts de Belgique 41*, 1–60.

Gilbert, P. (1886). Sur les produits composés d'un grand nombre de facteurs et sur le reste de la série de binet. *Annales de la Société scientifique de Bruxelles 10*, 191–200.

Gorenflo, R., Kilbas, A. A., Mainardi, F., and Rogosin, S. V. (2014). *Mittag-Leffler Functions, Related Topics and Applications.* (Springer).

Gorenflo, R., Kilbas, A. A., Mainardi, F., and Rogosin, S. V. (2020). *Mittag-Leffler Functions, Related Topics and Applications*, second edn. (Springer).

Gorenflo, R. and Mainardi, F. (1997). *Fractional Calculus: Integral and Differential Equations of Fractional Order.* Cism Courses and Lectures, No. 378, vol. Fractals and Fractional Calculus in Continuum, Mechanics, A. Carpinteri and F. Mainardi (Editors). (Springer Verlag, Wien and New York).

Gorenflo, R. and Mainardi, F. (2012). Parametric subordination in fractional diffusion processes. *arXiv:1210.8414*.

Gross, B. (1950). Note on the inversion of the Laplace transform. *Philosophical Magazine A 41*, 317, 534–544.

Gross, B. and Levi, B. (1946). Sobre el calculo de la transformacion inversa de Laplace. *Mathematicae Notae, Boletin del Instituto de Matematica, Rosario 6*, 213–224.

Grossman, D. (1991). Letters to the editor. *The American Mathematical Monthly 98*, 3, 233–234.

Gudermann, C. (1845). Additamentum ad functionis $\Gamma(a) = \int_0^\infty e^{-x}x^{a-1}dx$ theoriam. *Journal für die reine und angewandte Mathematik 29*, 209–212.

Hadamard, J. (1892-1893). *Essai sur l'étude des functions données par leur développement de Taylor & Étude sur les propriétés des functions entières et en particulier d'une function considérée par Riemann.* Extraits du Journal de Mathématiques pures et appliquées, 4ᵉ Série, vol. VIII and IX. (Gauthier-Villars et fils, Paris).

Hanneken, J. W., Narahari Achar, B. N., Puzio, R., and Vaught, D. M. (2009). Properties of the Mittag-Leffler function for negative alpha. *Physica Scripta T136*, 014037.

Hardy, G. H. (1928). Gösta Mittag-Leffler. *Journal of the London Mathematical Society s1-3*, 2 (April), 156–160.

Hardy, G. H. (1948). *Divergent Series.* (Oxford University Press, London).

Hardy, G. H. (1978). *Ramanujan*, 3rd edn. (Chelsea Publishing Company, New York).

Hardy, G. H. (2006). *A Course of Pure Mathematics*, 10th edn. (Cambridge University Press).

Hardy, G. H., Littlewood, J. E., and Polya, G. (1999). *Inequalities*, 2nd edn. (Cambridge University Press, Cambridge, U.K.).

Hardy, G. H. and Wright, E. M. (2008). *An Introduction to the Theory of Numbers*, 6th edn. (Oxford University Press, London).

Haubold, H. J., Mathai, A. M., and Saxena, R. K. (2011). Mittag-Leffler functions and their applications. *Journal of Applied Mathematics 2011*, Article ID 298628.

Hermite, C. (1895). Sur la function $\log \Gamma(a)$. *Journal für die reine und angewandte Mathematik 115*, 201–208.

Hermite, C. (1901). Extrait de Quelques Lettres de Monsieur Ch. Hermite à monsieur M. S. Pincherle. *Annali di Matematica Pura ed Applicata (also in Oeuvre Complète de Charles Hermite) 5*, 57–72.

Hilfer, R. and Seybold, H. J. (2006). Computation of the generalized Mittag-Leffler function and its inverse in the complex plane. *Integral Transforms and Special Functions 17*, 9, 637–652.

Hilfer, R. e. (2000). *Applications of Fractional Calculus in Physics.* (World Scientific, Singapore).

Jensen, J. L. W. V. (1891). Gammafunktiones theori: i elementaer fremstilling. i. *Nyt tidsskrift for matematik 2*, Afdeling B, 33–55.

Jensen, J. L. W. V. and Gronwall, T. H. (1916). An elementary exposition of the theory of the Gamma function. *Annals of Mathematics 17,* 3 (March), 124–166.

Jordan, C. (1909). *Cours d'Analyse de l'école Polytechnique*, 3rd edn. Vol. I: Calcul Différentiel. (Gauthier-Villars, Paris).

Jordan, C. (1913). *Cours d'Analyse de l'école Polytechnique*, 3rd edn. Vol. II: Calcul Intégral. (Gauthier-Villars, Paris).

Karniadakis, G. E. (2019). *Handbook of Fractional Calculus with Applications.* Vol. 3: Numerical Methods. (De Gruyter, Berlin).

Kittel, C. (1986). *Introduction to Solid State Physics*, 6th edn. (John Wiley & Sons, New York).

Klein, F. (1933). *Vorlesungen über die Hypergeometrische Funktion.* (Verlag von Julius Springer, Berlin).

Lanczos, C. (1964). A precision approximation of the Gamma function. *Journal of the Society for Industrial and Applied Mathematics (SIAM), Series B: Numerical Analysis 1,* 86–96.

Landau, E. (1974). *Handbuch der Lehre von der Verteilung der Primzahlen*, 3rd edn. (Chelsea Publishing Company, New York).

Laplace, P.-S. (1814). *Théorie Analytique des Probabilités*, 2nd edn. (Courcier, Paris).

Laugwitz, D. and Rodewald, B. (1987). A simple chacterization of the Gamma function. *The American Mathematical Monthly 94,* 6, 534–536.

Lerch, M. (1887). Note sur la function $R(w,x,s)=\sum_{k=0}^{\infty}\frac{e^{2k\pi ix}}{(w+k)^s}$. *Acta Mathematica 11,* 19–24.

Lerch, M. (1894). Další studie v oboru Malmsténovských řad. *Rozpravy Česke akademie, II. tř. 3,* 8. 28, 1–63.

Madelung, O. (1981). *Introduction to Solid-State Theory.* (Springer-Verlag, Berlin).

Mainardi, F. (2014). On some properties of the Mittag-Leffler function $E_\alpha(-t^\alpha)$, completely monotone for $t>0$ with $0<\alpha<1$. *Discrete and Continuous Dynamical Systems Series B 19,* 7 (September), 2267–2278.

Mainardi, F. (2020). Why the Mittag-Leffler function can be considered the Queen function of the Fractional Calculus? *Entropy (MDPI) 22,* 12, 1359/1–29.

Mainardi, F. (2021). *Fractional Calculus and Waves in Linear Viscoelasticity: An Introduction to Mathematical Models*, 2nd edn. (World Scientific Publishing, London, U.K.).

Mainardi, F. and Consiglio, A. (2020). The pioneers of the Mittag-Leffler functions in dielectrical and mechanical relaxation processes. *WSEAS Transactions on Mathematics 19,* 289–300.

Mainardi, F., Masina, E., and González-Santander, J. L. (2023). A note on the Lambert W function: Bernstein and Stieltjes properties for a creep model in linear viscoelasticity. *Symmetry 15,* 1654.

Markushevich, A. I. (1985). *Theory of Functions of a Complex Variable.* Vol. I–III. (Chelsea Publishing Company, New York).

Marsaglia, G. and Marsaglia, J. C. W. (1990). A new derivation of Stirling's approximation to $n!$. *The American Mathematical Monthly 97,* 9, 826–829.

Mascheronius, L. (1790). *Adnotationes ad calculum integralem Euleri in quibus nonnulla problemata ab Eulero proposita resolvuntur.* (Ex Typographia Petri Galeatii, Ticini).

Miller, K. S. and Samko, S. G. (1997). A note on the complete monotonicity of the generalized Mittag-Leffler function. *Real Analysis Exchange 23,* 2, 753–755.

Mittag-Leffler, M. G. (1903a). Sur la nouvelle function $E_\alpha(x)$. *Comptes Rendus hebdomadaires de Séances de l'Academié des Sciences, Paris 137,* 554–558.

Mittag-Leffler, M. G. (1903b). Une généralisation de l'intégrale de Laplace-Abel. *Comptes Rendus hebdomadaires de Séances de l'Academié des Sciences, Paris 136,* 537–539.

Mittag-Leffler, M. G. (1905). Sur la représentation analytique d'une branche uniforme d'une function monogène (cinquième note). *Acta Mathematica 29,* 101–181.

Morse, P. M. and Feshbach, H. (1978). *Methods of Theoretical Physics.* (McGraw-Hill Book Company, New York).

Nanjundiah, T. S. (1969). Note on the Beta and Gamma functions. *The American Mathematical Monthly 76,* 4, 411–413.

Nemes, G. (2013). Generalization of Binet's Gamma function formulas. *Integral Transforms and Special Functions 24,* 8, 595–606.

Newman, M. E. J. (2018). *Networks*, 2nd edn. (Oxford University Press, Oxford, U.K.).

Nielsen, N. (1956). *Die Gammafunktion: Band I. Handbuch der Theorie der Gammafunktion und Band II. Theorie des Integrallogarithmus und verwandter Transzendenten.* (B. G. Teubner, Leipzig 1906; republished by Chelsea, New York).

Ogilvy, C. S. (1951). The Beta function. *The American Mathematical Monthly 58,* 7 (August-September), 475–479.

Olver, F. W. J., Lozier, D. W., Boisvert, R. F., and Clark, C. W. (2010). *NIST Handbook of Mathematical Functions.* (Cambridge University Press, New York).

Paris, R. B. (2002). Exponential asymptotics of the Mittag-Leffler function. *Proceedings: Mathematical, Physical and Engineering Sciences 458,* 2028 (December), 3041–3052.

Paris, R. B. (2014). Comments on Exactification of Stirling's approximation for the logarithm of the Gamma function. *arXiv:1406.1320*.

Paris, R. B. and Kaminski, D. (2001). *Asymptotics and Mellin-Barnes Integrals.* Encyclopedia of Mathematics and its Applications, vol. 85. (Cambridge University Press, U. K.).

Pérez-Marco, R. (2022). On the definition of the Euler Gamma function. *L'Enseignement Mathématique 68,* 1/2, 135–160.

Podlubny, I. (1999). *Fractional Differential Equations.* Mathematics in Science and Engineering, vol. 198. (Academic Press, San Diego).

Pollard, H. (1946). The representation of e^{-x^λ} as a Laplace integral. *Bulletin of the American Mathematical Society 52,* 908–910.

Pollard, H. (1948). The completely monotonic character of the Mittag-Leffler function $e_a(-x)$. *Bulletin of the American Mathematical Society 54,* 1115–1116.

Popov, A. Y. and Sedletskii, A. M. (2013). Distribution of roots of Mittag-Leffler functions. *Journal of Mathematical Sciences 190,* 2 (April), 209–409.

Raabe, J. L. (1840). Angenäherte Bestimmung der Factorenfolge $1.2.3.4.5...n = \Gamma(1+n) = \int x^n e^{-x} dx$, wenn n eine sehr grosse Zahl ist. *Journal für die reine und angewandte Mathematik 25,* 146–159.

Rademacher, H. (1973). *Topics in Analytic Number Theory.* (Springer-Verlag, Berlin).

Rainville, E. D. (1971). *Special Functions*, reprinted edn. (Chelsea Publishing Company, New York).

Remmert, R. (1996). Wielandt's Theorem about the Γ-function. *The American Mathematical Monthly 103,* 3, 214–220.

Riemann, B. (1860). Über die Anzahl der Primzahlen unter einer gegebenen Grösse. *Monatsberichte der Koeniglichen Preussischen Akademie der Wissenschaften zu Berlin aus dem Jahre 1859*, 671–680.

Riemann, B. (1876). Versuch einer allgemeinen Auffassung der Integration und Differentiation. *Bernhard Riemann's gesammelte Mathematische Werke und Wissenschaftlicher Nachlass, edited by H. Weber and R. Dedekind, Leipzig Chapter XIX,* 331–344.

Riordan, J. (1968). *Combinatorial Identities.* (John Wiley & Sons, New York).

Ruben, H. (1976). A note on the Trigamma function. *The American Mathematical Monthly 83,* 8, 622–623.

Rubinstein, M. O. (2012). Identities for the Riemann zeta function. *Ramanjuan Journal 27,* 29–42.

Rubinstein, M. O. (2013). Identities for the Huwitz zeta function, Gamma function, and L-functions. *Ramanjuan Journal 32,* 421–464.

Sandev, T. and Tomovski, Z. (2019). *Fractional Equations and Models, Theory and Applications.* Developments in Mathematics, vol. 61. (Springer, Switzerland).

Sansone, G. and Gerretsen, J. (1960). *Lectures on the Theory of Functions of a Complex Variable.* Vol. 1 and 2. (P. Noordhoff, Groningen).

Schneider, W. R. (1996). Completely monotone generalized Mittag-Leffler functions. *Expositiones Mathematicae 14,* 3–16.

Seybold, H. J. and Hilfer, R. (2005). Numerical results for the generalized Mittag-Leffler function. *Fractional Calculus and Applied Analysis 8,* 2, 127–139.

Seybold, H. J. and Hilfer, R. (2008). Numerical algorithm for calculting the generalized Mittag-Leffler function. *SIAM Journal on Numerical Analysis 47,* 1, 69–88.

Simon, T. (2015). Mittag-Leffler functions and complete monotonicity. *Integral Transforms and Special Functions 26,* 1, 36–50.

Srinivasan, G. K. (2007). The Gamma Function: An Eclectic Tour. *The American Mathematical Monthly 114,* 4, 297–315.

Srivastava, H. M. (1994). A certain family of sub-exponential series. *International Journal of Mathematical Education in Science and Technology 25,* 2, 211–216.

Stirling, J. (1730). *Methodus differentialis: sive tractatus de summatione et interpolatione serierum infinitarum.* (Londini: Typis Gul. Bowyer; impensis G. Strahan).

Stokes, G. G. (1857). On the Discontinuity of Arbitrary Constants which appear in Divergent Developments. *Transactions of the Cambridge Philosophical Society 10,* 1 (May), 105–128.

Tannery, J. and Molk, J. (1893). *Éléments de la Théorie des Functions Elliptiques.* Thome I, (Gauthier-Villars et fils, Paris).

Titchmarsh, E. C. (1948). *Introduction to the Theory of Fourier Integrals*, 2nd edn. (Oxford University Press, Ely House, London W. I).

Titchmarsh, E. C. (1964). *The Theory of Functions.* (Oxford University Press, Amen House, London).

Titchmarsh, E. C. and Heath-Brown, D. R. (1986). *The Theory of the Zeta-function*, 2nd edn. (Oxford Science Publications, Oxford).

Ulgenes, D. P. H. (2025). Series and product representions of Gamma, Pseudogamma and inverse Gamma functions. *Maple Transactions 5,* 1, article 16090.

Van Mieghem, P. (1993–2026). *Characteristic Coefficients of a Complex Function.* (unpublished).

Van Mieghem, P. (1996). The asymptotic behavior of queueing systems: Large deviations theory and dominant pole approximation. *Queueing Systems 23*, 27–55.

Van Mieghem, P. (2014). *Performance Analysis of Complex Networks and Systems.* (Cambridge University Press, Cambridge, U.K.).

Van Mieghem, P. (2021). Binet's factorial series and extensions to Laplace transforms. *arxiv: 2102.04891.*

Van Mieghem, P. (2022). The origin of the fractional derivative and fractional non-Markovian continuous time processes. *Physical Review Research 4,* 2, 023242.

Van Mieghem, P. (2023a). *Graph Spectra for Complex Networks*, 2nd edn. (Cambridge University Press, Cambridge, U.K.).

Van Mieghem, P. (2023b). Properties of a triangular matrix and its linear differential equation. Delft University of Technology, report20230703 (www.nas.ewi.tudelft.nl/people/Piet/TUDelftReports.html).

Van Mieghem, P. and Liu, Q. (2019). Explicit non-Markovian SIS mean-field epidemic threshold for Weibull and Gamma infections but Poisson curings. *Physical Review E 100,* 2 (August), 022317.

Watson, G. N. (1929). Theorems stated by Ramanujan (V): Approximation connected with e^x. *Proceedings of the London Mathematical Society s2-29,* 1, 293–308.

Watson, G. N. (1995). *A Treatise on the Theory of Bessel Functions*, Cambridge Mathematical Library edn. (Cambridge University Press, Cambridge, U.K.).

Webster, R. (1997). Log-convex solutions to the functional equation $f(x+1) = g(x)f(x)$: Γ-type functions. *Journal of Mathematical Analysis and Applications 209*, 605–623.

Whittaker, E. T. and Watson, G. N. (1996). *A Course of Modern Analysis*, Cambridge Mathematical Library edn. (Cambridge University Press, Cambridge, U.K.).

Widder, D. V. (1946). *The Laplace transform.* (Princeton University Press, Princeton).

Wilf, H. S. (1994). *generatingfunctionology*. (Academic Press, Inc.).

Wiman, A. (1905a). Über den Fundamentalsatz in der Teorie der Funktionen $e_a(x)$. *Acta Mathematica 29*, 191–205.

Wiman, A. (1905b). Über die Nullstellen der Funktionen $e_a(x)$. *Acta Mathematica 29*, 217–234.

Wong, R. and Zhao, Y.-Q. (2002). Exponential Asymptotics of the Mittag-Leffler function. *Constructive Approximation 18*, 355–385.

Zeng, C. and Chen, Y. (2015). Global Padé approximations of the generalized Mittag-Leffler function and its inverse. *Fractional Calculus & Applied Analysis 18,* 6, 1492–1506.

Index